Artificial Intelligence

and

Democratic Values

CAIDP AI Index 2026

Volume 1

Center for AI and Digital Policy

CAIDP.org

This report is available online at:
https://www.caidp.org/reports/caidp-index-2026/
ISBN 979-8-9956958-1-3

TABLE OF CONTENTS

EXECUTIVE SUMMARY

Purpose and Scope

The Center for AI and Digital Policy (CAIDP) ***Artificial Intelligence and Democratic Values Index*** (CAIDP AI Index) is the first global survey to assess progress toward trustworthy AI based on detailed narrative reports, combined with a methodology that produces ratings and rankings for national AI policies and practices.

The **CAIDP AI Index** has these objectives: (1) to document AI policies and practices, based on publicly available sources, (2) to establish a methodology for the evaluation of AI policies and practices, based on global norms, (3) to provide a basis for comparative evaluation, (4) to provide the basis for future evaluations, and (5) to ultimately encourage all countries to make real the promise of AI that is trustworthy, human-centric, and provides broad social benefit to all.

Artificial Intelligence and Democratic Values focuses on human rights, rule of law, and democratic governance metrics. Endorsement and implementation of the OECD/G20 AI Principles and the UNESCO Recommendation on the Ethics of Artificial Intelligence are among the primary metrics. Opportunities for the public to participate in the formation of national AI policy, adoption of the right to algorithmic transparency, and creation of independent agencies to address AI challenges are also among the metrics. Patents, publications, investment, and employment impacts are important metrics for the AI economy, but they are not considered here.

The ***CAIDP AI Index*** is published annually and will evolve as country practices change and new issues emerge.

The 2026 Edition

The 2026 edition of the report updates and expands the previous reports. To reflect some of the key discussions and developments in the AI policy landscape, our efforts to enhance usability, and our expansion as an organization, we have also made the following changes for ***CAIDP AI Index 2026***.

- We updated the metric concerning the UNESCO Recommendation on the Ethics of AI. Recognition for full implementation (Y) of the Recommendation requires completion of the Readiness Assessment Methodology (RAM) and additional steps.
- We added 10 new country reports to the Index, bringing our total to 90 countries. The new countries for the 2026 edition are: Benin, Ecuador, Greece, Jordan, Mozambique, Namibia, Oman, Romania, Slovakia, and Tanzania.
- We introduced a section on AI Literacy to weigh policies promoting education and capacity-building for students, workers, and civil servants. We focus on programs designed to prepare citizens to live and participate in a world shaped by AI.

- The scores for country reports previously published were reviewed and revised based on developments during the past year concerning AI policies and practices.
- The number of researchers participating in the project has grown significantly. More than 500 CAIDP Research Group members representing more than 90 countries contributed updates from 2025. This brings our total contributors above 2000 researchers representing 130 countries.

Acknowledgements

Editor-in-Chief: April Yoder (2025–2026), Karine Caunes (2024), Marc Rotenberg (2021–2023)

Assistant Editors: Merve Hickok, Grace Thomson, Ren Bin Lee Dixon

Index Graphics: Spencer Michaels, Christabel Randolph, Alireza Saebi, Nidhi Sinha

Management: Marc Rotenberg, Merve Hickok, Christabel Randolph, Selim Alan

Policy Clinic Team: Merve Hickok, Marc Rotenberg, Grace Thomson, Tamiko Eto, Clea Strydom, Jason Raymond, Varsha Swelal; Dila Sen, Humeyra Cirit

Special Editorial and Research Assistance in 2026: Abhinaya Murthy, Aditi Rukhaiyar, Ameneh Dehshiri, Aysu Dericioğlu Egemen, Brenda Maina, Candice Alder, Carolina Judith Medina Guzmán, Cecilia García Podoley, Christabel Randolph, Idil Ilayda Kaner, Janani Sathiyaseelan, Jésica Edith Tapia Reyes, Luisa Fernanda Jiménez Mahecha, Marcela Campos Jabor, Maria Lungu, McDonald Moyo, Nayyara Rahman, Nazareen Ebrahim, Pritika Magima, Ren Bin Lee Dixon, Ruoyi Lu, Tatiana Zasheva, Tim Sowa, Tran Vu Ha Minh, Zoë Aikman

2025 Contributors: Aakash Gurung, Aakash Shah, Aaron Margolis, Aaron Promise Mbah, Abdelrahman Yacoub, Abdulhaleem Salisu Shehu, Abdullah elbi, AbdulQuddus Mohammed, Abdulrosheed Olalekan Fadipe, Abhinaya Murthy, Abubakar Sadiq Muhammad, Adelle Engmann, Adithya Rajendra Kumar, Aditi Bharti, Aditi Rukhaiyar, Aditya Thomas, Adolfo Arreola Garcia, Agustina Brizio, Ahmad Amr Ali Abdeltawab, Ahmad Haidar, Aigerim Amirova, Aiko Shimizu, Aiymgul Kachyke, Akanksha Bisoyi, Akanksha Ray, Akmaral Orazaly, Al-Imran Ruhul Islam, Alberto González Pulido, Alessandro Ciceri, Alexey Sidorenko, Alia Yofira Karunian, Alice Liu, Altai Ismailovi, Amalia Peralta, Amanda Camilotti, Amelie Soares , Ameneh Dehshiri, Amy Darajati Utomo, Ana Lucía Martín Paredes, Ana Paula Simoes Lureiro, Ana Toskic Cvetinovic, Anabel Karina Arias, Ananya Mukherjee, Anda Bologa, Andras Hlacs, Andrea Iwaki Motta, Andrea Kainz, Andrea Palumbo, Andrea Thomas, Angela Onikepe, Angelo Leone, Angesom Teklu, Ani Meliksetyan, Anisa Abeytia, Annita Mwagiru, Anuujin Sanjaajamts, Archana Raghavan, Archit Lohani, Arianna Valentini Céspedes, Arife KaraÇelik, Arthit Suriyawongkul, Asela Waidyalankara, Ashish Rauniyar, Ashley Pusey, Atdhe Lila, Ayse Nur Dil Barutcu, Aysegul Guzel Turgut, Aysu Dericioğlu Egemen, Azeem Sajjad, Bagus Jatmiko, Benedicta Frema Boamah, Betul Kubra Ekinci, Bilguun Erdenebat, Blessing James Ademulegun, Buddhi Withanage, Camila Beltran Reyes, Camille Garrison, Camille Moura, Candice Carmel Alder, Carola Frey-Tventarnii, Carolina Changoluisa, Carolina Medina Guzmán, Caroline Lancelot, Cecilia Garcia Podoley, Chaeri Park, Chazara Clark-Smith, Chelsea Horne, Chidimma Cynthia Obidiegwu, Chilufya Theresa Mulenga, Chimdessa Fekadu Tsega, Chimezie Precious

Erugo, Chioma Phibe Nwaodike, Chithra Madhusudhanan, Chitoo Veronica Mbama, Chloe Messdaghi, Christian Stiegler, Christine Custis, Chuma Akana, Cindy Friedman, Cing Hoih, Clea Strydom, Cosmas Ondino, Cristiana Firullo, Cynthia Akosua Antwi, Daire Maria Ni Uanachain, Danait Girma, Daniel Gómez González, Daniela Camberos, Darleen Langen, Darmain Segaran, Darshatha Damith Gamage, Davit Totadze, Deborah Arthur, Demet Ceylan Demircan, Deuk Yeon Lee, Devvrat Vaishnav, Dhara Atul Mungra, Diana Gutierrez, Diana Duran Alvarez, Diego Lisoni Caro, Dimitra Eleni Pavlidou, Divin Lereh, Dominic Mathew, Dona Mathew, Doreen Aglago-Cofie, Dorothy Ankunda, Douglas Barrios Anderson, Dvija Mehta, Dyllan Brown-Bramble, Ebele Eucharia Okafor, eileen carter, Elana Banin, Elias Kassaye Gebregziabher, Elikplim Sabblah, Ella Duus, Ella Howard, Emma Jane O'Connell, Emma Jones, Engy Said, Erika Martins de Oliveira, Ermelinda Kanushi, Ethan Ward, Evans Odongo Agembo, Fabian Ulmer, Fabienne Tarrant, Facundo Fernando Bonaura, Faith Njoki Kabata, Farah Alqattan, Farah Zainal, Farhan Yusuf, Fathima Amra Ismail, Federica Paolucci, Federico Dante De Falco, Femi Gbolahan, Feride Bahar Erdogan, Feven Alemu Beyecha, Francesca Aurora Sacchi, Gabriel Akinremi, Galina Stoeva, Geetha Raju, Geneve Phillip Durham, George Mealy, Georgiana Birdan, Gijs Reusken, Giovana Fleck, Giovanna Jaramillo Gutierrez, Giulia Consonni, Gretel Schaj, Gülce Kazaz, Gulraiz Iqbal, Ha Minh Tran Vu, Habiba Sarhan, Haleh Asgarinia, Hamilton Tamburayi Katsvairo, Hamna Khalid, Hanani Thamsanqa Hlomani, Hande Ulker Pehlivan, Hanhee Yoon, Heather Leson, Hebah Foda, Hemchandra Betchoo, Hicran Erol, Huayizi Chen, Hussain Rezai, Ian Chiang, Ian Hillary Oboka, Ian Olwana Ouma, Iaroslava Kharkova, Ida Nganga, Igor Vikhrov, Imane Kabori, Immaculate Nabuggwawo, Immaculate Wanjiru Kimani, Ina Kokinova, Indras Ghosh, Induja Mony Jalajadevi, Inès Belhadj, İrem Çınar, Irene Warima Kabua, Isabela Rosal Santos, Isabella de Vivo, Isabella Pereira, Isadora Assuncao, Ishu Gupta, Jameela Sahiba, James Edward Griffith, Jan Llenzl Dagohoy, Janani Sathiyaseelan, Janice Lam, Janki Nawale, Jason Praise Solomon, Jason Raymond, Jasper Mokua Michieka, Javier Ibáñez-Guzmán, Jean Mattos Duarte, Jennifer Park, Jésica Tapia Reyes, Jessa Henderson, Jessica Torres De La Piedra, Joana Maria, Joanita Nagaba, João Rocha-Gomes, Joaquin Collao Jul, John Oluwaseye Adebayo, Jonathan Lim, Joshua Kitili, Juan Carlos Montoya, Judith Chiamaka Ohaeri, Julia Mykhailiuk, Julia Oliveira Rosa, Juliana Coelho dos Santos, Juliett Suárez Ferreira, Julio Cesar Espinosa, Kaden Madson, Karen Garcia Alvarado, Karma Peiró, Karolina Romoli-Walczak, Karryl Sagun Trajano, Katherin Corredor Páez, Kathleen Anderson, Kedar Bhasme, Kemfon Ibanga Josephneke, Kenzie Burchell, Keron Alleyne, Kevin Harerimana, Keyan Simpson, Khandsuren Lkhamsuren, Khensani Xivuri, Kirtan Padh, Koichi Tsukioka, Kollin Napier, Kparobor Akpomiemie, Kristen Hutchens, Kristina Horn, Kristina Kempkey, Kristina Khutsishvili, Krystal Jackson, Kseniia Guliaeva, Kyla Williams Tate, Kyle Gracey, Lakshmi Prasath Muniandi, Lam Dieu Hien Phan, Lamia Ben Ayach, Laura Edwards, Laura Romero Villamizar, Leah Wambui Maina, Leire Beltran Sagaseta, Lia Chkhetiani, Lidia Leontjeva, Ligia Lage Vieira, Lionceau Clovis Agre, Lizzette Soria

Sotelo, Logesh Rajendran, Lone Lottering, Lorena Barić, Lorena Pereira, Lovet Ndialle, Lucia Flores Echaiz, Lucia Kobzová, Ludovico Jr. Delloro, Luis Martinez, Luisa Jiménez Mahecha, Luisa Maciel Perez, Luke Keller, Lynn Monzer, Mandy Lit, Marcel Mir Teijeiro, Marcela Campos Jabor, Maria Beatriz Torquato Rego, Maria Paola Silva, María Serrano, Mariana Gonzalez Carrillo, Mariane Piccinin Barbieri, Marie Roschelle Quintero, Marietjie Botes, Marin Milenkoski, Mario George Mackay, Marta Burgos González, Martha Njeri Mugambi, Masoomeh Bakhshi-Jafroodi, Matias Carpio, Matteo Mastracci, Matthew Sutcliffe, Matthias Hudobnik, McDonald Moyo, MD Khorshed Alam, Meem Arafat Manab, Melodena Stephens, Mert Özyiğit, Meyda Nento, Michael Dugeri, Michael Tettey, Michał Kubiak, Miguel Angel Jesus Correa, Mohsin Dhali, Monika Manolova, Monique Nichole Morrison, Monserrat Lopez Perez, Montserrat Legorreta Luna, Muhammad Deckri Algamar, Muhammad Sadiq, Mukilan Deivarajan Suresh, Münüse Gülşah Gökçe, Myrnelle Cinco, Nada Alwadi, Nadjiba Boukemidja, Nam Phuong Ngo, Nancy Eke-Agu, Nanda Min Htin, Natalia Maj, Nazareen Ebrahim, Nelson Matendo Nkari, Nerushka Bowan, Nguyen Tran, Nicole Bennett, Nikunj Doshi, Nimmi Patel, Noah Levy, Norman Valdez, Nur Sena Sevindi, Ojus Tyagi, Olesia Chizheva, Olesia Ved, Olubukola Familoni, Omolola Oluyemisi Haastrup, Onyinye Onuh, Orkhan Damirli, Orland Tubola, Oscar Obando Chaves, Osvaldo Ramirez Hurtado, Oussama Elmerrahi, Owolabi Paul Adelana, Paige Benton, Paloma Baytelman, Pascal Thibeault, Patrick Herbert, Patrycja Niemczyk, Paula Luvini, Paula Roko, Pauline Baron, Piyal Uddin, Piyali Mitra, Pınar Saruhan, Pooja Puranik, Pranjal Dwivedi, Preethivirajsingh Mohabeer, Princess Nsoko-Nkwor, Priscila Chaves, Pritika Marguerite Magima, Pritika Reddy, Qidi Zhao, Rachael Olaitan Aborishade, Rachele Carli, Rae Baker, Raghav Kapoor, Raghda Elhalawany, Rajesh Kumar Shakya, Rakshitha Siva, Ramiz Arslan, Randon Taylor, Ranjana Kushwaha, Raymond Francis Sarmiento, Rebecca Duke Wiesenberg, Renato Manuel Berrino, Richard Holman Matanta, Richard Oshoma Abdulahi, Rishita Chaudhary, Risper Onyango, Rita Orji, Robert Hamlett, Rodolfo Cordova Alcaraz, Romana Afroze, Rommel Abilio Infante Asto, Roshan Prakash Melwani, Rukiya Deetjen Ruiz, Ruoyi Lu, Sadman Rahman, Şafak Etike, Saima Tariq Khan, Samagi Manusha Rajakaruna, Sandeep Sharma, Sania Green - Reynolds, Sara Concetta Santoriello, Sarah Chantal Delporte, Sarah O'Connell, Saseeta Rakha, Sebastian Chik, Sebastian Winter, Sebastiao Espindola, Sebnem Yardimci Geyikci, Setu Rajput, Sevinj Novruzova, Seyide Direk, Sezer Selvi, Shagufta Sen, Shahtaj nazmeen, Sharathappriyaa V, Shazade Jameson, Shelly Hall, Sher Afgun Khan Afgun, Sherlie Grand Pierre, Shila Nhemi, Shirley Genga, Shorefunmi Bola-Saliu, Shruti Sharma, Shuvarthi Bhattacharjee, Simeen Mirza, Siti Hasliah Salleh, Slava (Veaceslav) Balan, Snezana Nikcevic, Sodiq Ajala, Sofiia Klymchuk, Sophie Strassmann, Soraya Belghazi, Spencer Michaels, Srabonty Das Gupta, Sreya Francis, Stephanie Grasser, Steven Gonzalez Guerrero, Sudeshna Mukherjee, Sukaasini Latchumanan, Sukhdeep Kaur, Supriya Kulkarni, Suraj Juddoo, Suriya Balakrishnan Nair, Susan Morrissey, Susanne Ogolla, Syed Kasim Shah, Tahniat Khan, Takehito Akima, Tamiris Garbelotti, Tanushree Unakal, Tarcila Matos

Renner, Tarek Chaouch, Tatiana Pilon, Tatjana Titareva, Tayrone Chiavone, Taysir Mathlouthi, Tazqia Al-Djufri, Teerawat Potikul, Tetiana Avdieieva, Thi Kim Hang Ho, Tiago Filipe Sousa, Tiffany Petricini, Trisha Ray, Tugce Levi, Unggul Sagena, Unnikrishnan Nagarajan, Uran Esengeldiev, Uzoma Mkparu, Vanessa (Tram) Bui, Vanessa Maidoh, Vasiliki Pechlivani, Victoria Di Baggio Vega, Victoria Martina Copeland, Vidda Guzzo Faustino, Vivek Kumar, Vivek Rana, Vladimir Choi, Vrikson Acosta, Waridah Makena, Warren Alan Bowles, Wisdom Obinna, Wm. Matthew Kennedy, xiaolin xu, Yaru Wang, Yasmin Zeidan, Yejin Lee, Yesim Keskin, Yihwan Cho, Yohan Veppumthara, Yoshita Sharma, Yuvraj Tokas, Zamara Rodriguez Duenas, Zeynep Meryem Ünal, Zhanat Murzakulova, Zhen Wen Lim, Zoë Aikman, Zoi Roupakia

2024 Contributors: Aaron Schultz, Aashna Kothiyal, Abiodun Solanke, Abiola Joseph Azeez, Abraham Alonzo Guiyab, Aditi Rukhaiyar, Adriana Mardueno Villasenor, Adrianna Tan, Agita Pasaribu, Agus Azka Haria Fitra, Aimeé Pujadas Clavel, Aisha Najera, Aishvarya Rajesh, Aishwarya Gurung, Akanksha Ray, Akintunde Agunbiade, Alejandro Carlos Baltazar Ruiz, Aleksandra Iugunian, Aleksandra Wójtowicz, Alexandra Krastins Lopes, Alhassan Sulemana Alhassan, Alireza Saebi, Ally Tang, Allyson Kapin, Altynay Junusova, Amanda Leal, Amanda Maria Horzyk, Amelia Ayang Sabrina, Amir Noy, Ana Camerano, Ana Carolina César, Ana Carolina Sousa Dias, Ana Segovia, Anabel Karina Arias, Anahita Valakche, Ananda Gautam, Ananya Reddi, Andras Hlacs, Andrea Palumbo, Andrés Eduardo Rodriguez Gomez, Andressa Girotto Vargas, Andrew McIntyre, Andy Sanchez, Angie Orejuela, Anmol Bharuka, Anmol Mathur, Anna Pellegatta, Anna Teresa Aguilar, Antonella Perini, Antonia Dionicia Bogado Rodas, Anushka Jain, Anushka Sachdev, April Yoder, Ariana Gamarra, Ariel Riera, Arpit Gupta, Artem Kobrin, Aruzhan Seitmagambet, Arvin Obnasca, Atika Chahby, Aysu Dericioğlu Egemen, Bahaa El-Taweal, Bahadır Ozgenc, Bakyt Omurzakov, Barbara Simão, Bartlett Morgan, Begyira Donald Sharp, Belen Alondra Bringas Machicado, Bendjedid Rachad Sanoussi, Berenice Fernández Nieto, Bernard M. Nyaga, Beşir Orak, Biljana Molan Klisarova, Bridget Roddy, Bridgette Ndlovu, Bruna Sellin Trevelin, Caio Cesar Vieira Machado, Caitlin Goldenberg, Camila Alejandra Beltran Reyes, Camila Hidalgo, Camille Ford, Caragh Aylett-Bullock, Carissa Mears, Carla Verdiane Nkusi, Carlos Eduardo Torres Giraldez, Carlos Guerrero Argote, Carmen Ng, Carolina Judith Medina Guzmán, Caroline Lancelot, Caryn Swart, Casandra Rusti, Challace Pahlevan-Ibrekic, Chelsea Cunningham, Chialuka Prisca-Mary Onuoha, Christian Stiegler, Christina Huntzinger, Christina Varytimidou, Christine Gibson, Clara Balon, Claudia Ramly, Collins Otoo, Confidence Osein, Cristian Andres Mejia Caballero, Cynthia Picolo, Cyril Ikechukwu Obika, Damian Flisak, Daniel Acheampong, Daniela Dora Eilberg, Daniele Sabato, Daniella Esi Darlington, David Backovsky, Davit Totadze, Denisa Reshef Kera, Deniz Celikkaya, Desiree Valentine, Diana Nyarang'o, Dila Sen, Diná Santana Santos, Dmytro Chumachenko, Donald Erepamowei Sikpi, Doruk Ozulu, Douglas Barrios Anderson, Dumitrita

Rogojinaru, Eddie Liywalii, Eduarda Costa, Elena Sertore, Eleonore Shalomita Hana, Elina Margarita Castillo Jiménez, Elise Sakura Christiane Farge Di Maria, Eliza Aspen, Elizabeth Rothman, Ellie Copeland, Elliot Gianfranco Mejía Trujillo, Emiliano Mora, Emma Haywood, Emma Lamberton, Emmy Chirchir, Enelton Satria, Enzo Dunayevich, Eren Yenigun, Erica Yen Yi Liaw, Erika Del Carmen Fuchs, Erva Akin, Erva Çolak, Essa Mohamedali, Esther Montoya Martinez van Egerschot, Eugenia Gonzalez Ehlinger, Eustace Aroh, Ezgi Turgut Bilgiç, Faith Mutune, Faith Oronga Amatika-Omondi, Fazilet Caglayan, Federico Dante De Falco, Fitri Lestari, Foteini-Maria Vassou, Francesco Saraceno, Frederick Antonovics, Freyja Van den Boom, Galiya Yelubayeva, Gamze Busra Kaya, George Lwanda, Ghofran Vilquin, Gianmarco Gabriele Marchionna, Giovana Fleck, Giovanna Jaramillo-Gutierrez, Gisele Waters, Gisli Ragnar Gudmundsson, Gizem Yilmaz, Ha Minh Tran Vu, Haissam Khan, Halyna Padalko, Hans Christian Arents, Helen Titilola Olojede, Helga M. Brøgger, Henok Russom, Heramb Podar, Hesam Nourooz Pour, Hilal Durek, Hilmy Hanif, Himanshu Joshi, Hina Bashir, Humeyra Cirit, Iakovina Kindylidi, Idil Ilayda Kaner, Imane Hmiddou, Irem Atik, Irene Unceta Mendieta, Isabel Neubert, Isabelle Anzabi, Jae Seung Lee, James Mwole, James Townsend, Jamil AlKhatib, Jan Peter Ganter de Otero, Jane Chinyere Ezirigwe, Janet Lim, Jason Raymond, Jeanette Wong, Jenna Manhau Fung, Jeremy Ng, Jerome Igbokwe, Jésica Edith Tapia Reyes, Jessica Santos, Jessie Yang, Jiayi Wu, Jihyun Kwon, Jinqian Li, Johannes Nordin, Jorge Montiel, Jose Carmelo Cueto, Julia Małgorzata Kalus, Julia Meltzer, Júlia Mendonça, Julia Mykhailiuk, Julia Pomares, Junaid Aziz, Justine Gluck, Kadian Alicia Davis-Owusu, Kardelen Filorinalı Kargın, Karla Patricia Ramirez Sanchez, Kennedy Okong'o, Kerrie Hooper, Kewa Jiang, Khaoula Chehbouni, Kimberli Jeffery, Kirtan Shyamal Padh, Krerkkaiwan Kajohnkunchonlathorn, Krishna Kumar Saha, Kristiina Juurmaa, Krystal Jackson, Kuan-Wei Chen, Kwetchi Takam Ezekiel, Kyoung Yang Kim, Kyoungsic Min, Lara Salgueiro Oliveira, Laura Betancourt, Laura Centeno Casado, Laura Gacho, Lavina Rao Rao, Lea Hani, Leda Kuneva, Lena Patricia Nieper, Lia Chkhetiani, Lillie Coney, Linda Shum, Lindsey Washburn, Lisa Wymer, Logan O'Shaughnessy, Lorena Almaraz De La Garza, Luciana Longo, Luis Jacob Retanan, Luis Vasquez, Luisa Abreu Dall'Agnese, Luisa Verónica Arroyo Revatta, Luise Eder, Luka Begiashvili, Lynn Pickering, Madalina Nicolai, Mansur Omonov, Manuel Gustavo Isaac, Maral Niazi, Marcel Mir Teijeiro, Marcell Petho, Marcelo Pasetti, Marco Bani, Maria Lungu, Maria ("Mary") Lang, Maria Hadjicosta, Maria Lungu, Mariana Avelar, Mariana Correa, Marianela Sarabia, Mariette Awad, Marisa Stones, Mark Mulobi, Marta Basystiuk, Marta Bienkiewicz Watson, Mary Oduor, Mashael Alzaid, Mateusz Łabuz, Md Ferdows Hossen, Megan McLaughlin, Melike Akkaya, Melodena Stephens, Mennatullah Hendawy, Mercy Chinazom Matthew, Mert Cuhadaroglu, Mgechikwere Nana Nwachukwu, Minna Adlan, Mireia Pérez Carretero, Mohammad Ghasemi, Morgan McMurray, Mostafa Abdelaziz Ashhab Elkadi, Mouloud Khelif, Mubarak Raji, Muhammad Sadiq, Murtala Makama Dattijo, Mykyta Petik, Nabil Fancy, Nada Alwadi, Nam Phuong Ngo, Nana Afua Brantuo, Nana Khechikashvili, Nandita Sampath, Natalia Alarcon, Natascia Arcifa,

Natasha Karner, Nate McFaul, Nazareen Ebrahim, Neesha Patel, Nelly Rousseau, nesibe kiris, Niel Swanepoel, Nika Pranata, Niki Agrotou, Nikolas Schmidt, Ninsiima Melissa, Noah-Levin Damschke, Noemi Mejia, Noora Arajärvi Parsons, Nurulazmi Lamanda, Nuurrianti Jalli, Ojus Tyagi, Olga Grygorovska, Omer Han Bilgin, Ona Oshen, Onome Cynthia Anakanire, Ornela Sollaku, Oyinkan Adebimpe, Öznur Uğuz, Pablo Fuentes Nettel, Paola Katherine Galvez Callirgos, Paola Katherine Gálvez Callirgos, Parishrut Jassal, Parul Anand, Patricio Espinoza, Patrick Armstrong, Patrycjusz Szubryt, Paula Guedes Fernandes da Silva, Paula Marques Rodrigues, Paula Martins, Paulina Denhi Gomez Escalona, Pauline Baron, Pavlo Riepin, Perpetua Ndidiamaka Ogwuche, Pervin Harvinder SK, Peter Mmbando, Petra Müllerová, Pierre du Pasquier, Pius Kwao Gadosey, Prabhu Pradhan, Prakash Gupta, Pranav Bhaskar Tiwari, Precious Osegbo, Priya Goswami, Ranjini Syam, Raymond Amumpaire, Raymond Francis Sarmiento, Rémy Ibarcq, Renan Canaan, Rhea Subramanya, Ritika R Ranka, Rizka Khairunissa Herdiani, Romilla Syed, Rommel Abilio Infante Asto, Rose Boutboul, Rotimi Owolabi, Ruoyi Lu, Ruwani Sandeepani De Silva Uyanage, Rwamamara Lionel, Ryan Burns, Saba Tiku Beyene, Sabire Sanem Yilmaz, Saleh Laslom, Samantha González, Samantha Khoo, SaMee Harden, Samuel Goodger, Sandra Makumbirofa, Santiago Molina, Sara Concetta Santoriello, Sarah D'Andrea, Sechaba Keketsi, Sedef Akinli Kocak, Septiani Eka Wahyu Pratiwi, Serçin Kutucu, Serign Modou Bah, Shakhboz Juraev, Sharinee Lalit Jagtiani, Sharon Kutukwa, Sharon Ndinda Nzuki, Shauna Blackmon, Shezaad Dastoor, Shihori Maeda, Shiva Kanwar, Shuvarthi Bhattacharjee, Sidney Frank Engelbrecht, Snezana Nikcevic, Soledad Guilera, Srija Chakraborty, Stacey Berry, Stacy Owino, Stella Anne Teoh Ming Hui, Stella Lee, Stella Yiantet Letuya, Stephanie Forbes, Stephanie Muthoni Muthungu, Stephie-Lea Tabujara, Sukaasini Latchumanan, Suljo Corsulic, Sumair Gul, Supriya Kulkarni, Susan Ibukun Akinade, Susana Aires Gomes, Sven Ronen Spangenberg, Taina Silveira Baylao, Talha Bin Tariq, Tania Maria Skrapaliori, Tapiwa Ronald Cheuka, Taposh Dutta Roy, Taria Khaoma, Tatevik Davtyan, Tatia Mosidze, Tatiana Levi, Tatiana Zasheva, Tatjana Titareva, Temitayo Fajobi, Thiago Moraes, Thierry Grimm, Thomas Cassara, Thomas Linder, Thummim Iyoha - Osagie, Tiago Filipe Sousa Gonçalves, Timothy Sowa, Tlhaloso Mpopo, Tobias Wirth, Tom Utum, Tomiris Amirova, Tooba Kazmi, Tori Rousay, Tracy Vivian Odipo, Tsepo Ramoholi, Tulika Avni Sinha, Tyler Lance Jaynes, Uchenna Anyamele, Ulysse Richard, Uzma Alam, Uzoma Mkparu, Valeria Allen Gimenez, Valerie Elefante, Veronica Cretu, Veronica Tan, Veronika Klymova, Vicente Antonio Arias Gonzalez, Vicente Martinez Fernandez, Victor Bohorquez, Victoria Hendrickx, Vidhi Sharma, Vishmi Fernand, Vivek S. Rana, Vladimir Cortes, Waceera Kabando Kabando, Waridah Makena, Worawan Klinsawai, Xitshembhiso Russel Mulamula, Yasmin Hussien, Yerdaulet Rakhmatulla, Yiğitcan Çankaya, Yilmaz Sahin, Yolanda Botti-Lodovico, Youssef Horchani, Yuli Rubinsky, Yulia Askhadulina, Zeynep Ogretmen Kotil, Zoë Aikman, Zoe Yi Soh, Zoi Roupakia

2023 Contributors: Abdelrahman Hassanein, Abdulkarim Althiyabi, Abdulmecit İçelli, Abiodun Solanke, Abiola Joseph Azeez, Adeeba Asri, Adriana Mardueno Villasenor, Ahmed Esmat, Ahmet Bilal Aytekin, Aicha Jeridi, Aimeé Pujadas Clavel, Aishwarya Gurung, Akintunde Agunbiade, Alain Otaegui, Alain Perdomo, Alejandro Carlos Baltazar Ruiz, Alejandro Tamez, Alessandro Tufano, Alessio Azzutti, Alexa Hanson, Alexandra Krastins Lopes, Alexandre Aubert, Alfredo Collosa, Alhassan Sulemana Alhassan, Alina Antonie, Alireza Saebi, Allyson Kapin, Amanda Leal, Ameen Jauhar, Amina Akhmetbekova, Amir Noy, Amira Dhalla, Ana Camerano, Ana Carolina César, Ana Segovia, Anahita Valakche, Ananda Gautam, Ananya Reddi, Anastasia Nefeli Vidaki, Anca Radu, Andras Hlacs, André Fernandes, Andrea Sanchez Aguilar, Andréane Sabourin Laflamme, Andrew McIntyre, Anil Sena Bayindir, Anindita Bose, Anmol Mathur, Anna Pellegatta, Anna Teresa Aguilar, Anushka Sachdev, Apolline Rolland, April Yoder, Ariel Riera, Arohi Kashyap, Arpit Gupta, Artem Kobrin, Aruzhan Seitmagambet, Aryashree Kunhambu, Ashwini Natesan, Atika Chahby, Axel Beelen, Axel Ebermann, Ayca Ariyoruk, Ayesha Abduljalil, Ayomide Owoyemi, Aziz Soltobaev, Bahaa El-Taweal, Bahadır Ozgenc, Bakyt Omurzakov, Baris Cihan Canturk, Bartlett Morgan, Begyira Donald Sharp, Belen Sanchez Hidalgo, Bendjedid Rachad Sanoussi, Benjamin Faveri, Berenice Fernández Nieto, Bernard M. Nyaga, Beşir Orak, Bhavna Sharma, Billy Kipkemboi Boruett, Blessing Udo, Brenda Maina, Bridget Chimbga, Busra Memisoglu, Caitlin Kearney, Cameron Benton, Camila Hidalgo, Camille Ford, Carissa Mears, Carla Vazquez Wallach, Carla Verdiane Nkusi, Carleen Boyer, Carlos Eduardo Torres Giraldez, Carlos Ferreira, Carmen Ng, Carolin von Bredow, Caroline Friedman Levy, Casandra Rusti, Catriona Gray, Cezary Gesikowski, Chananya Groner, Charitarth Bharti, Charlotte Esler, Chelsea Cunningham, Cherie Oyier, Chhavi Sharma, Chi Zhang, Chinmoy Rajpal, Chris Kihereko, Christabel Randolph, Christina Huntzinger, Christina Zarogianni, Christine Axsmith, Christopher Musodza, Çiğdem Akin, Clara Balon, Clara Linda Harris Hawking, Clarence Tocchio, Claudia Moreno Arcila, Claudio Ndeleva Mutua, Clea Strydom, Cody Rutherford, Colin Phillips, Collins Okoh, Confidence Osein, Cynthia Chepkemoi, Damian Flisak, Daniel Acheampong, Daniela Constantin, Daniela Dora Eilberg, Daniela Duta, Daniele Sabato, Daniella Esi Darlington, Daphne Wanjiku, Dariel Suarez, David Backovsky, Denisa Reshef Kera, Denise Schalet, Deniz Celikkaya, Desiree Valentine, Devin Almonor, Diana Mocanu, Diana Nyarang'o, Dila Sen, Dimple Shah, Diná Santana Santos, Dirk Brand, Dirk Johannes Brand, Divya Dwivedi, Dominique Greene-Sanders, Duduetsang Mokoele, Dumitrita Rogojinaru, Duong Hoang, Ebere Lisa Iroegbu, Eddie Liywalii, Edla Aittokallio, Eduarda Costa, Ekene Chuks-Okeke, Elan Silver, Elena Turci, Elina Margarita Castillo Jiménez, Ellie Copeland, Emiliano Mora, Emma Haywood, Emma Lamberton, Emmy Chirchir, Emsie Erastus, Enelton Satria, Enton Dimni, Erva Akin, Essa Mohamedali, Eva Sachar, Evangelia-Maria Kostopoulou, Evelina Ayrapetyan, Fabio Seferi, Faith Gitonga, Faith Mutune, Faith Oronga Amatika-Omondi, Fatemeh Babaeian, Fidis Muriithi, Fiona Melzer, Foteini-Maria Vassou, Francis Ngige, Freyja Van den Boom, Frincy Clement, Gabriel Toscano, Gabriella Maia, Galiya

Yelubayeva, Gamze Büşra Kaya, Geetanjali Bisht, Georgeanela Flores Bustamante, Georgios Theodotou, Gergana Tzvetkova, Germán López Ardila, Ghofran Vilquin, Gianmarco Gabriele Marchionna, Gisele Waters, Giuseppe Claudio Cicu, Gizem Yardimci, Gizem Yilmaz, Hamid Mukhtar, Hans Christian Arents, Harri Williams, Harshana Ghoorhoo, Helen Robinson, Helen Titilola Olojede, Helena Broj, Helga M. Brøgger, Héloïse Rivoal, Henok Russom, Heramb Podar, Hesam Nourooz Pour, Hoore Jannat, Humeyra Cirit, Iakovina Kindylidi, Ibrahim Yeku, Idil Ilayda Kaner, Idil Kula, Ifeoma Nwafor, Ikran Abdirahman, İlayda Süer, Ilona Poseliuzhna, Immaculate Odwera, Inés Fernández Gallego, Ingrid Soares, Insun Park, Irem Atik, Isabel Neubert, Ismael Kherroubi Garcia, Iulia Gabriela Popescu, Iva Sisul Brdar, Iwona Karkliniewska, Jacquelyn (Jacky)) Omotalade, Jae Seung Lee, James Townsend, Jane Chinyere Ezirigwe, Janelle Radcliffe, Janet Lim, Janhvi Patel, Javed Sajad, Jaya Vijayen, Jayati Dev, Jean Linis-Dinco, Jeanette Wong, Jenna Manhau Fung, Jenny Greve, Jens Meijen, Jeremy Ng, Jeremy Pesner, Jessie Yang, Jinqian Li, João Filipe Santiago Silva, Joel Christoph, Jorge Montiel, Jose E Diaz Azcunaga, Joseph Wehbe, Josephine Kaaniru, Joshua Mounsey, Joyce Grant, Juan Armando Becerra Gutiérrez, Julia Andrusiak, Julia Henriques Souza, Julia Meltzer, Júlia Mendonça, Julia Sterling, Julian Buchwalter, Julian Theseira, Julieta Carballo, Jumpei Komoda, Kara Doriani O'Shee, Kardelen Filorinalı Kargın, Kasia Polanska, Katleho Mokoena, Katria Tomko, Kayle Hatt, Kesang Yuden, Ketevani Kukava, Kewa Jiang, Kholofelo Kugler, Krerkkaiwan Kajohnkunchonlathorn, Krishna Kumar Saha, Kristiina Juurmaa, Kristina Fort, Kritika Roy, Kuan-Wei Chen, Kyoungsic Min, Lara Salgueiro Oliveira, Laura Gacho, Laura Greco, Lea Hani, Leah Park, Leandri Pretorius , Leila Doty, Lena Patricia Nieper, Lesa Lawrence, Leslie Salgado, Lillie Coney, Livia Crepaldi Wolf, Logan O'Shaughnessy, Lorena Almaraz De La Garza, Luciana Longo, Lucy Mwaura, Luis Jacob Retanan, Luis Vasquez, Luisa Abreu Dall'Agnese, Luise Eder, Lynn Pickering, Maciej Piszcz, Madeleine Matsui, Madhubhashini Rathnayaka, Maia Levy Daniel, Maitreya Shah, Malek Ben Achour, Manail Anis, Manal Siddiqui, Mansur Omonov, Manuel Gustavo Isaac, Maram Aldaiel, Marcelo Pasetti, Marcia Lindsey, Marcin Rabiza, Marco Bani, Maria Hadjicosta, Maria Lungu, Maria Tran, Mariam Shainidze, Marian Adly, Mariana Correa, Marine Lipartia, Marisa Stones, Marta Bienkiewicz Watson, Martina Malcheva, Martyna Kalvaityte, Marwa Soudi, Mary Ebeling, Marzia Coltri, Mashael Alzaid, Matias Banchero, Mayara Carneiro, Medhavi Mishra, Meg McNulty, Megan Cansfield, Megan McLaughlin, Melike Akkaya, Melissa Hamilton, Melvin W, Mennatullah Hendawy, Michael Nayebare, Michael Shoag, Michaela Sullivan-Paul, Michelle Calabro, Michelle Tan, Minna Adlan, Miranda McClellan, Mirela Buturovic, Mohammad Ghasemi, Moin Khan, Mollie Howard, Monique Munarini, Mostafa Abdelaziz Ashhab Elkadi, Mouloud Khelif, Mubarak Raji, Muneeba Rizvi, Musa Bitrus Ndahi, Nabil Fancy, Nana Afua Brantuo, Nana Khechikashvili, Nana Sakvarelidze, Nandita Sampath, Nasiruddeen Muhammad, Nata Kapanadze, Natalia Martins Fritzen, Natalie Milman, Natascia Arcifa, Natasha Karanja, Nate McFaul, Nayyara Rahman, Nelly Rousseau, Ngozichukwuamaka Yvonne Onyemenam,

Niharika Gujela, Nika Pranata, Niki Agrotou, Nikolaos Prodromos, Nikolas Schmidt, Nino Taganashvili, Ninsiima Melissa, Nkechi Emmanuella Agugoesi, Nkurunziza Christophe, Noah-Levin Damschke, Noemi Mejia, Noora Arajärvi Parsons, Nurcan Uyar, Nuurrianti Jalli, Obichi Obiajunwa, Olga Grygorovska, Olgu Özdemir Ertürk, Olivia Corcoran, Ona Oshen, Ori Freiman, Ornella Belfiori, Oscar Lopez, Paige Lord, Pamela De La Rosa, Paola Katherine Gálvez Callirgos, Parul Anand, Pascal Musyoki, Patrick Armstrong, Patrick Folinsbee, Patrick Lin, Patrycjusz Szubryt, Paula Garuz Naval, Paula Guedes Fernandes da Silva, Paula Hernandez, Paulina Denhi Gomez Escalona, Perpetua Ndidiamaka Ogwuche, Pervin Harvinder SK, Peter Mmbando, Petra Müllerová, Petruta Pirvan, Philip Usigbe, Pierre du Pasquier, Pius Kwao Gadosey, Polycarp Okumu, Preskal Tadrous, Priya Goswami, Radhika Bajpai, Radoslaw Komuda, Rajalakshmi Balasubramanian, Ranjini Syam, Raúl Vizcarra Chirinos, Raymond Amumpaire, Rebecca Aspetti, Rebecca Bonnevie, Rebecca Distler, Reed Shaw, Reeneth Santos, Rehema Baguma, Rémy Ibarcq, Remy Takang Arrey, Ren Bin Lee Dixon, Renan Canaan, Renee Black, Roch Glowacki, Roselyn Martin, Rotimi Owolabi, Ruby, Ruofei Wang, Rwamamara Lionel, Ryan Burns, Ryan Watkins, Saba Elizbarashvili, Saba Tiku Beyene, Sabire Sanem YILMAZ, Sahr Muhammedally, Saimum Reza Talukder, Sajid Mahmud, Samantha González, Samantha Khoo, Samantha Ndiwalana, SaMee Harden, Sandra Makumbirofa, Santiago Molina, Sara Fratti, Sarah D'Andrea, Sarah Fu, Sarah Sinclair, Scarleth Herrera, Sebastián Dueñas Müller, Sechaba Keketsi, Sedef Akinli Kocak, Selin Ozbek Cittone, Serçin Kutucu, Serif Onur Bahcecik, Serign Modou Bah, Shahneela Shahbaz, Sharinee Lalit Jagtiani, Sharon Bassan, Sharon Ndinda Nzuki, Sharvari Dhote, Shauna Blackmon, Sheila Casserly, Shiva Kanwar, Shokhrukh Nuraliev, Shreenithi Annadurai, Sidney Frank Engelbrecht, Simone Maria Parazzoli, Siqi Zhao, Sofia Calado, Sofia Surla, Soham Bose, Soledad Guilera, Solomon Karanja Meru, Sophie Hale, Stacey Berry, Stacy Owino, Stefanos Vitoratos, Stella George, Stella Lee, Stella Yiantet Letuya, Stephan Sonnenberg, Stephanie Forbes, Stephanie Muthoni Muthungu, Stephie-Lea Tabujara, Sujitha Subramanian, Suljo Corsulic, Sumit Kumar, Supriya Kulkarni, Susan Ibukun Akinade, Svetluša Surová, Svitlana Tarasenko, Tabani Moyo, Taina Silveira Baylao, Talha Bin Tariq, Tamiko Eto, Tamilla Triantoro, Tatiana Zasheva, Teddy Odira, Tessa Darbyshire, Thiago Moraes, Thomas Linder, Thummim Iyoha Osagie, Timothy Sowa, Tlhaloso Mpopo, Tobias Wirth, Tom Utum, Tomiris Amirova, Tori Rousay, Uchenna Anyamele, Uchenna Mgbaja, Ulysse Richard, Uzma Alam, Valéria Silva, Valerie Edna, Valerie Elefante, Varsha Sewlal, Veronika Klymova, Victor Bohorquez, Vikash Madduri, Vishakha Agrawal, Vishmi Fernando, Vladimir Cortes, Vu Thach Thao Nguyen, Waridah Makena, Wathagi Ndungu, William Bello, Winnie Kungu, Worawan Klinsawai, Yasin Tokat, Yelyzaveta Markova, Yiğitcan Çankaya, Yishu Mao, Yogendrasingh Pawar, Youssef Horchani, Yulia Askhadulina, Yuval Sinay, Zanele Sokatsha, Zina Akrout, Zoe Yi Soh, Zurab Karchava

2022 Contributors: Abraham Kuuku Sam, Aleksei Gudkov, Alketa Hotaj, Ananya Ramani, Ashwini Natesan, Ayesha Dawood, Avinash Dadhich, Belen Bricchi, Benjamin Faveri, Bisma Shoaib, Busra Memisoglu, Caitlin Power, Cari Miller, Caroline Friedman Levy, Chukwuyere Izuogu, Çiğdem Akın, Claudia Wladdimiro Quevedo, Dalila Hoover, Deo Shao, Divya Diwedi, Doreen Aoko Abiero, Emily Mehalek, Eric Wamugu Maina, Fabienne Graf, Fatuma Ibrahim, Grace Thomson, Hande Çağla Yılmaz, Harsha Agrawal, Helen Samukeliso Sithole, Ibrahim Sabra, Idil Kula, Ingrid Soares, Jaya Vijayen, João Santiago Silva, Jens Meijen, Jordan Richard Schoenherr, Julian Theseira, Leandri Pretorius, Liliane Obrecht, Livia Crepaldi Wolf, Lucy Wanjiku Mwaura, Magdalena Rzaca, Mahak Rathee, Mai Kato, Manail Anais, Mariam Jamal, Mark Sayre, Martyna Kalvaityte, Matthew Ogbeifun, Mokesioluwa Fanoro, Nasir Muftic, Nicole Lemke, Nikolas Kyriacou, Ori Freiman, Osaye Ekomwereren, Osman Gazi Güçlütürk, Peter Graham, Peter Tsamwa, Sajid Mahmud, Samuel Chukwuebuka Uzoigwe, Sanduni Wickramasinghe, Selina Onyando, Shouhui Zhou, Soraya Weiner, Tamar Kankava, Tina Lassiter, Victor Famubode, Yussuf Abdirahim, Davor Ljubenkov, Elisa Elhadj, Joel Kumwenda, Jonathan Kurniawan, Mélissa M'Raidi-Kechichian, Nidhi Sinha, Mercy Chinazom Godwin, Zhuozheng (Rick) Cai, Aditya Kumar Pal, Allan Ochola, Angel Arroyo, Ashkan Alinaghian, Ayca Ariyoruk, Bhredipta Socarana, Brenda Wangechi Maina, Canan Erez, Catherine Setiawan, Cherie Adhiambo Oyier, Christof Wolf-Brenner, Amir Mikail, Cody D. Rutherford, Deyanira Juliet Murga, Duță Daniela, Edwin Concepcion, Evelina Ayrapetyan, Florence Anyango Ogonjo, Gvantsa Baidoshvili, Ibrahim Said, Ibrahim Yeku, Jacob Odame-Baiden, Jamie Wong, Jodi Masters-Gonzales, John Mathew, Kerstin Waxnegger, Lesa Lawrence, Luca Nannini, Luz Elena Gonzalez, Marian Ela C. Ebillo, Emaediong Akpan, Jackline Akello, Marine Lipartia, Mario Emmanuel Rodriguez Trejo, Mark Anthony Perez, Melody Musoni, Nayyara Rahman, Nazam Laila, Nicholas Kisundu, Fredrick Ishengoma, Nojus Antanas Bendoraitis, Olajide Olugbade, Omolola Oviroh, Parisa Osivand Pour, Parul Shukla, Pierrinne Leukes, Prarthana Vasudevan, Ruofei Wang, Saba Elizbarasvhili, Samantha De Soysa, Sandra Lattner, Sandra Omayi Musa, Shahira El Alfy, Shilpa Ramachandran, Shradhanjali Sarma, Sivaramakrishnan R Guruvayur, Takaya Terakawa, Umut Pajaro Velasquez, Unyime Paul Akpabio, Vaibhav Garg, Vakhtang Chkhenkeli, Stella Wamae, Varsha Sewlal, Vasileios Rovilos, Mohamed Elbashir, Victor Werimo, Wellington Galassi, Maha Jouini, Yasin Tokat, and Yogendrasingh Pawar

2021 Contributors: Adeboye Adegoke, Temofe Akaba, Selim Alan, Zelal Binici, Afi Blackshear, Bridget Boakye, Giuliano Borter, Stephanie Cairns, Shangamitra Chakraborty, Lyantoniette Chua, Roberto López Dávila, Pam Dixon, Divya Dwivedi, Elisa Elhadj, Pete Furlong, August Gweon, Lida Haghnegahdar, Gokce Cobansoy Hizel, Jason K. Johnson, Eddan Katz, Dongwoo Kim, Writankar Kundu, Thobekile Matimbe, Natalia Menendez, Kushang Mishra, Charles Kajoloweka, Mélissa M'Raidi-Kechichian, Alex Moltzau, Tamra Moore, Oarabile Mudongo, Somaieh Nikpoor, Diana Kemunto Nyakundi, Ubongabasi Edidiong Obot, Marcos López

Oneto, Ananya Ramani, Rishi Ray, Rachel Stockton, Anirbar Sen, Niovi Vavoula, Sherry Wu, Khatia Zukhubaia, Larissa Zutter

2020 Contributors: Mariame Almoghanam, Francesca Bignami, Anne Carblanc, Karine Caunes, Emilio De Capitani, Pam Dixon, Paula Soumaya Domit, Giovanni De Gregorio, Douglas Frantz, Merve Hickok, Regina Iminova, Lorraine Kisselburgh, Rebecca Leeper, Wonki Min, Pablo Molina, Maria Helen Murphy, Stephanie Perrin, Bilyana Petkova, Oreste Pollicino, Nguyen Dinh Quy, Cristos Velasco, Wendell Wallach, and Larissa Zutter

Reviewers since 2020: Dr. Elena Abrusci, Brunel University London; Prof. Syed Ishtiaque Ahmed, University of Toronto; Tetiana Avdieieva, Digital Security Lab Ukraine; Bruno Bioni, Data Privacy Brazil; Isabelle Buscke, Brussels Office, Federation of German Consumer Organisations; Darren Grayson Chng, Singapore Personal Data Protection Commission; Prof. Renée Cummings, University of Virginia; Prof. Giovanni De Gregorio, Católica Global School of Law; Prof. Virginia Dignum, Umea University; Dr. Hlengiwe Dube, University of Pretoria; Marina Garrote, Data Privacy Brazil; Armando Guio Español, Berkman Klein Center, Harvard University; Dr. Gry Hasselbalch, DataEthics.eu, InTouchAI.eu; Dr. Masao Horibe, Professor Emeritus at Hitotsubashi University, Former Chairman of Personal Information Protection Commission (PPC), Japan; Bulelani Jili, Harvard University; Leyla Keyser, IT Law Research Center, Istanbul Bilgi University (with Işıl Selen Denemeç, Zümrüt müftüoğlu, and Dr. Atilla Aydin); Prof. Dr. Katja Langenbucher, Goethe University; Prof. Seth Lazar, Australian National University; Prof. Justin Longo, University of Regina; Prof. Wonki Min, Technology Ambassador, Government of Korea; Prof. Valsamis Mitsilegas, Queen Mary University of London; Dr. Pablo G. Molina, Georgetown University; Julia Mendonça, Data Privacy Brazil; Prof. Maria Murphy, Maynooth University; Angella Ndaka, Centre for Epistemic Justice Foundation; Dr. Irena Nesterova, Researcher at the Institute of Legal Science, Faculty of Law, University of Latvia, member of the UNESCO Ad Hoc Expert Group for the Recommendation on the Ethics of Artificial Intelligence; Prof. Oreste Pollicino, Bocconi University; Prof. Viviana Polisena, Universidad Católica de Córdoba; David Roldán Martínez, International WoMenX in Business for Ethical AI; Prof. Carolina Rossini, University of Massachusetts, Amherst; Prof. Emma Ruttkamp-Bloem, University of Pretoria; Prof. Edward Santow, University of Technology Sydney; Prof. Dr. Ingrid Schneider, Universität Hamburg, Department of Informatics; Prof. Giovanni Sartor, University of Bologna, European University of Florence; Prof. Ben Shneiderman, University of Maryland; Takaya Terakawa, Technica Zen; Dr. Niovi Vavoula, Université du Luxembourg; Dr. Cristos Velasco; Dr. Kutoma Wakunuma, De Montfort University; Prof. Hannes Werthner, Digital Humanism; Prof. Michael Zimba, Malawi University of Science and Technology

CAIDP Board of Directors:

Lorraine Kisselburgh, Chair

Thank you to the following funders for helping make this report possible with their support of the Center for AI and Digital Policy

Foundation Support
Craig Newmark Charitable Foundation
Heising-Simons Foundation
McCourt School of Public Policy
Minderoo Foundation
Omidyar Network
The Patrick J. McGovern Foundation
Planet Heritage Foundation

Individuals
Anonymous
Mohamed Elbashir Ahmed
Steve Bunnell
Ivan and Sharon Fong
Oscar and Judith Gandy
Joi Ito
Leonard Kennedy
The Leah Foundation
Somaieh Nikpoor
Estate of Charles Rotenberg
Michael and Karen Rotenberg
Bruce Schneier
The Barbara Simons Fund
David Stern
Washington Community Charitable Foundation

Paul Wolfson

Endorsements

The CAIDP AI and Democratic Values report is a trusted global reference.

– Council of Europe Secretary General Alain Berset (2026)

For anyone who wants to engage AI governance seriously, as a researcher, practitioner, educator, or advocate, the CAIDP Index is as close to an essential resource as this field has produced.

– Carolina Rossini, University of Massachusetts, Amherst (2026)

The Index offers a valuable resource for policymakers, researchers, and civil society working to ensure that artificial intelligence is developed and deployed in ways that are trustworthy, human-centric, and beneficial to society.

– David Roldán Martínez, International WoMenX in Business for Ethical AI (2026)

The CAIDP Index and the work of the CAIDP in general demonstrate the value of civil society engagement as a factor in policy, in this case as an essential evidence-based contribution to the regulation and governance of AI.

– Hannes Werthner, Digital Humanism (2026)

The Index is essential for informing anyone curious about the key trends on how nations are responsibly adopting AI.

– Michael Zimba, Malawi University of Science and Technology (2026)

The wealth of information in the CAIDP Index is unparalleled and the rigor is uncanny. What an excellent tool for researchers and policy makers!

– Niovi Vavoula, Université du Luxembourg (2026)

The CAIDP Artificial Intelligence and Democratic Values Index is a critical resource for practitioners and technology adoption leaders

– Pablo Molina, Drexel University (2026)

[CAIDP's study is] a significant milestone that opens the door for us to consider the long-term material impacts of technology, ensuring that our pursuit of innovation consistently protects the intellectual and spiritual well-being of all societies.

– Syed Ishtiaque Ahmed, University of Toronto (2026)

As the development and use of AI continues to increase exponentially, governments and civil society have struggled to keep pace. [...] This report is essential reading for anyone looking to grasp the key trends that link how nation states are responding to the rise of AI, and for anyone looking to understand the key developments in specific states. It is both thoroughly researched and the product of deep and broad expertise.

– *Edward Santow, Human Technology Institute, UTS (2025)*

The AIDV Index is an incredibly impressive and valuable compendium. The country reports themselves offer an extremely valuable collection of case studies.

– *Justin Longo, University of Regina (2025)*

The AI and Democratic Values 2024 report is a well-developed and highly technical work that provides essential insights into the intersection of artificial intelligence and democratic governance. This report serves as a crucial tool for understanding how nations can effectively promote and safeguard democratic values in the age of AI. It is an invaluable resource for countries seeking to strengthen democratic principles while leveraging AI responsibly.

– *Armando Guio Español, Berkman Klein Center (2025)*

It is always encouraging to see the progress in AIDV and its contribution to the formation of AI policies around the world.

– *Wonki Min, KAIST (2025)*

The report seeks to shed light on the broad landscape of AI regulation and governance. Accordingly, it seeks to explicate both the challenges and opportunities surrounding the adoption of AI systems, making the AIDV a useful reference document that points towards more inclusive digital futures.

– *Bulelani Jili, Harvard University (2025)*

I endorse this report for its clear and diverse inclusion of countries in both the Global North and the Global South. The report gives an important insight into AI-related activities globally. This is important because AI does not operate in a vacuum, nor does it operate in silos: its impact is intertwined across the globe.

– *Kutoma Wakunuma, De Montfort University (2025)*

The CAIDP 2024 Report continues its tradition of excellence in providing essential insights into the global relationship among democratic values, ethics and the accelerated pace of AI innovation and development. It is an essential tool to understand the nature of AI and its relationship to the public good. At UNESCO, with our global standard on the ethics of AI, we are proud to partner with CAIDP to achieve a technological transition that is fair and sustainable.

– *Former UNESCO Assistant Director-General for Social and Human Sciences at Gabriela Ramos (2024)*

CAIDP's report sheds light on the current landscape of international regulations, national laws and policies emerging from the fascinating and complex field of AI. The report emphasizes the imperative to protect fundamental rights and uphold the rule of law and will serve as a useful roadmap for shaping AI governance and to foster and advance a crucial dialogue among diverse stakeholders in the international arena."

– *Cristos Velasco, DHBW Cooperative State University (2024)*

CAIDP's flagship report provides insightful information how the AI principles are translated into practice across 75 countries.

– *Wonki Min, SUNY Korea (2023)*

It is a great exercise of comparative digital law in action, providing a comprehensive perspective on the evolution of AI policy also contributing to underline the constitutional relationship between AI & democracy. Digital Democracy and rule of law in the digital context will be the real challenges for the next decade.

– *Oreste Pollicino, Bocconi University (2023)*

A very worthwhile analysis.

– *Stuart Russell, University of Berkeley, California (2022)*

An impressive body of work.

– *Alessandro Acquisti, Carnegie Mellon University (2022)*

An excellent report, a very useful tool for the community and a reference to see nexus between AI and Democratic Values.

– *Leyla Keser, Istanbul Bilgi University (2022)*

A very insightful report on AI laws and policies that addresses significant international developments and provides a terrific comparison of AI laws and policies across 50 countries.

– *Lee J. Tiedrich, Duke University (2022)*

This report is unique in that it compares AI policies and practices around the world and should inspire progress toward trustworthy and human-centric AI. The report also documents the important role of digital rights and consumer groups in this endeavor.

– *Ursula Pachl, European Consumer Association (BEUC) (2022)*

It's rare to read a document that has the potential to influence national policies on artificial intelligence around the world. The country descriptions and evaluations are thoughtfully organized, well-written, and carefully documented. The country rankings enable readers to have a broad understanding of who the leaders are and why, while pointing out what still needs to be done. The AI and Democratic Values index gives me hope that AI policy efforts can improve human rights, social justice, and dignity.

– Ben Shneiderman, author, Human-Centric AI (Oxford 2022)

In this historical moment of global interest negotiation and competition on AI we are in right now, we urgently need a shared narrative serving the human interest and democracy. With its comprehensive overview of key global and local power actors and initiatives The Artificial Intelligence and Democratic Values Index is an essential instrument for the facilitation of this shared global dialogue on AI now and in many years to come.

– Gry Hasselbalch, DataEthics.eu, InTouchAI.eu (2022)

This is a very ambitious and important exercise. The outcome is impressive—detailed and rigorous.

– Valsamis Mitsilegas, Queen Mary University of London (2022)

This comparative study sheds critical light on the adoption of Artificial Intelligence in democratic societies. AI and Democratic Values is an indispensable reference source for regulators, reporters, academics, and practitioners

– Dr. Pablo G. Molina, Georgetown University and founder, International Applied Ethics and Technology Association (iaeta.org) (2022)

A Year in Review

2025: Progress, Backsliding, and Pushing Forward

The *AI and Democratic Values Index* (***CAIDP Index***), published annually by the Center for AI and Digital Policy provides a comprehensive review of AI policies and practices worldwide. In this sixth edition, we highlight increased support for the international AI treaty and progress in various countries toward comprehensive AI legislation and the creation of supervisory authorities. We also noted backsliding, particularly with attempts by the US federal government to interfere with state legislation on AI and influence the implementation of the EU AI Act. Despite these setbacks, AI governance continues to move forward, led by international organizations including the United Nations and UNESCO and, increasingly, middle powers, including countries from the Global South.

Progress

In the roughly 20 AI Intelligence and Human Rights, Democracy, and the Rule of Law opened for signature, 45 countries have endorsed this first international treaty on AI. Bosnia and Herzegovina, Canada, Japan, Liechtenstein, Switzerland, Ukraine, and Uruguay signed in 2025, and Armenia signed in early 2026.[1] The mandate of the Committee on Artificial Intelligence (CAI), which drafted the Convention, concluded in 2025.[2] The Treaty extends to South America, North America, Europe, and East Asia. The final meeting reflected the global commitment to the treaty's principles and expanded global outreach. Ecuador joined the meeting as a CAI observer for the first time and committee members engaged in a Thematic Hearing on "Connecting Asia, the Global South, and Europe on Responsible AI in the Public Sector."[3] The Declaration on AI and Global Governance[4] at the conclusion of the Lisbon Forum organized with the GPAI Expert Community[5] reflected efforts to connect Europe, Africa, the Middle East, Latin America, and Asia in discussions on AI governance challenges. A proposed Steering Committee on New and Emerging Technologies (CDNET) to complete work on implementation tools for the Framework Convention

---

[1] Council of Europe Treaty Office, *Chart of Signatures and Ratifications of Treaty 225* (Mar. 31, 2026), https://www.coe.int/en/web/Conventions/full-list/?module=signatures-by-treaty&treatynum=225

[2] Council of Europe, *Committee on Artificial Intelligence (CAI)* (2026), https://www.coe.int/en/web/artificial-intelligence/cai

[3] Committee on Artificial Intelligence, *14th Plenary Meeting*, Agenda Items 4, 9 (Nov. 3–5, 2025), https://rm.coe.int/cai-2025-13-14th-meeting-report/488029e106

[4] Lisbon Forum, *Lisbon Declaration on AI and Global Governance* (Oct. 28–29, 2025), https://rm.coe.int/2025-lisbon-declaration/488029b11d

[5] Council of Europe North-South Centre, *Lisbon Forum Declaration on AI and Global Governance Now Published* (Dec. 10, 2025), https://www.coe.int/en/web/north-south-centre/-/lisbon-forum-declaration-on-ai-and-global-governance-now-published

and to offer advice to the Committee of Ministers on AI governance could further increase the treaty's global support.[6]

Several countries formalized legal frameworks to guide the development, deployment, and use of artificial intelligence and to answer questions about liability for harms. China's revised cybersecurity law[7] defines liability and specifies penalties for individuals and for network operators responsible for "serious" failures such as data breaches or infrastructure failure. Individuals and operators can also be liable for disseminating or failing to stop the spread of prohibited information. In the European Union, Italy charged forward, delivering the first national framework implementing the EU AI Act.[8] Slovenia[9] and Hungary, too,[10] passed national acts to create a more robust framework for enforcing the regional act. Even as the European Commission's Digital Omnibus loosens requirements in the name of simplification,[11] national governments are developing regulations to provide clarity for companies and protections for their citizens.

Vietnam[12] and Peru[13] both passed comprehensive, risk-based AI legislation near the end of 2025. Like the EU AI Act, the Vietnamese and Peruvian legal frameworks require safeguards for high-risk systems and prohibit certain AI applications such as social scoring in the Vietnamese case[14] and mass surveillance and

---

[6] Ibid, Agenda Items 10–11

[7] National People's Congress China Daily, *AI Support at Heart of Cybersecurity Revision* (Oct. 30, 2025), http://en.npc.gov.cn.cdurl.cn/2025-10/30/c_1136472.htm

[8] Gazzetta Ufficiale, *Provisions and Delegations to the Government on Artificial Intelligence [Disposizioni e deleghe al Governo in matria di intelligenza artificiale]*, Law No. 132 of 23 September 2025, Official Gazette of the Italian Republic, no. 223 (Sept. 23, 2025), https://www.gazzettaufficiale.it/eli/id/2025/09/25/25G00143/sg

[9] Uradni List, *3035. Act on the Implementation of the Regulation (EU) laying down harmonised rules on artificial intelligence (ZIUDHPUDI),* Official Gazette, page 9899 (Oct. 31, 2025), https://www.uradni-list.si/glasilo-uradni-list-rs/vsebina/2025-01-3035

[10] Nemzeti Jogszabálytár, *Decree 344/2025 on the Implementation of the Act Executing the European Union Regulation on Artificial Intelligence in Hungary [Korm. rendelet Az Európai Unió mesterséges intelligenciáról szóló rendeletének magyarországi végrehajtásáról]* (Oct. 31, 2025), https://njt.hu/jogszabaly/2025-344-20-22

[11] European Commission, *Simplification—Digital Package and Omnibus* (2026), https://ec.europa.eu/info/law/better-regulation/have-your-say/initiatives/14855-Simplification-digital-package-and-omnibus_en

[12] Vietnam News Agency (VNA), *NA Passes First-Ever Law on Artificial Intelligence* (Dec. 10, 2025), https://en.vietnamplus.vn/na-passes-first-ever-law-on-artificial-intelligence-post334057.vnp

[13] Presidencia del Consejo de Ministros, *Supreme Decree No. 115-2025-PC, that approves the Regulation of the Law No. 31814, Law Promoting the Use of Artificial Intelligence for the Country's Economic and Social Development [Decreto Supremo que aprueba el Reglamento de la Ley No 31814, Ley que promueve el uso de la inteligencia artifical en favor del desarrollo económico y social del país]* (Sept. 9, 2025), https://www.gob.pe/institucion/pcm/normas-legales/7133522-115-2025-pcm

[14] Ministry of Science and Technology, *Details of the Draft Document QPLL* (Sept. 29, 2025), https://mst.gov.vn/van-ban-phap-luat/du-thao/2294.htm; *Luật Trí Tuệ Nhân Tạo* [full text,

biometric identification (with exceptions) in the Peruvian case.[15] These prohibitions offer a step toward realizing the Global Call for AI Red Lines launched during the UN General Assembly and supported by over 90 organizations, including CAIDP, and AI and governance experts.[16] Most of the provisions of the Peruvian law entered into force at the end of 2025 while Vietnamese law entered into force in March 2026, months after the Korean AI Basic Law promulgated in January 2025.[17] Like the EU law, the Peruvian law has rolling deadlines, including progressive steps over three years toward full implementation of the article on Algorithmic Transparency. Taiwan's AI Basic Act enacted in January 2026 defines principles and a risk-based framework for future legislation.[18]

Implementation of global frameworks and governance mechanisms progressed in 2025, too. UNESCO has supported more than 40 countries through the report stage of Readiness Assessment Methodology (RAM).[19] Countries from the Global South and global majority continue to lead in the completing the RAM. Asia saw the greatest number of completions since the launch of the 2025 Index, with 10 countries—Bangladesh, Cambodia, India, Laos, Malaysia, Maldives, Philippines, Thailand, Timor-Leste, and Vietnam—joining Indonesia. Botswana, Chad, Mauritius, Namibia, São Tomé and Príncipe, Tanzania, and Zimbabwe joined African countries identified last year, with Rwanda nearing completion. Egypt, listed under Arab States and Africa, and Oman joined Morocco and Saudi Arabia in the Middle East. Eight countries from Latin America and the Caribbean also published reports since April 2025, including Peru, Paraguay,[20] Jamaica, Ecuador, Curaçao, Cuba, Colombia,

---

Vietnamese] (2025), https://mic.mediacdn.vn/document/2025/10/2/250925duthao-luatai-v10-1759393446665893284209.pdf

[15] Presidencia del Consejo de Ministros, *Law Promoting the Use of Artificial Intelligence for the Country's Economic and Social Development*, Art. 23 (Sept. 9, 2025), https://cdn.www.gob.pe/uploads/document/file/8619777/7133522-decreto-supremo-n-115-2025-pcm.PDF?v=1757422328

[16] AI Red Lines, *Global Call for AI Red Lines* (2025), https://red-lines.ai/#call

[17] Ministry of Science and ICT, *A New Chapter in the Age of AI: Basic Act on AI Passed at the National Assembly's Plenary Session* (Dec. 26, 2024), https://www.msit.go.kr/eng/bbs/view.do?sCode=eng&mId=4&mPid=2&pageIndex=&bbsSeqNo=42&nttSeqNo=1071&searchOpt=ALL&searchTxt=

[18] Presidential Order, *Artificial Intelligence Basic Act* (Jan. 8, 2025), https://www.linkedin.com/posts/luisalbertomontezuma_taiwan-activity-7443304394474979328-ajVp; Office of the President, *Formulate a Basic Law on Artificial Intelligence* (Jan. 14, 2026), https://www.president.gov.tw/Page/294/50131

[19] UNESCO Global AI Ethics and Governance Observatory, *Global Hub* (Oct. 2025), https://www.unesco.org/ethics-ai/en/global-hub; as of 3/31/26, 43 RAM reports are accessible in the Digital Library, though not all of those countries are listed on the Hub as completed. Colombia, Democratic Republic of Congo, and Paraguay have published reports but have no profile. Rwanda has a profile, but the report is not available.

[20] UNESCO Office Montevideo and Regional Bureau for Science in Latin America and the Caribbean, *Paraguay: Evaluation of the Status of Artificial Intelligence Preparation [Paraguay: Evaluación del estadio de preparación de la inteligencia artificial]* (2025), https://unesdoc.unesco.org/ark:/48223/pf0000396524

Antigua and Barbuda. The Netherlands became the first European country to complete the process in 2025, with Moldova following in early 2026. Countries have also participated in programs to equip civil servants with basic knowledge about the uses and risks of AI technologies through UNESCO initiative such as the Supervising AI by Competent Authorities program[21] and the AI Literacy Program for Civil Servants.[22]

Beyond the RAM, many countries have taken action to implement the UNESCO Recommendation[23] and OECD AI Principles[24] through policies to integrate AI literacy and digital skills into national education programs and support for community-based initiatives. Namibia's national digital strategy emphasizes integrating digital literacy into the national education curriculum, training teachers, and developing community-based digital literacy and AI training programs to empower citizens to know when and how to use AI systems effectively.[25] Argentina's Secretariat of Education and Ministry of Human Capital launched a beta version of a program (PAIDEIA) to prepare students in primary and secondary education to "understand and actively participate in a world increasingly mediated by technology."[26] The program applies cognitive and durable skills to coding and computer science as well using and developing AI systems. Finland's Ministry of Education and Culture's Recommendation on AI also centers AI literacy with attention to legal and ethical issues.[27] Finland's Recommendation identifies competencies students can develop through AI literacy education and the safe use of AI tools in learning. These programs also build citizens' capacity to engage in AI policymaking.

Japan continues to model action-oriented governance globally, particularly through the country's leadership engaging with the G7 Hiroshima AI Process (HAIP). The OECD and G7 released the HAIP Reporting Framework in early 2025, and 9 of

---

[21] UNESCO, *Expanding Capacity Building for Competent Authorities on AI: National Trainings across the EU* (Dec. 9, 2025), https://www.unesco.org/en/articles/expanding-capacity-building-competent-authorities-ai-national-trainings-across-eu

[22] Christabel Randolph and Merve Hickok, *UNESCO's AI Literacy Trianing for Civil Servants: Empowering Ethical AI Governance around the World*, UNESCO Global AI Ethics and Governance Observatory (Dec. 2, 2025), https://www.unesco.org/ethics-ai/en/articles/unescos-ai-literacy-training-civil-servants-empowering-ethical-ai-governance-around-world-0

[23] UNESCO Digital Library, *UNESCO's Recommendation on the Ethics of Artificial Intelligence: Key Facts*, p. 14 (2023), https://unesdoc.unesco.org/ark:/48223/pf0000385082.page=14

[24] OECD AI Policy Observatory, *Building Human Capacity and Preparing for Labour Market Transformation (Principle 2.4)* (2026), https://oecd.ai/en/dashboards/ai-principles/P13

[25] Minister of Information and Communications Technology, *National Digital Strategy Namibia 2025–2028*, pp. 61–63 (2025), https://mict.gov.na/documents/869282/6500577/NDS4+040825.pdf/aeb1c2a6-63a9-f04e-7f78-3e15549eefcf?t=1754320341496

[26] Ministerio de Capital Humano, *Paideia: Argentine Program of Education Innovation with Artificial Intelligence [Paideia: Programa Argentino de Innovación de la Educación con Inteligencia Artificial]*, https://www.argentina.gob.ar/capital-humano/educacion/paideia

[27] Ministry of Education and Culture, *Recommendations for AI* (Sept. 18, 2024), https://okm.fi/en/project?tunnus=OKM021:00/2024

the 25 companies who have submitted reports as of January 2026, are Japanese.[28] Japan explicitly drew on the HAIP to formulate the national AI Act, which is intended to "improve the quality of life of the Japanese people and boost the development of the national economy by promoting AI research, development, and utilization."[29] Japan's integration of the HAIP in the national AI Act exemplifies how countries can integrate international frameworks and mechanisms into national legislation to ensure citizens' safety and to build trust.

The AI Action Summit in early 2025 marked a clear shift in the global AI policy agenda from a narrow focus on catastrophic risk at the 2023 Bletchley Safety Summit to a broader emphasis on deployment, economic opportunity, and global governance. The conference also featured many new participants from the Global South. More than 60 countries and major international organizations endorsed a political declaration emphasizing that AI should be "open, inclusive, transparent, ethical, safe, secure and trustworthy" and aligned with human rights and sustainability goals.[30] Surprisingly, the United States and the United Kingdom chose not to endorse the conference declaration.

The Vatican has also been prominent in discussions on AI governance. Pope Francis's message to the AI Action Summit stressed the power of the heart over algorithms and reiterated that "the voices of all stakeholders should be taken into account, including the poor, the powerless and others who often go unheard in global decision-making processes"[31] in discussions on AI governance. The Vatican endorsed the Statement on Inclusive and Sustainable AI.[32] Pope Leo emphasized human dignity, too, telling AI governance leaders "Acknowledging and respecting what is uniquely characteristic of the human person is essential to the discussion of any adequate ethical framework for the governance of AI."[33] This effort, requires

---

[28] OECD AI Policy Observatory, *G7 Reporting Framework—Hiroshima AI Process (HAIP) International Code of Conduct for Organizations Developing Advanced AI Systems* (2026), https://transparency.oecd.ai/reports

[29] Public Relations Office Government of Japan, *Act on Promotion of Research and Development, and Utilization of Artificial Intelligence-Related Technology Now in Full Effect*, Policy Related News Series, Vol. 209 (Nov. 2025), https://www.gov-online.go.jp/hlj/en/november_2025/november_2025-08.html

[30] Élysée, *Statement on Inclusive and Sustainable Artificial Intelligence for People and the Planet* (Feb. 11, 2025), https://www.elysee.fr/en/emmanuel-macron/2025/02/11/statement-on-inclusive-and-sustainable-artificial-intelligence-for-people-and-the-planet

[31] Holy See, *Message of the Holy Father to the President of Republic of France on the Occasion of the 'Sommet pour l'Action sur l'Intelligence Artificielle'* (Feb. 6, 2025), https://www.vatican.va/content/francesco/en/messages/pont-messages/2025/documents/20250207-messaggio-summit-parigi-ia.html

[32] Élysée, *Statement on Inclusive and Sustainable Artificial Intelligence for People and the Planet* (Feb. 11, 2025), https://www.elysee.fr/en/emmanuel-macron/2025/02/11/statement-on-inclusive-and-sustainable-artificial-intelligence-for-people-and-the-planet

[33] Holy See, *Message of Pope Leo XIV to Participants in the Second Annual Conference on Artificial Intelligence, Ethics, and Corporate Governance* (Jun. 17, 2025), https://www.vatican.va/content/leo-xiv/en/messages/pont-messages/2025/documents/20250617-messaggio-ia.html

"widespread participation that gives everyone the opportunity to be heard with respect, even the most humble"[34] to restore "confidence in the human ability to guide the development of these technologies [...] a confidence that today is increasingly eroded by the paralyzing idea that [AI] development follows an inevitable path." Actors from politics, business, communities, and religious and other institutions must place this joint responsibility, Pope Leo argued, "before any partisan interest or profit, which is increasingly concentrated in the hands of a few."

Backsliding

The US federal administration's swift revocation of the 2024 Executive Order on Safe, Secure, and Trustworthy AI[35] marked the beginning of a steady decline in the country's leadership in trustworthy artificial intelligence and global governance. Administration leaders invoked a race with China and security threats from unnamed adversaries to rationalize an innovation-at-all costs narrative that opposed well-established values for AI governance, including fairness, transparency and accountability, and reliability.[36] This narrative has material effects. With the withdrawal from UNESCO[37] and opposition to the AI Scientific Panel and the Paris Declaration, the United States has increasingly acted on the world stage as an adversary to the AI governance principles it helped establish.

The United States has threatened progress in the European Union, too. The proposed AI Digital Omnibus would grant companies easier access to data for training models[38] without the AI Act protections designed to build trust.

---

[34] Holy See, *Address of His Holiness Pope Leo XIV to Participants in the Conference "Artificial Intelligence and Care of Our Common Home" Organized by the Centesimus Annus Pro Pontifice Foundation and the Strategic Alliance of Catholic Research Universities* (Dec. 5, 2026), https://www.vatican.va/content/leo-xiv/en/speeches/2025/december/documents/20251205-conferenza.html

[35] Federal Register, *Executive Order 14110: Safe, Secure, and Trustworthy Development and Use of Artificial Intelligence* (Oct. 30. 2023), https://www.federalregister.gov/documents/2023/11/01/2023-24283/safe-secure-and-trustworthy-development-and-use-of-artificial-intelligence

[36] Trump White House, *Removing Barriers to American Leadership in Artificial Intelligence* (Jan. 23, 2025), https://www.whitehouse.gov/presidential-actions/2025/01/removing-barriers-to-american-leadership-in-artificial-intelligence/

[37] Trump White House, *Withdrawing the United States from International Organizations, Conventions, and Treaties That Are Contrary to the Interests of the United States: Memorandum for the Heads of Executive Departments and Agencies* (Jan. 7, 2026), https://www.whitehouse.gov/presidential-actions/2026/01/withdrawing-the-united-states-from-international-organizations-conventions-and-treaties-that-are-contrary-to-the-interests-of-the-united-states/

[38] European Commission, *Proposal for a Regulation of the European Parliament and of the council amending Regulation (EU) 2024/1689 and (EU) 2018/1139 as regards the simplification of the implementation of harmonised rules on artificial intelligence (Digital Omnibus on AI)* (Nov. 19, 2025), https://eur-lex.europa.eu/legal-content/EN/TXT/?uri=CELEX:52025PC0836

Pushing Forward

Resistance to the Trump Administration has united actors in the United States and globally. California,[39] New York,[40] and Texas,[41] for example, moved forward with new AI safeguards. The US Senate defeated a proposed moratorium on state laws, 99–1.[42] G20 leaders moved forward with their summit in November despite a US boycott over claims South Africa—which was shunned in international organizations in the 1960s–early 1990s over its apartheid regime—was persecuting the white minority.[43] The G20 Declaration highlighted a broader range of global voices in the discussions on AI, data governance, and innovation for sustainable development.[44] By contrast, the US State Department media note announcing assumption of the 2026 G20 presidency promises to "return the G20 to focusing on its core mission of driving economic growth and prosperity to produce results."[45]

Many national governments moved to combat deepfakes and AI-generated nonconsensual sexual material,[46] especially involving children. Italy's national AI law criminalizes harmful uses of artificial intelligence, including deepfakes,[47] making users generating and sharing undressed or otherwise harmful content without consent liable to criminal charges. Italy's data protection office (*Garante*) also warned companies that own products used to create these images, including Grok and ChatGPT, of their obligation "to design, develop, and make available applications and platforms […] to ensure that users can use them in compliance with privacy

---

[39] California Legislative Information, *AB-316 Artificial Intelligence: Defenses,* Chapter 672 (Oct. 14, 2025), https://leginfo.legislature.ca.gov/faces/billNavClient.xhtml?bill_id=202520260AB316; California Legislative Information, *SB-53 Artificial Intelligence Models: Large Developers* (Sept. 29, 2025), https://leginfo.legislature.ca.gov/faces/billTextClient.xhtml?bill_id=202520260SB53

[40] New York State Senate, *Assembly Bill A6453/A Relates to the Training and Use of Artificial Intelligence Frontier Models* (Dec. 19, 2025), https://www.nysenate.gov/legislation/bills/2025/A6453/amendment/A

[41] Senate Research Center, *CSHB 149: Texas Responsible Artificial Intelligence Governance Act* (May 20, 2025), https://capitol.texas.gov/tlodocs/89R/analysis/html/HB00149S.htm; [full text] https://capitol.texas.gov/tlodocs/89R/billtext/pdf/HB00149F.pdf

[42] 119th Congress, *H.R.1—An Act to Provide for Reconciliation Pursuant to Title II of H. Con. Res. 14*, Actions, SA 2814 (Jul. 1, 2025), https://www.congress.gov/bill/119th-congress/house-bill/1/all-actions

[43] John Eligon, *Can the World Move on Without the US? G20 Leaders Gave It a Shot*, New York Times (Nov. 23, 2025), https://www.nytimes.com/2025/11/23/world/africa/g20-united-states.html

[44] G20 South Africa 2025, *G20 South Africa Summit: Leaders' Declaration* (Nov. 22, 2025), https://dirco.gov.za/wp-content/uploads/2025/11/2025-G20-Summit-Declaration.pdf

[45] US Department of State, *United States Assumes Presidency of the Group of 20* (Dec. 1, 2025), https://www.state.gov/releases/2025/12/united-states-assumes-presidency-of-the-group-of-20

[46] 119th Congress, *S.146—TAKE IT DOWN Act* (May 19, 2025), https://www.congress.gov/bill/119th-congress/senate-bill/146

[47] Gazzetta Ufficiale, *Provisions and Delegations to the Government on Artificial Intelligence [Disposizioni e deleghe al Governo in matria di intelligenza artificiale], Law No. 132 of 23 September 2025*, Art. 26, Official Gazette of the Italian Republic, no. 223 (Sept. 23, 2025), https://www.gazzettaufficiale.it/eli/id/2025/09/25/25G00143/sg

regulations."[48] Brazil's government gave xAI, the company that owns Grok, 30 days to prevent sexualized deepfakes, while Indonesia and Malaysia imposed at least temporary bans on Grok.[49] China's Cyberspace Administration published an instrument to standardize the labeling of AI-generated content,[50] a move that complements earlier restrictions. While Grok and xAI have been named most prominently because of the "spicy" mode on the image generation tool, the United Kingdom's Department for Science, Innovation and Technology (DSIT) Secretary of State emphasized the broader problem: "The platforms that host such material must be held accountable—including X." These actions reflect progress in defining liability for users and for providers.

Lost funding has not deterred the United Nations from fulfilling its role as a site for international dialogue and norm-making for AI governance. The 2025 General Assembly formally established the International Scientific Panel on AI, with members selected through early 2026.[51] The first Global Dialogue on AI Governance also took place,[52] though informally, offering what BRICS leaders called "an inclusive platform within the United Nations for states and stakeholders to discuss the critical issues concerning AI facing humanity today."[53]

The Global Dialogue and International Scientific Panel demonstrate the centrality of the United Nations in furthering global AI governance. The AI Red Lines movement also gathered attention at the General Assembly. The next UN Secretary-General will have to navigate an expanded, urgent role amid opposition from a former partner. AI governance will certainly remain a crucial topic in the interactive dialogues with candidates for Secretary-General in 2026.[54]

---

[48] Garante per la Protezione dei Dati Personali, *Press Release—Deefakes, the Guarantor Warns: Fundamental Rights and Freedoms at Risk [Communicato Stampa—Deepfake, il Garante avverte: A rishio diritti e libertà fondamentali]* (Jan.8, 2026), https://garanteprivacy.it/home/docweb/-/docweb-display/docweb/10207147

[49] Justin Hendrix and Ramsha Jahangir, *Tracking Regulator Responses to the Grok "Undressing" Controversy* (Jan. 16, 2026), https://www.techpolicy.press/tracking-regulator-responses-to-the-grok-undressing-controversy/

[50] Cyberspace Administration of China, *Notice on Issuing the "Measures for Identifying Artificial Intelligence-Generated and Synthetic Content"* [关于印发 "人工智能生成合成内容标识办法" 的通知] (Mar. 14, 2025), https://www.cac.gov.cn/2025-03/14/c_1743654684782215.htm

[51] UN Independent International Scientific Panel on AI, *Independent International Scientific Panel on AI* (2025), https://www.un.org/independent-international-scientific-panel-ai/en

[52] UN Secretary-General, *Statement Attributable to the Spokesperson for the Secretary-General—on the General Assembly Decision on New Artificial Intelligence Governance Mechanisms within the United Nations* (Aug. 26, 2025), https://www.un.org/sg/en/content/sg/statements/2025-08-26/statement-attributable-the-spokesperson-for-the-secretary-general-the-general-assembly-decision-new-artificial-intelligence-governance-mechanisms-within-the-united

[53] BRICS Brasil 2025, *BRICS Leaders' Statement on the Global Governance of Artificial Intelligence*, Presidency Documents (Jul. 6, 2025), https://brics.br/en/documents/presidency-documents

[54] President of the 80th Session, *Statement of the President of the United Nations General Assembly, HE Ms. Annalena Baerbock, on the Dates for the Interactive Dialogues with Candidates for the*

The UN Secretary-General report on lethal autonomous weapons systems (LAWS) demonstrated wide consensus that weapons systems with no human control challenge international humanitarian law.[55] Discussions in the UN General Assembly in 2025 pushed the idea of a treaty forward with only 5 countries opposing and 5 abstaining on a resolution to hold informal discussions on emerging normative proposals around LAWS.[56] Countries demonstrated similar support to continue the dialogue on artificial intelligence in the military domain more broadly. 118 delegations also supported[57] a draft resolution demanding human control and oversight over command, control, and communications systems of nuclear weapons.[58] These resolutions encourage the Convention on Certain Conventional Weapons (CCW) Group of Governmental Experts (GGE) on LAWS to continue toward an international instrument by the end of 2026. The global movement for AI red lines and UN efforts converge here, against allowing machines to decide and act to end a human life.

The BRICS Leaders' Statement on the Global Governance of Artificial Intelligence reflects the Global South's leadership in the push to implement AI governance based on shared values and national sovereignty.[59] At the Global AI for Africa Summit, leaders backed the African Union AI Continental Strategy with plans to bolster data and technological sovereignty and the establishment of the Africa AI Council.[60] The ASEAN provided concrete guidelines and action steps "to create the ideal conditions for responsible AI to flourish in the region" with the Responsible AI Roadmap[61] and the expansion of the Guide on AI Governance and Ethics to address Generative AI.[62] Progress in the G20 on data governance, non-discrimination in AI,

---

Position of Secretary-General (Jan. 14, 2026), https://www.un.org/pga/80/2026/01/14/tatement-of-the-president-of-the-united-nations-general-assembly-h-e-ms-annalena-baerbock-on-the-dates-for-the-interactive-dialogues-with-candidates-for-the-position-of-secretary-general/

[55] UN Secretary-General, *Lethal Autonomous Weapons Systems: Report of the Secretary-General*, United Nations Digital Library (2024), https://digitallibrary.un.org/record/4059475?v=pdf

[56] UN General Assembly, *Report of the First Committee*, 80th Session, Agenda items 90–106. 121, 137 (Nov. 13, 2025), https://docs.un.org/en/A/80/534

[57] Ibid

[58] UN General Assembly, *Possible Risks of the Integration* of Artificial Intelligence into Command, Control and Communications Systems of Nuclear Weapons, 80th Session First Committee, Agenda Item 99 (Oct. 16, 2025), https://docs.un.org/en/A/C.1/80/L.56

[59] BRICS Brasil 2025, *BRICS Leaders' Statement on the Global Governance of Artificial Intelligence*, Presidency Documents (Jul. 6, 2025), https://brics.br/en/documents/presidency-documents

[60] Global AI Summit on Africa, *The Africa Declaration on Artificial Intelligence* (Apr. 4, 2025), https://c4ir.rw/docs/Africa Declaration on Artificial Intelligences.pdf

[61] ASEAN, *ASEAN Responsible AI Roadmap (2025–2030)*, https://asean.org/book/asean-responsible-ai-roadmap-2025-2030/

[62] ASEAN Malaysia 2025, *Chairman's Statement of the 47th ASEAN Summit* (Oct. 26, 2025), https://asean.org/wp-content/uploads/2025/11/CHAIRMANS-STATEMENT-47TH-ASEAN-SUMMIT.pdf

and innovation for sustainable development owes much to the three consecutive years of leadership by countries in the Global South.[63]

The G7 continues to offer tools for AI developers and deployers to align their practices with AI principles as part of the commitment to "drive innovation and adoption of secure, responsible, and trustworthy AI that benefits people, mitigates negative externalities, and promotes our national security."[64] Beyond the Hiroshima Process reporting, the G7 this year moved forward with an AI Adoption Roadmap outlining "the G7's shared vision and practical steps to help our small and medium-sized enterprises (SMEs) access, understand and adopt AI in ways that drive value and productivity."[65] Canada's G7 Presidency achieved the key outcome of developing the SME AI Adoption Blueprint.[66]

This year has demonstrated again the value of international institutions and multilateralism to move closer to a "world where technology promotes broad social inclusion based on fundamental rights, democratic institutions, and the rule of law,"[67] even as authoritarianism threatens. International institutions and individual countries have pushed implementation of AI principles forward while demonstrating the benefits of global cooperation. The stakes are too high to upset a growing global consensus for AI governance.

Key Developments 2025

- 45 countries support the Framework Convention on Artificial Intelligence and Human Rights, Democracy, and the Rule of Law
- Japan and China pass comprehensive AI legislation; Peru and Vietnam enact laws defining red lines against AI applications that undermine human rights and democratic values; South Korea's 2025 law comes into effect
- US states adopt innovative AI safeguards while the executive branch withdraws from leadership in AI governance
- EU countries, including Italy, Slovenia, and Hungary, move forward the AI Act, and appoint oversight authorities
- UN establishes the Global Dialogue on AI Governance and an Independent International Scientific Panel on AI
- Countries at AI Action Summit in Paris endorse declaration on "inclusive and sustainable AI" as inequality and sustainability emerge as key governance themes

---

[63] G20 South Africa 2025, *G20 South Africa Summit: Leaders' Declaration* (Nov. 22, 2025), https://dirco.gov.za/wp-content/uploads/2025/11/2025-G20-Summit-Declaration.pdf

[64] G7 Canada, *G7 Leaders' Statement on AI for Prosperity*, p. 1 (Jun. 17, 2025), https://g7.canada.ca/assets/ea689367/Attachments/NewItems/pdf/g7-summit-statements/ai-en.pdf

[65] G7 2025 Kananaskis, *Implementing the G7 AI Adoption Roadmap* (Dec. 9, 2025), https://g7.canada.ca/en/news-and-media/news/implementing-the-g7-ai-adoption-roadmap/

[66] Government of Canada, *The SME AI Adoption Blueprint* (Dec. 9, 2025), https://ised-isde.canada.ca/site/ised/en/sme-ai-adoption-blueprint

[67] Cener for AI and Digital Policy, *Home*, https://www.caidp.org/

- UNESCO promotes AI readiness, focusing on the Global South and reaching Europe with the Netherlands and Moldova
- UNESCO begins training of AI Supervisory Authorities
- The South Africa G20 presidency moves forward on inclusive AI governance to protect against discrimination and establishes an institute to assist with AI governance
- ASEAN publishes the ASEAN Responsible AI Roadmap (2025–2030) and the Expanded ASEAN Guide on AI Governance and Ethics, strengthening alignment with foundational AI governance frameworks
- African nations support data and technological sovereignty and establish AI Council

2026 Recommendations

- Global support for the International AI Treaty
- Prohibitions on AI systems that undermine human rights and democratic values
- Human oversight of AI systems across the lifecycle
- Bridge the AI safety and fairness agendas
- Implementation and enforcement of AI governance frameworks, such as the EU AI Act and the Hiroshima AI Process
- Algorithmic transparency, including the accountability to contest adverse outcomes
- Liability rules for developers, deployers, and users of AI systems

THE GLOBAL AI POLICY LANDSCAPE[68]

As a field of research, AI policy is rapidly maturing. In the last several years, national governments and international organizations have adopted policy frameworks that explicitly regulate "Artificial Intelligence."[69] This has occurred at a rapid pace. While government funding for work on Artificial Intelligence goes back to the mid-1950s, it was many years before governments examined the consequences of this research. That gap is now closing. Governments around the world confront important decisions about AI priorities, AI ambitions, and AI risks. Much of this report concerns the current policies and practices of national governments.

In addition to national governments, many intergovernmental organizations are pursuing AI policies and initiatives. This section provides an overview of these organizations, listed in a simple A to Z. We also note the important work of technical associations and civil society organizations. This section briefly summarizes these activities, as of early 2026.

Council of Europe

The Council of Europe (COE) is the continent's leading human rights organization.[70] The COE comprises 47 member states, 27 of which are members of the European Union. All COE member states have endorsed the European Convention of Human Rights, a treaty designed to protect human rights, democracy, and the rule of law. Article 8 of the Convention concerning the right to privacy has influenced the development of privacy law around the world.

The COE Convention 108 (1981) was the first binding international instrument protecting the individual against abuses that may accompany the collection and processing of personal data and regulating the transborder flow of personal data.[71] In 2018, the Council of Europe amended Convention 108 and opened for signature and ratification the COE Modernized Convention 108+.[72] Article 9(1)(c) specifically addresses AI decision-making. As the COE explains, the "modernised Convention extends the catalogue of information to be transmitted to data subjects when they

---

[68] [Editorial note: For the 2025 edition of the ***AI and Democratic Values Index***, we removed much of the background text that was typically included in the Global Policy Landscape section and focused on recent developments. These materials are still available in the 2024 edition of AIDV, which is available online at https://www.caidp.org/reports/aidv-2023/

[69] Marc Rotenberg, *Human Rights Alignment: The Challenge Ahead for AI Lawmakers,* in H. Werthner, et al., *Introduction to Digital Humanism*, Springer (Dec. 21, 2023), https://doi.org/10.1007/978-3-031-45304-5_38

[70] Council of Europe, *Who We Are,* https://www.coe.int/en/web/about-us/who-we-are

[71] Council of Europe Treaty Office, *Details of Treaty No. 108,* https://www.coe.int/en/web/conventions/full-list/-/conventions/treaty/108

[72] Council of Europe, Data Protection, *Modernisation of Convention 108*, https://www.coe.int/en/web/data-protection/convention108/modernised

exercise their right of access. Furthermore, data subjects are entitled to obtain knowledge of the reasoning underlying the data processing, the results of which are applied to her/him. This new right is particularly important in terms of profiling of individuals."[73] Forty-six states have signed the protocol amending the Privacy Convention and 33 have ratified.[74]

The development of the Framework Convention on AI emerged in response to growing concerns over the artificial intelligence. The Council of Europe initiated discussions on AI governance in 2019, leading to the drafting of a binding legal framework. In 2024, the Council of Europe adopted the Framework Convention on Artificial Intelligence and Human Rights, Democracy, and the Rule of Law. The AI Treaty aims to ensure that the development and use of AI technologies align with fundamental human rights, democratic values, and the rule of law, addressing risks such as misinformation, algorithmic discrimination, and threats to public institution.

At the time of publication, over 40 countries, counting the individual member states of the European Union, had endorsed the treaty, including Canada, Japan, United States, Ukraine, and Uruguay.[75] Leading experts in international law and AI have also endorsed the AI Treaty.[76]

The Committee on AI (CAI), which drafted the Treaty, completed the mandate in 2025, though a new body with an advisory role is now established.[77] The Steering Committee for New and Emerging Digital Technologies (CDNET) will coordinate the work of the Council of Europe in the field of new and emerging technologies, ensuring that this work supports innovation, and give its legal and policy expertise.[78] CDNET also serves as custodian of the Framework Convention until the Conference of the Parties is established. The HUDERIA methodology,[79] a tool for actors in the public and private sectors to assess the risk and impact assessment of AI systems with an eye to human rights, democracy, and the rule of law, also facilitates implementation of the Treaty and its principles.

---

[73] Council of Europe, Data Protection, *Modernisation of Convention 108: Overview of the novelties* https://rm.coe.int/modernised-conv-overview-of-the-novelties/16808accf8

[74] Council of Europe, Treaty Office, *Chart of Signatures and Ratifications of Treaty 223* (Feb. 2, 2026), https://www.coe.int/en/web/conventions/full-list?module=signatures-by-treaty&treatynum=223

[75] Council of Europe Treaty Office, *Chart of Signatures and Ratifications of Treaty 225* (Apr. 7, 2026), https://www.coe.int/en-GB/web/conventions/full-list2?module=signatures-by-treaty&treatynum=225

[76] CAIDP, *Council of Europe AI Treaty*, https://www.caidp.org/resources/coe-ai-treaty/

[77] Council of Europe, *Committee on Artificial Intelligence (CAI)* (2026), https://www.coe.int/en/web/artificial-intelligence/cai

[78] Council of Europe, Steering Committee for New and Emerging Digital Technologies (CDNET), https://www.coe.int/en/web/artificial-intelligence/cdnet

[79] Council of Europe, *HUDERIA—Risk and Impact Assessments of AI Systems* (2026), https://www.coe.int/en/web/artificial-intelligence/huderia-risk-and-impact-assessment-of-ai-systems

European Court of Human Rights

The European Court of Human Rights has generated an abundance of case law interpreting Article 8 of the European Convention on Human Rights on the right to private and family life. This case law will establish governance norms for AI systems. The opinions of the Court on privacy and data protection are widely regarded by other courts. The Court has addressed privacy challenges in relation to telephone conversations, computers, mass surveillance, bulk interceptions of telecommunications and the internet.[80] The Court have deemed these aspects of personal data protection of fundamental importance to a person's enjoyment of their right to respect for private and family life.

In early 2025, the European Court of Human Rights sponsored a judicial seminar on Protecting Human Rights in a World of Artificial Intelligence, Algorithms and Big Data.[81] A background paper prepared for the session outlined relevant European Court of Human Rights caselaw in three categories: Freedom of expression, right to a fair trial, and prohibition on discrimination.[82] The paper further detailed specific areas of intersection between the case law of the Court of Human Rights and AI systems, such as facial recognition, the right to be forgotten, bias in the system, reasoned decision-making, and gender-based cyber violence.

European Union

The European Union (EU) was established by the Treaty of Masstricht in 1993. The European Union is a supranational political and economic union of 27 member states. Many institutions in the European Union now play a significant role in the development of AI policies and practices.

The EU AI Act, a comprehensive regulation for the governance of AI, was formally adopted in March 2024 by the European Parliament, with a large majority of 523-46 votes in favor of the legislation. The risk-based framework notably includes a category of prohibited AI systems, such as subliminal manipulation, social scoring, scraping of facial images for facial recognition, biometric categorization, and real-time biometric identification. At the time of publication, nearly 200 national supervisory authorities have been selected to ensure the enforcement and implementation of the AI Act, with data protection agencies given a prominent role.

---

[80] For an overview of the case law, see European Court of Human Rights, *Mass Surveillance* (Jan. 2022), https://www.echr.coe.int/documents/fs_mass_surveillance_eng.pdf; *Personal Data Protection* (Jan. 2022), https://www.echr.coe.int/Documents/FS_Data_ENG.pdf

[81] European Court of Human Rights, *Opening of the Judicial Year 2025* (Jan. 31, 2025), https://www.echr.coe.int/w/opening-of-the-judicial-year-2025-1

[82] European Court of Human Rights, *Protecting Human Rights in a World of Artificial Intelligence, Algorithms and Big Data* (Nov. 19, 2024), https://www.echr.coe.int/documents/d/echr/seminar-background-paper-2025-eng

European Commission

The European Commission plays an active role in developing the EU's overall strategy and in designing and implementing EU policies. The Commission is the initiator of EU legislation. AI was identified as a priority when the new Commission, under the Presidency of Ursula von der Leyen, was established in late 2019.[83] At that time, von der Leyen recommended new rules on Artificial Intelligence that respect human safety and rights.[84] Von der Leyen's proposal followed remarks at the G20 summit in 2019 by Chancellor Angela Merkel, who called on the European Commission to propose comprehensive regulation for artificial intelligence: "It will be the job of the next Commission to deliver something so that we have regulation similar to the General Data Protection Regulation that makes it clear that artificial intelligence serves humanity."

European Parliament

The European Parliament is co-legislator, together with the Council of the European Union. The Parliament has convened hearings and adopted resolutions to outline the elements of EU legislation.[85] One resolution urged the Commission to establish legal obligations for artificial intelligence and robotics, including software, algorithms, and data. A second would make those operating high-risk AI systems strictly liable for any resulting damage. A third resolution on intellectual property rights makes clear that AI should not have legal personality: only people may claim IP rights.

Two committees in the European Parliament have taken the reins for the EU AI Act. The Committee on the Internal Market and Consumer Protection (IMCO Committee) is responsible for the legislative oversight and scrutiny of EU rules on the single market, including the digital single market, customs, and consumer protection.[86] The Committee on Civil Liberties, Justice, and Home Affairs (LIBE Committee) "is responsible for the majority of legislation and democratic oversight of policies that enable the European Union to offer its citizens an area of freedom, security and justice (Article 3 TEU). While doing so, we ensure, throughout the EU, the full respect of and compliance with the EU Charter of Fundamental Rights, in conjunction with the European Convention on Human Rights."[87]

---

[83] CAIDP Update 1.3, *European Commission Proposes Four Options for Ethical AI* (Aug. 2, 2020), https://dukakis.org/center-for-ai-and-digital-policy/center-for-ai-policy-update-european-commission-proposes-four-options-for-ethical-ai/

[84] European Commission, *A Union that Strives for More: The First 100 Days* (Mar. 6, 2020), https://south.euneighbours.eu/news/union-strives-more-first-100-days/

[85] CAIDP Update 1.12, *European Parliament Adopts Resolutions on AI* (Oct. 24, 2020), https://dukakis.org/center-for-ai-and-digital-policy/caidp-update-european-parliament-adopts-resolutions-on-ai/

[86] European Parliament, *About IMCO, Welcome,* https://www.europarl.europa.eu/committees/en/imco/about

[87] European Parliament, *LIBE, About,* https://www.europarl.europa.eu/committees/en/libe/about

The influential LIBE Committee has also highlighted concerns about AI and fundamental rights and AI in criminal justice.[88] In February 2020, the Committee held a hearing on Artificial Intelligence and Criminal Law and examined the benefits, risks, and ethical and rights implications of AI, predictive policing, and facial recognition. LIBE worked in association with the United Nations Interregional Crime and Justice Research Institute (UNICRI), the European Union Agency for Fundamental Rights (FRA), and the Council of Europe (COE). In November 2020, LIBE issued an opinion concerning AI and the application of international law.[89]

Two Councils

The European Council defines the EU's overall political direction and priorities.[90] Its members are the heads of state or government of the 27 EU member states, the European Council President, and the President of the European Commission. The European Council is not one of the EU's legislating institutions, so does not negotiate or adopt EU laws. This is the prerogative of the Council of the European Union (Council), composed of representatives of member states' ministers.

Court of Justice of the European Union

Although the Court of Justice has yet to rule directly on the EU AI Act, the Court will play a significant role as AI policies evolve and AI law is adopted.[91] The Court issued several judgements that implicate AI systems. In its *Ligue des Droits Humains* Judgment of June 2022, concerning the Passenger Name Records Directive, the Court of Justice reaffirmed the primacy of a human-centered approach to AI. The Court of Justice ruled that machine learning techniques may be incompatible with the protection of fundamental rights.[92] The Court observed that the opacity of artificial intelligence might make it impossible to understand the reason why a given program arrived at a positive match.[93] As the Advocate General had earlier observed,

---

[88] CAIDP Update 1.8 *LIBE Committee of EU Parliament Examines AI Practices, Data Protection* (Sept. 9, 2020), https://dukakis.org/center-for-ai-and-digital-policy/caidp-update-libe-committee-of-eu-parliament-examines-ai-practices-data-protection/

[89] European Parliament, Committee on Civil Liberties, Justice and Home Affairs, *On Artificial Intelligence: Questions of Interpretation and Application of International Law in so far as the EU Is Affected in the Areas of Civil and Military Uses and of State Authority Outside the Scope of Criminal Justice (2020/2013 (INI))* (Nov. 23, 2020), https://www.europarl.europa.eu/doceo/document/LIBE-AD-652639_EN.pdf

[90] European Council, https://www.consilium.europa.eu/en/european-council/

[91] CAIDP Update 1.1, *EU Privacy Decision Will Have Global Consequences* (Jul. 19, 2020), https://dukakis.org/news-and-events/center-for-ai-and-digital-policy-update-eu-privacy-decision-will-have-global-consequences/

[92] Court of Justice of the European Union, *Ligue des droits humains v. Conseil des Ministres*, (Jun. 21, 2022), https://curia.europa.eu/juris/document/document.jsf?text=&docid=261282&pageIndex=0&doclang=EN&mode=req&dir=&occ=first&part=1&cid=13059170

[93] Court of Justice of the European Union, *Ligue des droits humains,* para. 194

algorithms "must function transparently and that the result of their application must be traceable."[94] The Court added in *Ligue des droits humains* that the use of pre-determined criteria also precludes the use of systems that modify "the assessment criteria on which the result of the application of that process is based as well as the weighting of those criteria."[95] These holdings could have far-reaching significance for the use of AI techniques by law enforcement agencies and the future interpretation of the EU AI Act.[96]

The Court has also addressed the transparency of automated processing. In the SCHUFA case, the Court of Justice held that the generation of credit scores by credit reference agencies are within the scope of the prohibition on automated decision-making in Article 22(1) of the GDPR.[97] The practical consequence is that automated decision-making enabled by AI systems will be subject to the GDPR. In another case, the Court found that individuals subject to adverse decisions are entitled to receive meaningful information about the basis of the decision.[98] In the Court's words: "The explanation provided must enable the data subject to understand and challenge the automated decision."

Judgments of the Court concerning data transfers will also impact the development of AI systems. In the 2020 *Schrems II* judgment, the Court struck down the Privacy Shield framework that permitted the transfer of personal data from the European Union to the United States.[99] The *Schrems II* judgment will likely limit the collection and use of personal data for AI systems.

European Data Protection Board

The European Data Protection Board (EDPB) is an independent European body that contributes to the consistent application of data protection rules throughout the European Union and promotes cooperation among the EU's data protection authorities (DPAs).[100]

---

[94] Opinion of the Advocate General, *Ligue des droits humains*, para. 228

[95] Court of Justice of the European Union, *Ligue des droits humains,* para. 194

[96] *CAIDP Statement to the European Data Protection Board on Facial Recognition and Law Enforcement* (Jul. 18, 2022), https://www.caidp.org/app/download/8405683363/CAIDP-Statement-EDPB-FRT-01082022.pdf; Marc Rotenberg, *CJEU PNR Decision Unplugs the "Black Box,"* European Data Protection Law Review (2022)

[97] Court of Justice of the European Union, *Judgment—SCHUFA Holding, Case C-634/21 (Scoring)* (Dec. 7, 2023), https://curia.europa.eu/juris/liste.jsf?num=C-634/21

[98] Court of Justice of the European Union, Press Release No. 22/25, *Automated Credit Assessment: The Data Subject Is Entitled to an Explanation as to How the Decision Was Taken in Respect of Him or Her* (Feb. 27, 2025), https://curia.europa.eu/jcms/upload/docs/application/pdf/2025-02/cp250022en.pdf

[99] CJEU, *The Court of Justice Invalidates Decision 2016/1250 on the Adequacy of the Protection Provided by the EU-US Data Protection Shield* (Jul. 16, 2020), https://curia.europa.eu/jcms/upload/docs/application/pdf/2020-07/cp200091en.pdf

[100] EDPB, *Who We Are,* https://edpb.europa.eu/about-edpb/about-edpb_en

The EDPB issued several reports in 2024 concerning Artificial Intelligence. A ChatGPT Taskforce expressed preliminary views on lawfulness, fairness, transparency, data accuracy, and the rights of data subjects.[101] A July 2024 statement from the EDPB addressed the role of DPAs in the AI Act framework.[102] The Statement described the relationship between the AI Act and the GDPR as "complementary and mutually reinforcing instruments" and recommended a "prominent role" for DPAs due to their experience and expertise in developing guidelines and carrying out enforcement actions on AI-related issues with respect to the processing of personal data. For these reasons and others, the EDPB recommended that DPAs be designated as Market Surveillance Authorities (MSA) under the EU AI Act.

In December 2024, the EDPB issued an opinion on AI Models.[103] The report addressed four questions raised by the Irish DPA: (1) when and how an AI model can be considered anonymous; (2) how controllers can demonstrate the appropriateness of legitimate interest as a legal basis in the development and (3) deployment phases; and (4) what are the consequences of the unlawful processing of personal data in the development phase of an AI model on the subsequent processing or operation of the AI model. Among the key findings: "the determination of whether an AI model is anonymous should be assessed, based on specific criteria, on a case-by-case basis"[104]; AI models are very likely to require such a thorough evaluation of the risks of Identification"[105]; "When there is the finding of an infringement, SAs may impose corrective measures, such as ordering controllers, taking into account the circumstances of each case, to take actions in order to remediate the unlawfulness of the initial processing";[106] and the lawfulness of the processing should be assessed case-by-case.[107]

In 2025, the European Data Protection Board (EDPB) played a central role in clarifying the application of EU data protection law to artificial intelligence, with a particular focus on generative AI and cross-border enforcement. The EDPB advanced coordinated enforcement through the establishment and ongoing work of a task force on generative AI, aimed at ensuring consistent supervisory responses among national data protection authorities.[108] In parallel, the EDPB continued to issue guidance and

---

[101] EDPB, *Report of the Work Undertaken by the ChatGPT Taskforce* (May 23, 2024), https://www.edpb.europa.eu/system/files/2024-05/edpb_20240523_report_chatgpt_taskforce_en.pdf

[102] EDPB, *Statement 3/2024 on Data Protection Authorities' Role in the Artificial Intelligence Act Framework* (Jul. 16, 2024), https://www.edpb.europa.eu/system/files/2024-07/edpb_statement_202403_dpasroleaiact_en.pdf

[103] EDPB, *Opinion 28/2024 on Certain Data Protection Aspects Related to the Processing of Personal Data in the Context of AI Models* (Dec. 17, 2024), https://www.edpb.europa.eu/system/files/2024-12/edpb_opinion_202428_ai-models_en.pdf

[104] Ibid, p. 14

[105] Ibid, p. 16

[106] Ibid, p. 32

[107] Ibid, p. 33

[108] EDPB, *Report of the Work Undertaken by the ChatGPT Taskforce* (2025), https://edpb.europa.eu/system/files/2025-04/edpb_report_chatgpt_taskforce_en.pdf

binding decisions under the GDPR's consistency mechanism, reinforcing a harmonized, rights-based approach to AI governance across the European Union.[109]

European Data Protection Supervisor

The European Data Protection Supervisor is the European Union's independent data protection authority.[110] The EDPS responsibilities include the mission to "monitor and ensure the protection of personal data and privacy when EU institutions and bodies process the personal information of individuals." This AI work of the EDPS builds on earlier initiatives, including Recommendations on the AI Act (2023) and the EDPS-EDPB Joint Opinion on the AI Act (2021).[111]

The European Data Protection Supervisor undertook many projects related to AI in 2024. In June, the EDPS published Orientations on "generative Artificial Intelligence and personal data protection" to provide EU institutions, bodies, offices and agencies with practical advice and instructions on the processing of personal data when using generative AI systems, to facilitate their compliance with the requirements of the data protection legal framework.[112] The Guidelines emphasize the general principles of data protection that should help EU institutions comply with the data protection requirements. The European Data Protection Supervisor also supported the work of the Global Privacy Assembly and the G7 DPA Privacy Roundtable, discussed above.

In 2025, the European Data Protection Supervisor (EDPS) advanced work on artificial intelligence through a series of detailed supervisory guidelines and policy initiatives aimed at safeguarding fundamental rights. Most notably, the EDPS issued revised guidance on generative AI, providing practical instructions for EU institutions on lawful data processing, transparency, and risk mitigation, including heightened scrutiny of web scraping and training data practices.[113] The EDPS also released comprehensive guidance on AI risk management, offering lifecycle-based

---

[109] EDPB, *Annual Report 2025*, https://edpb.europa.eu/system/files/2026-01/edpb_annual_report_2025_en.pdf

[110] EDPS, *About*, https://edps.europa.eu/about-edps_en

[111] EDPS, *EDPS Opinion 44/2023 on the Proposal for Artificial Intelligence Act in the Light of Legislative Developments* (Oct. 23, 2023), https://www.edps.europa.eu/data-protection/our-work/publications/opinions/2023-10-23-edps-opinion-442023-artificial-intelligence-act-light-legislative-developments_en; EDPS, *EDPS-EDPB Joint Opinion on the Proposal for a Regulation of the European Parliament and of the Council Laying Down Harmonised Rules on Artificial Intelligence (Artificial Intelligence Act)* (Jun. 18, 2021), https://www.edps.europa.eu/node/7140_en

[112] EDPS, *Generative AI and the EUDPR. First EDPS Orientations for Ensuring Data Protection Compliance when Using Generative AI Systems* (Jun. 3, 2024), https://www.edps.europa.eu/system/files/2024-06/24-06-03_genai_orientations_en.pdf

[113] EDPS, *EDPS Unveils Revised Guidance on Generative AI, Strengthening Data Protection in a Rapidly Changing Digital Era* (Oct. 28, 2025), https://www.edps.europa.eu/press-publications/press-news/press-releases/2025/edps-unveils-revised-guidance-generative-ai-strengthening-data-protection-rapidly-changing-digital-era_en

frameworks to identify and mitigate risks to personal data and fundamental rights, with particular attention to fairness, accuracy, data minimization, and security.[114]

G7

The Group of Seven (G7) is an inter-governmental political forum consisting of Canada, France, Germany, Italy, Japan, the United Kingdom, and the United States. The members represent the wealthiest liberal democracies. The group is officially organized around shared values of pluralism and representative government. The G7 is also the incubator for significant work on AI policy.

In 2024, Artificial Intelligence was a prominent feature in the G7 Leaders' Communique. G7 leaders said they would "pursue an inclusive, human-centered, digital transformation that underpins economic growth and sustainable development, maximizes benefits, and manages risks, in line with our shared democratic values and respect for human rights."[115] The G7 leaders expressed support for the Hiroshima AI process and the development of a reporting framework for monitoring the International Code of Conduct for Organizations Developing Advanced AI Systems.

G7 leaders called attention to labor issues and said they would launch an action plan on the use of AI in the world of work. "We ask our Labor Ministers to develop the action plan, envisaging concrete actions to fully leverage the potential of AI to enable decent work and workers' rights and full access to adequate reskilling and upskilling, while addressing potential challenges and risks to our labor markets." The G7 leaders endorsed the UN General Assembly Resolution on Seizing the Opportunities of Safe, Secure and Trustworthy AI Systems for Sustainable Development and said they would work "towards closing digital divides, including the gender digital divide, and achieving digital inclusion."[116]

The G7 leaders also recognized the impact of AI on the military domain and the need for a framework for responsible development and use. They stated, "We welcome those who have endorsed the Political Declaration on Responsible Military Use of AI and Autonomy (REAIM) and the REAIM Call to Action, and we encourage more States to do so to ensure that military use of AI is responsible, complies with applicable international law, particularly international humanitarian law, and enhances international security."[117]

Under Canada's G7 Presidency in 2025, G7 leaders again focused on artificial intelligence, dedicating a statement to AI for Prosperity.[118] The leaders committed to

---

[114] EDPS, *Guidance for Risk Management of Artificial Intelligence Systems* (Nov. 11, 2025), https://www.edps.europa.eu/data-protection/our-work/publications/guidelines/2025-11-11-guidance-risk-management-artificial-intelligence-systems_en

[115] G7 Italia, *Apulia G7 Leaders' Communiqué*, p. 21 (Jun. 15, 2024), https://www.g7italy.it/wp-content/uploads/Apulia-G7-Leaders-Communique.pdf

[116] Ibid, p. 22

[117] Ibid, p. 23

[118] G7 Canada, *G7 Leaders' Statement on AI for Prosperity* (Jun. 17, 2025), https://g7.canada.ca/assets/ea689367/Attachments/NewItems/pdf/g7-summit-statements/ai-en.pdf

collaboration on integrating AI into the public sector to enhance services "while respecting human rights and privacy, as well as promoting transparency, fairness, and accountability."[119] They also laid out action steps to support SMEs to adopt AI, manage the energy challenges posed by AI, and partner with emerging markets and developing countries to increase access to AI.[120]

France assumed the G7 presidency for 2026.

G7 Privacy Roundtable

In recent years, privacy officials from G7 nations have gathered to host seminars and issue statements in conjunction with the Leaders' Summit. The G7 Privacy officials issued several statements related to AI at the 4th G7 Data Protection and Privacy Authorities Roundtable in Rome, Italy. First, the 2024 Communique emphasized "the challenges that artificial intelligence (AI) poses to privacy, data protection, and other fundamental rights and freedoms."[121] In a second statement on AI and Children, the G7 Privacy officials warned about "potential violations of privacy and data protection linked to the use of AI systems which could have serious implications for children and young people."[122] The privacy officials identified several threats, including AI-based decision-making, manipulation and deception, and the use of children's data for training AI models. The officials recommended greater emphasis on privacy by design, privacy impact assessments, transparency, and digital literacy.

A third statement explored the role of Data Protection Agencies in fostering trustworthy AI.[123] The G7 Privacy officials welcomed the recognition of DPAs in many of the international instruments for AI, including the OECD AI Principles (2019) and the UNESCO Recommendation on AI Ethics (2021), as well as the Bletchley Declaration (2023) and the Seoul Declaration (2024). The privacy officials explained, "many AI technologies, including generative AI, are based on the processing of personal data, which can subject natural persons to unfair stereotyping, bias and discrimination even when not directly processing their respective personal data." The G7 privacy officials emphasized, "Current privacy and data protection laws apply to the development and use of generative AI products, even as different

---

[119] Ibid, p. 1

[120] Ibid, pp. 2–3

[121] G7 Privacy, Roundtable of G7 Data Protection and Privacy Authorities, *G7 DPAs' Communiqué: Privacy in the Age of Data* (Oct. 11, 2024), https://www.edps.europa.eu/system/files/2024-10/g7_dpas_rome_roundtable_draft_communique_en.pdf

[122] G7 Privacy, Roundtable of G7 Data Protection and Privacy Authorities, *Statement on AI and Children* (Oct. 11, 2024), https://www.edps.europa.eu/system/files/2024-10/statement_on_ai_and_children_en.pdf

[123] G7 Privacy, Roundtable of G7 Data Protection and Privacy Authorities, *Statement on the Role of Data Protection Authorities in Fostering Trustworthy AI* (Oct. 11, 2024), https://www.edps.europa.eu/system/files/2024-10/draft_statement_dpas_governance_trustworthy_ai_en.pdf

jurisdictions continue to develop AI-specific laws and policies." The statement concluded, "Many data protection overarching principles can be transposed into broader AI governance frameworks"; "DPAs supervise a core component of AI"; "DPAs can help address problems at their source"; and "DPAs have experience."

The 2025 Privacy Roundtable Statement stressed data protection and privacy as necessary conditions for innovation, arguing "When individuals have confidence that their data is protected and used lawfully and responsibly, trust exists; where trust exists, innovation is embraced."[124] The data protection and privacy reiterated how organizations can prioritize privacy to support market confidence and individuals' trust and to protect children while offering specific actions for prioritizing privacy.

G20

The G20 is an international forum made up of 19 countries, the African Union, and the European Union, representing the world's major developed and emerging economies.[125] Together, the G20 members represent 85% of global GDP, 75% of international trade, and two-thirds of the world's population. According to the OECD, because of its size and strategic importance, the G20 has a crucial role in setting the path for the future of global economic growth.

The 2025 Leaders' Declaration highlighted advances by the Indian and Brazilian presidencies in facilitating a broader range of global voices in the discussions on AI, data governance, and innovation for sustainable development as exemplified by the Solidarity, Equality, and Sustainability theme promoted by the South African G20 presidency.[126] The G20 also progressed on ensuring more equitable access to AI development and participatory governance through organizations such as the Technology Policy Assistance Facility established with UNESCO to support countries in developing AI policies based on research and best practices. The United States did not attend the meetings in South Africa, alleging the country was violating the rights of white Afrikaners and acting aggressively toward the United States and allies.[127]

The United States assumed the G20 presidency in December 2025, promising "the best is yet to come" at the 2026 summit in Miami.[128] The United States has not invited South Africa to the meeting, citing "politics of grievance" from South Africa

---

[124] Office of the Privacy Commissioner of Canada, *2025 G7 Data Protection and Privacy Authorities Roundtable Statement* (Jun. 19, 2025), https://www.priv.gc.ca/en/opc-news/speeches-and-statements/2025/js-dc-g7_20250619/

[125] OECD, *OECD and G20*, https://www.oecd.org/en/about/oecd-and-g20.html

[126] G20 South Africa 2025, *G20 South Africa Summit: Leaders' Declaration* (Nov. 22, 2025), https://dirco.gov.za/wp-content/uploads/2025/11/2025-G20-Summit-Declaration.pdf

[127] Trump White House, *Addressing Egregious Actions of the Republic of South Africa* (Feb. 7, 2025), https://www.whitehouse.gov/presidential-actions/2025/02/addressing-egregious-actions-of-the-republic-of-south-africa/

[128] G20 America 2026, *Miami 2026*, https://g20.org/

and the country's "radical agendas that have nothing to do with economic growth"—namely discussions on climate change, diversity and inclusion, and aid dependency.[129]

Global Privacy Assembly

The Global Privacy Assembly is the global network of privacy officials and experts. The Global Privacy Assembly meets annually to discuss emerging privacy issues and to adopt resolutions. In recent years, the focus of the GPA has moved toward AI.[130]

In October 2023, the Assembly adopted a resolution on Generative AI Systems. The signatories stressed that they were particularly "concerned by the release—often with insufficient pre-deployment assessment—of generative AI systems to the wider public, which may present risks and potential harms to data protection, privacy and other fundamental human rights if not properly developed and regulated."[131]

The GPA Working Group on Ethics and Data Protection in Artificial Intelligence presented a report at the October 2024 meeting in Jersey.[132] The Working Group set out a 2024–2025 workplan to (1) establish a clear definition of what constitutes a meaningful human review of automated decisions, (2) report on Generative AI-related work, (3) assess AI auditing capabilities, and (4) examine discrimination in insurance pricing and decisions. In 2025, the GPA adopted a resolution proposed by the Working Group that includes steps organizations can take to ensure meaningful human oversight while acknowledging that "in some circumstances meaningful human oversight may not be possible"[133] such as when the scale and timeframe of automated decisions make oversight of individual decisions impractical. The Working Group announced plans to develop a checklist for organizations formulating internal policies on the use of generative AI and compiling DPA's issued AI-related guidelines.[134]

---

[129] US Secretary of State Marco Rubio, *America Welcomes a New G20*, US Embassy & Consulates in South Africa (Dec. 4, 2025), https://za.usembassy.gov/america-welcomes-a-new-g20/

[130] CAIPD Update 1.15, *Privacy Commissioners Adopt Resolutions on AI, Facial Recognition* (Oct. 19, 2020), https://dukakis.org/center-for-ai-and-digital-policy/caidp-update-privacy-commissioners-adopt-resolutions-on-ai-facial-recognition/

[131] Global Privacy Assembly, *Resolution on Generative AI* (Oct. 2023), https://globalprivacyassembly.org/wp-content/uploads/2023/10/5.-Resolution-on-Generative-AI-Systems-101023.pdf

[132] CNIL, EDPS, PIPC, *Working Group on Ethics and Data Protection in Artificial Intelligence* (Oct. 2024), https://globalprivacyassembly.org/wp-content/uploads/2024/11/8.-Ethics_and_Data_Protection_in_AI_Working_Group_Annual_Report.pdf

[133] Global Privacy Assembly, *Resolution on Meaningful Human Oversight of Decisions Involving AI Systems* (Sept. 2025), https://globalprivacyassembly.com/wp-content/uploads/2025/10/GPA-Resolution-Human-Oversight-of-Automated-Decisions.pdf

[134] Global Privacy Assembly, *Working Group on Ethics and Data Protection in Artificial Intelligence Report* (Jul. 2025), https://globalprivacyassembly.com/wp-content/uploads/2025/10/AIWG-Annual-Report-2025.pdf

OECD

The OECD is an international organization that "works to build better policies for better lives."[135] The goal of the OECD is to "shape policies that foster prosperity, equality, opportunity and well-being for all." The OECD has led the global effort to develop and establish the first government-endorsed framework for AI policy. This was a result of a concerted effort by the OECD and member states to develop a coordinated international strategy. The OECD AI Principles also build on earlier OECD initiatives such as the OECD Privacy Guidelines, a widely recognized framework for transborder data flows and the first global framework for data protection.[136] OECD policy frameworks are not treaties, do not have legal force, and are not directly applicable to OECD member states. However, there are many instances of countries adopting national laws based on OECD policies and a clear convergence of legal norms, particularly in the field of data protection.

OECD publishes regular reports and research aligned with implementation of the AI Principles. A report on AI in government published in 2025 "aims to provide governments with the necessary elements for effective AI use and to identify areas where AI is having an impact and where gaps remain. Future efforts will build on the growing evidence base of policies and use cases, seeking to further assist governments putting in place the enablers, guardrails and engagement mechanisms needed for a strategic and trustworthy approach to AI."[137] The organization also partnered with the European Commission to develop a draft framework to guide leaders and policymakers in creating AI Literacy materials and standards toward competence that "empowers learners to understand AI and make decisions about its use in meaningful and ethical ways."[138]

Global Partnership on AI

The Global Partnership on Artificial Intelligence (GPAI) emerged from the OECD Recommendation on Artificial Intelligence.[139] GPAI activities are intended to foster the responsible development of AI grounded in "human rights, inclusion, diversity, innovation, and economic growth."[140] The GPAI aims to "bridge the gap between theory and practice on AI by supporting cutting-edge research and applied activities on AI-related priorities." GPAI formalized an integration partnership with OECD "for harnessing the potential of AI for Good and for All"[141] in 2024. According

---

[135] OECD, *Who We Are*, https://www.oecd.org/about/

[136] OECD, *OECD Guidelines on the Protection of Privacy and Transborder Flows of Personal Data* (1981, updated Feb. 12, 2002), https://www.oecd.org/en/publications/2002/02/oecd-guidelines-on-the-protection-of-privacy-and-transborder-flows-of-personal-data_g1gh255f.html

[137] OECD, *Governing with Artificial Intelligence: The State of Play and Way Forward in Core Government Functions* (Sept. 18, 2025), https://doi.org/10.1787/795de142-en

[138] OECD, European Commission, and code.org, *Empowering Learners for the Age of AI: An AI Literacy Framework for Primary and Secondary Education* (2025), https://ailiteracyframework.org

[139] GPAI, *The Global Partnership on Artificial Intelligence*, https://gpai.ai

[140] GPAI, *About GPAI*, https://gpai.ai/about/

[141] India, Ministry of Electronics & IT, *6th Meeting of the GPAI Ministerial Council Held on 3rd July 2024 at New Delhi [New Delhi Declaration]* (Jul. 3, 2024), https://pib.gov.in/PressReleasePage.aspx?PRID=2030534

to a press statement, "The OECD and the Global Partnership on Artificial Intelligence (GPAI) are joining forces to advance an ambitious agenda for implementing human-centric, safe, secure and trustworthy Artificial Intelligence (AI) embodied in the principles of the OECD Recommendation on AI."[142] The OECD further said, "The existing GPAI Expert Support Centres—in Paris, France, Montreal, Canada, and Tokyo, Japan—will continue to take an active role in shaping and advancing the new partnership's work."[143]

As of January 2026, GPAI's members include Argentina, Australia, Belgium, Brazil, Canada, Chile, Colombia, Costa Rica, Czech Republic, Denmark, Estonia, Finland, France, Germany, Greece, Hungary, Iceland, India, Ireland, Israel, Italy, Japan, Korea, Latvia, Lithuania, Luxembourg, Mexico, the Netherlands, New Zealand, Norway, Poland, Portugal, Senegal, Serbia, Singapore, Slovakia, Slovenia, Spain, Sweden, Switzerland, Türkiye, the United Kingdom, and the United States.[144]

OECD AI Policy Observatory

The OECD AI Policy Observatory launched in February 2020 provides extensive data and multi-disciplinary analysis on artificial intelligence across a wide range of policy areas.[145] According to the OECD, the AI Policy Observatory is based on multidisciplinary, evidence-based analysis and Global multi-stakeholder partnerships.

Collaboration with United Nations

In September 2024, the OECD and the United Nations announced a collaborative effort to strengthen global AI governance. This partnership focuses on conducting regular, science-based assessments of AI risks and opportunities.[146] The collaboration seeks combines the OECD's technical expertise with the UN's global reach to inform policy decisions and promote responsible AI development worldwide. OECD Deputy Secretary General Knudsen said, "The speed of AI technology development and the breadth of its impact requires diverse policy ecosystems to work more cohesively. And in real time. I am delighted that the OECD and the UN will link their efforts to help governments improve the quality and timeliness of their policy response to AI's opportunities and its risks." UN Under-Secretary General Gill said,

---

[142] OECD, *GPAI and OECD Unite to Advance Coordinated International Efforts for Trustworthy AI* (Jul. 3, 2024), https://www.oecd.org/en/about/news/speech-statements/2024/07/GPAI-and-OECD-unite-to-advance-coordinated-international-efforts-for-trustworthy-AI.html

[143] Ibid

[144] OECD AI Policy Observatory, *About the Global Partnership on Artificial Intelligence (GPAI)* (2026), https://oecd.ai/en/about/about-gpai

[145] OECD, *AI Policy Observatory*, https://www.oecd.ai/

[146] OECD, *OECD and UN Announce Next Steps in Collaboration on Artificial Intelligence* (Sept. 22, 2024), https://www.oecd.org/en/about/news/press-releases/2024/09/oecd-and-un-announce-next-steps-in-collaboration-on-artificial-intelligence.html

"We will work with all stakeholders, including leading scientists and academic centres from around the globe, to realise this goal."

United Nations

The Secretary-General of the United Nations has led a global campaign for AI governance over many years. Speaking in February 2025 at the AI Action Summit in Paris, Secretary-General António Guterres said, "We must all work together so that artificial intelligence can bridge the gap between developed and developing countries—not widen it. It must accelerate sustainable development—not entrench inequalities."[147] Guterres emphasized the need to ensure technology serves humanity, not the other way around. "The creation of an Independent International Scientific Panel on AI will be central to translating this vision into reality," said the Secretary-General. He also called attention to the growing impact of AI on the climate and sustainability. He concluded, "It is in all our interests for Governments and technology leaders to commit to global guardrails, share best practices and shape fair policy and business models."

The UN accomplished several milestones toward this goal in 2024. The Pact for the Future, adopted by World Leaders at the Summit for the Future in September 2024, outlined several goals for AI aligned with the Sustainable Development Goals, including enhancing international governance of AI for the benefit of humanity.[148] The Pact also warned of the "risks of existing and potential risks associated with the military applications of artificial intelligence."[149] The Global Digital Compact, also adopted at the Summit for the Future, established the first universal agreement on the governance of AI. The Compact reiterated calls for building AI capacity in developing nations.

The UN implemented these initiatives in 2025 with the establishment of International Scientific Panel on AI[150] and the first informal Global Dialogue on AI at the 80th General Assembly.[151] The panel and dialogue reflect the UN's centrality in furthering global AI governance, a role validated by the BRICS Leaders' recognition of the Global Dialogue as "an inclusive platform within the United Nations for states

---

[147] United Nations, *Secretary-General, at Action Summit, Urges Working Together so Artificial Intelligence Expedites Sustainable Development, Not Creates World of "Haves and Have-Nots"* (Feb. 11, 2025), https://press.un.org/en/2025/sgsm22548.doc.htm

[148] United Nations, *Pact for the Future* A/RES/79/1 (Sept. 22, 2024), https://www.un.org/en/summit-of-the-future/pact-for-the-future

[149] Ibid, p. 19

[150] UN Independent International Scientific Panel on AI, *Independent International Scientific Panel on AI* (2025), https://www.un.org/independent-international-scientific-panel-ai/en

[151] UN Secretary-General, *Statement Attributable to the Spokesperson for the Secretary-General—on the General Assembly Decision on New Artificial Intelligence Governance Mechanisms within the United Nations* (Aug. 26, 2025), https://www.un.org/sg/en/content/sg/statements/2025-08-26/statement-attributable-the-spokesperson-for-the-secretary-general-the-general-assembly-decision-new-artificial-intelligence-governance-mechanisms-within-the-united

and stakeholders to discuss the critical issues concerning AI facing humanity today."[152]

UNESCO

In 2020, UNESCO embarked on a project to develop a global standard for Artificial Intelligence. UNESCO Director General Audrey Azoulay stated, "Artificial intelligence can be a great opportunity to accelerate the achievement of sustainable development goals. But any technological revolution leads to new imbalances that we must anticipate."[153]

The UNESCO Recommendation sets out about a dozen principles, five Action Goals, and eleven Policy Actions. Notable among the recommendations is the emphasis on Human Dignity, Inclusion, and Diversity. UNESCO also expresses support for Human Oversight, Privacy, Fairness, Transparency and Explainability, and Safety and Security, among other goals. Understandably, UNESCO is interested in the scientific, educational, and cultural dimensions of AI, the agency's program focus.

The UNESCO Recommendation was adopted in 2021. This is the first global agreement on the Ethics of Artificial Intelligence.[154] UNESCO Director General Audrey Azoulay stated, "The world needs rules for artificial intelligence to benefit humanity. The recommendation on the ethics of AI is a major answer. It sets the first global normative framework while giving member states the responsibility to apply it at their level. UNESCO will support its 193 member states in its implementation and ask them to report regularly on their progress and practices."

UNESCO explained, "The Recommendation aims to realize the advantages AI brings to society and reduce the risks it entails. It ensures that digital transformations promote human rights and contribute to the achievement of the Sustainable Development Goals, addressing issues around transparency, accountability and privacy, with action-oriented policy chapters on data governance, education, culture, labour, healthcare and the economy." The key achievements of the UNESCO AI Recommendation include:

1. Protecting data. The UNESCO Recommendation calls for action beyond what tech firms and governments are doing to guarantee individuals more protection by ensuring transparency, agency and control over their personal data.
2. Banning social scoring and mass surveillance. The UNESCO Recommendation explicitly bans these uses of AI systems.

---

[152] BRICS Brasil 2025, *BRICS Leaders' Statement on the Global Governance of Artificial Intelligence*, Presidency Documents (Jul. 6, 2025), https://brics.br/en/documents/presidency-documents

[153] UNESCO, *Artificial Intelligence with Human Values for Sustainable Development,* https://en.unesco.org/artificial-intelligence

[154] UNESCO, *Recommendation on the Ethics of Artificial Intelligence* (2021), https://unesdoc.unesco.org/ark:/48223/pf0000380455

3. Monitoring and Evaluation. The UNESCO Recommendation establishes new tools that will assist in implementation, including Ethical Impact Assessments and a Readiness Assessment Methodology (RAM).
4. Protecting the environment. The UNESCO Recommendation emphasizes that AI actors should favor data-, energy-, and resource-efficient AI methods to help ensure that AI becomes a more prominent tool in the fight against climate change and on tackling environmental issues.

The Recommendation aims to provide a basis to make AI systems work for the good of humanity, individuals, societies, and the environment and ecosystems, and to prevent harm. It also aims at stimulating the peaceful use of AI systems. The Recommendation provides a universal framework of values and principles of the ethics of AI. It sets out four values: respect, protection, and promotion of human rights and fundamental freedoms and human dignity; environment and ecosystem flourishing; ensuring diversity and inclusiveness; living in peaceful, just, and interconnected societies.

Further, the Recommendation outlines 10 principles—proportionality and do no harm, safety and security, fairness and non-discrimination, sustainability, right to privacy and data protection, human oversight and determination, transparency and explainability, responsibility and accountability, awareness and literacy—backed by more concrete policy actions on how they can be achieved. The Recommendation also introduces red lines to unacceptable AI practices. For example, it states that "AI systems should not be used for social scoring or mass surveillance purposes."

The Recommendation focuses not only on values and principles but also on their practical realization via concrete eleven policy actions. UNESCO encourages Member States to introduce frameworks for ethical impact assessments, oversight mechanisms, etc. Member States should ensure that harms caused through AI systems are investigated and redressed by enacting strong enforcement mechanisms and remedial actions to make certain that human rights, fundamental freedoms, and the rule of law are respected.

UNESCO developed the Readiness Assessment Methodology (RAM) to assist Member States in implementing the Recommendation.[155] As of early 2026, UNESCO has supported more than 40 countries in assessing their readiness for AI integration along 5 dimensions: legal, social/cultural, scientific/educational, economic, and technical and infrastructural. Country profiles on UNESCO's Global Hub[156] include key insights and recommendations from the reports, revealing key action steps and common challenges as well as useful approaches to governing AI aligned to the Recommendation.

---

[155] UNESCO Digital Library, *Readiness Assessment Methodology: A Tool for the Recommendation on the Ethics of Artificial Intelligence* (2023), https://unesdoc.unesco.org/ark:/48223/pf0000385198

[156] UNESCO Global AI Ethics and Governance Observatory, *Global Hub* (2026), https://www.unesco.org/ethics-ai/en/global-hub

UNESCO's Supervising AI by Competent Authorities[157] project, with the assistance of the Center for AI and Digital Policy, provided hands-on training to "equip national authorities with the tools, knowledge, and peer support needed to supervise AI systems effectively." In 2025, the program delivered sessions to more than 700 civil servants across 12 EU Member States. The program also produced an AI Supervision Toolkit.[158] CAIDP also assisted UNESCO on the AI Literacy Program for Civil Servants,[159] as part of the AI Ethics Experts without Borders (AIEB) network. The program aims to advance national implementation of the UNESCO Recommendation by preparing government official and public services with the "knowledge, tools, and competencies that can help guide AI adoption in ways that uphold human rights, democracy, and the rule of law." The European Commission Directorate-General for International Partnerships (DG INTPA) and the Patrick J. McGovern Foundation also support the project.

UN High Commissioner for Human Rights

In the Roadmap for Digital Cooperation, the UN Secretary-General stated, "To address the challenges and opportunities of protecting and advancing human rights, human dignity and human agency in a digitally interdependent age, the Office of the United Nations High Commissioner for Human Rights will develop system-wide guidance on human rights due diligence and impact assessments in the use of new technologies, including through engagement with civil society, external experts and those most vulnerable and affected."[160]

In September 2021, the UN High Commissioner for Human Rights Michelle Bachelet called for a moratorium on the sale and use of AI that pose a serious risk to human rights until adequate safeguards are put in place.[161] She also called for a ban on AI applications that do not comply with international human rights

---

[157] UNESCO, *Expanding Capacity Building for Competent Authorities on AI: National Trainings across the EU* (Dec. 9, 2025), https://www.unesco.org/en/articles/expanding-capacity-building-competent-authorities-ai-national-trainings-across-eu

[158] UNESCO and Dutch Authority for Digital Infrastructure, *AI Supervision Training Toolkit: Building Capacity and Skills for AI Supervisory Authorities* (2025), https://unesdoc.unesco.org/ark:/48223/pf0000396566

[159] Christabel Randolph and Merve Hickok, *UNESCO's AI Literacy Trianing for Civil Servants: Empowering Ethical AI Governance around the World*, UNESCO Global AI Ethics and Governance Observatory (Dec. 2, 2025), https://www.unesco.org/ethics-ai/en/articles/unescos-ai-literacy-training-civil-servants-empowering-ethical-ai-governance-around-world-0

[160] UN Secretary General, *Report: Roadmap for Digital Cooperation* (Jun. 2020), https://www.un.org/en/content/digital-cooperation-roadmap/assets/pdf/Roadmap_for_Digital_Cooperation_EN.pdf)

[161] UN Human Rights, Office of the High Commissioner, *Artificial Intelligence Risks to Privacy Demand Urgent Action—Bachelet* (Sept. 15, 2021), https://www.ohchr.org/EN/NewsEvents/Pages/DisplayNews.aspx?NewsID=27469&LangID=E; see also *UN Urges Moratorium on AI that Violates Human Rights,* CAIDP Update 2.34 (Sept. 15, 2021), https://www.caidp.org/app/download/8343909663/CAIDP-Update-2.34.pdf

law. "Artificial intelligence can be a force for good, helping societies overcome some of the great challenges of our times. But AI technologies can have negative, even catastrophic, åeffects if they are used without sufficient regard to how they affect people's human rights," Bachelet said.

The High Commissioner's statement accompanied the release of a new report on The Right to Privacy in the Digital Age. The UN Report details how AI systems rely on large datasets, with information about individuals collected, shared, merged, and analyzed in multiple and often opaque ways. The UN Report finds that data used to guide AI systems can be faulty, discriminatory, out of date, or irrelevant. Long-term storage of data also poses risks, as data could in the future be exploited in as yet unknown ways.[162]

UN Special Rapporteur

An extensive 2018 report by a UN Special Rapporteur explored the implications of artificial intelligence technologies for human rights in the information environment, focusing in particular on rights to freedom of opinion and expression, privacy, and non-discrimination.[163] The Report of the Special Rapporteur on the Promotion and Protection of the Right to Freedom of Opinion and Expression defines key terms "essential to a human rights discussion about artificial intelligence"; identifies the human rights legal framework relevant to artificial intelligence; and presents preliminary guidelines to ensure that human rights are considered as AI systems evolve. The report emphasizes free expression concerns and notes several frameworks, including the International Covenant on Civil and Political Rights and the UN Guiding Principles on Business and Human Rights.

Among the Recommendations, the Special Rapporteur proposed "Companies should make all artificial intelligence code fully auditable and should pursue innovative means for enabling external and independent auditing of artificial intelligence systems, separately from regulatory requirements. The results of artificial intelligence audits should themselves be made public." The report emphasizes the need for transparency in the administration of public services. "When an artificial intelligence application is being used by a public sector agency, refusal on the part of the vendor to be transparent about the operation of the system would be incompatible with the public body's own accountability obligations," the report advises.

---

[162] Human Rights Council, *The Right to Privacy in the Digital Age, Report of the United Nations High Commissioner for Human Rights* (Sept. 13, 2021), https://www.ohchr.org/EN/HRBodies/HRC/RegularSessions/Session48/Documents/A_HRC_48_31_AdvanceEditedVersion.docx

[163] UN Special Rapporteur, *Report of the Special Rapporteur on the promotion and protection of the right to freedom of opinion and expression,* A/73/348 (Aug. 29, 2018), https://freedex.org/wp-content/blogs.dir/2015/files/2018/10/AI-and-FOE-GA.pdf

An article published in 2024 by the CAIDP founder urged the creation of a Special Rapporteur for AI and Human Rights.[164] Rotenberg wrote, "to effectively navigate the intricate landscape of AI and human rights, there is a pressing need for the creation of a UN Special Rapporteur on AI and Human Rights. This role would not only complement existing efforts but also provide the agility, authority, and competence required to address emerging challenges and safeguard human rights in the digital age."

UN and Lethal Autonomous Weapons

In 2024, the UN Secretary-General urged the Security Council to act decisively to establish international guardrails for artificial intelligence, warning that delays could heighten risks to global peace and security.[165] "Every moment of delay in establishing international guardrails increases the risk for us all," Mr. Guterres said. "No country should design, develop, deploy or use military applications of AI in armed conflict that violate international humanitarian and human rights laws." "Recent conflicts have become testing grounds for AI military applications," he said, citing AI use in autonomous surveillance, predictive policing, and even reported life-and-death decisions.

Particularly alarming, he underscored, is the potential integration of AI with nuclear weapons and the advent of quantum-AI systems that could destabilize global security. "The fate of humanity must never be left to the 'black box' of an algorithm," he stated, stressing the importance of human control over decisions involving the use of force.

The Secretary-General's remarks follow a long history of efforts to establish rules to limit the use of AI systems in warfare. In fact, one of the first AI applications to focus the attention of global policymakers was the use of AI for warfare.[166] In 2016, the United Nations established the Group of Governmental Experts (GGE) on Lethal Autonomous Weapons Systems (LAWS) following a review of the High Contracting Parties to the Convention on Certain Conventional Weapons (CCW).[167] In November

---

[164] Marc Rotenberg, *The Imperative for a UN Special Rapporteur on AI and Human Rights*, Journal of AI Law and Regulation, Volume 1, Issue 1, pp. 110–112 (2024), DOI: https://doi.org/10.21552/aire/2024/1/13

[165] United Nations, *Humanity's Fate Can't Be Left to Algorithms, UN Chief Tells Security Council* (Dec. 19, 2024), https://un.dk/humanitys-fate-cant-be-left-to-algorithms-un-chief-tells-security-council/

[166] The Computer Professionals for Social Responsibility (CPSR), a network of computer scientists based in Palo Alto, California, undertook early work on this topic in the 1980s. CPSR History, http://cpsr.org/about/history/; See also David Bellin and Gary Chapman, *Computers in Battle Will They Work?* (1987)

[167] United Nations, *2018 Group of Governmental Experts on Lethal Autonomous Weapons Systems (LAWS),* https://www.unog.ch/80256EE600585943/(httpPages)/7C335E71DFCB29D1C1258243003E8724

2019,[168] the CCW High Contracting Parties endorsed 11 Guiding Principles for LAWS.[169] But concerns about the future of regulation of lethal autonomous weapons remain. At present, some countries believe that current international law "mostly suffices" while others believe new laws are needed.[170] Human Rights Watch provided an important overview of country positions on the future of banning fully autonomous weapons in August 2020.[171] Concerns over killer reports also arose at the 75th UN Assembly in October 2020. Pope Francis warned that lethal autonomous weapons systems would "irreversibly alter the nature of warfare, detaching it further from human agency." He called on states to "break with the present climate of distrust" that is leading to "an erosion of multilateralism, which is all the more serious in light of the development of new forms of military technology."[172] The Permanent Representative of the Holy See to the UN called for a ban on autonomous weapons in 2014.[173]

At the 2022 UN General Assembly, 70 countries endorsed a joint statement on autonomous weapons systems. The joint statement urged "the international community to further their understanding and address these risks and challenges by adopting appropriate rules and measures, such as principles, good practices, limitations and constraints. We are committed to upholding and strengthening compliance with International Law, in particular International Humanitarian Law,

---

[168] Meeting of the High Contracting Parties to the Convention on Prohibitions or Restrictions on the Use of Certain Conventional Weapons Which May Be Deemed to Be Excessively Injurious or to Have Indiscriminate Effects, *Final Report* (Dec. 13, 2019), https://undocs.org/Home/Mobile?FinalSymbol=CCW%2FMSP%2F2019%2F9&Language=E&DeviceType=Desktop

[169] Group of Governmental Experts on Emerging Technologies in the Area of Lethal Autonomous Weapons System, *Report of the 2019 session of the Group of Governmental Experts on Emerging Technologies in the Area of Lethal Autonomous Weapons Systems* (Sept. 25, 2019), https://documents-dds-ny.un.org/doc/UNDOC/GEN/G19/285/69/PDF/G1928569.pdf?OpenElement

[170] Dustin Lewis, *An Enduring Impasse on Autonomous Weapons*, Just Security (Sept. 28, 2020), https://www.justsecurity.org/72610/an-enduring-impasse-on-autonomous-weapons/

[171] Human Rights Watch, *Stopping Killer Robots: Country Positions on Banning Fully Autonomous Weapons and Retaining Human Control* (Aug. 10, 2020), https://www.hrw.org/report/2020/08/10/stopping-killer-robots/country-positions-banning-fully-autonomous-weapons-and#

[172] Address of His Holiness Pope Francis to the Seventy-fifth Meeting of the General Assembly of the United Nations, *The Future We Want, the United Nations We Need: Reaffirming our Joint Commitment through Multilateralism* (Sept. 25, 2020), https://reachingcriticalwill.org/images/documents/Disarmament-fora/unga/2020/25Sept_HolySee.pdf

[173] Statement by H.E. Archibishop Silvano M. Tomasi, Permanent Representative of the Holy See to the United Nations and Other International Organizations in Geneva at the meeting of Experts on Lethal Autonomous weapons systems of the High Contracting Parties to the Convention, *On Prohibitions or Restrictions on the Use of Certain Conventional Weapons which May Be Deemed to be Excessively Injurious or to Have Indiscriminate Effect* (May 13, 2014), https://www.unog.ch/80256EDD006B8954/(httpAssets)/D51A968CB2A8D115C1257CD8002552F5/$file/Holy+See+MX+LAWS.pdf

including through maintaining human responsibility and accountability in the use of force."[174]

At the 78th UN General Assembly First Committee in 2023, 164 states voted in favor[175] of resolution L.56[176] on autonomous weapons systems. The Resolution emphasizes the "urgent need for the international community to address the challenges and concerns raised by autonomous weapons systems," and mandated the UN Secretary-General to prepare a report reflecting the views of member and observer states on autonomous weapons systems. The report presented in July 2024 summarized views from more than 40 Member and Observer States and 28 civil society organizations.[177] The statements reflect a breadth of perspectives and concerns over defining LAWS and whether international law is sufficient for governing these weapons or a new instrument should be developed. Questions about the best approach remain but the Secretary-General noted the urgency "to take preventative action on this issue" and reiterated his call for "the conclusion by 2026, of a legally binding instrument to prohibit lethal autonomous weapons that function without human control or oversight and that cannot be used in compliance with international humanitarian law."[178] He concluded, "The autonomous targeting of humans by machines is a moral line that must not be crossed."

The 80th UN General Assembly furthered support for an international instrument to ensure the development and use of lethal autonomous weapons (LAWS) align with international humanitarian law. Negotiations for a treaty pushed forward with only 5 countries opposing and 5 abstaining on a resolution to extend conversations on beyond the GGE to informal discussions on emerging normative proposals at the General Assembly.[179] Countries also overwhelmingly supported ongoing dialogue on artificial intelligence in the military domain more broadly. 118 delegations supported[180] a draft resolution demanding human control and oversight

---

[174] UN General Assembly, First Committee, *Joint Statement on Lethal Autonomous Weapons Systems First Committee, 77th United Nations General Assembly Thematic Debate – Conventional Weapons* (Oct. 21, 2022), https://estatements.unmeetings.org/estatements/11.0010/20221021/A1jJ8bNfWGlL/KLw9WYcSnnAm_en.pdf

[175] Isabelle Jones, *164 States Vote against the Machine at the UN General Assembly*, Stop Killer Robots (Nov. 1, 2023), https://www.stopkillerrobots.org/news/164-states-vote-against-the-machine/

[176] UN General Assembly, *Resolution L56: Lethal Autonomous Weapons* (Oct. 12, 2023), https://reachingcriticalwill.org/images/documents/Disarmament-fora/1com/1com23/resolutions/L56.pdf

[177] UN General Assembly, *Lethal Autonomous Weapons Systems: Report of the Secretary-General* (Jul. 1, 2024), https://docs.un.org/en/a/79/88

[178] Ibid, p. 18

[179] UN General Assembly, *Report of the First Committee*, 80th Session, Agenda items 90–106. 121, 137 (Nov. 13, 2025), https://docs.un.org/en/A/80/534

[180] Ibid

over the command, control, and communications systems of nuclear weapons.[181] These resolutions encourage the Convention on Certain Conventional Weapons (CCW) Group of Governmental Experts (GGE) on LAWS to continue toward an international instrument by the end of 2026.

The Vatican

Pope Francis served as a leading figure in AI policy. In addition to his statements on autonomous weapons, in November 2020 the Pope warned that AI could exacerbate economic inequalities around the world if a common good is not pursued. "Artificial intelligence is at the heart of the epochal change we are experiencing. Robotics can make a better world possible if it is joined to the common good. Indeed, if technological progress increases inequalities, it is not true progress. Future advances should be oriented towards respecting the dignity of the person and of Creation."[182]

Earlier in 2020, the Pope endorsed the Rome Call for AI Ethics.[183] The goal of the Rome Call is to "support an ethical approach to Artificial Intelligence and promote a sense of responsibility among organizations, governments and institutions." The Pope said, "The Call's intention is to create a movement that will widen and involve other players: public institutions, NGOs, industries and groups to set a course for developing and using technologies derived from AI." The Pope also said that the Rome Call for Ethics is the "first attempt to formulate a set of ethical criteria with common reference points and values, offering a contribution to the development of a common language to interpret what is human."[184]

The key principles of the Rome Call are 1) Transparency: AI systems must be explainable; 2) Inclusion: the needs of all human beings must be taken into consideration so that everyone can benefit and all individuals can be offered the best possible conditions to express themselves and develop; 3) Responsibility: those who design and deploy the use of AI must proceed with responsibility and transparency; 4) Impartiality: do not create or act according to bias, thus safeguarding fairness and human dignity; 5) Reliability: AI systems must be able to work reliably; 6) Security and privacy: AI systems must work securely and respect the privacy of users. These principles are described as "fundamental elements of good innovation."

The Pope delivered several speeches on Artificial Intelligence in 2024. Speaking to the G7 leaders in June, the Pope described AI as an "exciting and

---

[181] UN General Assembly, *Possible Risks of the Integration* of Artificial Intelligence into Command, Control and Communications Systems of Nuclear Weapons, 80th Session First Committee, Agenda Item 99 (Oct. 16, 2025), https://docs.un.org/en/A/C.1/80/L.56

[182] Vatican News, *Pope's November Prayer Intention: That Progress in Robotics and AI "Be Human"* (Nov. 2020), https://www.vaticannews.va/en/pope/news/2020-11/pope-francis-november-prayer-intention-robotics-ai-human.html

[183] *Rome Call AI Ethics*, https://romecall.org

[184] Pontifical Academy for Life, *Rome Call for Ethics* (Feb. 28, 2020), http://www.academyforlife.va/content/pav/en/events/intelligenza-artificiale.html

fearsome" tool, "that can autonomously adapt to the task assigned to it and, if designed this way, can make choices independent of the person in order to achieve the intended goal."[185] The Pope called attention to inherent bias in AI systems, particularly those that may be deployed for judicial decision-making, and explained that algorithms are neither objective nor neutral. The Pope also urged world leaders to reconsider the development and use of lethal autonomous weapons and ultimately ban their use. He said, "No machine should ever choose to take the life of a human being."

Speaking to the World Forum in Davos in 2025, the Pope said that AI must promote and never violate human dignity.[186] The Pope said, AI raises "fundamental questions about ethical responsibility, human safety, and the broader implications of these developments for society." He warned against the risk that AI will be used to advance the "technocratic paradigm," which subordinates human dignity and fraternity to "the pursuit of efficiency, as though reality, goodness, and truth inherently emanate from technological and economic power."

Pope Leo XIV stressed to participants at the Second Annual Rome Conference on Artificial Intelligence "the need to weigh the ramifications of AI in light of the 'integral development of the human person and society' (Note *Antigua et Nova*, 6). This entails taking into account the well-being of the human person not only materially, but also intellectually and spiritually […] Ultimately, the benefits or risks of AI must be evaluated precisely according to this superior ethical criterion."[187] In a later speech, he called for widespread participation and unity "before any partisan interest or profit, which is increasingly concentrated in the hands of a few"[188] in a joint responsibility to protect this essential humanity against "the paralyzing idea that [AI] development follows an inevitable path." Pope Leo also challenged leaders to act against the perceived inevitability of an AI-enabled arms race.[189] He noted that artificial intelligence in weapons and other applications "is a tool that requires

---

[185] The Holy See, *Address of His Holiness Pope Francis* (Jun. 14, 2024), https://www.vatican.va/content/francesco/en/speeches/2024/june/documents/20240614-g7-intelligenza-artificiale.html

[186] Vatican News, *Pope to World Economic Forum: AI Must Promote and Never Violate human Dignity* (Jan. 23, 2025), https://www.vaticannews.va/en/pope/news/2025-01/pope-francis-sends-message-to-davos.html

[187] Holy See, *Message of Pope Leo XIV to Participants in the Second Annual Conference on Artificial Intelligence, Ethics, and Corporate Governance* (Jun. 19, 2025), https://www.vatican.va/content/leo-xiv/en/messages/pont-messages/2025/documents/20250617-messaggio-ia.html

[188] Holy See, *Address of His Holiness Pope Leo XIV to Participants in the Conference "Artificial Intelligence and Care of Our Common Home" Organized by the Centesimus Annus Pro Pontifice Foundation and the Strategic Alliance of Catholic Research Universities* (Dec. 5, 2026), https://www.vatican.va/content/leo-xiv/en/speeches/2025/december/documents/20251205-conferenza.html

[189] Holy See, *Address of Pope Leo XIV to Members of the Diplomatic Corps Accredited to the Holy See* (Jan. 9, 2026), https://www.vatican.va/content/leo-xiv/en/speeches/2026/january/documents/20260109-corpo-diplomatico.html

appropriate and ethical management, together with regulatory frameworks focused on the protection of freedom and human responsibility."

Technical Societies

Technical societies have also played a leading role in the articulation of AI principles. The IEEE led several initiatives, often in cooperation with government policymakers, to develop and promote Ethically Aligned Design (EAD).[190] The initial report, A Vision for Prioritizing Human Well-being with Autonomous and Intelligent Systems, was published in 2015. The IEEE published the second edition in 2017.[191] In 2019 the IEEE issued a Positions Statement on Artificial Intelligence, concluding that "AI systems hold great promise to benefit society, but also present serious social, legal and ethical challenges, with corresponding new requirements to address issues of systemic risk, diminishing trust, privacy challenges and issues of data transparency, ownership and agency."[192]

In November 2022, the IEEE Board approved the IEEE Standard for Operator Interfaces of Artificial Intelligence. The standard defines a set of operator interfaces frequently used in AI applications and highlights various types of operators, such as those related to basic mathematics, neural networks, and machine learning.[193] As of 2025,[194] the IEEE Standards Association includes the 7000 standards series to address transparency, privacy, algorithmic bias, and accountability in AI and other autonomous systems and a personal credentialing and product certification program to be compatible with the EU AI Act. The South African Chair's Summary following the World Summit on the Information Society (WSIS) High-Level Event 2025, co-sponsored by International Telecommunications Union (ITU), UNESCO, and several other UN entities, called on countries to empower digital and education ministers through political support and funding to carry out the "skills development, infrastructure investment, and knowledge sharing" needed to "ensure that all countries can benefit equitably."[195]

---

[190] IEEE Standards Association, *Autonomous and Intelligent Systems (AIS)*, https://standards.ieee.org/initiatives/autonomous-intelligence-systems/

[191] IEEE Standards Association, *IEEE Releases Ethically Aligned Design, Version 2 to Show "Ethics in Action" for the Development of Autonomous and Intelligent Systems (A/IS)* (Dec. 12, 2017), https://standards.ieee.org/news/2017/ead_v2.html

[192] IEEE, *Artificial Intelligence* (Jun. 24, 2019), https://globalpolicy.ieee.org/wp-content/uploads/2019/06/IEEE18029.pdf

[193] IEEE Standards Association, *IEEE Standard for Artificial Intelligence (AI) Model Representation, Compression, Distribution, and Management* (Mar. 2022), https://standards.ieee.org/ieee/2941/10363/

[194] IEEE Standards Association, *IEEE Standards Commitment to Advancing AI Governance Includes Impactful Contributions to New International AI Standards Exchange* (Jul. 11, 2025), https://standards.ieee.org/news/ieee-standards-commitment-to-advancing-ai-governance-includes-impactful-contributions-to-new-international-ai-standards-exchange/

[195] WSIS+20 High-Level Event 2025, *Chair's Summary: WSIS Accelerating Vision & Empowerment* (Jul. 7–11, 2025),

ACM, an international society of computer scientists and professionals, has also contributed to the global AI policy landscape.[196] In 2017 ACM released a Statement on Algorithmic Transparency and Accountability, identifying key principles to minimize bias and risks in algorithmic decision-making systems, including transparency, accountability, explainability, auditability, and validation.[197] In 2020, in response to growing concerns about the use of facial recognition technologies in public spaces, ACM released another statement addressing the unique issues of biometric data systems and the potential bias and inaccuracies that have significant consequences for violation of human rights.[198] A 2023 Statement addressed emerging challenges of generative AI and proposed several new principles, including Limits and guidance on Deployment and Use, Ownership, Personal Data Control, and Correctability.[199]

ACM supported efforts in the United States and Europe to ensure AI use aligned with the organization's code of Ethics and Professional Conduct. In a statement on the inappropriate use of AI Chatbots,[200] the organization expressed deep concerns about chatbots and similar tools that "engage in interactions of a sensual, sexual, or suggestive nature with minors" and urged "investigation and oversight of any such design or deployment." An ACM brief called on policymakers and legislators to draft laws against technology-facilitated domestic violence, human trafficking, and child exploitation.[201] The ACM U.S. Technology Policy Committee's recommendations to include AI education and workforce training were included in the AI Action Plan.[202] The Europe Technology Policy Committee submitted a policy

---

https://www.itu.int/net4/wsis/forum/2025/Files/outcomes/WSIS20HighLevelEvent2025-ChairsSummary.pdf

[196] Association for Computing Machinery, www.acm.org/public-policy

[197] ACM U.S. Public Policy Council, *Statement on Algorithmic Transparency and Accountability* (Jan. 12, 2017), https://www.acm.org/binaries/content/assets/publicpolicy/2017_usacm_statement_algorithms.pdf

[198] ACM U.S. Technology Policy Committee, *Statement on Principles and Prerequisites for the Development, Evaluation and Use of Unbiased Facial Recognition Technologies* (Jun. 30, 2020), https://www.acm.org/binaries/content/assets/public-policy/ustpc-facial-recognition-tech-statement.pdf

[199] ACM U.S. Technology Policy Committee, *Statement on Principles for the Development, Deployment and Use of Generative AI Technologies* (Jun. 27, 2023), https://www.acm.org/binaries/content/assets/public-policy/ustpc-approved-generative-ai-principles

[200] ACM U.S. Technology Policy Committee, *Statement on the Inappropriate Use of AI Chatbots* (Sept. 18, 2025), https://www.acm.org/binaries/content/assets/public-policy/usacm/ai/ai-chatbot-statmentsept-2025.pdf

[201] Jody R. Westby and Simson L. Garfinkel, *Technology Policy Can Curb Domestic Violence, Human Trafficking, and Crimes against Children*, ACM Technology Policy Council, TechBriefs, Winter 2025, no. 12 (2025), https://dl.acm.org/doi/epdf/10.1145/3717828

[202] ACM U.S. Technology Policy Committee, *Comments in Response to the Office of Science and Technology Policy Request for Information on the Development of an Artificial Intelligence (AI) Action Plan* (Mar. 14, 2025), https://www.acm.org/binaries/content/assets/public-policy/acm---ustpc-comments-on-ai-action-plan-final-submitted.pdf

brief to EU AI Act policymakers with recommendations to "enable true environmental accountability in the EU AI Act."[203]

Legal Societies

Legal societies are also playing a more prominent role in the AI policy landscape. The European Law Institute has published several reports and recommendations on AI law and policy.[204] In 2024, the International Bar Association endorsed the Council of Europe Framework Convention on Artificial Intelligence, Human Rights, Democracy and the Rule of Law.[205]

Civil Society

Civil society organizations engage in AI policymaking through a variety of means. Whether researching AI-related justice issues like facial recognition in public spaces, submitting comments, or sitting in the rooms where policymakers make decisions on legislation, civil society organizations hold government leaders accountable for rights-violating AI uses and help develop more just policies. This section highlights some recent engagement by civil society organizations.

Global Efforts

International organizations such as the G20 provide channels for representatives from civil society. In 2024, for example, Brazil launched the G20 Social to "increase the participation of non-governmental actors in G20 activities and decision-making processes."[206] The São Luís Declaration, a joint statement on inclusive AI by the civil, labor, think-tank, and women engagement groups, emerged from meetings leading up to the Social Summit held in connection with the Leaders' Summit.[207]

Other global meetings and venues such as the AI Safety Summit also include civil society organizations. The Sixth Athens Roundtable on AI and the Rule of Law in December 2024 brought together over 130 leading AI decision-makers in person

---

[203] ACM Europe Technology Policy Committee, *Policy Brief on the EU AI Act for True Environmental Accountability* (Jun. 2, 2025), https://www.acm.org/binaries/content/assets/public-policy/europe-tpc/acm_climate_disclosure_final.pdf

[204] See, for example, European Law Institute, *ELI Guiding Principles and Model Rules on Algorithmic Contracts* (2022), https://www.europeanlawinstitute.eu/projects-publications/current-projects/current-projects/eli-guiding-principles-and-model-rules-on-algorithmic-contracts/

[205] International Bar Association, *The IBA Is the First Association of Legal Practitioners to Endorse the Council of Europe Framework Convention on Artificial Intelligence* (Oct. 2, 2024), https://www.ibanet.org/The-IBA-is-the-first-association-of-legal-practitioners-to-endorse-the-Council-of-Europe-Framework-Convention-on-Artificial-Intelligence

[206] G20 Brasil 2024, *G20 Social* (2024), https://g20.gov.br/en/g20-social

[207] T20, Brasil 2024, *São Luís Declaration: A Joint Statement from Engagement Groups to the G20 States on Artificial Intelligence* (Sept. 19, 2024), https://www.t20brasil.org/en/news/82/sao-luis-declaration-a-joint-statement-from-engagement-groups-to-the-g20-states-on-artificial-intelligence

and more than 800 participants online. The event emphasized the urgency of establishing robust, globally coordinated accountability mechanisms to safeguard the rule of law, democratic values, and global well-being. Key discussions centered on enforcing global standards for AI risk management, building an interoperable ecosystem for AI evaluations, and ensuring transparency and accessible information flows.

Africa

In Africa, civil society organizations have actively engaged with continental initiatives surrounding AI governance, exemplified by an open letter published in July 2024 addressing the African Union (AU).[208] The letter was coordinated by Pollicy and the Media Institute of Southern Africa (MISA) and supported by a coalition comprising various organizations including Mozilla, Paradigm Initiative, African Internet Rights Alliance, and the Women of Uganda Network. The letter was issued in response to the endorsement of the Continental AI Strategy and the African Digital Compact by African ICT and Communications Ministers, applauding the AU for its visionary leadership and collective effort.

Signatories expressed significant concerns regarding the effectiveness of implementation at the national level, emphasizing that the transformative potential of these continental frameworks depends heavily on their translation into binding domestic policies. The coalition highlighted the importance of protecting African societies from potential negative impacts of AI, such as misinformation and AI-enabled propaganda, stressing the critical necessity of robust policies addressing data governance and fundamental rights. Recommendations included calling upon African Heads of State to actively participate in discussions during the AU Assembly, urging AU Chairperson Moussa Faki Mahamat to underscore the importance of the AU Data Policy Framework, and advocating for strengthened partnerships between civil society, communities, and AU bodies like AUDA-NEPAD. The letter strongly emphasized collaborative governance as essential, underlining civil society's willingness to partner with political leaders to ensure AI development truly benefits all Africans, reflecting a collective vision toward digital transformation rooted firmly in human rights and ethical standards.

Americas

Civil society organizations continue to advocate for individual rights in their struggle against facial recognition in public spaces. In Argentina, the Civil Rights Association (*Asociación por los Derechos Civiles*, ADC) organized against facial

---

[208] Pollicy, Media Institute of Southern Africa (MISA), & Kristophina Shilongo, *We Submitted an Open Letter to the African Union. Here Is What We Learned about AI Policymaking in Africa,* Mozilla (Aug. 12, 2024), https://foundation.mozilla.org/en/blog/we-submitted-an-open-letter-to-the-african-union-here-is-what-we-learned-about-ai-policymaking-in-africa/

recognition in 2020–2023.[209] A judge ruled the system implemented in Buenos Aires unconstitutional in 2022.[210] The Coalition Rights on the Net (*Coalizão Direitos na Rede*) in cited several examples of arrests after mistaken identity in a 2024 open letter advocating for the Brazilian AI regulation to ban facial recognition technologies.[211] This action continues research informing the public on facial recognition in public spaces by Brazilian think tank Instituto Igarapé[212] and a 2018 class action by the Brazilian Institute for Consumer Protection before the São Paulo Court of Justice in 2018.[213]

In Mexico, the multi-stakeholder National Alliance on Artificial Intelligence (ANIA, *Alianza Nacional de Inteligencia Artificial)* revitalized the charge for a national AI strategy with the Proposal for a National AI Strategy for Mexico, 2024–2030.[214] ANIA brings together experts from academia, private enterprise, government, and NGOs to conduct roundtables, offer public webinars and workshops, research AI use cases, and collaborate with other regional leaders.

In the United States, a variety of civil society organizations, including the Center for AI and Digital Policy, engage in policymaking and public advocacy. The Algorithmic Justice League, led by Dr. Joy Buolamwini, [215] unites these areas with testimony to Congress and statements to legislators, organizing and capacity-building through public campaigns, and media such the documentary *Coded Bias*.[216] Buolamwini's study, *Gender Shades*, co-authored with former Google AI Ethics co-lead Timnit Gebru,[217] prompted her advocacy and activism. Gebru's organization, Distributed AI Research Institute, centers research that aims to "mitigate/disrupt/eliminate/slow down harms caused by AI technology, and cultivate spaces to accelerate imagination and creation of new technologies and tools to build

---

[209] Asociación por los Derechos Civiles, *ConMiCaraNo,* https://conmicarano.adc.org.ar/

[210] Future of Privacy Forum, *Judge Declares Fugitive Facial Recognition System Unconstitutional* (Sept. 30, 2022), https://fpf.org/blog/judge-declares-buenos-aires-fugitive-facial-recognition-system-unconstitutional/; See also Judicial Branch of the City of Buenos Aires, *Judicial Decision on the Administrative and Tribute Litigation No. 4* (Sept. 7, 2022), https://www.cels.org.ar/web/wp-content/uploads/2022/09/reconocimientofacialsentencia070922.pdf

[211] Coalizão Direitos Na Rede (Coalition Rights on the Net), *Open Letter: Advocating for Brazilian AI Regulation that Protects Human Rights* (Jul. 8, 2024), https://direitosnarede.org.br/2024/07/08/open-letter-advocating-for-brazilian-ai-regulation-that-protects-human-rights/

[212] Instituto Igarapé, *Facial Recognition in Brazil* (Accessed Feb. 3, 2025), https://igarape.org.br/infografico-reconhecimento-facial-no-brasil/

[213] Instituto Brasileiro de Defesa do Consumidor, *ViaQuatro* (Aug. 30, 2018), https://idec.org.br/sites/default/files/acp_viaquatro.pdf

[214] ANIA, *About ANIA [Sobre ANIA]* (2024), https://www.ania.org.mx/

[215] Algorithmic Justice League, *About* (2025), https://www.ajl.org/about

[216] Algorithmic Justice League, *Policy/Advocacy* (2025), https://www.ajl.org/library/policy-advocacy

[217] Joy Buolamwini and Timnit Gebru, *Gender Shades: Intersectional Accuracy Disparities in Commercial Gender Classification*, Proceedings of Machine Learning Research 81, no. 1 (2018), https://proceedings.mlr.press/v81/buolamwini18a/buolamwini18a.pdf

a better future."[218] DAIR centers research on the real, present harms of AI technologies in society.

Asia

AI Safety Asia (AISA), co-founded in 2024 by Lyantoniette Chua, Senior Policy Advisor and Fellow at the Center for AI and Digital Policy is a global non-profit organization dedicated to the international governance of safe AI, beginning with initiatives in Southeast Asia. By promoting safe and governed AI in Asia, AISA aims to mitigate catastrophic risks, reduce adverse societal impacts, and unlock equitable socioeconomic development for 60% of the global population. Through programs such as the AISA Governance Academy and the Southeast Asia Policy Observatory, AISA seeks to enhance AI safety governance, upskill civil servants, incubate junior researchers, and foster cross-disciplinary collaboration. AISA has conducted AI Safety Roundtables in Manila, Jakarta, Singapore, Kuala Lumpur, and Bangkok, convening stakeholders to discuss regional opportunities, challenges, and strategies for AI safety governance.

AISA joined with Asia Society France at the AI Action Summit in Paris, France, in February 2025, to convene a high-level dialogue in Paris on the sidelines of the AI Action Summit titled From Oil to Water: Governing AI as a Global Public Good—A Europe-Asia Dialogue.

Europe

Civil society organizations across Europe have issued an urgent collective response to the third draft of the EU Code of Practice for General Purpose AI (GPAI) models, warning that crucial protections for fundamental rights and child safety have been significantly weakened. The joint letter,[219] endorsed by influential groups including the Ada Lovelace Institute, ARTICLE 19, 5Rights Foundation, and The Future Society, criticized the relegation of fundamental rights concerns such as child sexual abuse material (CSAM) and non-consensual intimate image abuse (NCII) to voluntary measures listed in an appendix, rather than central obligations. According to these groups, this approach drastically shifts responsibility for assessing risks from GPAI model providers to downstream system developers, undermining the accountability envisioned by the EU AI Act. Furthermore, the coalition asserts that by reducing fundamental rights issues to mere "potential considerations," the code becomes incompatible with established international frameworks, such as the International Scientific Report on the Safety of Advanced AI, which explicitly

---

[218] DAIR, *Research Philosophy* (2024), https://www.dair-institute.org/research/

[219] *Civil Society's Urgent Warning: The EU's Code of Practice for General Purpose AI Final Draft Cannot Abandon Fundamental Rights, and Protections for Children*, Center for Democracy & Technology (Mar. 28, 2025), https://cdt.org/insights/joint-civil-society-letter-urging-the-eu-institutions-to-protect-fundamental-rights-in-the-code-of-practice-for-general-purpose-ai-final-draft/; [pdf] https://cdt.org/wp-content/uploads/2025/03/Joint-Letter-CSO-CoP.pdf

acknowledges these harms. They also highlighted inadequacies in current industry safety frameworks cited by the drafters, pointing out that prominent frameworks from companies including Meta and Google notably fail to address critical risks related to human rights, discrimination, and child safety.

Civil society organizations urged EU policymakers to ensure the code aligns firmly with the original intent of the AI Act, maintaining mandatory protections rather than permitting companies to selectively recognize or ignore systemic risks. The signatories concluded that without these revisions, the code risks normalizing a regulatory landscape in which safeguarding fundamental rights and protecting vulnerable groups become discretionary rather than obligatory moral imperatives.

COUNTRY REPORTS

Argentina

In 2025, Argentina's data protection agency released guidelines on transparency in artificial intelligence while the Secretariat of Education's AI literacy program, PAIDEIA, launched in the beta stage.

National AI Strategy

The former Ministry of Science, Technology and Productive Innovation (MINCYT) published the National Strategy for Artificial Intelligence in 2019.[220] The plan includes two priority initiatives: Digital Agenda Argentina 2030[221] and the National Strategy for Science, Technology, and Innovation, Argentina Innovates 2030.[222]

With the goal of positioning Argentina as a regional leader on AI, the ten-year strategy seeks to transform the country through AI, leveraging the technology in pursuit of developmental objectives built on the UN's sustainable Development Goals (SDGs). The Strategy aims to minimize the potential risks of AI development and implementation for Argentinean society by protecting personal data and individual privacy through guidelines for the design of AI systems consistent with ethical and legal principles. The strategy also proposes to analyze the impact in the production scheme, to measure effects on labor force, and to prevent automate systems from reproducing or reinforcing discriminatory or exclusionary stereotypes. The cross-cutting themes in the Strategy are Ethics and regulation, Communication and awareness building, and International co-operation.

As the wide range of topics indicates, the Strategy requires a whole-of-government effort that brings together different government ministries under the leadership of the Digital Agenda Executive Roundtable (*Mesa Ejecutiva Agenda Digital*), which ended in 2019. This effort is supported by twenty different government agencies, as well as a Multi-sectoral Committee of Artificial Intelligence and a Scientific Committee of experts.

The 2019 National AI Strategy for Argentina set out ambitious goals that built on other national strategies developed under the former President Mauricio Marci, just before President Alberto Fernandez was elected in December 2019. The Executive

---

220 OECD AI Policy Observatory, *AI National Plan* (2023), https://oecd.ai/en/dashboards/countries/Argentina

221 Presidency of the Nation, *Argentina Digital Agenda — Decree 996/2018* (Nov. 2, 2018), https://www.boletinoficial.gob.ar/detalleAviso/primera/195154/20181105

222 Presidency of the Nation, *National Artificial Intelligence Plan Argentina (Plan Nacional de Inteligencia Artificial de la República de Argentina*) (2019), https://oecd-opsi.org/wp-content/uploads/2021/02/Argentina-National-AI-Strategy.pdf

Branch under the Fernandez administration published the Productive Development Plan Argentina 4.0 in April 2021.[223] An initiative of the now-dissolved National Ministry of Productive Development, aims to promote the incorporation of technologies 4.0—including AI—in the national production chain. The Argentina 4.0 Plan does not refer to the National AI Strategy. In November 2021, the Secretariat for Strategic Affairs also adopted the Artificial Intelligence Program.[224] The aim is to foster the responsible use of technology, such as artificial intelligence, "that contribute to consolidating Argentine technological sovereignty in the 4.0 revolution."[225]

Local Strategies

The Autonomous City of Buenos Aires launched Future City: AI Strategy in August 2021.[226] The Plan outlines three objectives:

- Use AI for the city's development
- Use AI for the benefit of the citizens
- Use cross-cutting tools to ensure the city's sustainability

Under this strategy, the Buenos Aires government has established Buenos Aires AI Lab (*BA Laboratorio IA*), which provides opportunities for training and professional development to youth and serves as a hub for facilitating R&D and application of AI. As with the National Plan, one of the key aspects of the Buenos Aires strategy is that it aims to foster mechanisms and tools for the development and use of AI technology that respects fundamental values and human rights.[227]

The Ministry of Government of the Province of Buenos Aires (*Gobierno de la Provincia de Buenos Aires*) released guidelines for using generative AI in the public administration.[228] The guidelines outline eight key principles to ensure the safe and appropriate use of AI in public administration, including 1) Authorization for Use; 2) Ethical Behavior; 3) Transparency of the tools; 4) Confidentiality and Data Protection;

---

[223] Ministry of Productive Development, *Productive Development Plan: Argentina 4.0* (Apr. 2021), https://www.argentina.gob.ar/sites/default/files/plan_de_desarrollo_productivo_argentina_4.0.vf__2.pdf

[224] Secretary for Strategic Affairs, *Resolution 90/2021* (Nov. 26, 2021), https://www.argentina.gob.ar/normativa/nacional/resoluci%C3%B3n-90-2021-357421

[225] National Executive Branch, *Decree 970/2020* (Dec. 1, 2020), https://www.argentina.gob.ar/normativa/nacional/decreto-970-2020-344786/texto

[226] Buenos Aires Secretary of Innovation and Digital Transformation, *Future City: Artificial Intelligence Plan for the City [Ciudad Futuro: Plan Estratégico Inteligencia Artificial]* (Aug. 13, 2021), http://buenosaires.gob.ar/jefaturadegabinete/innovacion/noticias/ciudad-futuro-el-primer-plan-de-inteligencia-artificial-de

[227] Ibid

[228] Gobierno de la Provinicia de Buenos Aires, *Resolution 4/2025 that Approves the "Guidelines for the Use of Generative Artificial Intelligence in Public Administration of the Province of Buenos Aires [Resolución 4/2025 de la Subsecretaría de Gobierno Digital del Ministerio de Gobierno que Aprueba las "Diatrices de uso de Inteligencia Artificial Generativa en la Administración Pública de la Provincia de Buenos Aires"]*, Boletín 29939 (Feb. 14, 2025), https://normas.gba.gob.ar/ar-b/resolucion/2025/4/493209

5) Information Security in not entering sensitive information in AI systems; 6) Evaluation of AI Outputs to ensure human oversight for accuracy, relevance, data sensitivity, bias, and legal compliance; 7) Prohibited Uses to mandate human intervention in decision-making, especially in matters affecting rights, accreditation, verifications, civil applications, or legal investigations; and 8) Training and Continuous Learning for public officials. The Undersecretariat of Digital Government is responsible for overseeing this initiative to protect human rights, foster responsible innovation, and ensure human oversight in AI applications across Buenos Aires' public sector.

Public Participation

Since the development of Argentina's national AI strategy, the government has taken further steps to increase public participation in AI-related initiatives. The Agency for Access to Public Information (*Agencia de Acceso a la Información Pública*, AAIP) launched a program in 2023[229] focused on AI transparency and the protection of personal data. To further support these efforts, the AAIP published the Guide to Transparency and the Protection of Personal Data in Responsible AI,[230] outlining AI's potential risks and challenges and offering guidance on implementing responsible AI principles throughout the AI lifecycle. These efforts reflect Argentina's aim to ensure AI development aligns with societal values and effectively addresses public concerns.

The AAIP oversaw a public consultation on a proposed update of the Personal Data Protection Law (*Ley 25.326*) in 2022.[231] A webpage on the project reported more than 123 participants from civil society, organizations, universities, and public and private institutions submitted opinions and documents and includes summaries of roundtables to inform the bill submitted in 2023.[232] The site references informational documents, including the UNESCO Recommendation on the Ethics of AI.

---

[229] Agencia de Acceso a la Información Pública, *Program for Transparency and Protection of Personal Data in the Use of Artificial Intelligence — Resolution 161/2023* (Sept. 4, 2023), https://www.argentina.gob.ar/noticias/programa-de-transparencia-y-proteccion-de-datos-personales-en-el-uso-de-la-inteligencia

[230] Agencia de Acceso a la Información Pública, *Guide for Transparency and Personal Data Protection in Responsible AI* (Sept. 16, 2024), https://www.argentina.gob.ar/sites/default/files/aaip-argentina-guia_para_usar_la_ia_de_manera_responsable.pdf

[231] Agencia de Acceso a la Información Pública, *Comenzó la consulta pública sobre la propuesta de anteproyecto de Ley de Protección de Datos Personales* (Sept. 12, 2022), https://www.argentina.gob.ar/noticias/comenzo-la-consulta-publica-sobre-la-propuesta-de-anteproyecto-de-ley-de-proteccion-de

[232] Agencia de Acceso a la Información Pública, *Personal Data Protection Bill [Proyecto de Ley de Protección de Datos Personales]*, https://www.argentina.gob.ar/aaip/datospersonales/proyecto-ley-datos-personales

Argentina's Library of Congress (*Biblioteca del Congreso de la Nación*) dedicated an edition of the Legislative Dossier to artificial intelligence.[233] The Dossier includes bills, international frameworks such as the OECD AI Principles, and judicial decisions related to AI. The issue provides basic information such as the definition of artificial intelligence and impacts as well as relevant legislation.

Data Protection

Article 43 of Argentina's Constitution guarantees an individual's access to personal data in private and public registries, and exercise agency over how that data is used. Argentina's Personal Data Protection Law (PDPL) follows international standards regarding basic personal data rules and had even been deemed adequate by the European Commission.[234] A new proposal put forward by the former administration to reform the PDPL and related legislation has been stalled in the National Congress. The purpose of this reform is not only for the country to keep its international status as a jurisdiction that provides an adequate level of protection, particularly after the passing of the European General Data Protection Regulation, but also to keep its data protection regime up to date with the technological and legal developments that have taken place in recent years. Two new bills to modernize the PDPL, one in the Chamber of Deputies (*Cámara de Diputados*),[235] and one in the Senate (*Senado*)[236] may succeed in modernizing the law though they remain under committee consideration with no plenary vote recorded in either chamber.

The Agency for Access to Public Information (*Agencia de Acceso a la Información Pública*, AAIP) is Argentina's data protection authority. Although the AAIP enjoys functional autonomy by law, the agency remains under the National Executive Branch from a structural perspective; an aspect that, with the absence of proper mechanisms, has led civil society groups to question the impartiality and independence of the appointment process of its Executive Director.[237]

---

[233] Biblioteca del Congreso de la Nación, *Dossier legislative: Inteligencial artificial,* Vol. XIII, no. 312 (Apr. 2025), https://bcn.gob.ar/uploads/Dossier-312-Legis-nacional-Inteligencia-Artificial-Abril-2025.pdf

[234] DLA Piper, *Data Protection Laws of the World – Argentina* (Jan. 28, 2025), https://www.dlapiperdataprotection.com/index.html?c=AR

[235] Honorable Cámara de Diputados de la Nación, *Personal Data Protection Regime [Proyecto de ley 1948-D/2025 — Régimen de protección de datos personales]* (2025), https://www4.hcdn.gob.ar/dependencias/dsecretaria/Periodo2025/PDF2025/TP2025/1948-D-2025.pdf

[236] Honorable Senado de la Nación Argentina, *Personal Data Protection Regime [Expediente S-0644/2025—Régimen de protección de datos personales]* https://www.senado.gob.ar/parlamentario/comisiones/verExp/644.25/S/PL

[237] Association for Civil Rights, *Observations of the ADC on the Proposed Candidate for the Director of the Agency of Access to Public Information* (Mar. 17, 2021), https://adc.org.ar/2021/03/17/observaciones-de-la-adc-a-la-candidatura-propuesta-para-la-direccion-de-la-agencia-de-acceso-a-la-informacion-publica/

Following an extended public consultation, the AAPI took up 80 for the revised Personal Data Protection Law articles in the final proposal and modified 43 based on public comments. The reform package was presented to Argentina's government for review before introduction to the National Congress of Argentina in June 2023.[238]

AAIP issued a resolution that established the Transparency and Protection of Data Program on the use of AI.[239] This program has the objective of enhancing regulatory frameworks and strengthening state powers necessary for the development and use of AI, based on the assumption that the state is responsible for guaranteeing the effective exercise of citizens' rights regarding transparency and the protection of personal data. The Resolution included the formation of the Observatory on AI, an initiative to foster governance and community engagement. A week after this Resolution was created, Argentina's President Chief of Staff issued Administrative Decision No. 750/2023, which created an Inter-Ministerial roundtable on AI.[240]

As a member of the Ibero-American Network for the Protection of Personal Data (*Red Iberoamericana de Protección de Datos*, RIPD), which comprises 16 data protection authorities of 12 countries, the AAPI endorsed the General Recommendations for the Processing of Personal Data in Artificial Intelligence[241] and the accompanying Specific Guidelines for Compliance with the Principles and Rights that Govern the Protection of Personal Data in Artificial Intelligence Projects.[242]

In May 2023, the RIPD data protection authorities initiated a coordinated action regarding ChatGPT, developed by OpenAI, on the basis that it may entail risks for the rights and freedoms of users in relation to the processing of their personal data. Concerns regarding the risk of misinformation. "ChatGPT does not have knowledge and/or experience in a specific domain, so the precision and depth of the response may vary in each case, and/or generate responses with cultural, racial or gender biases, as well as false ones."[243]

---

[238] National Executive Branch, *The National Executive Branch Sent the Personal Data Protection Bill to Congress* (Jun. 30, 2023), https://www.argentina.gob.ar/noticias/el-poder-ejecutivo-nacional-envio-al-congreso-el-proyecto-de-ley-de-proteccion-de-datos

[239] Agencia de Acceso a la Información Pública, *Resolution No 161/23* (Sept. 2023), https://www.boletinoficial.gob.ar/detalleAviso/primera/293363/20230904

[240] M. OFarrell Mairal, *Agency of Access to Public Information Creates Transparency and Personal Data Protection Program for Using Artificial Intelligence,* Lexology (Oct. 5, 2023), https://www.lexology.com/library/detail.aspx?g=d4c2fb3b-4502-46af-8831-783ee5946503

[241] Ibero-American Network for the Protection of Personal Data (RIPD), *General Recommendations for the Processing of Personal Data in Artificial Intelligence* (Jun. 21, 2019), https://www.redipd.org/en/document/guide-general-recommendations-processing-personal-data-ai-en.pdf

[242] Ibero-American Network for the Protection of Personal Data (RIPD), *Specific Guidelines for Compliance with the Principles and Rights that Govern the Protection of Personal Data in Artificial Intelligence Projects* (Jun. 21, 2019), https://www.redipd.org/en/document/guide-specific-guidelines-ai-projects-en.pdf

[243] Ibero-American Network for the Protection of Personal Data (RIPD), *Authorities from the Ibero-American Network for the Protection of Personal Data Begin a Coordinated Action in Relation to*

Argentina's AAIP has been a member of the Global Privacy Assembly (GPA) since 2018.[244] AAIP co-sponsored the 2018 GPA Resolution on AI and Ethics, [245] 2020 Resolution on AI and Accountability,[246] 2022 Resolution on Facial Recognition Technology, [247] and the 2023 Resolution on Generative AI. [248]

Argentina is also party to Council of Europe Convention 108 since June 2019. Argentina became the 23rd state to ratify 108+ on data protection in 2023.[249]

Building on these initiatives, in July 2024 the Argentine Congress proposed the establishment of a Federal Observatory on AI[250] to guide ethical AI policy, enhance transparency, and ensure that AI systems respect constitutional rights. The Center for AI and Digital Policy (CAIDP)[251] praised Argentina's efforts, particularly the AAIP guidelines' focus on transparency and data protection, which is aligned with global standards such as the OECD AI Principles and UNESCO Recommendation on AI Ethics.

Algorithmic Transparency

Argentina ratified the Council of Europe Convention 108+, granting Argentines the right to information on algorithmic logic as well as access to

---

ChatGPT (May 8, 2023), https://www.redipd.org/noticias/autoridades-red-iberoamericana-de-proteccion-de-datos-personales-inician-accion-chatgpt

244 Global Privacy Assembly, *List of Accredited Members* (2026), https://globalprivacyassembly.org/participation-in-the-assembly/list-of-accredited-members/

245 Global Privacy Assembly, *Declaration on Ethics and Data Protection in Artificial Intelligence* (Oct. 23, 2018), https://globalprivacyassembly.org/wp-content/uploads/2018/10/20180922_ICDPPC-40th_AI-Declaration_ADOPTED.pdf

246 Global Privacy Assembly, *Resolution on Accountability in the Development and Use of Artificial Intelligence* (Oct. 2020), https://globalprivacyassembly.org/wp-content/uploads/2020/10/FINAL-GPA-Resolution-on-Accountability-in-the-Development-and-Use-of-AI-EN-1.pdf

247 Global Privacy Assembly, *Resolution on Principles and Expectations for the Appropriate Use of Personal Information in Facial Recognition Technology* (Oct. 2022), https://globalprivacyassembly.org/wp-content/uploads/2022/11/15.1.c.Resolution-on-Principles-and-Expectations-for-the-Appropriate-Use-of-Personal-Information-in-Facial-Recognition-Technolog.pdf

248 Global Privacy Assembly, *Resolution on Generative Artificial Intelligence Systems* (Oct. 2023), https://globalprivacyassembly.org/wp-content/uploads/2023/10/5.-Resolution-on-Generative-AI-Systems-101023.pdf

249 Council of Europe, *Argentina Becomes the 23rd State to Ratify Convention 108+ on Data Protection* (Apr. 17, 2023), https://www.coe.int/en/web/human-rights-rule-of-law/-/argentina-ratifies-convention-108-during-the-privacy-symposium

250 Chamber of Deputies of Argentina, *Bill No. 3900-D-2024 — Establishing the Federal Observatory on Artificial Intelligence* (currently under deliberation) (Jul. 24, 2024), https://digitalpolicyalert.org/event/21749-introduced-bill-establishing-the-federal-observatory-on-artificial-intelligence-bill-no-3900-d-2024

251 Center for AI and Digital Policy (CAIDP), *Comments to the Agency for Access to Public Information (AAIP) of the Government of the Argentine Republic - Consultation on AAIP Guide to Using Artificial Intelligence Responsibly* (Oct. 29, 2024), https://files.constantcontact.com/dfc91b20901/9606ce1c-1bdb-4e26-8ee9-60314f3d40ac.pdf

correction.[252] The AAIP[253] published a preliminary draft of the Guide on Transparency and Personal Data Protection for Responsible AI,[254] which outlines challenges and risks associated with AI while providing recommendations for implementing these principles throughout the AI lifecycle, in 2024. The guide promotes algorithmic transparency in the AI lifecycle and ensures data subjects are provided with relevant information regarding the AI system. Public consultation on this guide was opened after its publication.[255]

The AAPI provided interpretation guidelines of the Personal Data Protection Act[256] through Resolution No. 4/2019[257] in which it recognizes that, under the right of access enshrined in the current data protection law, data subjects have the right to request from data controllers an explanation about the logic used by any system that reaches decisions solely based on automated processing of data and which can affect citizens or have pernicious legal effects on them.[258]

The AAPI's proposed reform Act includes the right of citizens to get information about "the existence of automated decision systems, including those that create digital profiles," as well as "meaningful information" about the logic applied by those systems.[259] A formal right to object to a decision based solely on automatic processing methods is also included in the proposal.[260]

The RED Specific Guidelines for Compliance with the Principles and Rights that Govern the Protection of Personal Data in Artificial Intelligence Projects also

---

[252] Boletín Oficial, *Ley 27699, Proteccion of People with Respect to the Automatic Treatment of Personal Data [Protección de las Personas con Respecto al Tratamiento Automatizado de Datos de Carácter Personal]* (Nov. 20, 2022), https://www.boletinoficial.gob.ar/detalleAviso/primera/276783/20221130

[253] Data Guidance, *Argentina: AAIP Publishes Guide on Transparency and Personal Data Protection in AI* (Sept. 17, 2024), https://www.dataguidance.com/news/argentina-aaip-publishes-guide-transparency-and

[254] Agencia de Acceso a la Información Pública, *Guide for Private and Public Entities for Transparency and Protection of Personal Data for Responsible AI [Guía para entidades públicas y privadas en materia de Transparencia y Protección de Datos Personales para una Inteligencia Artificial responsable: Versión preliminar]* (Sept. 27, 2024), https://www.argentina.gob.ar/sites/default/files/aaip-argentina-guia_para_usar_la_ia_de_manera_responsable.pdf

[255] Agencia de Acceso a la Información Pública, *AAIP Guide for the Responsible Use of AI* (2024), https://www.argentina.gob.ar/noticias/guia-de-la-aaip-para-usar-la-inteligencia-artificial-de-manera-responsable

[256] UNODC, *Personal Data Protection Act [Translation]* (2000), https://sherloc.unodc.org/cld/uploads/res//uncac/LegalLibrary/Argentina/Laws/Argentina%20Personal%20Data%20Protection%20Act%202000.pdf

[257] Government of Argentina, *Resolution 4 /2019, Guiding Criteria and Indicators for Best Practices in the Application of Law No. 25.326* (2019), https://www.argentina.gob.ar/normativa/nacional/resoluci%C3%B3n-4-2019-318874

[258] *Annex I of Resolution 4/2019* (IF-2019-01967621-APN-AAIP), http://servicios.infoleg.gob.ar/infolegInternet/anexos/315000-319999/318874/norma.htm.

[259] Ibid, Article 28 (h) of the draft Bill

[260] Ibid, Article 32 of the draft Bill

provide guidance on algorithmic transparency such that "The information provided regarding the logic of the AI model must include at least basic aspects of its operation, as well as the weighting and correlation of the data, written in a clear, simple and easily understood language, it will not be necessary to provide a complete explanation of the algorithms used or even to include them. The above always looking not to affect the user experience."[261]

Data Scraping

The AAIP and 16 of its international counterparts released a follow-up joint statement laying out expectations for industry related to data scraping and privacy in 2024.[262] The initial statement in August 2023 established that "[social media companies] and other websites are responsible for protecting personal information from unlawful scraping."[263] The follow-up, which incorporates feedback from engagement with companies such as Alphabet (YouTube), X Corp (formerly Twitter), and Meta Platforms (Facebook, Instagram, WhatsApp), adds that organizations must "comply with privacy and data protection laws when using personal information, including from their own platforms, to develop AI large language models"; must regularly review and update privacy protection measures to keep pace with technological advances; and ensure that permissible data scraping, such as that for research, "is done lawfully and following strict contractual terms."[264]

Data scraping generally involves the automated extraction of data from the web. The statements responded to increasing incidents involving data scraping, particularly from social media companies and the operators of other websites that host publicly accessible data. Scraped personal information can be exploited for targeted cyberattacks, identity fraud, monitoring, profiling and surveillance purposes, unauthorized political or intelligence gathering purposes, unwanted direct marketing or spam.

---

[261] Ibero-American Network for the Protection of Personal Data (RED), *Specific Guidelines for Compliance with the Principles and Rights that Govern the Protection of Personal Data in Artificial Intelligence Projects* (Jun. 2019), p. 17, www.redipd.org/en/document/guide-general-recommendations-processing-personal-data-ai-en.pdf

[262] ICO, *Global Privacy Authorities Issue Follow-Up Joint Statement on Data Scraping after Industry Engagement* (Oct. 28, 2024), https://ico.org.uk/about-the-ico/media-centre/news-and-blogs/2024/10/global-privacy-authorities-issue-follow-up-joint-statement-on-data-scraping-after-industry-engagement/

[263] Regulatory Supervision Information Commissioner's Office of the United Kingdom, J*oint Statement on Data Scraping and the Protection of Privacy* (Aug. 24, 2023), https://www.priv.gc.ca/en/opc-news/speeches-and-statements/2023/js-dc_20230824/

[264] ICO, *Global Privacy Authorities Issue Follow-Up Joint Statement on Data Scraping after Industry Engagement* (Oct. 28, 2024), https://ico.org.uk/about-the-ico/media-centre/news-and-blogs/2024/10/global-privacy-authorities-issue-follow-up-joint-statement-on-data-scraping-after-industry-engagement/

Use of AI in Public Administration

Argentina is among the Latin American and the Caribbean countries that have been utilizing AI in the public sector. Some of the implementations include Prometea,[265] developed in 2017, which is a virtual assistant that anticipates case solutions (based on previous cases and solutions) and helps assemble judicial case files. Laura, an application developed by the Ministry of Finance of the Province of Cordoba, automates tasks in bureaucratic procedures.[266] One of the most controversial uses was the use of AI by the Government of the Province of Salta, which implemented a system to predict teenage pregnancy and school dropout using machine learning algorithms trained on personal, environmental, and health data collected in low-income districts of Salta City in 2016 and 2017.[267] The program came to an end following the 2019 national and state elections when the new administration terminated several programs, including the use of algorithms to predict pregnancy.[268]

The Undersecretariat of Information Technologies enacted a resolution delineating a comprehensive framework of ethical principles and recommendations for the design, development, implementation, and use of AI projects in the public service.[269] The Ministry of Justice launched the National Comprehensive Program of Artificial Intelligence in Justice.[270] The National Program aims to apply AI in strategic areas that directly impact the lives of citizens, especially in justice and public administration, health, and education and training.[271] The Ministry of Security furthered the use of AI systems in the justice system with the creation of the Artificial Intelligence Unit Applied to Security (UIAAS)[272] to enhance crime prevention and investigation using AI tools like social media monitoring, facial recognition, and predictive algorithms.

---

[265] OECD, *The Strategic and Responsible Use of Artificial Intelligence in the Public Sector of Latin America and the Caribbean: AI Use Cases in LAC governments* (Mar. 2022), https://oecd-opsi.org/wp-content/uploads/2022/03/lac-ai.pdf

[266] Ibid

[267] Wired, *The Case of the Creepy Algorithm That 'Predicted' Teen Pregnancy* (Feb. 16, 2022), https://www.wired.com/story/argentina-algorithms-pregnancy-prediction/

[268] TNI, *What Artificial Intelligence is Hiding, Microsoft and Vulnerable Girls in Northern Argentina* (Feb. 23, 2023), https://www.tni.org/en/article/what-artificial-intelligence-is-hiding

[269] Jefatura de Gabinete de Ministros Subsecretaría de Tecnologías de la Información, *Resolution 2/2023: Guidelines for Reliable Artificial Intelligence* (Jun. 1, 2023), https://www.boletinoficial.gob.ar/detalleAviso/primera/287679/20230602

[270] Government of Argentina, *Resolution 149/2024* (May 6, 2024), https://www.argentina.gob.ar/normativa/nacional/resoluci%C3%B3n-149-2024-398921

[271] World Law Group, *The National Artificial Intelligence Program: Harnessing Technology for the Common Good in Argentina*, (Oct. 21, 2024), https://www.theworldlawgroup.com/membership/news/news-the-national-artificial-intelligence-program-harnessing-technology-for-the-common-good-in-argentina-1

[272] Government of Argentina, *Resolution 710/2024* (Jul. 26, 2024), https://www.boletinoficial.gob.ar/detalleAviso/primera/311381/20240729

Facial Recognition

Several documented cases of facial recognition technology use have been reported in various cities and localities as well as at the provincial level in the country. Facial recognition systems being deployed include: for the identification and capture of fugitives (in the Autonomous City of Buenos Aires);[273] for the identification of missing persons and of people with criminal backgrounds (town of Tigre, Buenos Aires province);[274] for the use of the police to surveil mass gatherings (Mendoza province);[275] or for the prevention and prosecution of crimes (Salta province).[276] The program in the City of Buenos Aires in particular was denounced by the UN Special Rapporteur for the Right of Privacy as a technology whose "proportionality" was questionable when compared to the "serious privacy implications" for people not related to any crime and for not carefully updating and checking for accuracy.[277] Human Rights Watch also denounced the system, noting the illegal exposure of minor's personal information.[278] The City legislature approved a bill in 2020 to authorize the use of AI for the purpose of capturing fugitives.[279] But it has been alleged that this fact does not alter the unconstitutional character of the Buenos Aires program.[280] The increasing and unaccountable use of this technology led to the creation of a national campaign by the Association for Civil Rights (*Asociación por los Derechos Civiles*), a well-known Argentinian human rights organization. With the

---

[273] Al Sur, *Facial Recognition in Latin America: Trends in the Implementation of a Perverse Technology* (2021), p. 11, https://www.alsur.lat/sites/default/files/2021-11/ALSUR_Reconocimiento_facial_en_Latam_ES.pdf

[274] Ambito, *Tigre Implemented a New Facial Recognition System* (May 10, 2019), https://www.ambito.com/municipios/municipios/tigre-lanzo-un-nuevo-sistema-reconocimiento-facial-n5030978

[275] El Sol, *Facial Recognition: More Than 100 People with Arrest Warrants Found* (May 20, 2019), https://www.elsol.com.ar/reconocimiento-facial-hallaron-a-mas-de-100-personas-con-pedido-de-captura

[276] Salta Government, Secretary of Press and Communication, *Facial Recognition Cameras Lead to the Detention of a Person with an Arrest Warrant* (Jun. 19, 2019), https://www.salta.gob.ar/prensa/noticias/las-camaras-de-reconocimiento-facial-permitieron-detener-a-una-persona-con-pedido-de-captura-64939

[277] OHCHR, *Statement to the Media by the United Nations Special Rapporteur on the Right to Privacy, on the Conclusion of His Official Visit to Argentina, 6–17 May 2019* (May 17, 2019), https://www.ohchr.org/en/statements/2019/05/statement-media-united-nations-special-rapporteur-right-privacy-conclusion-his?LangID=E&NewsID=24639

[278] Human Rights Watch, *Argentina Publishes Personal Data of Boys and Girls Accused of Crimes Online* (Oct. 9, 2020), http://www.hrw.org/es/news/2020/10/09/argentina-publica-en-linea-datos-personales-de-ninos-y-ninas-acusados-de-cometer

[279] Asociación por los Derechos Civiles, *The Buenos Aires Legislature Should Reject the Use of Facial Recognition Technology for the Surveillance of Public Space* (Oct. 21, 2020), https://adc.org.ar/2020/10/21/la-legislatura-portena-debe-rechazar-el-uso-de-la-tecnologia-de-reconocimiento-facial-para-la-vigilancia-del-espacio-publico/

[280] iProfessional, *Now they come for your face? This Expert Warns of the Dangers of Facial Recognition* (May 10, 2020), https://www.iprofesional.com/tecnologia/338236-reconocimiento-facial-advierten-sobre-peligros-en-argentina

slogan: "Con mi Cara No" ("Not with my face"), the organization aims to raise awareness about the dangers facial recognition technologies pose to citizens, particularly when their data is included within opaque and unaccountable systems.[281] A trial judge declared the implementation of the Fugitive Facial Recognition System (SRFP, for its name in Spanish) by the Government of the City of Buenos Aires unconstitutional.[282]

The Court of Appeals of the City of Buenos Aires confirmed the unconstitutionality of the use of the Fugitive Facial Recognition System implemented by the Buenos Aires City Government in 2023.[283] The use of SRFP is currently suspended. Meanwhile on the national level, the Artificial Intelligence Applied to Security Unit (UIAAS) plans to use facial recognition to predict future crimes.[284] AI may be used to cross-reference databases related to the commission of crimes (such as records of convictions, arrests, etc.) with others related to criminals and other persons of interest to generate predictions. UIAAS also expects to deploy facial recognition software to identify "wanted persons," patrol social media, and analyze real-time security camera footage to detect suspicious activities.

Environmental Impact of AI

In Argentina, the role of AI in environmental management is becoming increasingly important as it has the potential to address critical challenges such as climate change, pollution, and biodiversity loss. AI is already being used in initiatives such as the United Nations Environment Programme's (UNEP)[285] World Environment Situation Room (WESR),[286] which uses AI for real-time analysis of environmental data. This includes monitoring CO2 levels, sea level rise and other key indicators to inform policy decisions and improve environmental sustainability. AI also plays an important role in monitoring emissions, optimizing the use of renewable energy and

---

[281] Association for Civil Rights, *Con Mi Cara No: Facial Recognition in Buenos Aires*, https://conmicarano.tedic.org/

[282] Future of Privacy Forum, *Judge Declares Fugitive Facial Recognition System Unconstitutional* (Sept. 30, 2022), https://fpf.org/blog/judge-declares-buenos-aires-fugitive-facial-recognition-system-unconstitutional/; See also Judicial Branch of the City of Buenos Aires, *Judicial Decision on the Administrative and Tribute Litigation No. 4* (Sept. 7, 2022), https://www.cels.org.ar/web/wp-content/uploads/2022/09/reconocimientofacialsentencia070922.pdf

[283] CELS, *The Court of Appeals of the City of Buenos Aires Confirmed the Unconstitutionality of the Use of the Fugitive Facial Recognition System (SRFP) Implemented by the Buenos Aires City Government* (Apr. 29, 2023), https://www.cels.org.ar/web/en/2023/04/the-court-of-appeals-of-the-city-of-buenos-aires-confirmed-the-unconstitutionality-of-the-use-of-the-fugitive-facial-recognition-system-srfp-implemented-by-the-buenos-aires-city-government/

[284] Government of Argentina, *Resolution 710/2024* (Jul. 26, 2024), https://www.boletinoficial.gob.ar/detalleAviso/primera/311381/20240729

[285] United Nations Environment Programme, *How Artificial Intelligence Is Helping Tackle Environmental Challenges* (Nov. 7, 2022), https://www.unep.org/news-and-stories/story/how-artificial-intelligence-helping-tackle-environmental-challenges

[286] UN Environment Programme, *World Environment Situation Room* (2024), https://wesr.unep.org/

supporting smarter environmental policies. However, there is growing concern about the environmental costs of AI, particularly in terms of data processing and e-waste. Argentina's growing focus on AI for sustainability must balance the technology's environmental impact with its benefits.

Lethal Autonomous Weapons

Argentina has been very critical about the development and use of lethal autonomous weapons systems, particularly those without significant human involvement. Argentina has set out a strong position in public statements as well as within international organizations, including during meetings regarding the Convention on Conventional Weapons. Within the framework of those meetings, Argentina stressed the need "to preserve meaningful human control at all phases of the development and use" of weapons systems.[287] On behalf of the Group of Latin American and Caribbean Countries, Argentina raised several concerns over fully autonomous weapons, including the risks of reprisal, retaliation, and terrorism.[288] And Argentina has called for a "preemptive prohibition of the development of lethal autonomous systems."[289]

At the 78th UN General Assembly First Committee in 2023, Argentina voted in favor of the resolution on autonomous weapons systems, along with 163 other states. The resolution stressed the "urgent need for the international community to address the challenges and concerns raised by autonomous weapons systems," and mandated the UN Secretary-General to prepare a report reflecting the views of member and observer states on autonomous weapons systems and ways to address the related challenges and concerns they raise from humanitarian, legal, security, technological and ethical perspectives and on the role of humans in the use of force.

Argentina was one of the 70 countries that endorsed a 2022 joint statement on autonomous weapons systems at the UN General Assembly meeting. In this joint statement, States urged "the international community to further their understanding and address these risks and challenges by adopting appropriate rules and measures, such as principles, good practices, limitations and constraints. We are committed to upholding and strengthening compliance with International Law, in particular

---

[287] United Nations, *Statement to the Convention on Conventional Weapons Group of Governmental Experts on Lethal Autonomous Weapons Systems* (Mar. 26, 2019), https://documents.unoda.org/wp-content/uploads/2020/09/CCW_GGE.1_2019_3_E.pdf

[288] Human Rights Watch, *Stopping Killer Robots: Country Positions on Banning Fully Autonomous Weapons and Retaining Human Control* (Aug. 10, 2020), https://www.hrw.org/report/2020/08/10/stopping-killer-robots/country-positions-banning-fully-autonomous-weapons-and

[289] United Nations, *Statement to the Convention on Conventional Weapons Fifth Review Conference* (Dec.12, 2016), https://meetings.unoda.org/ccw-revcon/convention-certain-conventional-weapons-fifth-review-conference-2016

International Humanitarian Law (IHL), including through maintaining human responsibility and accountability in the use of force."[290]

Argentina endorsed, along with more than 30 other Latin American and Caribbean states, the Belén Communiqué,[291] which calls for "urgent negotiation" of a binding international treaty to regulate and prohibit the use of autonomous weapons to address the grave concerns raised by removing human control from the use of force. Argentina voted in favor of UN General Assembly Resolution 78/241[292] on lethal autonomous weapons systems, which takes up several of the points expressed in Resolution L.56. Most recently, the country also supported a draft resolution that calls for open informal consultations that shall be open to the full participation of all member states and observer states, international and regional organizations, the International Committee of the Red Cross, and civil society, including the scientific community and industry.[293]

AI Literacy

Argentina's Secretariat of Education (*Secretaría de Educación*), under the Ministry of Human Capital (*Ministerio de Capital Humano*), oversees the Argentine Program of Education Innovation with Artificial Intelligence (PAIDEIA) to "develop key skills in students from primary to secondary education, preparing them to understand and actively participate in a world increasingly mediated by technology."[294] PAIDEIA emphasizes human cognition skills across three objectives including computational thinking, application of AI, and development of AI.[295] The program aligns with a humanist vision of AI as a tool for human development, "derived from the ancient Greek *paideia* meaning education, formation, or growth."[296] The program has launched in a beta phase with plans for gradual national

---

[290] United Nations (UN) General Assembly, First Committee, *Joint Statement on Lethal Autonomous Weapons Systems First Committee, 77th United Nations General Assembly Thematic Debate – Conventional Weapons* (Oct. 21, 2022), https://estatements.unmeetings.org/estatements/11.0010/20221021/A1jJ8bNfWGlL/KLw9WYcSnnAm_en.pdf

[291] Latin American and the Caribbean Conference of Social and Humanitarian Impact of Autonomous Weapons, *Communiqué* (Feb. 24, 2023), https://www.rree.go.cr/files/includes/files.php?id=2261&tipo=documentos

[292] UN General Assembly, *Lethal Autonomous Weapons, Resolution 78/241* (Dec. 22, 2023), https://digitallibrary.un.org/record/4033027?ln=en&v=pdf

[293] UN General Assembly, *Lethal Autonomous Weapons, Resolution L.77* (Nov. 5, 2024), https://digitallibrary.un.org/record/4065061?ln=en&v=pdf

[294] Ministerio de Capital Humano, *Paideia: Argentine Program of Education Innovation with Artificial Intelligence [Paideia: Programa Argentino de Innovación de la Educación con Inteligencia Artificial]*, https://www.argentina.gob.ar/capital-humano/educacion/paideia

[295] Ibid, *Reach of the PAIDEIA Program [Alcance del Programa PAIDEIA]*, https://www.argentina.gob.ar/capital-humano/educacion/paideia/alcance-del-programa-paideia

[296] Ibid, *Vision of Artificial Intelligence in Education [Visión de la INteligencia Artificial en Educación*, https://www.argentina.gob.ar/capital-humano/educacion/paideia/vision-de-la-inteligencia-artificial-en-educacion

implementation.[297] The Secretariat of Education released a guide for AI in education as part of the PAIDEIA program.[298]

Human Rights

According to the Freedom House 2024 report, Argentina is "Free," receiving overall a score of 85/100.[299] The Freedom House country report highlights concerns regarding democratic backsliding. Citizens continue to use social media to mobilize protests on political and social issues although new president Javier Milei has implemented restrictions on protests. The President has also dismantled several ministries, reducing their number from 20 to 8. Some controversial reforms have also been engaged regarding women, gender, and Indigenous affairs, raising concerns regarding the protection of vulnerable groups. On the other hand, in June 2024 the Argentine Chamber of Deputies proposed a bill to create an AI regulatory framework with INTI oversight, penalties for non-compliance, and an advisory board for ethical and human rights impact assessment.[300]

In the international arena, Argentina has shown a strong commitment to the protection of human rights, including international and regional initiatives that pertain to AI.

OECD / G20 AI Principles

As part of the G20 and as a prospective member to the OECD, Argentina has endorsed the OECD/G20 AI Principles.[301] According to an OECD report, several policies of Argentina's national AI strategy align with the G20 AI principles. These include the comprehensive, human-centered and human rights-focus nature, which aligns with the Principles for Responsible Stewardship of Trustworthy AI (Section 1). Argentina's investment initiatives, the focus on conditions for AI development, educational plans and international engagements implement Section 2 of the G20 AI Principles (National Policies and International Co-operation for Trustworthy AI).[302]

---

[297] Ibid, *Content and Resources [Contenidos. Y Recurso—PAIDEIA]*, https://www.argentina.gob.ar/capital-humano/educacion/paideia/contenidos-y-recursos-paideia

[298] Ministerio de Capital Humano, Secretaría de Educación, *Guide for the Integration of Artificial Intelligence in Education [Guía para la Integración de las Inteligencias Artificiales en Eduación]* (2025), https://www.argentina.gob.ar/sites/default/files/documento_guia_de_integracion_vf_digital.pdf

[299] Freedom House, *Freedom in the World 2025: Argentina* (2025), https://freedomhouse.org/country/argentina/freedom-world/2025

[300] Chamber of Deputies, *Proposed Bill that Establishes the Legal Regime Applicable for the Responsible Use of AI* (Jun. 10, 2024), https://www.diputados.gov.ar/comisiones/permanentes/clgeneral/proyecto.html?exp=3003-D-2024

[301] OECD, *OECD Takes First Step in Accession Discussions with Argentina, Brazil, Bulgaria, Croatia, Peru and Romania* (Jan. 225, 2022), https://www.oecd.org/newsroom/oecd-takes-first-step-in-accession-discussions-with-argentina-brazil-bulgaria-croatia-peru-and-romania.htm

[302]OECD/CAF, *The Strategic and Responsible Use of Artificial Intelligence in the Public Sector of Latin America and the Caribbean*, OECD Public Governance Reviews (2022),

In October 2023, the OECD's The State of the Implementation of the OECD AI Principles Four Years On noted that Argentina's Ethics Principles for the Development of AI reflect the five value-based OECD AI Principles and the five recommendations to national governments.[303]

Council of Europe AI Treaty

Argentina contributed as an Observer State in the negotiations of the Council of Europe Framework Convention on AI, Human Rights, Democracy and the Rule of Law.[304] However, Argentina has signed this first legally binding treaty on AI and human rights.[305]

UNESCO Recommendation on AI Ethics

Argentina is a UNESCO member and adopted the UNESCO Recommendation on the Ethics of AI during the 41st General Conference in November 2021.

Argentina is part of the Regional Council comprising national and local governments from Latin America and the Caribbean, established through collaboration between UNESCO and the Latin American Development Bank (CAF) in support of ethical AI.[306] Argentina signed the Santiago Declaration to Promote Ethical Artificial Intelligence that resulted from the first council meeting.[307] The declaration establishes fundamental principles that should guide public policy on AI. These include proportionality, security, fairness, non-discrimination, gender equality, accessibility, sustainability, privacy, and data protection. Argentina also signed the Declaration of Cartagena de Indias,[308] which establishes the commitment to promote

---

https://doi.org/10.1787/1f334543-en; OECD Policy Observatory, *AI in Argentina, AI Policies in Argentina* (2024), https://oecd.ai/en/dashboards/national/argentina

[303] OECD, *The State of Implementation of the OECD AI Principles Four Years On* (Oct. 2023), https://www.oecd-ilibrary.org/science-and-technology/the-state-of-implementation-of-the-oecd-ai-principles-four-years-on_835641c9-en

[304] Council of Europe, *Framework Convention on Artificial Intelligence* (2026), https://www.coe.int/en/web/artificial-intelligence/the-framework-convention-on-artificial-intelligence

[305] Council of Europe Treaty Office, *Chart of Signatures and Ratifications of Treaty 225* (Jan. 21, 2025), https://www.coe.int/en/web/conventions/full-list?module=signatures-by-treaty&treatynum=225

[306] UNESCO, *Chile Will Host the First Latin American and Caribbean Ministerial and High Level Summit on the Ethics of Artificial Intelligence* (Sept. 25, 2023), https://www.unesco.org/en/articles/chile-will-host-first-latin-american-and-caribbean-ministerial-and-high-level-summit-ethics

[307] High-Level Ministerial Summit on the Ethics of Artificial Intelligence, *Declaration of Santiago "To Promote Ethical Artificial Intelligence in Latin America and the Caribbean"* (Oct. 2023), https://minciencia.gob.cl/uploads/filer_public/40/2a/402a35a0-1222-4dab-b090-5c81bbf34237/declaracion_de_santiago.pdf

[308] Digital Policy Alert, *Argentina: Signed Cartagena de Indias Declaration for Governance, the Construction of AI Ecosystems and the Promotion of AI Education in an Ethical and Responsible Manner in Latin America and the Caribbean* (Aug. 9, 2024), https://digitalpolicyalert.org/event/22018-signed-cartagena-de-indias-declaration-for-governance-the-

governance frameworks and AI ecosystems for a safe, inclusive, ethical and responsible development of AI within the Latin American region.

Despite a record of international collaboration, Argentina refused to sign the UN Pact for the Future, which included the Global Digital Compact, a comprehensive framework for global governance of digital technology and artificial intelligence in 2024.[309]

Argentina has taken a clear step to implement the UNESCO Recommendation on the Ethics of Artificial Intelligence by beginning the Readiness Assessment Methodology (RAM).[310]

Evaluation

Argentina's comprehensive, ambitious, and human-centered national strategy reflects the country's interest in matching socioeconomic development with strong human rights commitments in the design and development of AI. Despite the initial enthusiasm that surrounded the launching of the Plan, the new government seems to have abandoned this path. It remains to be seen how the Productive Development Plan Argentina 4.0 will be operationalized in practice. Despite significant progress in modernizing data protection law and bringing it up to international standards, the deployment of facial recognition systems has raised widespread concern that AI could be used for purposes that violate individual rights. This threat is especially concerning given recent reports of democratic backsliding. The non-signing of the Pact for the Future or the adoption of the Council of Europe's AI Treaty could represent a significant change of foreign policy to the detriment of human rights safeguards.

Argentina has the resources and the infrastructure to pursue regional leadership. Argentina has signed the UNESCO Recommendation on AI Ethics and, with the collaboration of the CAF, is on its path to its implementation. Argentina has also been calling for a prohibition and regulation of autonomous weapons.

---

construction-of-ai-ecosystems-and-the-promotion-of-ai-education-in-an-ethical-and-responsible-manner-in-latin-america-and-the-caribbean

[309] UN, *World Leaders Pledge Bold Action to Protect Present, Future Generations amid Climate Crisis, Conflicts Gripping Globe, as General Assembly Adopts Pact for Future* (Sept. 22, 2024), https://press.un.org/en/2024/ga12627.doc.htm

[310] UNESCO Global AI Ethics and Governance Observatory, *Global Hub* (Oct. 2025), https://www.unesco.org/ethics-ai/en/global-hub

Australia

In 2025, Australia updated the National AI Policy along with the related impact assessment and procurement tools. The country also furthered implementation of the Guidelines for the responsible use of AI in the public service.

National AI Strategy

The Australian government published a Roadmap for AI in 2019 to "help develop a national AI capability to boost the productivity of Australian industry, create jobs and economic growth, and improve the quality of life for current and future generations."[311] Australia's AI Technology Roadmap is intended to help guide future investment in AI and provide a pathway to ensure Australia captures the full potential of AI.[312] The Roadmap identifies three domains of AI development and application where AI could transform Australian industry based on existing strengths and comparative advantages, opportunities to solve Australian problems, and opportunities to export solutions to the rest of the world. These domains are Health, Aging, and Disability; Cities, Towns, and Infrastructure (including connected and automated vehicle technology); and National Resources and Environment (building on strengths related to mining and agriculture). The Roadmap also elaborates necessary foundations for realizing these goals in terms of data governance, ethics, trust research, skills, and infrastructure, while underscoring its complementarity with the OECD AI Principles.

Australia also published an AI Ethics Framework in 2019 to "help guide businesses and governments looking to design, develop, and implement AI in Australia."[313] Key goals are to achieve better outcomes, reduce the risk of negative impact, and practice the highest standards of ethical business and good governance. The eight AI Ethics Principles are Human, social, and environmental wellbeing; Human-centered values; Fairness; Privacy protection and security; Reliability and safety; Transparency and explainability; Contestability and Accountability.[314]

---

[311] Commonwealth Scientific and Industrial Research Organisation (CSIRO), *Artificial Intelligence Roadmap* (Nov. 2019), https://www.csiro.au/en/research/technology-space/ai/Artificial-Intelligence-Roadmap

[312] Commonwealth Scientific and Industrial Research Organisation (CSIRO), *Artificial Intelligence Roadmap: Solving Problems, Growing the Economy and Improving Our Quality of Life* (2019), https://www.csiro.au/en/research/technology-space/ai/artificial-intelligence-roadmap

[313] Department of Industry, Science, Energy and Resources, *Towards an Artificial Intelligence Ethics Framework* (Nov. 7, 2019), https://www.industry.gov.au/news/towards-artificial-intelligence-ethics-framework

[314] Hon. Karen Andrews MP, Archived Content, *Businesses Ready to Test AI Ethics Principles* (Nov. 7, 2019), https://www.minister.industry.gov.au/ministers/karenandrews/media-releases/businesses-ready-test-ai-ethics-principles

In 2020, the Australian Department of Industry, Science and Resources (DISR) launched a public consultation to inform the AI Action Plan,[315] which was published in 2021 as part of the Digital Economy Strategy.[316] This plan aims to position Australia as a leader in developing and adopting "trusted, secure, and responsible AI," enhancing business productivity, building a skilled workforce, and addressing national challenges."[317] The Plan created the National AI Centre (NAIC) and four Capability Centres to support SMEs and promote ethical AI.[318] DISR subsequently took over administration of the NAIC to strengthen government-industry engagement on AI.[319]

Australia's AI Strategy includes coordinated initiatives for responsible AI development and adoption. After a public consultation launched in 2023 on safe and responsible AI in Australia, including the need to regulate AI,[320] the government's interim response in January 2024 concluded that the current regulatory framework likely did not sufficiently address known risks presented by AI systems. [321]

Following the AI Strategy and interim public consultation response, DISR established a temporary AI Expert Group to advise on measures for testing, transparency, and accountability in high-risk AI contexts and ensure system safety.[322] Comprising 12 industry, academic, and legal experts, the group operated through September 2024, with plans for a permanent group expected.[323]

---

[315] Department of Industry, Science and Resources, *Australia's AI Action Plan: Discussion Paper* (Jun. 22, 2021), https://consult.industry.gov.au/australias-ai-action-plan-discussion-paper

[316] Department of Industry, Science and Resources—Digital Economy Branch, *Australia's Digital Economy* (Mar. 30, 2022), https://web.archive.org.au/awa/20220816053410mp_/https://digitaleconomy.pmc.gov.au/

[317] Department of Industry, Science and Resources, *An Action Plan for Artificial Intelligence in Australia* (Jun. 18, 2021), https://www.industry.gov.au/news/action-plan-artificial-intelligence-australia

[318] Department of Industry, Science and Resources, *National Artificial Intelligence Centre*, https://www.industry.gov.au/science-technology-and-innovation/technology/national-artificial-intelligence-centre

[319] Commonwealth Scientific and Industrial Research Organisation (CSIRO), *National Artificial Intelligence Centre* (Jul. 2024), https://www.csiro.au/en/work-with-us/industries/technology/National-AI-Centre

[320] Department of Industry, Science and Resources, *Responsible AI in Australia: Have Your Say* (Jun. 1, 2023), https://www.industry.gov.au/news/responsible-ai-australia-have-your-say

[321] Department of Industry, Science and Resources, *The Australian Government's Interim Response to Safe and Responsible AI Consultation* (Jan. 17, 2024), https://consult.industry.gov.au/supporting-responsible-ai

[322] Ministers for the Department of Industry, Science and Resources, *The Hon Ed Husic MP: New Artificial Intelligence Expert Group* (Feb. 14, 2024), https://www.minister.industry.gov.au/ministers/husic/media-releases/new-artificial-intelligence-expert-group

[323] Department of Industry, Science and Resources, *New Expert Group Will Help Guide the Future of Safe and Responsible AI in Australia* (Feb. 14, 2924), https://www.industry.gov.au/news/new-expert-group-will-help-guide-future-safe-and-responsible-ai-australia

DISR launched the Voluntary AI Safety Standard in September 2024[324] to establish consistent practices and help organizations manage AI risks, aligning with the broader objectives of the AI Action Plan. The standard provides guidance on the safe and responsible use of AI through ten voluntary guidelines emphasizing testing, transparency, and accountability. The AI Impact Navigator, a tool launched by the NAIC in October 2024, will assist users in implementing the Voluntary AI Safety Standard and in evaluating the impact of their AI systems.[325]

DISR also laid a path toward regulating the use of AI in high-risk settings with the release of a proposal paper outlining definitions, principles, and mandatory guardrails, including for the use of general-purpose AI systems.[326] These guardrails reference international AI regulations and include five of the most agreed-upon goals in the AI policy realm: fairness, accountability, transparency, rule of law, and fundamental rights. In the public consultation of the proposed guardrails, the Australian Human Rights Commission highlighted that 'the proposed guardrails do not specify the need for high-risk AI to comply with human rights obligations' and recommended it to explicitly include human rights considerations.[327] Public consultation has since closed but how and if DISR implements feedback remains to be seen.[328]

The government has also acted to strengthen existing laws in areas that will help to address known harms with AI. This includes the implementation of privacy law reforms, including an in-principle agreement to require non-government entities to conduct a privacy impact assessment to identify and manage, minimize or eliminate risks, which is already a requirement for government entities and an amendment to the Online Safety Act 2021.[329] A proposed amendment to existing communications

---

[324] Department of Industry, Science and Resources, *Voluntary AI Safety Standard* (Sept. 5, 2024), https://www.industry.gov.au/publications/voluntary-ai-safety-standard

[325] Department of Industry, Science and Resources, *AI Impact Navigator Will Help Australian Companies Tackle the Challenges of AI* (Oct. 21, 2024), https://www.industry.gov.au/news/ai-impact-navigator-will-help-australian-companies-tackle-challenges-ai

[326] Department of Industry, Science and Resources, *Introducing Mandatory Guardrails for AI in High-Risk Settings: Proposals Paper* (Sept. 5, 2024), https://consult.industry.gov.au/ai-mandatory-guardrails

[327] Australian Human Rights Commission, *Guardrails for High-Risk AI, Submission by the Commission* (Oct. 30, 2024), https://humanrights.gov.au/our-work/legal/submission/guardrails-high-risk-ai

[328] Department of Industry, Science and Resources, *Introducing Mandatory Guardrails for AI in High-Risk Settings: Proposals Paper, Published Responses*, https://consult.industry.gov.au/ai-mandatory-guardrails/submission/list

[329] Parliament of Australia, *Online Safety Amendment (Social Media Minimum Age) Bill 2024* (Nov. 29, 2024), https://www.aph.gov.au/Parliamentary_Business/Bills_Legislation/Bills_Search_Results/Result?bId=r7284

legislation targeted misinformation and disinformation[330] but was withdrawn in committee over censorship concerns.[331]

Australia launched the National AI Plan in December 2025 to deliver three goals (1) capture the opportunity by building smart infrastructure, backing domestic AI capability, and attracting global investment (2) spread the benefits through widespread AI adoption, supporting and training Australian workers, and improved public services (3) keep Australians safe with legislative and regulatory frameworks that mitigate AI harms, while promoting widespread responsible practices and international engagement that upholds Australia's values.[332] The Plan signals the government's confidence that existing legal frameworks "can apply to AI and other emerging technologies" to mitigate AI harms, while committing to continue assessing adequacy.[333] The confidence in existing frameworks suggests a preference for regulatory flexibility over the adoption of a comprehensive AI-specific framework.

While the AI Strategy and bodies like the Australian Human Rights Commission recognize the need for regulation, Australia lacks a specific federal framework for AI. The Senate formed the Select Committee on Adopting Artificial Intelligence to explore potential opportunities and challenges. [334] Based on these findings, the Committee recommends that the Australian Government implement comprehensive AI legislation to regulate high-risk uses, support sovereign AI development, address workplace risks, ensure fair compensation for creatives, and prioritize transparency, privacy, and sustainable infrastructure growth.[335]

Bringing greater transparency to AI adoption in the country, the National AI Centre launched the Responsible AI Index 2025 to benchmark organizational governance maturity across fairness, safety, and transparency metrics.[336] The Index assessed 500 Australian organizations and provided a public dashboard for tracking

---

[330] Parliament of Australia, *Communications Legislation Amendment (Combatting Misinformation and Disinformation) Bill 2024* (Oct. 23, 2024), https://www.aph.gov.au/Parliamentary_Business/Bills_Legislation/bd/bd2425/25bd014

[331] Parliament of Australia, *Communications Legislation Amendment (Combatting Misinformation and Disinformation) Bill 2024 [Provisions]: Report* (Nov. 2024), https://www.aph.gov.au/Parliamentary_Business/Committees/Senate/Environment_and_Communications/MisandDisinfobill/Report

[332] Department of Industry Science and Resources, *National AI Plan* (Dec. 2, 2025), https://www.industry.gov.au/publications/national-ai-plan

[333] Ibid, pp. 28-30

[334] Parliament of Australia, *Select Committee on Adopting Artificial Intelligence (AI)* (Mar. 26, 2024), https://www.aph.gov.au/Parliamentary_Business/Committees/Senate/Adopting_Artificial_Intelligence_AI

[335] Senate Committees, *Select Committee on Adopting Artificial Intelligence: Final Report* (Nov. 26, 2024), https://www.aph.gov.au/Parliamentary_Business/Committees/Senate/Adopting_Artificial_Intelligence_AI/AdoptingAI/Report

[336] National Artificial Intelligence Centre, *Responsible AI Index 2025: National Benchmark Report* (Aug. 26, 2025), https://www.industry.gov.au/news/australias-national-benchmark-responsible-ai-adoption-now-available

national progress on responsible AI adoption, with 12 percent of organizations classified as "leading" and 17 percent in the "emerging" category.

International Cooperation on Safe and Responsible AI

Australia and Singapore signed an updated Memorandum of Understanding (MOU) in 2024. The MOU renews a shared commitment to "further cooperate and leverage on Australia and Singapore's comparative strengths to ensure the participants can realize the benefits of AI and minimise any risks, and also shape the development of international AI-related frameworks, policies and standards" from a 2020 MOU.[337]

Australia participated in the first AI Safety Summit and endorsed the Bletchley Declaration in November 2023.[338] Australia thus committed to participate in international cooperation efforts on AI "to promote inclusive economic growth, sustainable development and innovation, to protect human rights and fundamental freedoms, and to foster public trust and confidence in AI systems to fully realise their potential." Endorsing parties affirmed that "for the good of all, AI should be designed, developed, deployed, and used, in a manner that is safe, in such a way as to be human-centric, trustworthy and responsible."

At the subsequent AI Safety Summit—which was renamed the AI Action Summit—Australia joined 90+ countries in endorsing the Statement on Inclusive and Sustainable AI Statement and co-led the AI safety, trust, and governance initiatives.[339]

Public Participation

The Australian government has responded to recommendations and feedback from the public in national efforts on AI and digital policy, including the AI Ethics Framework and Supporting Responsible AI discussion paper.[340] The Australian

---

[337] Department of Industry, Science and Resources, *Memorandum of Understanding on Cooperation on Artificial Intelligence between Australia and Singapore* (Dec. 16, 2024), https://www.industry.gov.au/publications/memorandum-understanding-cooperation-artificial-intelligence-between-australia-and-singapore

[338] UK Department for Science, Innovation & Technology, Foreign, Commonwealth & Development Office, Prime Minister's Office, *The Bletchley Declaration by Countries Attending the AI Safety Summit* (Nov. 2, 2023), https://www.gov.uk/government/publications/ai-safety-summit-2023-the-bletchley-declaration/the-bletchley-declaration-by-countries-attending-the-ai-safety-summit-1-2-november-2023

[339] Department of Industry, Science and Resources, *Australia Signs Paris AI Action Summit Statement* (Feb. 14, 2025), https://www.industry.gov.au/news/australia-signs-paris-ai-action-summit-statement

[340] Department of Industry, Science and Resources, *Australia's Artificial Intelligence Ethics Principles: How We Developed the Principles* (Dec. 2, 2025), https://www.industry.gov.au/publications/australias-artificial-intelligence-ethics-principles; Department of Industry, Science and Resources, *Supporting Responsible AI: Discussion Paper* (Jan. 17, 2023), https://consult.industry.gov.au/supporting-responsible-ai

Government opened a public consultation[341] as part of a review of the Online Safety Act 2021.[342]

The Australian government published an interim response to the Safe and Responsible AI in Australia discussion paper in early 2024.[343] The government committed to developing a regulatory environment that builds community trust and promotes AI adoption. Subsequently, DISR issued a formal request for comment on Mandatory Guardrails for AI in High-risk Settings.[344] The views were asked on the proposed guardrails, high-risk AI definition, and regulatory options for mandating the guardrails.

The Therapeutic Goods Administration of the Department of Health and Aged Care[345] and Department itself issued separate calls for public consultation in September 2024 as they reviewed their legislative frameworks to "clarify and strengthen legislation and regulation for AI in Australia's health care settings.[346] In particular, they sought feedback on stakeholders, what about AI to regulate, and how to prevent AI harms and enable the benefits. The Joint Committee of Public Accounts and Audit adopted an inquiry into the use and governance of artificial intelligence systems by public sector entities in September 2024.[347]

The Australian Treasury opened a formal request for comment for a Review of AI and the Australian Consumer Law (ACL).[348] The review will examine the ACL's fitness to address consumer harms and support responsible use of AI. Inputs are sought to determine if provisions on product safety, prohibitions on businesses for

---

[341] Department of Infrastructure, Transport, Regional Development, Communications and the Arts, *Consultation Open for the Online Safety Act Review* (Apr. 29, 2024), https://www.infrastructure.gov.au/department/media/news/consultation-open-online-safety-act-review

[342] Federal Register of Legislation, *Online Safety Act 2021* (Oct. 14, 2024), https://www.legislation.gov.au/C2021A00076/latest/text

[343] Department of Industry, Science and Resources, *The Australian Government's Interim Response to Safe and Responsible AI Consultation* (Jan. 17, 2024), https://www.industry.gov.au/news/australian-governments-interim-response-safe-and-responsible-ai-consultation

[344] Department of Industry, Science and Resources, *Safe and Responsible AI in Australia: Proposals Paper for Introducing Mandatory Guardrails for AI in High-Risk Settings* (Sept. 2024), https://consult.industry.gov.au/ai-mandatory-guardrails

[345] Department of Health and Aged Care, *Consultation: Clarifying and Strengthening the Regulation of Artificial Intelligence (AI)* (Sept. 12, 2024), https://www.tga.gov.au/resources/consultation/consultation-clarifying-and-strengthening-regulation-artificial-intelligence-ai

[346] Department of Health and Aged Care, *Safe and Responsible Artificial Intelligence in Health Care: Legislation and Regulation Review* (Sept. 13, 2024), https://consultations.health.gov.au/medicare-benefits-and-digital-health-division/safe-and-responsible-artificial-intelligence-in-he

[347] Joint Committee of Public Accounts and Audit, *Inquiry into the Use and Governance of Artificial Intelligence Systems by Public Sector Entities* (Sept. 12, 2024), https://www.aph.gov.au/Parliamentary_Business/Committees/Joint/Public_Accounts_and_Audit/PublicsectoruseofAI

[348] Treasury, *Review of AI and the Australian Consumer Law* (Oct. 25, 2024), https://treasury.gov.au/consultation/c2024-584560

deceptive conduct and unfair contract terms, prohibitions on false representations, etc., sufficiently safeguard and provide remedies against consumer harm. The review also seeks input on whether there are ambiguities around ACL's application to AI and how it can be addressed.

Data Protection

In February 2023, the Australian Attorney-General's Department released its review of the Privacy Act 1988,[349] a significant step initiated in 2020.[350] The review was initiated after the Australian Competition and Consumer Commission (ACCC) recommended reforms to "ensure consumers are adequately informed, empowered and protected, as to how their data is being used and collected."[351] The Privacy Act Review Report[352] includes 116 recommendations based on 30 "key themes and proposals" from stakeholders. "The proposed reforms are aimed at strengthening the protection of personal information and the control individuals have over their information. Stronger privacy protections would support digital innovation and enhance Australia's reputation as a trusted trading partner," according to the Attorney-General's Department.

The Office of the Australian Information Commissioner welcomed the release of the report. "This is an important milestone as we move towards further reform of Australia's privacy framework," said Angelene Falk, the Australian Information Commissioner and Privacy Commissioner at the time of the report's release. "As the privacy regulator we see the proposal to introduce a positive obligation that personal information handling is fair and reasonable, as a new keystone of the Australian privacy framework. This shifts the burden from individuals, who are currently required to safeguard their privacy by navigating complex privacy policies and consent requirements, and places more responsibility on the organisations who collect and use personal information to ensure that their practices are fair and reasonable in the first place."

The Government released its Response to the Privacy Act 1988 Review, noting 10 proposals, agreeing in-principle to 68, and fully agreeing to 38 proposals, including that "further consideration should be given to enhanced risk assessment requirements in the context of facial recognition technology and other uses of biometric information and that this work should be coordinated with the

[349] Attorney-General's Department, *Privacy Act 1988* (Nov. 30, 2024), https://www.legislation.gov.au/Series/C2004A03712
[350] Attorney-General's Department, *Privacy Act Review Issues Paper* (Oct. 30, 2020), https://www.ag.gov.au/rights-and-protections/publications/review-privacy-act-1988-cth-issues-paper
[351] Australian Competition and Consumer Commission, *Digital Platforms Inquiry—Final Report* (Jul. 26, 2019), https://www.accc.gov.au/about-us/publications/digital-platforms-inquiry-final-report
[352] Department of the Attorney-General, *Review of the Privacy Act 1988* (Feb. 16, 2023), https://www.ag.gov.au/integrity/consultations/review-privacy-act-1988

Government's ongoing work on Digital ID and the National Strategy for Identity Resilience.[353]

These reforms build on Australia's efforts to enact regulation for accountability and data protection. The Privacy APP Code 2017 requires agencies subject to the Privacy Act (Australian Government Agencies—Governance) to conduct privacy impact assessments for all "high privacy risk projects."[354] Furthermore, section 33D of the Privacy Act grants the Commissioner authority to require an agency to provide impact assessments whenever an activity or function may have a "significant impact on the privacy of individuals."[355]

The Privacy Act also builds on efforts to address privacy concerns raised by facial recognition and other biometric technology. The Federal Attorney-General referred positively to the 2022 Human Technology Institute (HIT) report outlining a Facial Recognition Model Law to address "threats to Australians' privacy and other human rights" using a risk-based approach.[356] The Federal Attorney-General noted that the Model Law is "a way of striking the right balance, endorsing, in principle, a risk assessment approach to regulating facial recognition and other biometric technologies."[357]

The Parliament approved a bill to amend the Privacy Act 1988 to implement reforms that include "creating a statutory tort for serious invasions of privacy" and amending the Criminal Code Act 1995 to introduce two offenses for doxxing.[358] The Privacy and Other Legislation Amendment Bill 2024 received Royal Assent in December 2024, with most provisions commencing immediately.[359] Further obligations such as automated decision-making transparency will be phased in through December 2026.[360]

---

[353] Attorney-General's Department, *Government Response to the Privacy Act Review Report* (Sept. 28, 2023), https://www.ag.gov.au/rights-and-protections/publications/government-response-privacy-act-review-report

[354] Office of the Australian Information Commissioner, *When Do Agencies Need to Conduct a Privacy Impact Assessment?* (Sept. 14, 2020), https://www.oaic.gov.au/privacy/privacy-guidance-for-organisations-and-government-agencies/government-agencies/australian-government-agencies-privacy-code/when-do-agencies-need-to-conduct-a-privacy-impact-assessment

[355] Office of the Australian Information Commissioner, *Chapter 10: Directing a Privacy Impact Assessment* (Jan. 2023), https://www.oaic.gov.au/about-the-OAIC/our-regulatory-approach/guide-to-privacy-regulatory-action/chapter-10-directing-a-privacy-impact-assessment

[356] Human Law Technology, *A Blueprint for Regulation of Facial Recognition Technology* (Sept. 27, 2022), https://www.uts.edu.au/news/2022/09/blueprint-regulation-facial-recognition-technology

[357] Human Technology Institute, *Facial Recognition Technology: Towards a Model Law* (2026), https://www.uts.edu.au/human-technology-institute/projects/facial-recognition-technology-towards-model-law

[358] Parliament of Australia, *Privacy and Other Legislation Amendment Bill 2024* (Nov. 18, 2024), https://www.aph.gov.au/Parliamentary_Business/Bills_Legislation/bd/bd2425/25bd016

[359] Parliament of Australia, *Privacy and Other Legislation Amendment Act 2024* (Dec. 10, 2024), https://www.legislation.gov.au/C2024A00128/asmade/text

[360] Office of the Australian Information Commissioner, *Chapter 1: APP 1 Open and Transparent Management of Personal Information*, https://www.oaic.gov.au/privacy/australian-privacy-

The Office of the Australian Information Commissioner is a member of the Global Privacy Assembly (GPA).[361] Australia supported the 2022 GPA Resolution on Principles and Expectations for the Appropriate Use of Personal Information in Facial Recognition Technology[362] but did not support the 2018 Declaration on Ethics and Data Protection in Artificial Intelligence,[363] 2020 Resolution on Accountability in the Development and Use of Artificial Intelligence,[364] or 2023 Resolution on Generative Artificial Intelligence Systems.[365]

Algorithmic Transparency

Australian law does not explicitly protect a right to algorithmic transparency. However, the transparency and explainability, contestability, and accountability principles in the AI Ethics Framework provide guidance for organizations[366] and the 2024–2025 National Action Plan[367] charges the Attorney-General's Department and DISR to "create transparency in the use of automated decision making and artificial intelligence." The Australian Government released the Transparency of Automated Decision Making (AU0024) to address public concerns about the transparency and integrity of government decisions made using automated systems and AI.[368]

These actions follow reports and advice from various agencies to establish algorithmic transparency as a right. In early 2019, the Australian Human Rights Commission called for an AI Policy Council to guide companies and regulators

---

principles/australian-privacy-principles-guidelines/chapter-1-app-1-open-and-transparent-management-of-personal-information

361 Global Privacy Assembly, *List of Accredited Members* (2026), https://globalprivacyassembly.com/participation-in-the-assembly/list-of-accredited-members/

362 Global Privacy Assembly (GPA), *Resolution on Principles and Expectations for the Appropriate Use of Personal Information in Facial Recognition Technology* (Oct. 2022), https://globalprivacyassembly.com/wp-content/uploads/2022/11/15.1.c.Resolution-on-Principles-and-Expectations-for-the-Appropriate-Use-of-Personal-Information-in-Facial-Recognition-Technolog.pdf

363 International Conference of Data Protection & Privacy Commissioners, *Declaration on Ethics and Data Protection in Artificial Intelligence* (Oct. 23, 2018), https://globalprivacyassembly.com/wp-content/uploads/2018/10/20180922_ICDPPC-40th_AI-Declaration_ADOPTED.pdf

364 Global Privacy Assembly (GPA), *Adopted Resolution on Accountability in the Development and Use of Artificial Intelligence* (Oct. 2020), https://globalprivacyassembly.com/wp-content/uploads/2020/11/GPA-Resolution-on-Accountability-in-the-Development-and-Use-of-AI-EN.pdf

365 Global Privacy Assembly (GPA), *Resolution on Generative Artificial Intelligence Systems* (Oct. 2023), https://globalprivacyassembly.com/wp-content/uploads/2023/10/5.-Resolution-on-Generative-AI-Systems-101023.pdf

366 Department of Industry, Science and Resources, *Australia's AI Ethics Principles* (Accessed Dec. 8, 2024), https://www.industry.gov.au/publications/australias-artificial-intelligence-ethics-principles/australias-ai-ethics-principles

367 Open Government Partnership, *Australia's Third Open Government Partnership National Action Plan 2024–2025* (Accessed Dec. 8, 2024), https://www.opengovpartnership.org/wp-content/uploads/2023/12/Australia_Action-Plan_2023-2025_December.pdf

368 Open Government Partnership, *Australia: Transparency of Automated Decision Making (AU0024)*, https://www.opengovpartnership.org/members/australia/commitments/AU0024/

regarding artificial intelligence technology. "When companies use AI decision-making systems, they must build them in a way that allows a person to understand the basis of decisions that affect them. This is fundamental to ensuring accountability and will be really important for all companies that use AI," Human Rights Commissioner Ed Santow said.[369] The Australian Human Rights Commission published the Human Rights and Technology report recommending the government to legislate a right to contest automated decisions and introduce a requirement for transparency in AI systems used for significant decisions.[370]

The Office of the Australian Information Commissioner (OAIC) emphasized the need for greater transparency in its 2023 response to the Safe and Responsible AI in Australia discussion paper.[371] The OAIC highlighted that "most Australians view transparency about the use of AI and the right to request information on how AI decisions are made as essential preconditions for its responsible deployment in decision-making."

The Senate Economics References Committee's Interim Report on international digital platforms in 2023 included a chapter on algorithmic transparency.[372] Chapter 6 outlines risks such as misinformation, echo chambers, and biases in automated decision-making, stressing the need for algorithm disclosure, especially in content curation and targeted ads. The report called for regulation to ensure algorithms promote fairness, accuracy, privacy, and user safety.[373] An interim report released by the Senate Select Committee on Adopting Artificial Intelligence in October 2024[374] recommended that mandatory transparency and disclosure

---

[369] James Eyers, *Call for "AI Policy Council" to Govern How Algorithms Use Personal Information*, Financial Review (Mar. 15, 2020), https://www.afr.com/technology/call-for-ai-policy-council-to-govern-how-algorithms-use-personal-information-20190315-h1cej1

[370] Australian Human Rights Commission, *Human Rights and Technology: Final Report*, p. 14 (Mar. 2021), https://humanrights.gov.au/resource-hub/by-resource-type/publications/technology-and-human-rights/final-report-summary-human-rights-and-technology

[371] Office of the Australian Information Commissioner, *OAIC Submission to the Department of Industry, Science and Resources—Safe and Responsible AI in Australia Discussion Paper* (Sept. 21, 2023), https://www.oaic.gov.au/engage-with-us/submissions/oaic-submission-to-the-department-of-industry-science-and-resources-safe-and-responsible-ai-in-australia-discussion-paper

[372] Parliament of Australia, *Influence of International Digital Platforms* (Nov. 2023), https://www.aph.gov.au/Parliamentary_Business/Committees/Senate/Economics/Digitalplatforms/Report

[373] Parliament of Australia, *Influence of International Digital Platforms: Chapter 6: Algorithmic Transparency* (Nov. 2023), https://www.aph.gov.au/Parliamentary_Business/Committees/Senate/Economics/Digitalplatforms/Report/Chapter_6_-_Algorithmic_transparency

[374] Parliament of Australia, *Interim Report* (Oct. 2024), https://www.aph.gov.au/Parliamentary_Business/Committees/Senate/Adopting_Artificial_Intelligence_AI/AdoptingAI/Interim_report

requirements for AI-generated content should be developed before the 49th Parliament election.[375]

In its formal response to the Select Committee's report, issued in August 2025, the Australian government accepted in principle the recommendation for mandatory transparency requirements. The Minister for Communications announced that the government would direct the Australian Communications and Media Authority (ACMA) to develop a mandatory code of practice for digital platforms under the Online Safety Act. This code will require designated platforms to provide clear, plain-language explanations of how their recommender systems and content moderation algorithms work, and to produce annual transparency reports on their operation and impact. A draft of the code is expected to be released for public consultation in early 2026.[376]

The OAIC released a report and recommendations on Automated decision-making and public reporting under the Freedom of Information Act, assessing "how transparent Australian Government agencies are about their use [Automated Decision Making] ADM in decision making process."[377] The report assesses how agencies authorized to use ADM disclose their use of ADM as required under the Freedom of Information Act 1982 (FOI Act). The report found that 17 percent of agencies reviewed disclosed the use of ADM, while 2 percent were identified as "likely using ADM" but did not disclose this publicly. As a result, the report recommended agencies to disclose their status regardless of whether they're actually using ADM, and expanded on the information that should be published if ADM is used.

Data Scraping

The Office of the Australian Information Commissioner (OAIC) and 16 of its international data protection and privacy counterparts released a follow-up joint statement laying out expectations for industry related to data scraping and privacy.[378] The initial statement in August 2023 established that "[social media companies] and other websites are responsible for protecting personal information from unlawful

---

[375] Parliament of Australia, *Select Committee on Adopting Artificial Intelligence* (Oct. 2024), https://parlinfo.aph.gov.au/parlInfo/download/committees/reportsen/RB000493/toc_pdf/SelectCommitteeonAdoptingArtificialIntelligence(AI).pdf

[376] Department of Infrastructure, Transport, Regional Development, Communications and the Arts, *Government Response to Select Committee Report on Digital Platforms* (Aug. 2025), https://www.infrastructure.gov.au/department/media/publications/algorithmic-transparency-response-2025

[377] Office of the Australian Information Commissioner, *Automated Decision-Making and Public Reporting under the Freedom of Information Act* (Jan. 21, 2026), https://www.oaic.gov.au/freedom-of-information/information-commissioner-decisions-and-reports/foi-reports/Automated-decision-making-and-public-reporting-under-the-Freedom-of-Information-Act

[378] Office of the Australian Information Commissioner, *Concluding Joint Statement on Data Scraping and the Protection of Privacy* (Oct. 2024), https://www.priv.gc.ca/en/opc-news/speeches-and-statements/2024/js-dc_20241028/

scraping."[379] The follow-up, which incorporates feedback from engagement with companies such as Alphabet (YouTube), X Corp (formerly Twitter), and Meta Platforms (Facebook, Instagram, WhatsApp), adds that organizations must "comply with privacy and data protection laws when using personal information, including from their own platforms, to develop AI large language models"; must regularly review and update privacy protection measures to keep pace with technological advances; and ensure that permissible data scraping, such as that for research, "is done lawfully and following strict contractual terms."[380]

The OAIC's press release places the joint initiative in the context of national efforts to hold industry accountable to privacy laws, noting "The joint initiative, with its focus on the implications of AI large language models, coincides with the OAIC recently publishing detailed guidance on privacy and developing and training generative AI models.[381] These guidelines establish that the collection and use of publicly available data for AI training must meet a "fair and reasonable" standard, defining specific criteria for lawful data collection practices. The OAIC identified six factors to consider when determining whether web scraping is fair, including the sensitivity of information, intended purpose, risk of harm to individuals, and whether individuals intentionally made the information public. The guidelines provide practical guidance for organizations developing AI systems with scraped data and represent a concrete implementation step for Australia's commitment to ethical AI development while ensuring consistent privacy protections across technologies and data sources.

Privacy Commissioner Carly Kind's statement in August 2024 on the OAIC decision against investigating Clearview AI cited the initial joint statement on data scraping to reiterate companies' obligations to protect personal information.[382] In 2021, the Australian Information Commissioner declared that Clearview AI must "cease collecting images from individuals in Australia" and delete images it previously collected because the company had violated the Privacy Act and several Australian Privacy Principles in its use of facial recognition technology to collect images and biometric templates from people in Australia. Kind reiterated that "the

---

[379] Office of the Australian Information Commissioner, *Global Expectations of Social Media Platforms and Other Sites to Safeguard against Unlawful Data Scraping* (Aug. 24, 2023), https://www.oaic.gov.au/news/media-centre/global-expectations-of-social-media-platforms-and-other-sites-to-safeguard-against-unlawful-data-scraping

[380] Office of the Australian Information Commissioner, *Global Privacy Authorities Issue Follow-up Joint Statement on Data Scraping* (Oct. 29, 2024), https://www.oaic.gov.au/news/media-centre/global-privacy-authorities-issue-follow-up-joint-statement-on-data-scraping

[381] Office of the Australian Information Commissioner, *Guidance on Privacy and Developing and Training Generative AI Models* (Oct. 23, 2024), https://www.oaic.gov.au/privacy/privacy-guidance-for-organisations-and-government-agencies/guidance-on-privacy-and-developing-and-training-generative-ai-models

[382] Office of the Australian Information Commissioner, *Statement on Clearview AI* (Aug. 21. 2024), https://www.oaic.gov.au/news/media-centre/statement-on-clearview-ai

determination against Clearview AI still stands" and promised guidance for companies seeking to develop and train AI models.

Use of AI in Public Administration

After a scandal involving an automated system issuing unlawful and inaccurate debt notices, the Australian government has worked to establish principles and procedures for the responsible use of AI systems in public administration. These efforts culminated in the Policy for the Responsible Use of AI in Government, which took effect on September 1, 2024.[383]

From 2016 through 2019, the Australian government agency Services Australia used an automated debt recovery system known as Robodebt to calculate overpayments and issue debt notices to welfare recipients. Robodebt was an automated data matching system that compared the records of Services Australia payment compliance program with averaged income data from the Australian Taxation Office. Robodebt was suspended in 2019 following investigations by the Commonwealth Ombudsman,[384] two Senate Committees,[385] and several legal challenges, [386] including class-action lawsuits against the government. In one such suit, Federal Court Justice Bernard Murphy condemned the Government in his ruling and approved a $1.8 billion settlement, including repayments of debts paid, wiping of outstanding debts, and legal costs.[387] In October 2022, the newly elected government effectively forgave the debts of 197,000 people that were still under review.[388]

---

[383] Digital Transformation Agency, *Policy for the Responsible Use of AI in Government* 1.1 (Sept. 1, 2024), https://www.digital.gov.au/policy/ai/policy

[384] Commonwealth Ombudsman, *Lessons Learnt about Digital Transformation and Public Administration: Centrelink's Online Compliance Intervention* (Jul. 2017), https://www.ombudsman.gov.au/__data/assets/pdf_file/0024/48813/AIAL-OCI-Speech-and-Paper.pdf

[385] Matthew Doran, *Centrelink Debt Recovery Program to Be Investigated at Senate Committee Today,* ABC News (Mar. 7, 2017), http://www.abc.net.au/news/2017-03-08/centrelink-debt-recovery-program-to-be-investigated/8334072; Parliament of Australia, *Centrelink's Compliance Program*, https://www.aph.gov.au/Parliamentary_Business/Committees/Senate/Community_Affairs/Centrelinkcompliance

[386] Victoria Legal Aid, *Learning from the Failures of Robodebt—Building a Fairer, Client-centred Social Security System* (Nov. 14, 2023), https://www.legalaid.vic.gov.au/learning-from-the-failures-of-robodebt; Luke Henriques-Gomes, *Centrelink Cancels 40,000 Robodebts, New Figures Reveal: Robodebt Faces Landmark Legal Challenge over 'Crude' Income Calculations*, The Guardian (Feb. 6, 2019), https://www.theguardian.com/australia-news/2019/feb/06/robodebt-faces-landmark-legal-challenge-over-crude-income-calculations

[387] Rebecca Turner, *Robodebt Condemned as a "Shameful Chapter" in Withering Assessment by Federal Court Judge*, Australian Broadcasting Corporation (ABC News) (Jun. 11, 2021), https://www.abc.net.au/news/2021-06-11/robodebt-condemned-by-federal-court-judge-as-shameful-chapter/100207674

[388] Matthew Doran, *Robodebt Cases Dumped and Debts Wiped Amid Royal Commission into Controversial Scheme*, ABC News (Oct. 11, 2022), https://www.abc.net.au/news/2022-10-11/robodebt-reviews-wiped-government-clears-final-remnants-scheme/101523702

A Royal Commission into the Robodebt scheme issued a report[389] detailing the human impact of system and its errors, including false or incorrectly calculated debt notices; negative consequences on the physical and mental health of debt notice recipients, often among the most vulnerable; and the lawfulness of the scheme.

In August 2023, the House of Representatives passed a formal motion apologizing for the scheme on behalf of the Parliament. In November 2023, the Australian government formally accepted or accepted in principle all 56 recommendations in the Royal Commission report.[390] In the Government Response, the Prime Minister and Cabinet promised increased funding to implement the recommendations aligned to reforms aimed at "building trust in government, investing in a capable public sector, delivering strong institutions[,] and ensuring humans are put back at the centre of human services and service delivery."

In September 2023, the Digital Transformation Agency (DTA) and the Department of Industry, Science and Resources (DISR) established the Artificial Intelligence in Government Taskforce.[391] The Chief Executive Officer for the DTA cited the government's role in setting an example for the safe and ethical use of AI technologies, adding "We don't want to be left behind but we do want to protect government systems and ensure we're ultimately benefiting the wider Australian community." The Taskforce comprising representatives from multiple Australian Public Service (APS) agencies identified four key principles for implementing AI safely and responsibly in the public sector: "1) AI should be deployed responsibly in low-risk situations, 2) Transparency and explainability: tell when AI is used and why its use was warranted, 3) Privacy protection and security: use only public information, and 4) Accountability and human centred decision-making: final word should be with a human."[392]

The AI Taskforce released initial interim guidance on government use of publicly available generative artificial intelligence (AI) platforms.[393] According to the Guidance, one of the golden rules that Australian Public Service staff should consider for the responsible use of generative AI tools is: "you should be able to explain, justify

---

[389] Royal Commission into the Robodebt Scheme, *Report* (Jul. 2023), https://robodebt.royalcommission.gov.au/system/files/2023-09/rrc-accessible-full-report.PDF;

[390] Department of the Prime Minister and Cabinet, *Government Response to the Royal Commission into the Robodebt Scheme* (Nov. 13, 2023), https://www.pmc.gov.au/resources/government-response-royal-commission-robodebt-scheme

[391] Digital Transformation Agency, *The AI in Government Taskforce: Examining Use and Governance of AI by the APS* (Sept. 20, 2023), https://www.dta.gov.au/blogs/ai-government-taskforce-examining-use-and-governance-ai-aps

[392] Antonino Nielfi, *Ethical AI and the Australian Public Sector: Plotting an Unknown Course*, Parliament of Australia, Flagpost (Nov. 13, 2023), https://www.aph.gov.au/About_Parliament/Parliamentary_departments/Parliamentary_Library/FlagPost/2023/November/Ethical_AI_and_the_Australian_public_sector

[393] Digital Transformation Agency and Department of Industry, Science and Resources, *Interim Guidance on Government Use of Publicly Available Generative Artificial Intelligence* (Nov. 22, 2023), https://architecture.digital.gov.au/guidance-generative-ai

and take ownership of your advice and decisions." In addition, the Interim Guidance encouraged all APS staff to "read and understand Australia's AI Ethics Principles," and outlined steps to adhere to the Principles.[394]

The Australian government issued the Data and Digital Government Strategy: The data and vision for a world-class APS to 2030.[395] The Data and Digital Government Strategy is accompanied by an Implementation Plan and a Roadmap timeline. With the Strategy, the government commits to "improving and maintaining trust in its use of data and digital technology through adopting a whole-of-government Data Ethics Framework [...] and adopting AI technologies in safe, ethical and responsible ways."[396]

Building on the Strategy, the Data and Digital Ministers Meeting released a National Framework for using AI in government.[397] This framework establishes a consistent and holistic approach to AI assurance across all aspects of government.[398] Shortly after release of the national framework, the Digital Transformation Agency's Policy for the responsible use of AI in government took effect.[399] The policy outlines an adaptive whole-of-government framework that attempts to unite uses of AI in Public Administration and strengthen public trust in the Government's use of AI.[400] Adherence is mandatory for non-corporate Commonwealth entities, excepting national defense matters and members of the national intelligence community, who are encouraged to adopt elements of the policy that do not compromise national security.[401] Government agencies are also required to designate accountable official(s) to assume responsibility for their respective agency's implementation of the Policy.[402] These officials are tasked with ensuring AI tools align with the core principles outlined in the National Framework for the Assurance of AI in Government, including human oversight and privacy protection. Additionally, Section 5.2 of the policy requires agencies to publish public transparency statements detailing AI use cases, risk mitigation strategies, and human oversight mechanisms. The Senate Select

---

[394] Ibid

[395] Australian Government, *Data and Digital Government Strategy: The Data and Vision for a World-Class APS to 2030* (Dec. 15, 2023), https://www.dataanddigital.gov.au/sites/default/files/documents/2025-10/Data%20and%20Digital%20Government%20Strategy%20v1.0_0.pdf

[396] Ibid, p. 23

[397] Department of Finance, *National Framework for the Assurance of Artificial Intelligence in Government* (Jun. 21, 2024), https://www.finance.gov.au/government/public-data/data-and-digital-ministers-meeting/national-framework-assurance-artificial-intelligence-government

[398] Ibid

[399] Artificial Intelligence in Government, *Policy for the Responsible Use of AI in Government*, 2.0 (Dec. 1, 2025), https://www.digital.gov.au/policy/ai/policy

[400] Australian Government, Artificial Intelligence in Government, *Policy Aims*, https://www.digital.gov.au/policy/ai/aim

[401] Digital Transformation Agency, *Policy for the Responsible Use of AI in Government* (Sept. 1, 2024), https://architecture.digital.gov.au/responsible-use-of-AI-in-government

[402] Australian Government, Digital Transformation Agency, *Standards for Accountable Officials*, https://architecture.digital.gov.au/standard-accountable-officials

Committee on Adopting AI emphasized in its October 2024 interim report that such disclosures are critical to rebuilding public trust after systemic failures such as the Robodebt scandal.[403]

Australia's AI Plan for the Australian Public Service 2025 outlines the government's approach to leveraging AI to enhance public service delivery, framed around the goal of delivering "better services faster."[404] The plan provides a platform for fostering AI literacy and capacity-building among public servants, supported by designated Chief AI Officers tasked with driving AI adoption across agencies. To support the AI Plan for the Australian Public Service 2025, the Digital Transformation Agency (DTA) published three initiatives. First, the update to the Policy for the Responsible Use of AI in Government addresses the evolving capabilities of AI, strengthens accountability requirements, and mandates impact assessments for specified use cases.[405] Second, the AI Impact Assessment Tool introduces a standardized tool for managing impacts and risks against Australia's AI Ethics Principles.[406] Third, the Guidance on AI Procurement in Government comes with a checklist and detailed guidance expanding on each stage of the procurement process, highlighting AI-specific risks and considerations.[407]

Use of AI in Education

In May 2022, in a global investigative report on the education technology (EdTech) endorsed by for children's education during the pandemic, Human Rights Watch analyzed the technical and policy features of Minecraft: Education Edition used in Australia. Human Rights Watch found that the endorsements of this online learning platform put at risk or directly violated children's rights due their tracking abilities for advertising purposes.

Following the report, education ministers agreed to develop an evidence-based, best practice framework to guide schools in harnessing AI tools to support teaching and learning, and to establish a Taskforce to develop the framework. The Taskforce issued the Australian Framework for Generative AI in Schools in

---

[403] Parliament of Australia, *Chapter 5 —Automated Decision-making* (2024), https://www.aph.gov.au/Parliamentary_Business/Committees/Senate/Adopting_Artificial_Intelligence_AI/AdoptingAI/Report/Chapter_5_-_Automated_decision-making

[404] Australian Government, *AI Plan for the Australian Public Service 2025* (Nov. 12, 2025), https://www.digital.gov.au/policy/ai/australian-public-service-ai-plan-2025

[405] Digital Transformation Agency, *AI Policy Update: Strengthening Responsible Use across Government* (Jan. 12, 2025), https://www.dta.gov.au/articles/ai-policy-update-strengthening-responsible-use-across-government

[406] Digital Transformation Agency, *Artificial Intelligence Impact Assessment Tool* (Dec. 1, 2025), https://www.digital.gov.au/ai/impact-assessment-tool

[407] BuyICT, *Guidance on AI Procurement in Government* (Dec. 2, 2025), https://www.buyict.gov.au/sp?id=buyer&kb=KB0011755; Digital Transformation Agency, *AI Policy Overhauled with New Impact Assessment Tool and Procurement Guidance* (Dec. 2, 2025), https://www.dta.gov.au/media-releases/ai-policy-overhauled-new-impact-assessment-tool-and-procurement-guidance

November 2023.[408] The framework relies on key principles such as Human and Social Wellbeing, including the use of AI tools "in ways that respect and worker rights, including individual autonomy and dignity"; Transparency, including explainability: "vendors ensure that end users broadly understand the methods used by generative AI tools and their potential biases"; Fairness; Accountability; Privacy, Security and Safety: safe and ethical use of generative AI tools; best practice implementation of generative AI tools in the classroom to lift student outcomes; reducing workload burden and administration using generative AI tools; and establishing education-specific standards and governance to meet the needs of Australian schools.

The House of Representatives Standing Committee on Employment, Education, and Training tabled its report, Study Buddy or Influencer,[409] following its Inquiry into the Use of Generative Artificial Intelligence in the Australian Education System.[410] The Committee's 25 recommendations explore how Australian schools can maximize the opportunities presented by generative artificial intelligence (GenAI) while successfully mitigating the risks of using the emerging technology and ensuring adequate safeguards and guardrails are in place to prevent misuse. The recommendations focus on integrating GenAI into Australia's national curriculum as a study buddy for all students.

Use of AI in the Military

Although Australia participates in forums on AI use in the military, the Defence Digital Strategy and Roadmap 2024 did not explicitly include a plan for integrating AI systems.[411] The roadmap outlines a vision for a "secure, integrated, and scalable digital environment to compete and succeed in the digital age," but only briefly mentions the use of modern technologies such as AI and business process mining to inform and enhance defense operations.

Australia is also collaborating with international allies to develop advanced AI capabilities in defense. Under the AUKUS security pact, [412] Australia is working

---

[408] National AI in Schools Taskforce, *Australian Framework for Generative Artificial Intelligence in Schools* (Nov. 17, 2023), https://www.education.gov.au/schooling/resources/australian-framework-generative-artificial-intelligence-ai-schools

[409] House of Representatives Standing Committee on Employment, Education, and Training, *Study Buddy or Influencer* (Aug. 2024), https://www.aph.gov.au/Parliamentary_Business/Committees/House/Former_Committees/Employment_Education_and_Training/AIineducation/Report

[410] House of Representatives Standing Committee on Employment, Education, and Training, *Inquiry into the Use of Generative Artificial Intelligence in the Australian Education System* (May 24, 2023), https://www.aph.gov.au/Parliamentary_Business/Committees/House/Former_Committees/Employment_Education_and_Training/AIineducation

[411] Australian Government, Defence, *Defence Digital Strategy and Roadmap 2024* (Aug. 27, 2024), https://www.defence.gov.au/about/strategic-planning/defence-digital-strategy-roadmap-2024

[412] Louisa Brooke-Holland, *AUKUS Pillar 2: Advanced Military Capabilities*, House of Commons Library (Sept. 02, 2024), https://researchbriefings.files.parliament.uk/documents/CBP-9842/CBP-9842.pdf

alongside the United States and the United Kingdom on next-generation military technologies, including AI, quantum computing, and autonomous systems. Early AUKUS initiatives have explored AI-enabled battlefield decision-making and autonomous vehicle deployment. However, these developments necessitate strong ethical and legal oversight.

Moreover, Australian universities are collaborating with industry and government to train students in applied AI and other emerging technologies as part of Australia's Next Generation Graduates Program. The program seeks to create a skilled workforce that will enhance Australia's defense and aerospace sectors while leveraging engagement with Australia's First Nations peoples.[413]

Environmental Impacts of AI

Although AI technology is seen as a possible solution to clean energy and sustainability, Australia's national science agency, the Commonwealth Science and Industrial Research Organization (CSIRO),[414] has shifted focus to the potential environmental impacts of AI technology. This focus is expected to progress considering Australia's aim to "become a host for AI data [centers] in the Asia-Pacific region."[415] Australia's DISR commissioned a 2021 report to review the energy efficiency of data centers. The report found that "[g]rowth in edge computing (the use of data centres close to the point of use), which is required to support applications such as 5G, Artificial Intelligence and Virtual Reality, is likely to increase energy demand."[416] Beyond data center energy impacts, the DISR proposal for mandatory guardrails included environmental impacts among those organizations should consider in assessing whether the use of an AI system is "high-risk."[417]

Australia was one of 19 signatories to the Net Zero Government Initiative at the United Nations Framework Convention on Climate Change Conference of Parties

---

[413] Commonwealth Scientific and Industrial Research Organisation (CSIRO), *Applied AI and Digital Innovation for Defence and Aerospace Applications*, https://www.csiro.au/en/work-with-us/funding-programs/funding/next-generation-graduates-programs/awarded-programs/applied-ai-and-digital-innovation-for-defence-and-aerospace-applications

[414] Commonwealth Science and Industrial Research Organization (CSIRO), *AI for Clean Energy and Sustainability*, https://www.csiro.au/en/work-with-us/funding-programs/funding/Next-Generation-Graduates-Programs/Awarded-programs/AI-Clean-Energy-Sustainability

[415] Australian Trade and Investment Commission, *Artificial Intelligence and Data Centres: Helping to Power Digital Transformation*, https://international.austrade.gov.au/en/do-business-with-australia/sectors/technology/ai-and-data-centres#ref3

[416] Department of Industry, Science, Energy and Resources, *International Review of Energy Efficiency in Data Centres*, Commissioned Report by Fiona Brocklehurst, Ballarat Consulting (Sept. 2021), https://www.dcceew.gov.au/sites/default/files/documents/international-review-energy-efficiency-data-centres.pdf

[417] Department of Industry, Science and Resources, *Introducing Mandatory Guardrails for AI in High-Risk Settings: Proposals Paper* (Sept. 5, 2024), https://consult.industry.gov.au/ai-mandatory-guardrails

(COP27).[418] The Net Zero in Government Operations Strategy, which outlines the Australian government's approach to achieve net zero government operations by 2030, set energy efficiency and environmental standards for new data centers.[419] As part of these efforts, the Digital Transformation Agency established a new Data Centre Panel to "help promote sustainable practices across the data centre market [...] to reach net zero."[420]

Lethal Autonomous Weapons

At the 78th UN General Assembly First Committee in 2023, Australia voted in favor[421] of resolution L.56[422] on autonomous weapons systems, along with 163 other states. The Resolution emphasized the "urgent need for the international community to address the challenges and concerns raised by autonomous weapons systems" from humanitarian, legal, security, technological, and ethical perspectives and reflect on the role of humans in the use of force. The Resolution mandated the UN Secretary-General prepare a report reflecting the views of member and observer states on autonomous weapons systems.

In addition, Australia is a high contracting party to the convention on Certain Conventional Weapons (CCW). [423] Australia has also endorsed the United Nations General Assembly Resolution 78/241 on lethal autonomous weapons systems (LAWS).[424] Australia made a submission to the Secretary-General's Report on Lethal Autonomous Weapons Systems in May 2024,[425] indicating that the government "considers that the CCW is the most appropriate framework for multilateral discussions on LAWS" and reinforced its compromises to IHL and ethical standards.

---

[418] Department of Finance, *APS Net Zero Emissions by 2030* (Jan. 17, 2024), https://www.finance.gov.au/government/climate-action-government-operations/aps-net-zero-emissions-2030

[419] Department of Finance, *Net Zero in Government Operations Strategy* (2023), https://www.finance.gov.au/sites/default/files/2023-11/Net_Zero_Government_Operations_Strategy.pdf

[420] Digital Transformation Agency, *New Data Centre Panel* (May 18, 2023), https://www.dta.gov.au/blogs/new-data-centre-panel

[421] Stop Killer Robots, *164 States Vote against the Machine at the UN General Assembly* (Nov. 1, 2023), https://www.stopkillerrobots.org/news/164-states-vote-against-the-machine/

[422] General Assembly, Lethal Autonomous Weapons Systems, *Resolution L56* (Oct.12, 2023), https://reachingcriticalwill.org/images/documents/Disarmament-fora/1com/1com23/resolutions/L56.pdf

[423] UN Office for Disarmament Affairs, *High Contracting Parties and Signatories CCW* (Mar. 18, 2025), https://disarmament.unoda.org/en/our-work/conventional-arms/convention-certain-conventional-weapons/high-contracting-parties-and-signatories-ccw

[424] UN General Assembly, *Resolution A/RES/78/241 on Lethal Autonomous Weapons Systems* (Dec. 28, 2023), https://documents.un.org/doc/undoc/gen/n23/431/11/pdf/n2343111.pdf?fe=true

[425] UN Office for Disarmament Affairs, *Australia's Submission to the United Nations Secretary-General's Report on Lethal Autonomous Weapons Systems* (May 2024), https://docs-library.unoda.org/General_Assembly_First_Committee_-Seventy-Ninth_session_(2024)/78-241-Australia-EN.pdf

The submission also called attention to a joint proposal Australia co-sponsored to the Group of Governmental Experts on Emerging Technologies in the Area of Lethal Autonomous Weapons Systems (GGE).[426] The Proposal introduces a two-tier approach that involves total bans on specific LAWS (e.g., those designed to target civilians or civilian objects) and the imposition of restrictions on other systems to ensure adherence to IHL.[427]

In October 2022, Australia was one of 70 states that endorsed a joint statement on autonomous weapons systems at the United Nations General Assembly. The statement called for the recognition of the dangers of autonomous weapons systems, acknowledged the need for human oversight and accountability, and emphasized the importance of an international framework of rules and constraints.[428] In this joint statement, States declared that they "are committed to upholding and strengthening compliance with International Law, in particular International Humanitarian Law (IHL), including through maintaining human responsibility and accountability in the use of force."[429]

Australia participated in the Responsible Artificial Intelligence in the Military Domain (REAIM) international summit, co-hosted by the Netherlands and the Republic of Korea.[430] Government representatives, including Australia, agreed on a joint call for action on the responsible development, deployment, and use of artificial intelligence in the military domain, stressing that AI in the military domain should be "employed in full accordance with international legal obligations and in a way that does not undermine international security, stability and accountability."[431] Australia has also endorsed the resulting Political Declaration on Responsible Military Use of

---

[426] Ibid

[427] UN Office for Disarmament Affairs, Group of Governmental Experts on Emerging Technologies in the Area of Lethal Autonomous Weapons System, *Draft Articles on Autonomous Weapon Systems: Prohibitions and Other Regulatory Measures on the Basis of International Humanitarian Law (IHL)*, CCW/GGE.1/2024/WP.10 (Aug. 26, 2024), https://docs-library.unoda.org/Convention_on_Certain_Conventional_Weapons_-Group_of_Governmental_Experts_on_Lethal_Autonomous_Weapons_Systems_(2024)/CCW-GGE.1-2024-WP.10.pdf

[428] Stop Killer Robots, *70 States Deliver Joint Statement on Autonomous Weapons Systems at UN General Assembly* (Oct. 21, 2022), https://www.stopkillerrobots.org/news/70-states-deliver-joint-statement-on-autonomous-weapons-systems-at-un-general-assembly/

[429] UN General Assembly, First Committee, *Joint Statement on Lethal Autonomous Weapons Systems First Committee*, 77th UN General Assembly Thematic Debate – Conventional Weapons (Oct. 21, 2022), https://estatements.unmeetings.org/estatements/11.0010/20221021/A1jJ8bNfWGlL/KLw9WYcSnnAm_en.pdf

[430] Responsible Artificial Intelligence in the Military Domain, *REAIM 2023 Programme* (2023), https://reaim2023.org/

[431] Responsible AI in the Military Domain Summit, *REAIM Call to Action* (Feb. 16, 2023), https://www.government.nl/documents/publications/2023/02/16/reaim-2023-call-to-action

AI and Autonomy issued in November 2023.[432] Australian commissioners and experts joined the Global Commission on Responsible AI in the Military Domain in the Hague.[433]

The second edition of the REAIM was held in Seoul, South Korea, on September 9–10, 2024.[434] The Summit led to the adoption of the Blueprint for Action on the responsible use of AI in the military domain, endorsed by Australia.[435] This commitment builds on the groundwork from the 2023 summit in The Hague, proposing broader guidance for the ethical and responsible application of AI in military contexts, emphasizing compliance with international law, human accountability, reliability, human involvement, and data governance.

AI Literacy

Australia's updated National AI Plan[436] includes supporting and training workers as Action 5 under the goal to spread the benefits of AI.[437] The Government supports lifelong learning in the plan and detailed under the Jobs and Skills Australia 2025 Generative AI report.[438] Industry partners including Microsoft are assisting the expansion of AI Literacy Hubs across the country for all citizens to promote "inclusive AI upskilling across Australia." However, the plan does not define AI literacy.

The National AI Plan adds AI to the digital literacy competency already embedded in the education system.[439] The curriculum will build on the 2023 Framework for Generative Artificial Intelligence in Schools,[440] which "seeks to guide the responsible and ethical use of generative AI tools in ways that benefits students,

---

[432] US Department of State, *Political Declaration on Responsible Military Use of Artificial Intelligence and Autonomy*, Endorsing States (Feb. 12, 2024), https://www.state.gov/political-declaration-on-responsible-military-use-of-artificial-intelligence-and-autonomy/

[433] The Hague Centre for Strategic Studies, *Global Commission on Responsible Artificial Intelligence in the Military Domain (GC REAIM)* (Accessed Nov. 11, 2024), https://hcss.nl/gcreaim-commissioners/

[434] The Hague Centre for Strategic Studies, *Summit on Responsible Artificial Intelligence in the Military Domain (REAIM), Seoul, South Korea* (2024), https://hcss.nl/global-commission-on-responsible-artificial-intelligence-in-the-military-domain-gc-reaim-seoul-south-korea-conference-meeting-2/

[435] Ministry of Foreign Affairs, Republic of Korea, *Outcome of Responsible AI in Military Domain (REAIM) Summit* (Sept. 12, 2024), https://www.mofa.go.kr/eng/brd/m_5674/view.do?seq=321057

[436] Department of Industry, Science and Resources, *National AI Plan* (Dec. 2, 2025), https://www.industry.gov.au/publications/national-ai-plan

[437] Ibid, *Action 5: Support and Train Australians*, https://www.industry.gov.au/publications/national-ai-plan/spread-benefits#action-5-support-and-train-australians

[438] Jobs and Skills Australia, *Our Gen AI Transition—Implications for Work and Skills* (Aug. 14, 2025), https://www.jobsandskills.gov.au/publications/generative-ai-capacity-study-report

[439] Department of Education, *Australian Curriculum* (Aug. 5, 2025), https://www.education.gov.au/australian-curriculum

[440] National AI in Schools Taskforce, *Australian Framework for Generative Artificial Intelligence (AI) in Schools* (2023), https://www.education.gov.au/download/17416/australian-framework-generative-artificial-intelligence-ai-schools/35400/australian-framework-generative-ai-schools/pdf

schools and society." The Framework defines guidelines around key principles, including Transparency, which specifies that "School communities understand how generative AI tools work, how they can be used, and when and how these tools are impacting them," requiring vendors to ensure users "broadly understand" how the tools work and their biases.[441] The Framework also includes principles on Fairness and Accountability, as well as Privacy and Data Rights. However, the Framework does not specify how to use AI in those ways or specify vendors. The report on the Framework following the annual review in 2024 "found that the Framework not only correctly identified and made provision for existing challenges but also effectively predicted emerging and current risks, such as the increased use of AI deep fakes." The review noted updates to the national curriculum (Version 9.0)[442] to support teacher and school training on AI, elaborate on AI content, and identify potential research into AI's impact on learning outcomes.

Human Rights

Human rights are respected and protected through various laws in Australia at the federal, state and territory levels, through the Australian Constitution, and the common law.[443] Australia has formally agreed to be bound as a party to seven core international human rights treaties,[444] including as an original signatory to the Universal Declaration of Human Rights.[445] Under the Human Rights (Parliamentary Scrutiny) Act 2011, all bills and legislations are scrutinized against these covenants and conventions for compatibility with human rights and freedoms.[446]

Freedom House ranked Australia very highly (95/100) in 2025 and reported that, "Australia has a strong record of advancing and protecting political rights and civil liberties. Challenges to these freedoms include the threat of foreign political influence, harsh policies toward asylum seekers, discrimination against LGBT+ people, legal constraints on the press, and ongoing difficulties ensuring the equal

---

[441] Ibid, p. 6

[442] Australian Curriculum, Assessment and Reporting Authority, *The Australian Curriculum Version 9.0* (2026), https://www.australiancurriculum.edu.au/

[443] Attorney-General's Department, *Human Rights Protections*, https://www.ag.gov.au/rights-and-protections/human-rights-and-anti-discrimination/human-rights-protections

[444] Attorney-General's Department, *International Human Rights Systems*, https://www.ag.gov.au/rights-and-protections/human-rights-and-anti-discrimination/international-human-rights-system

[445] Department of Foreign Affairs and Trade, *Australia's Commitment to Human Rights*, https://www.dfat.gov.au/international-relations/themes/human-rights

[446] Attorney-General's Department, *Human Rights Scrutiny*, https://www.ag.gov.au/rights-and-protections/human-rights-and-anti-discrimination/human-rights-scrutiny

rights of First Nations Australians."[447] Although Australia has maintained its "free" ranking, the score declined from 97 in 2021.[448]

Australia also ranks highly in the Freedom House Freedom on the Net report, scoring 76/100 in 2024. Although Australia ranks highly in access to information, the report warns of proposed legislation that "sought to increase online surveillance and limit the security of encrypted communications."[449]

Human rights and Artificial Intelligence came together in the Australian Human Rights Commission 2023 report to DISR on Safe and Responsible AI. The report outlined human rights risks of AI, called for government action to "ensure human rights-centred design in the deployment of new and emerging technologies (including AI)," and supported human rights impact assessments to increase public trust.[450] The Commission reiterated the need to center human rights in the development and deployment of AI systems in a 2024 report the Senate Select Committee on Adopting Artificial Intelligence.[451] The report provided 6 recommendations to the Select Committee to "address risks that are not currently within the scope of the existing regulatory framework."

OECD / G20 AI Principles

Australia has endorsed the OECD and the G20 AI Principles. Australia has been actively aligning its policies with the OECD AI Principles, including through the AI Ethics Principles, the creation of NAIC, and the Voluntary AI Safety Standard. The OECD notes the Australia Roadmap for AI, the AI Ethics Framework, and the Australia's AI Standards Roadmap, "intended to identify priority areas for AI standards development and a pathway for Australian leadership on international standardization activities for AI."[452] The OECD also notes the work of Australia on trustworthy AI for health.

---

[447] Freedom House, *Freedom in the World 2025: Australia* (2025), https://freedomhouse.org/country/australia/freedom-world/2025

[448] Freedom House, *Freedom in the World 2021: Australia* (2021), https://freedomhouse.org/country/australia/freedom-world/2021

[449] Freedom House, *Freedom on the Net 2024: Australia* (2024), https://freedomhouse.org/country/australia/freedom-net/2024

[450] Australian Human Rights Commission, *The Need for Human Rights-Centred Artificial Intelligence* (Jul. 26, 2023), https://humanrights.gov.au/our-work/legal/submission/need-human-rights-centred-ai

[451] Australian Human Rights Commission, *Adopting AI in Australia* (May 15, 2024), https://humanrights.gov.au/sites/default/files/select_committee_on_adopting_artificial_intelligence_0.pdf

[452] OECD, G20 Digital Economy Task Force, *Examples of National AI Policies* (2020), https://www.mcit.gov.sa/sites/default/files/examples-of-ai-national-policies.pdf; OECD, *State of Implementation of the OECD AI Principles: Insights from National AI Policies* (Jun. 18, 2021), https://doi.org/10.1787/1cd40c44-en

Australia joined the Global Partnership on AI as a founding member in June 2020.[453] Minister Andrews stated, "Australia is committed to responsible and ethical use of AI. Membership of the GPAI will allow Australia to showcase our key achievements in AI and provide international partnership opportunities which will enhance our domestic capability." Andrews further stated, "Membership of the GPAI will build on the work the Government started at last year's National AI Summit, which brought together 100 AI experts to discuss the challenges and opportunities which AI will present for the Australian economy."

Australia actively participated in the G20 meetings held in Brazil in 2024 and it also endorsed the Maceio Ministerial Declaration of the Digital Economy Working Group (DEWG).[454] The declaration calls for international cooperation to establish transparent and reliable AI guidelines to ensure equitable global benefits, human rights, and ethical standards. The declaration also highlights AI's potential in addressing challenges like climate change and health crises, urging investments in digital infrastructure and skills to maximize its positive impact.

Council of Europe AI Treaty

Australia participated in the negotiation and drafting process for the Council of Europe Framework Convention on Artificial Intelligence and Human Rights, Democracy and the Rule of Law.[455] The first-of-its-kind global framework, the treaty aims to ensure "respect of human rights, the rule of law and democratic legal standards in the use of artificial intelligence (AI) systems."[456] Australia has not yet signed the treaty.[457]

---

[453] Hon Karen Andrews MP, Minister for Industry, Science and Technology, *Australia Joins Global Partnership on Artificial Intelligence* (Jun. 16, 2019), https://www.minister.industry.gov.au/ministers/karenandrews/articles/australia-joins-global-partnership-artifical-intelligence

[454] G7G20 Documents Database, *Digital Economy Working Group Maceio Ministerial Declaration* (2024), https://g7g20-documents.org/database/document/2024-g20-brazil-sherpa-track-digital-economy-ministers-ministers-language-g20-dewg-maceio-ministerial-declaration

[455] Council of Europe, *The Framework Convention on Artificial Intelligence* (2026), https://www.coe.int/en/web/artificial-intelligence/the-framework-convention-on-artificial-intelligence

[456] Council of Europe, *Council of Europe Framework Convention on Artificial Intelligence and Human Rights, Democracy and the Rule of Law* (2024), https://edoc.coe.int/en/artificial-intelligence/11926-council-of-europe-framework-convention-on-artificial-intelligence-and-human-rights-democracy-and-the-rule-of-law.html

[457] Council of Europe Treaty Office, *Chart of Signatures and Ratifications of Treaty 225* (Mar. 1, 2026), https://www.coe.int/en/web/Conventions/full-list/?module=signatures-by-treaty&treatynum=225

UNESCO Recommendation on AI Ethics

Australia endorsed the UNESCO Recommendation on the Ethics of AI in 2021.[458] The Australian AI Action Plan contains no explicit reference to the Recommendation because the Plan was published before.[459] However, the recent responsible use of AI in government policy[460] and the national data sharing framework signal progress.[461] A comprehensive assessment, such as the UNESCO Readiness Assessment Methodology (RAM) could help Australia evaluate the effectiveness of policy and further implement the UNESCO Recommendation. However, Australia has not initiated the RAM.[462]

Evaluation

Australia has set out an AI Roadmap and an AI Ethics Framework, in addition to endorsing the OECD/G20 AI Principles. The Government is implementing these values-based principles through practical initiatives such as the National Artificial Intelligence Center (NAIC) based on the country's AI Ethical Principles. The AI Ethical Principles are built upon OECD values.[463] Australia has encouraged public participation in the development of AI policy, joined the Global Partnership on AI, and has a strong record on human rights. Australia has independent agencies, including a national regulator for privacy and freedom of information[464] and a human rights commission that is engaged in AI oversight. Australia was also a cosponsor of the GPA resolution on Accountability in the development and use of AI.

Although Australia has endorsed the UNESCO Recommendations on the Ethics of AI, the Government has not conducted a comprehensive assessment to evaluate their effectiveness and further implement the UNESCO Recommendation. The modernization of Australia's data protection law to better protect human rights continues to progress, most recently with the amendments to the Privacy Act passing through Parliament. In addition, Australia's national AI strategy continues to develop,

---

458 UNESCO, *UNESCO Adopts First Global Standard on the Ethics of Artificial Intelligence* (Aug. 31, 2023), https://www.unesco.org/en/articles/unesco-adopts-first-global-standard-ethics-artificial-intelligence

459 Department of Industry, Science, Energy, *Australia's AI Action Plan* (Jun. 2021), https://wp.oecd.ai/app/uploads/2021/12/Australia_AI_Action_Plan_2021.pdf

460 Digital Transformation Agency, *Standard for Accountability*, https://architecture.digital.gov.au/standard-accountable-officials

461 Office of National Data Commissioner, *Introducing the DATA Scheme* (Apr. 2023), https://www.datacommissioner.gov.au/the-data-scheme

462 UNESCO AI Ethics and Governance Observatory, *Global Hub* (Oct. 2025) https://www.unesco.org/ethics-ai/en/global-hub

463 OECD Artificial Intelligence Papers, *The State of Implementation of the OECD AI Principles Four Years On*, No. 3 (Oct. 2023), https://www.oecd-ilibrary.org/science-and-technology/the-state-of-implementation-of-the-oecd-ai-principles-four-years-on_835641c9-en

464 Office of the Australian Information Commission, *Human Rights and Technology Discussion Paper* (Dec. 2019), https://tech.humanrights.gov.au/sites/default/files/2019-12/TechRights2019_DiscussionPaper.pdf

including the release of the Voluntary AI Safety Standard, which aims to "ensure that the development and deployment of AI systems in Australia in legitimate but high-risk settings is safe and can be relied on, while ensuring the use of AI in low-risk settings can continue to flourish largely unimpeded."[465] Concerns exist regarding facial recognition.

[465] National Artificial Intelligence Centre, *Voluntary AI Safety Standard* (Aug. 2024), https://www.industry.gov.au/sites/default/files/2024-09/voluntary-ai-safety-standard.pdf

Austria

In 2025, Austria progressed on implementing the EU AI Act by operationalizing advisory and supervisory bodies. The AI Advisory Board and AI Stakeholder Forum will contribute to the newly established AI Monitor. Austria signed the Statement on Inclusive and Sustainable AI at the AI Action Summit.

National AI Strategy

The Austrian Government presented the national AI strategy, Artificial Intelligence Mission Austria 2030 (AIM AT 2030), in August 2021.[466] The objectives are: Using AI for the common good; Positioning Austria as an international hub of AI research and innovation; Boosting Austria's economy through strategic AI use.

AIM AT 2030 is guided by the two cornerstones of the European AI strategy: an ecosystem for trust and an ecosystem for excellence. Austria not only supports increased cooperation at the European level, as proposed in the European Commission's White Paper on AI[467] and the EU AI Act,[468] but also intends to shape national AI ecosystems in line with the European AI strategy.[469]

The Austrian strategy is based on a human-centered approach to AI to ensure that resources are used to support fundamental European values and respect and guarantee fundamental rights such as privacy and the principle of equality. Citizens' involvement is also identified as key.[470] Regarding ethical principles, the strategy refers to the European High-Level Expert Group on AI's Ethics guidelines for trustworthy AI.[471] Accordingly, AI systems must fulfill three basic principles to be considered trustworthy. They must:

---

[466] Federal Ministry for Climate Action, Environment, Energy, Mobility, Innovation and Technology (BMK), *Artificial Intelligence Strategy of the Austrian Federal Government, AIM AT 2030* (2021), https://digital-skills-jobs.europa.eu/sites/default/files/2023-10/Artificial%20Intelligence%20Mission%20Austria%202030_AIM_AT_2030.pdf

[467] European Commission, *White Paper on Artificial Intelligence: A European Approach to Excellence and Trust*, COM(2020) 65 final (Feb. 19, 2020), https://ec.europa.eu/info/sites/default/files/commission-white-paper-artificial-intelligence-feb2020_en.pdf

[468] European Union, EUR-Lex, *Regulation (EU) 2024/1689 of the European Parliament and of the Council of 13 June 2024 Laying Down Harmonised Rules on Artificial Intelligence and Amending Regulations (EC) No 300/2008, (EU) No 167/2013, (EU) No 168/2013, (EU) 2018/858, (EU) 2018/1139 and (EU) 2019/2144 and Directives 2014/90/EU, (EU) 2016/797 and (EU) 2020/1828* (Jun. 13, 2024), https://eur-lex.europa.eu/eli/reg/2024/1689/oj/eng

[469] Federal Ministry for Climate Action, Environment, Energy, Mobility, Innovation and Technology (BMK), *AIM AT 2030*, p. 15 (2021), https://digital-skills-jobs.europa.eu/sites/default/files/202310/Artificial%20Intelligence%20Mission%20Austria%202030_AIM_AT_2030.pdf

[470] Ibid, p. 17

[471] High-Level Expert Group on Artificial Intelligence, *Ethics Guidelines for Trustworthy AI* (2019), https://digital-strategy.ec.europa.eu/en/library/ethics-guidelines-trustworthy-ai

- "be lawful by respecting all existing laws and regulations;
- respect ethical principles and values such as equality and fairness; and
- be robust, both in a technical sense and from a societal perspective."[472]

AIM AT 2030 also mentions the need to establish a clear legal framework that releases innovation in science and economy, reduces uncertainties and at the same time guarantees legal certainty. The Austrian Federal Government supported the creation of a Europe-wide legal framework for AI applications to avoid isolated national solutions, a project realized with the EU AI Act.

The AI Implementation Plan introduced in 2024 is an interim report that serves as a significant addition and clarification to the AIM AT 30 and aligns "with fundamental European values and the European legal framework.[473]

The Federal Ministry Republic of Austria Climate Action, Environment, Energy, Mobility, Innovation and Technology oversees AI policy under the broad category of "Innovation." The Ministry also engages with research, policy, governance, industrial competitiveness, sustainability, and international cooperation.[474]

Austria's 11-member AI Advisory Board established in early 2024 includes experts from academia, law, and ethics. This board advises the federal government on AI regulatory measures and ethical concerns and assists in incorporating EU AI Act guidelines into national policy frameworks.[475]

Austria is proactively planning for emerging and future AI-related technologies. The Federal Ministry of Innovation, Mobility and Infrastructure (BMIMI), established a national roadmap, Virtual Worlds and Immersive Technologies 2035, detailing recommendations for action with the goal for Austria to effectively develop in extended reality (XR).[476] The roadmap highlights the need for domestic policy and legal frameworks as well as advocacy for these frameworks at the EU level to foster EU development and sovereignty of this technology.

---

[472] Federal Ministry for Climate Action, Environment, Energy, Mobility, Innovation and Technology (BMK), *AIM AT 2030*, p. 22 (2021), https://digital-skills-jobs.europa.eu/sites/default/files/2023-10/Artificial%20Intelligence%20Mission%20Austria%202030_AIM_AT_2030.pdf

[473] Federal Ministry for Innovation, Mobility and Infrastructure (BMIMI), *Implementation Plan 2024*, https://www.bmimi.gv.at/themen/innovation/publikationen/ikt/ai/Umsetzungsplan-2024.html

[474] Federal Ministry for Innovation, Mobility and Infrastructure (BMIMI), *Innovation*, https://www.bmimi.gv.at/themen/innovation.html

[475] Digital Austria, *Topics, Artificial Intelligence, AI Advisory Board* (2026), https://www.digitalaustria.gv.at/Themen/KI/AI-Advisory-Board.html

[476] Federal Ministry of Innovation, Mobility and Infrastructure (BMIMI), *Roadmap of the Federal Ministry of Innovation, Mobility and Infrastructure* (Jun. 21, 2025), https://www.bmimi.gv.at/en/topics/innovation/publications/virtual-worlds.html

EU Digital Services Act

As an EU member state, Austria is bound by the EU Digital Services Act (DSA).[477] Austria's DSA Accompanying Act (*DSA-Begleitgesetz*) has been in force since February 2024.[478]

EU AI Act

As an EU member State, Austria is bound by the EU AI Act.[479] Austria's Service Desk for Artificial Intelligence (AI Service Desk) under the Regulatory Authority for Broadcasting and Telecommunications (RTR) supports implementation of the EU AI Act and serves as a public information hub.[480] Austria has not designated a single point of contact under the EU AI Act or specified the supervisory role of the AI Service Desk.[481] The Service Desk coordinates with the AI Advisory Board and AI Forum to advise the Federal Government on national AI strategy, policies, and implementation.[482]

Public Participation

Experts and other stakeholders were involved in the development of the national AI strategy through the AI Stakeholder Forum,[483] and through the

---

[477] *Regulation (EU) 2022/2065 of the European Parliament and of the Council of 19 October 2022 on a Single Market for Digital Services and Amending Directive 2000/31/EC (Digital Services Act)*, (Oct. 21, 2022), https://eur-lex.europa.eu/legal-content/EN/TXT/?uri=celex%3A32022R2065

[478] Federal Ministry, Judiciary, *Digital Services Act* (2024), https://www.bmj.gv.at/themen/EU-und-Internationales/Digital-Services-Act.html

[479] European Parliament, *Artificial Intelligence Act, European Parliament Legislative Resolution of 13 March 2024 on the Proposal for a Regulation of the European Parliament and of the Council on Laying Down Harmonised Rules on Artificial Intelligence (Artificial Intelligence Act) and Amending Certain Union Legislative Acts (COM(2021)0206 – C9-0146/2021 – 2021/0106(COD))*, P9_TA(2024)0138, https://www.europarl.europa.eu/RegData/seance_pleniere/textes_adoptes/definitif/2024/03-13/0138/P9_TA(2024)0138_EN.pdf

[480] Regulatory Authority for Broadcasting and Telecommunications, *Service Desk for Artificial Intelligence*, https://www.rtr.at/rtr/service/ki-servicestelle/KI-Servicestelle.en.html

[481] European Commission, *Market Surveillance Authorities under the AI Act* (Sept. 26, 2025), https://digital-strategy.ec.europa.eu/en/policies/market-surveillance-authorities-under-ai-act#1720699867912-2

[482] Rechtsinformationsystem des Bundes, *Consolidated Federal Law: KommAustria Act, Article 20c*, Federal Law Gazette I No. 6/2024 (Jan. 1, 2024), https://www.ris.bka.gv.at/NormDokument.wxe?Abfrage=Bundesnormen&Gesetzesnummer=20001213&Paragraf=20c; Rechtsinformationssytem des Bundes *Consolidated Federal Law: Telecommunications Act 2021, Article 194*, Federal Law Gazette I No. 6/2024 (Jan. 1, 2024), https://www.ris.bka.gv.at/NormDokument.wxe?Abfrage=Bundesnormen&Gesetzesnummer=20011678&Paragraf=194a

[483] Digital Austria, *AI Governance in Austria* (2026), https://www.digitalaustria.gv.at/themen/kuenstliche-intelligenz/ki-organisation.html

consultation among government entitites.[484] The strategy also provides for broad participation of civil society organizations, intermediaries, and citizens in the implementation of the measures. The Austrian Regulatory Authority for Broadcasting and Telecommunications (RTR) created an AI Service Desk to serve as a "point of contact and information hub for the general public" on AI. The service desk also supports implementation of the EU AI Act.[485] The AI Stakeholders Forum also promotes exchange between the Federal Government and the country's diverse stakeholders in the field of artificial intelligence. As of December 2024, the Forum consisted of 29 different associations and interest groups. The central task of the Forum is to connect participating organizations, disseminate information and news on AI, and promote synergies.[486]

The Federal Government endeavors to formulate its target provisions in close coordination and comprehensive agreement with the fundamental values and objectives of the European Union. Austria is thus also contributing to the promotion of Europe's industrial and technical performance regarding AI.

Data Protection

Austria is an EU Member State, so the General Data Protection Regulation (GDPR)[487] applies. The Austrian Data Protection Act (DSG) supplements the GDPR.[488]

Following the EU AI Act's policy framework on AI systems, Austria has strengthened its data privacy measures. New policies require detailed impact assessments for systems handling sensitive data or biometric recognition. These updates further Austria's commitment to protecting data privacy and human rights.[489]

According to Article 35(4) GDPR, national supervisory authorities shall compose and publish the list of processing operations that requires performing a data protection impact assessment. According to the list established by the Austrian Data Protection Authority (*Datenshutz behörde*, DPB), a data protection impact assessment is necessary in cases involving AI, including:

---

[484] Digital Austria, *Strategy for Artificial Intelligence* (2026), https://www.digitalaustria.gv.at/verwaltung/strategien/ki-aim-at-2030.html

[485] Regulatory Authority for Broadcasting and Telecommunications, *Service Desk for Artificial Intelligence*, https://www.rtr.at/rtr/service/ki-servicestelle/KI-Servicestelle.en.html

[486] Digital Austria, *AI Governance in Austria* (2026), https://www.digitalaustria.gv.at/themen/kuenstliche-intelligenz/ki-organisation.html

[487] *Regulation (EU) 2016/679 of the European Parliament and of the Council of 27 April 2016 on the Protection of Natural Persons with Regard to the Processing of Personal Data and on the Free Movement of Such Data* (Apr. 27, 2016), https://eur-lex.europa.eu/EN/legal-content/summary/general-data-protection-regulation-gdpr.html

[488] Austria Data Protection Authority, *Federal Act Concerning the Protection of Personal Data (Datenschutzgesetz - DSG)* (May 25, 2018), https://www.ris.bka.gv.at/Dokumente/Erv/ERV_1999_1_165/ERV_1999_1_165.html

[489] Austria Data Protection Authority, *Legal Sources [Rechtsquellen]*, https://dsb.gv.at/rechte-pflichten/rechtsquellen

- assessment or classification related to a person's work, economic situation, health, preferences and interests, reliability or behavior, location or movements and that is based solely on automated processing and may have negative legal, physical, or financial consequences
- evaluating the behavior and other personal aspects in cases that may be used by third parties to make automated decisions that have legal effects on the persons evaluated or similarly significantly affect them
- using or applying new or novel technologies or organizational solutions in processing data that make it difficult to assess the impact on data subjects and the social consequences, in particular through the use of artificial intelligence and the processing of biometric data beyond the real-time reproduction of facial images
- where the use of algorithms makes it possible to take decisions that significantly affect the data subject by going beyond the processing operations normally expected of a data subject[490]

Austria also transposed the EU Data Protection Law Enforcement Directive (LED)[491] with the DSG.[492] The Austrian National Council approved the Austrian Data Access Act (*Datenzugangsgesetz*, DZG) to implement the EU Data Governance Act in 2025.[493]

Austria is a member of the Council of Europe and ratified the Council of Europe's Convention 108+ for the protection of individuals regarding the processing of personal data.[494]

The Austrian DPA is a member of the Global Privacy Assembly (GPA) since 2002. The DPA did not endorse the 2018 GPA Declaration on Ethics and Data Protection in Artificial Intelligence,[495] the 2020 GPA Resolution on AI

---

[490] European Data Protection Board, *Regulation of the Data Protection Authority on Processing Operations for which a Data Protection Impact Assessment Is Required (DSFA-V)* (2018), https://edpb.europa.eu/sites/default/files/decisions/at_sa_dpia_final_decision.pdf

[491] European Parliament, *Directive (EU) 2016/680 of the European Parliament and of the Council of 27 April 2016 on the Protection of Natural Persons with Regard to the Processing of Personal Data by Competent Authorities for the Purposes of the Prevention, Investigation, Detection or Prosecution of Criminal Offences or the Execution of Criminal Penalties, and on the Free Movement of Such Data, and Repealing Council Framework Decision 2008/977/JHA* (Apr. 27, 2016), https://eur-lex.europa.eu/legal-content/EN/TXT/?uri=celex%3A02016L0680-20160504

[492] Federal Legal Information System, *Federal Act Concerning the Protection of Personal Data* (Datenschutzgesetz - DSG) (May 25, 2018), https://www.ris.bka.gv.at/Dokumente/Erv/ERV_1999_1_165/ERV_1999_1_165.html

[493] Parlament Österreich, *Data Access Act – DZG (127. D.B.) [Datenzugangsgesetz, DZG]* (Jun. 18, 2025), https://www.parlament.gv.at/gegenstand/XXVIII/I/127?selectedStage=100

[494] Council of Europe, *Modernised Convention for the Protection of Individuals with Regard to the Processing of Personal Data* (May 18, 2018), https://www.coe.int/en/web/data-protection/convention108-and-protocol

[495] Global Privacy Assembly, *Declaration on Ethics and Data Protection in Artificial Intelligence* (Oct. 23, 2018), https://globalprivacyassembly.org/wp-content/uploads/2018/10/20180922_ICDPPC-40th_AI-Declaration_ADOPTED.pdf

Accountability,[496] the 2022 GPA Resolution on Facial Recognition Technology,[497] or 2023 GPA Resolution on Generative AI Systems.[498]

Algorithmic Transparency

Austria is subject to the GDPR, as well as Convention 108+ since July 2022.[499] The Austrian Data Protection Act implements the GDPR, ensuring Austrians have a general right to obtain access to information about automated decision-making and to the factors and logic of an algorithm.[500]

The case of the Data Protection Authority's (DSB) ruling with a tool developed by the Austrian Public Employment Service (AMS) demonstrates the DPB's regulatory role and the checks in support of innovation. After the DSB banned an AMS algorithm designed to support AMS staff in their daily advising to job seekers in 2020, the Austrian Federal Administrative Court (BVwG) overruled the initial ruling on appeal in 2025.[501] The appeal successfully demonstrated the lack of reliance on automatic decisions, making the ruling that the AMS did not have a suitable legal basis for automated decision-making under Article 22 GDPR moot.[502] According to the BVwG, "Investigations demonstrated that advisors had a substantive role in the decision-making process, rather than a merely formal one. Furthermore, [...] the controller implemented clear internal guidelines and measures to comply with them. Therefore, the Court concluded that there was no automated decision-making."[503]

---

[496] Global Privacy Assembly, *Resolution on Accountability in the Development and Use of Artificial Intelligence* (Oct. 2020), https://globalprivacyassembly.org/wp-content/uploads/2020/10/FINAL-GPA-Resolution-on-Accountability-in-the-Development-and-Use-of-AI-EN-1.pdf

[497] Global Privacy Assembly, *Resolution on Principles and Expectations for the Appropriate Use of Personal Information in Facial Recognition Technology* (Oct. 2022), https://globalprivacyassembly.org/wp-content/uploads/2022/11/15.1.c.Resolution-on-Principles-and-Expectations-for-the-Appropriate-Use-of-Personal-Information-in-Facial-Recognition-Technolog.pdf

[498] Global Privacy Assembly, *Resolution on Generative Artificial Intelligence Systems* (Oct. 20, 2023), https://www.edps.europa.eu/system/files/2023-10/edps-gpa-resolution-on-generative-ai-systems_en.pdf

[499] See Recital 63 and Article 22 of the GDPR; Article 9 c) of the Convention 108+ as well as Recital 77, *Explanatory Report, Convention 108+*, p. 24, https://rm.coe.int/convention-108-convention-for-the-protection-of-individuals-with-regar/16808b36f1

[500] Federal Chancellery of Austria, *Federal Act concerning the Protection of Personal Data (DSG)* (2018), https://www.ris.bka.gv.at/Dokumente/Erv/ERV_1999_1_165/ERV_1999_1_165.html

[501] Austrian Press Agency (OTS), *Data Protection and Innovation Compatible – Court Rules in Favor of AMS after Years of Legal Battle over AMS Algorithm* (2025), https://www.ots.at/presseaussendung/OTS_20250904_OTS0060/datenschutz-und-innovation-vereinbar-gericht-gibt-ams-nach-jahrelangem-rechtsstreit-um-ams-algorithmus-recht

[502] GDPR Hub, *BGH / W256 2235360-1* (2025), https://gdprhub.eu/index.php?title=BVwG_-_W256_2235360-1

[503] Austrian Press Agency (OTS), *Data Protection and Innovation Compatible – Court Rules in Favor of AMS after Years of Legal Battle over AMS Algorithm* (2025), https://www.ots.at/presseaussendung/OTS_20250904_OTS0060/datenschutz-und-innovation-vereinbar-gericht-gibt-ams-nach-jahrelangem-rechtsstreit-um-ams-algorithmus-recht

Still, the CEO of the AMS says that the tool will not be revived as too much time has passed and the tool is likely outdated."

The Austrian Government launched the AI Monitor, a dashboard that provides information about Austria's usage and the opportunities, and challenges associated with AI. Publicly available at digitalaustria.gv.at, the AI Monitor allows citizens greater AI oversight. The Monitor will be updated annually and include a "broad range of quantitative data, alongside short introductions to lighthouse projects and initiatives within the Austrian AI ecosystem and contributions by the Austrian AI Advisory Board and the AI Stakeholder Forum."[504]

Facial Recognition

Since December 2017, 25 E-Gates, where border controls are supported by facial recognition, have been installed at the Vienna International Airport in Austria. Manual border controls have also remained in place.[505]

After a one-year test phase, facial recognition for law enforcement has been in regular operation in Austria since August 2020. The police can use digital image comparison only if they suspect the commission of an intentional judicially punishable act in the case of unknown perpetrators. The Ministry of the Interior published information about the use of the software after parliamentary inquiries.[506] Accordingly, the Federal Criminal Police Office may use the system to investigate intentional acts authorized by the judiciary regardless of the level of punishment for an offense.[507]

The Austrian supervisory authority (*Datenschutzbehörde, DSB*) issued a decision in 2023 against the facial recognition company Clearview AI, which maintains a database of over 30 billion facial images collected globally. The DSB found Clearview AI to have violated several GDPR provisions, citing a lack of lawfulness, fairness, and transparency in processing the complainant's personal data, and of illegally scanning and extracting uniquely identifying facial features of the complainant. Clearview AI was ordered to delete the complainant's personal data and designate a representative within the European Union[508] but not required to pay a

---

[504] OECD.AI, *AI Monitor* (Jul. 9, 2025), https://oecd.ai/en/dashboards/policy-initiatives/ai-monitor-2797

[505] Passenger Self Service, *Vienna International Airport Installs ABC eGates from Secunet* (Dec. 19, 2017), https://www.passengerselfservice.com/2017/12/vienna-international-airport-installs-abc-egates-secunet/

[506] Federal Ministry of the Interior, *Query Response Parliamentary Question No. 2648/J: "Findings from the Test Operation of the Face Recognition System (2662/AB)"* (Sept. 4, 2020), https://www.parlament.gv.at/PAKT/VHG/XXVII/AB/AB_02662/index.shtml#

[507] Florian Terharen-Schoenherr, *Facial Recognition Technology: Regulations and Use – Austria* (Apr. 6, 2021), https://iclg.com/briefing/16092-facial-recognition-technology-regulations-and-use-austria

[508] European Data Protection Board, *Decision by the Austrian SA against Clearview AI Infringements of Articles 5, 6, 9, 27 of GDPR* (May 12, 2023), https://www.edpb.europa.eu/news/national-news/2023/decision-austrian-sa-against-clearview-ai-infringements-articles-5-6-9-27_en

fine.[509] Austrian privacy group noby filed a criminal complaint in Austria in 2025, accusing Clearview of illegally collecting photos to build its facial recognition database violating GDPR provisions.[510]

The ID Austria mobile app (previously Digitales Amt) was officially launched in 2025. The app aims to improve services for digital ID, electronic signatures, and qualified signatures. The login process has also been updated to allow users to log in using fingerprint, facial recognition, or their device password/PIN.[511]

Automated Tax Fraud Detection

The Austrian Ministry of Finance operates the Predictive Analytics Competence Center (PACC), which automates review to facilitate a more efficient risk management structure with increased efficiency in auditing, fraud prevention, and tax collection.[512]

PACC reviewed approximately 6 million income tax disclosures and 1.4 million applications for COVID-19 assistance payments using machine learning to derive potential fraud scenarios from a variety of historical data sources. The reviews facilitate the selection of tax audits, plausibility checks of tax returns, and the evaluation of start-ups and applications of all kinds. The Ministry of Finance claims that this will make it easier to detect tax evasion, including customs fraud.[513]

According to a Ministry of Finance's press release on the 2023 annual balance sheet, PACC's analysis identified 375,000 cases out of 34 million as potentially implausible, targeting them for further investigation into tax evasion and fraud.[514] For the 2024 cycle, PACC's automated reviewed analyzed 6.6 million tax cases and reviewed 23.4 million compliance cases, helping to detect false declarations, fraud attempts, and fake companies.[515] This work generated about €354 million in extra tax

---

[509] NOYB, Clearview AI data use deemed illegal in Austria, however no fine issued (May 10, 2023), https://noyb.eu/en/clearview-ai-data-use-deemed-illegal-austria-however-no-fine-issued

[510] Reuters, *Clearview AI Faces Criminal Complaint in Austria for Suspected Privacy Violations* (Oct. 28, 2025), https://www.reuters.com/sustainability/society-equity/clearview-ai-faces-criminal-complaint-austria-suspected-privacy-violations-2025-10-28/

[511] ID Austria, *Simple Online Identification with ID Austria* (2025), https://www.id-austria.gv.at/en

[512] Federal Ministry of Finance, *Predictive Analytics Competence Center* (2022), https://www.bmf.gv.at/themen/betrugsbekaempfung/einheiten-betrugsbekaempfung/Predictive-Analytics-Competence-Center.html

[513] Austria Press Agency, *BMF/Special Unit PACC: Around 6 Million Employee Assessments and 1.4 Million Applications for COVID-19 Aid Payments Checked in 2021: Predictive Analytics in the Fight against Organized Tax Evasion and Tariff Fraud* (Sept. 11, 2022), https://www.ots.at/presseaussendung/OTS_20220911_OTS0004/bmfspezialeinheit-pacc-2021-rund-6-mio-arbeitnehmerveranlagungen-und-14-mio-antraege-auf-covid-19-hilfszahlungen-ueberprueft

[514] Regfollower, *Austria Generates Additional EUR 185 Million in Tax Income from AI* (Aug. 14, 2025), https://regfollower.com/austria-generates-additional-eur-185-million-in-tax-income-from-ai/

[515] Federal Ministry of Finance, *Last Year, the Ministry of Finance Generated €354 Million in Additional Tax Revenue Thanks to AI Methods [Finanzministerium lukrierte im Vorjahr 354 Mio. Euro Steuermehreinnahmen dank KI-Methoden]* (Aug. 13, 2025), https://www.bmf.gv.at/presse/pressemeldungen/2025/august/pacc-ki.html

revenue. In one example, a farm failed to report vehicle sales and was required to pay over €300,000 in back taxes. The Ministry said that PACC plans to expand the use of AI and generative models to make Austria's financial administration even more effective.

User Tracking and Website Analytics

Many websites use tracking technologies to track users and show them personalized advertisement. The Austrian Data Protection Authority (DSB), after declaring the use of Google Analytics illegal,[516] held a similar decision on the Facebook Login and Meta Pixel tools provided by Meta.[517] If these tools are used, data is inevitably transferred to the United States, where the data is at risk of intelligence surveillance. Such transfer would contravene the Court of Justice of the European Union's (CJEU) 2020 *Schrems II* decision as the transferred data would not grant a level of protection equivalent to that guaranteed by the General Data Protection Regulation (GDPR).[518] According to the European Center for Digital Rights (noyb), "there is no information if a penalty was issued or if the DSB is planning to also issue a penalty. The GDPR foresees penalties of up to €20 million or 4% of the global turnover in such cases, but data protection authorities seem unwilling to issue fines, despite controllers ignoring two CJEU rulings for more than two years."[519]

Environmental Impact of AI

Austria's 2024 AI Implementation Plan identifies Climate Neutrality and Sustainability as a Key Area from and through AI.[520] The Environmental Impact Assessment Directive requires environmental impact assessments for 89 types of projects, including industrial plants, expected to have "substantial adverse impact on the environment." However, the directive contains threshold values and criteria that could allow some AI systems or data centers to avoid assessment.[521]

---

[516] GDPRhub, *DSB (Austria) – 2021-0.586.257 (D155.027)* (Dec. 12, 2021), https://gdprhub.eu/index.php?title=DSB_(Austria)_-_2021-0.586.257_(D155.027)

[517] GDPRhub, *DSB (Austria) – 2022-0.726.643* (Mar. 3, 2023), https://gdprhub.eu/index.php?title=DSB_(Austria)_-_2022-0.726.643

[518] European Parliament, *The CJEU Judgment in the Schrems II Case* (2020), https://www.europarl.europa.eu/RegData/etudes/ATAG/2020/652073/EPRS_ATA(2020)652073_EN.pdf

[519] NOYB, *Austrian DSB: Meta Tracking Tools Illegal* (Mar. 16, 2023), https://noyb.eu/en/austrian-dsb-meta-tracking-tools-illegal

[520] Federal Ministry for Innovation, Mobility and Infrastructure (BMIMI), *Implementation Plan 2024 [Umsetzungsplan 2024]* (Aug. 2024), https://www.bmimi.gv.at/themen/innovation/publikationen/ikt/ai/Umsetzungsplan-2024.html

[521] Business Service Portal, *Environmental Impact Assessment* (Sept. 3, 2025), https://www.usp.gv.at/themen/betrieb-und-umwelt/betriebliches-standortmanagement/umweltvertraeglichkeitspruefung.html

Austria joined over 100 countries to discuss the link between AI and energy at the AI Action Summit in Paris. The discussions aimed to share knowledge, support investments in sustainable AI systems, promote global dialogue on AI and the environment, and welcome the creation of an observatory on AI's energy impact with the International Energy Agency. Austria signed the Statement on Inclusive and Sustainable Artificial Intelligence for People and the Planet, along with 62 other countries.[522]

Lethal Autonomous Weapons

Austria supports developing a legally binding instrument that would ban autonomous weapons and systems that are not meaningfully controlled by humans.[523] At the virtual conference, Safeguarding Human Control over Autonomous Weapon Systems in 2021, the Austrian Federal Minister for European and International Affairs of Austria situated the country as the vanguard of many disarmaments, non-proliferation, and arms control issues. [524] They also talked about the challenges of AI and questioned algorithms that make death or life decisions based on ethics, morality, and law and called for a legal norm in the form of a treaty to ensure human control.[525]

Austria has continued this stance through efforts at the UN and REAIM conferences.[526] The country emphasized the need for a legally binding instrument to regulate autonomous weapons and highlighted both humanitarian, environmental and security-risks in remarks during the informal consultations on lethal autonomous weapons systems (LAWS) at the UN General Assembly.[527] Austria's submission to the Secretary-General Report in response to Resolution 79/239 emphasized the interface of AI and autonomous weapons and the need to link military-AI governance

---

[522] Élysée, *Statement on Inclusive and Sustainable Artificial Intelligence for People and the Planet* (Feb. 11, 2025), https://www.elysee.fr/en/emmanuel-macron/2025/02/11/statement-on-inclusive-and-sustainable-artificial-intelligence-for-people-and-the-planet

[523] DW Akademie, *Austria Wants Ethical Rules on Battlefield Killer Robots* (Nov. 15, 2020), https://www.dw.com/en/austria-wants-ethical-rules-on-battlefield-killer-robots/a-55610965

[524] Austrian Federal Ministry for European and International Affairs, *Autonomous Weapons Systems (AWS)*, https://www.bmeia.gv.at/en/european-foreign-policy/disarmament/conventional-arms/autonomous-weapons-systems

[525] EventMaker, *Safeguarding Human Control over Autonomous Weapons Systems, Speakers: High-Level Opening Panel* (Sept. 15, 2021), https://eventmaker.at/bmeia/laws_conference_2021/speakers.html

[526] For example, see Government of Netherlands, *Call to Action on Responsible Use of AI in the Military Domain* (Feb. 16, 2023), https://www.government.nl/documents/publications/2023/02/16/reaim-2023-call-to-action; US Department of State, *Political Declaration on Responsible Military Use of Artificial and Autonomy, Endorsing States* (Nov. 27, 2024), https://www.state.gov/political-declaration-on-responsible-military-use-of-artificial-intelligence-and-autonomy/

[527] Republic of Austria, *Remarks by the Republic of Austria*, Open Informal Consultations held in accordance with General Assembly Resolution 79/62, New York (12–13 May 2025), https://unodaweb-meetings.unoda.org/public/2025-05/LAWS%20informal%20consultations%20-%20Humanitarian%20considerations%20-%20Austria.pdf

with international humanitarian law.[528] Prior to submitting the report, Austria hosted the first Vienna Conference on Autonomous Weapons Systems Humanity at the Crossroads: Autonomous Weapons Systems and the Challenge of Regulation,[529] where more than 1000 participants from 140 countries discussed international regulation of autonomous weapons and the challenges of a legally binding document.

AI Literacy

Austria's Artificial Intelligence Strategy, AIM at 2030, recognizes that AI will fundamentally change teaching and learning processes. As a result, the Federal Government aims to establish Digital Basic Education as a compulsory subject for children and adolescents. Federal Ministry for Climate Action, Environment, Energy, Mobility, Innovation and Technology (BMK) posits this education will be essential for successful digital transformation, AI topics and related ethical questions can be addressed as pedagogical concepts.[530] The BMK AI Implementation Plan 2024 aims to ensure 80 percent of the population develops digital skills by 2030, an increase from the 63 percent assessed to possess them in 2023.[531] The Implementation Plan identifies increasing AI skills among teachers, students, and researchers as essential to minimize potential risks associated with AI use.[532] The Digital Skills Initiative (DKO) established in 2022 offered a strategy for realizing this goal and asserting Austria's leadership in digital competency with particular attention to the "training of educators, leveraging and supporting best practices, showcasing the public sector as a role model, addressing the IT skills shortage, offerings in the field of AI literacy."[533]

---

[528] UN General Assembly, *UNGA Resolution 79/239, Submission by Austria* (2025), https://docs-library.unoda.org/General_Assembly_First_Committee_-Eightieth_session_%282025%29/79-239-Austria-en.pdf

[529] Federal Ministry Republic of Austria Europe and International Affairs, *2024 Vienna Conference on Autonomous Weapons Systems, Chair's Summary* (Apr. 30, 2024), https://www.bmeia.gv.at/fileadmin/user_upload/Zentrale/Aussenpolitik/Abruestung/AWS_2024/Chair_s_Summary.pdf

[530] Federal Ministry for Climate Action, Environment, Energy, Mobility, Innovation and Technology (BMK), *Artificial Intelligence Mission Austria 2030 (AIM AT 2030)*, p. 37 (2021), https://digital-skills-jobs.europa.eu/sites/default/files/202310/Artificial%20Intelligence%20Mission%20Austria%202030_AIM_AT_2030.pdf

[531] Federal Ministry for Innovation, Mobility and Infrastructure (BMIMI), *Implementation Plan 2024 [Umsetzungsplan 2024]*, p. 40 (Aug. 2024), https://www.bmimi.gv.at/themen/innovation/publikationen/ikt/ai/Umsetzungsplan-2024.html

[532] Ibid, p. 48

[533] European Union, *Austria—Digital Skills Initiative* (Aug. 6, 2025), https://digital-skills-jobs.europa.eu/en/about/national-coalitions/austria-digital-skills-initiative

Human Rights

The 2025 Freedom House report scores Austria highly for political rights and civil liberties with (93/100) and designates the country as "Free."[534] Austria was the 70th country to join the United Nations and is party to the most important international legal instruments for the protection and defense of human rights.[535]

With regards to AI policy, as part of the objectives contained in its AI Strategy, Austria states that it will deploy AI responsibly targeting the common good relying on fundamental human rights,[536] aligned with the Council of Europe 2020 Recommendation to member states and the international treaty.[537] Austria reiterated the need that "All uses of AI must remain firmly anchored in international humanitarian and human rights law. […] AI technology should […] be aimed at promoting human rights" at the UN Security Council open debate on AI and international peace and security. [538]

OECD AI Principles

Austria has endorsed the OECD and G20 AI Principles.[539] Regarding implementation, the OECD notes that Austria is actively involved in relevant international organizations, the EU, and other processes and its AI strategy specifically addresses "human-centered values and fairness, robustness, security and safety, inclusive growth, sustainable development and well-being, investing in AI R&D and providing an enabling policy environment for AI."[540]

Council of Europe AI Treaty

Austria contributed as a Council of Europe and EU member state in the negotiations of the Council of Europe Framework Convention on AI, Human Rights,

---

[534] Freedom House, *Freedom in the World 2025: Austria* (2025), https://freedomhouse.org/country/austria/freedom-world/2025

[535] United Nations, *The United Nations in Vienna,* https://www.unvienna.org/

[536] *AIM AT 2030*

[537] Council of Europe Committee of Ministers, *Recommendation CM/Rec(2020)1 of the Committee of Ministers to Member States on the Human Rights Impacts of Algorithmic Systems* (Apr. 8, 2020), https://search.coe.int/cm/pages/result_details.aspx?objectid=09000016809e1154

[538] Permanent Mission of Austria to the United Nations in New York, *Security Council Open Debate—Maintenance of International Peace and Security: Artificial Intelligence and International Peace and Security* (Sept. 25 2025), https://www.bmeia.gv.at/oev-new-york/news/statements-and-speeches/2025/09/security-council-open-debate-maintenance-of-international-peace-and-security-artificial-intelligence-and-international-peace-and-security

[539] OECD Legal Instruments, *Recommendation of the Council on Artificial Intelligence*, Adherents (May 3, 2024), https://legalinstruments.oecd.org/en/instruments/oecd-legal-0449#adherents

[540] OECD AI Observatory, *AI Mission Austria 2030*, https://oecd.ai/en/dashboards/policy-initiatives/http:%2F%2Faipo.oecd.org%2F2021-data-policyInitiatives-24233

Democracy and the Rule of Law. Austria endorsed the international treaty as part of the European Union.[541]

UNESCO Recommendation on AI Ethics

Austria endorsed the UNESCO Recommendation on AI, the first ever global agreement on the ethics of AI.[542] The Austrian Commission for UNESCO created an Advisory Board on the Ethics of Artificial Intelligence to implement the UNESCO Recommendation on the Ethics of Artificial Intelligence. The Advisory Board aims to exchange information on developments related to AI and the UNESCO Recommendation, advise the National Contact Point at the Austrian Commission for UNESCO, and support measures to raise awareness and initiate a dialogue in society on the ethical implications of AI.[543]

Austria's TU Wien Informatics program became the first to hold a UNESCO Chair on Digital Humanism in 2023. The UNESCO Chair was part of a collaboration among the Federal Ministry of Education, Science and Research; Federal Ministry of Climate Action, Environment, Energy, Mobility, Innovation and Technology; Federal Ministry of European and International Affairs; and the City of Vienna. Following UNESCO's AI ethics recommendations, the Chair seeks to lay scientific groundwork for meaningful regulations and foster an interdisciplinary approach to educating future IT professionals.

Austria has taken steps to implement the UNESCO Recommendation on the Ethics of Artificial Intelligence through initiatives tied to the 2021 national AI strategy and digital decade. The country's programs for digital skills development, environmental protection, and creation of an AI oversight body demonstrate this progress. Austria has not initiated the UNESCO Readiness Assessment Methodology (RAM).[544]

Evaluation

The Austrian AI strategy released in late 2021 follows the larger goals of the EU strategy, emphasizing both excellence and the protection of fundamental rights. Austria has emphasized public participation in the development of the national AI strategy and receives expert advice from the Austrian Council on Robotics and Artificial Intelligence, which has stressed the importance of human-centric AI.

---

541 Council of Europe Portal, *Chart of Signatures and Ratifications of Treaty 225* (Jan. 22, 2025), https://www.coe.int/en/web/conventions/full-list?module=signatures-by-treaty&treatynum=225

542 UNESCO, *UNESCO Member States Adopt the First Ever Global Agreement on the Ethics of Artificial Intelligence* (Nov. 2021), https://www.unesco.org/en/articles/unesco-member-states-adopt-first-ever-global-agreement-ethics-artificial-intelligence

543 Austrian Commission for UNESCO, *New Advisory Board on the Ethics of Artificial Intelligence at the Austrian UNESCO Commission* (Jul. 6, 2023), https://www.unesco.at/presse/artikel/article/neuer-fachbeirat-ethik-der-ki

544 UNESCO Global AI Ethics and Governance Observatory, *Global Hub* (Oct. 2025), https://www.unesco.org/ethics-ai/en/global-hub

Austria ranks highly for traditional human rights protection and is active at the OECD, although it is not a member of the Global Partnership on AI. Austria ratified the Modernized Convention 108 of the Council of Europe, which includes an important provision on algorithmic transparency. This should avoid any doubts in the future regarding the scope of application of the right to algorithmic transparency. With the adoption of the EU AI Act, Austria shall establish a national supervisory mechanism which, it is to be hoped, will be an independent and will take the protection of human rights seriously. Concerns persist regarding the use of AI techniques for facial surveillance and predictive policing.

Azerbaijan

In 2025, Azerbaijan approved the Artificial Intelligence Strategy 2025–2028 and developed technical standards related to AI risk assessments and model training.

National AI Strategy

In recent years, Azerbaijan has taken an active interest in Artificial Intelligence and new technologies to reform the country's economy and extend internal security and policing.[545] The Ministry of Economy of the Republic of Azerbaijan announced in February 2022 that the country would establish an AI Strategy.[546] In March 2024, Acting Chairperson of the Innovation and Digital Development Agency of the Ministry of Digital Development and Transport Inara Valiyeva declared that a roadmap for the national AI strategy "is now at the approval stage."[547] The Minister of Digital Development and Transport announced the draft Artificial Intelligence Strategy 2025–2028 for the development of artificial intelligence in October 2024.[548] The minister's announcement focused on investment for startups and research to "establish a platform for the exchange of ideas and experience among all participants in the innovation ecosystem." The minister emphasized the plan to integrate sustainable develop principles in the agenda but did not mention ethical or human rights principles such as transparency or fairness.

The Artificial Intelligence Strategy 2025–2028 was approved by Presidential Decree in March 2025.[549] The Strategy outlines a national roadmap for AI development, focusing on innovation, public services, and capacity building. The AI strategy includes key objectives on developing a skilled workforce; establishing data governance systems to facilitate AI development while enhancing data quality; and ensuring transparency, mitigating bias, and enhancing cybersecurity through regulatory frameworks.

---

[545] Vahid Aliyev, *Azerbaijan's Newfound Orientation towards Artificial Intelligence and Robots* (May 20, 2020), https://www.researchgate.net/profile/Vahid-Aliyev/publication/341598034_Azerbaijan's_newfound_orientation_towards_Artificial_Intelligence_and_Robots/links/5ec9301092851c11a8817e42/Azerbaijans-newfound-orientation-towards-Artificial-Intelligence-and-Robots.pdf

[546] Azernews, *Azerbaijan to Develop National Artificial Intelligence Strategy* (Feb. 11, 2022), https://www.azernews.az/nation/189013.html

[547] Trend News Agency, *Azerbaijan's National AI Strategy Nears Approval* (Mar. 19, 2024), https://en.trend.az/business/it/3876181.html

[548] Nazrin Abdul, Azernews, *Azerbaijan Announces Draft Strategy for Artificial Intelligence Development* (Oct. 10, 2024), https://www.azernews.az/business/232395.html

[549] Innovation and Digital Development Agency, "*Artificial Intelligence Strategy" Approved—What to Expect?* (Mar. 19, 2025), https://idda.az/en/news/artificial-intelligence-strategy-approved-what-to-expect

As part of the implementation of the Action Plan for the Artificial Intelligence Strategy of the Republic of Azerbaijan for 2025–2028,[550] the Azerbaijan Standardization Institute adopted several national AI standards in 2025 to establish common rules for the functioning of AI systems across various platforms and industries and to ensure interoperability. The standards address risk management, using machine learning, assessing the robustness of neural networks, and managing systems more broadly.

The Fourth Industrial Revolution Analysis and Coordination Centre (4SIM), in collaboration with the World Economic Forum Artificial Intelligence and Machine Learning Platform, played a key role in preparing the roadmap for shaping Azerbaijan's National Strategy on Artificial Intelligence.[551] In parallel, the Azerbaijan is also preparing a digital development strategy,[552] including a data management strategy,[553] with the support of the World Bank.

In 2016, Azerbaijan approved the Strategic Roadmap for Development of Telecommunications and Information Technologies in the Azerbaijan Republic (Roadmap).[554] The Roadmap focuses on Information Communication Technology (ICT) sector and sets three main strategic targets: 1) Improve governance structures and strengthen ICT; 2) Increase productivity and operational efficiency of the business environment; 3) Digitize government and social environment.[555]

---

[550] Azerbaijan Standardization Institute, *New State Standards in the Field of Artificial Intelligence Have Been Adopted in the Country [Ölkədə Süni İntellekt sahəsində yeni dövlət standartları qəbul edilib]* (Jun. 5, 2025), https://azstand.gov.az/az/xeberler/olkede-suni-intellekt-sahesinde-yeni-dovlet-standartlari-qebul-edilib; Azerbaijan Standardization Institute, *New State Standard in the Field of Artificial Intelligence Has Been Adopted in the Country (Ölkədə Süni İntellekt sahəsində yeni dövlət standartı qəbul edilib)* (Jul. 28, 2025), https://azstand.gov.az/az/xeberler/olkede-suni-intellekt-sahesinde-yeni-dovlet-standarti-qebul-edilib

[551] Ministry of Economy of the Republic of Azerbeijan, *A Roadmap for the Development of a National Strategy on Artificial Intelligence Has Been Presented* (Jan. 26, 2023), https://economy.gov.az/az/post/1131/suni-intellekt-uzre-milli-strategiyanin-hazirlanmasi-%20ucun-yol-xeritesi-teqdim-olunub
Qabil Asirov, *Roadmap for Azerbaijan's National AI Strategy Has Been Prepared*, Azernews (Apr. 13, 2023), https://www.azernews.az/nation/208630.html

[552] E-Government, *The Document on Digital Development Strategy of Azerbaijan Republic Was Prepared on the Basis of Initial Modules by World Bank Group* (Jun. 6, 2023), https://www.e-gov.az/en/news/read/881

[553] Report News Agency, *Azerbaijan, World Bank Developing Strategy* (Nov. 13, 2023), https://report.az/en/infrastructure/azerbaijan-world-bank-developing-strategy/

[554] President of the Republic of Azerbaijan, *Strategic Roadmap for Development of Telecommunications and Information Technologies in Azerbaijan Republic* (Dec. 6, 2016), https://monitoring.az/assets/upload/files/6683729684f8895c1668803607932190.pdf

[555] Asian Development Bank, *Country Diagnostics Azerbaijan: Country Digital Development Overview*, p. 2 (Jan. 2019), https://www.adb.org/sites/default/files/institutional-document/484586/aze-digital-development-overview.pdf

In February 2021, Azerbaijan adopted its 2030 Vision: National Priorities on Socio-Economic Development,[556] which highlighted the importance of technological transformation.

Both the Ministry of Economy and the Ministry of Digital Development and Transport have been active in areas related to AI adoption. The Ministry of Economy manages the Center for Analysis and Coordination of the Fourth Industrial Revolution established by Presidential Decree of January 6, 2021. The mandate of the center is to respond to global challenges and trends including artificial intelligence.[557] In October 2021, another Presidential Decree established the Innovation and Digital Development Agency under the Ministry of Digital Development and Transport. This decision aimed to "improve governance in the field of digitalization, innovation, high technologies and communications in the Republic of Azerbaijan."[558]

Public Participation

Azerbaijan has a law that promotes public participation in the formulation and implementation of government policies.[559] Membership of the public council is open to everyone and elections are held to elect members of the public council.[560] As a member of the Council of Europe (CoE), the country is committed to implement the Council of Europe (CoE) National Action Plan for the Promotion of Open Government 2020–2022, through the adoption and implementation of an expanded CoE Action Plan for Azerbaijan 2022–2025, which among several goals, is intended to enhance digitalization, improve civilian oversight, expand public participation, and prevent corruption.[561]

Azerbaijan is part of the Enlarged Partial Agreement of the North-South Centre of the Council of Europe (NSC) to encourage a bottom-up dialogue between

---

[556] President of Azerbaijan, *Order of the President of the Republic of Azerbaijan on approval of Azerbaijan 2030: National Priorities for Socio-Economic Development* (Feb. 2, 2021), https://president.az/en/articles/view/50474

[557] Ministry of Economy of the Republic of Azerbaijan, *Fourth Industrial Revolution* (2026), https://www.economy.gov.az/en/page/dorduncu-senaye-inqilabi

[558] Ministry of Digital Development and Transport of the Republic of Azerbaijan, *Innovation and Digital Development Agency Public Legal Entity* (2021), https://mincom.gov.az/en/ministry/structure/innovation-and-digital-development-agency-public-legal-entity

[559] President of the Republic of Azerbaijan, *Law of the Republic of Azerbaijan on Public Participation* (Nov. 22, 2013), https://www.icnl.org/wp-content/uploads/Azerbaijan_Azerparticipate.pdf

[560] Ministry of Digital Development and Transport of the Republic of Azerbaijan, *Elections to Public Council Under Ministry of Digital Development and Transport Held* (Oct. 21, 2024), https://mincom.gov.az/en/media-en/news/elections-to-public-council-under-ministry-of- digital-development-and-transport-held

[561] Council of Europe, *Action Plan for Azerbaijan 2022–2025* (Feb. 1, 2022), https://rm.coe.int/action-plan-azerbaijan-2022-2025-eng/1680a59aa3

civil society and other democratic governance actors around four goals: global development education, youth co-operation, women's empowerment, and migration.

Nevertheless, no record of public consultations in connection with developing AI initiatives, including the AI Strategy, which outlines broad stakeholder engagement, has been found.

Data Protection

Data Protection is enshrined in the Constitution of the Republic of Azerbaijan. Article 32, paragraph VIII establishes the right to inviolability of private life, and states that the "scope of the personal information, as well as the conditions of their processing, collection, passing, use and protection is prescribed by law.[562]

The country's data protection laws currently encompass three bodies of legislation:[563] 1) the Law on State Secrets,[564] 2) the 1998 Law on Data, Data Processing and Data Protection, and 3) the 2010 Law on Personal Data.[565]

The Personal Data Law provides for data subjects' right to be informed, right to access, right to rectification, right to erasure, right to object opt-out, and right not to be subject to automated decision-making (unless this is required by law). The Law regulates the collection, processing and protection of personal data, in the public and private sectors. The Law also covers cross-border transfer of personal data and the rights and obligations of public bodies and local authorities, individuals, and legal entities operating in this field.

The Ministry of Digital Development and Transport is the authority tasked with the implementation of this Law.[566] A 2018 Decree of the President of the Republic of Azerbaijan authorized the ministry, formerly known as the Ministry of Transport, Communications, and High Technologies,[567] to exercise the authority to avoid infringements of the provision of the law, ensure information security, verify compliance in collection, processing and protection of personal data, and keep registry on information resources related to personal data. The State Security Service, Ministry

---

[562] President of the Republic of Azerbaijan Ilham Aliyev, *Constitution of the Republic of Azerbaijan* (Nov. 27, 1995), https://president.az/en/pages/view/azerbaijan/constitution

[563] Council of Europe. *Action Plan for Azerbaijan 2022-2025* (Feb 1, 2022), https://rm.coe.int/action-plan-azerbaijan-2022-2025-eng/1680a59aa3

[564] Republic of Azerbaijan, *Law of 7 September 2004 No. 733-IIQ: Azərbaycan Respublikasının Ədliyyə Nazirliyi Hüquqi aktların vahid elektron bazası, Dövlət sirri haqqında Azərbaycan Respublikasının Qanunu* (Sept. 7, 2004), https://e-qanun.az/framework/5526; CIS Legislation, *Law of the Azerbaijan Republic of September 7, 2004 No. 733–IIQ: About the State Secret*, unofficial AI translation (Amended Dec. 19, 2023), https://cis-legislation.com/document.fwx?rgn=7434

[565] Milli Majlis [National Assembly] of the Republic of the Azerbaijan, *Law No 998-IIIQ:Fərdi* məlumatlar haqqında (May 11, 2010), https://meclis.gov.az/news-qanun.php?id=1201&lang=az

[566] President of the Republic of Azerbaijan Ilham Aliyev, *Decree No.1464 of the President of the Republic of Azerbaijan* (Oct. 11, 2021), https://mincom.gov.az/storage/pages/1015/34a816171817c6fe6feadf9e170def84.pdf

[567] E-gov, *Regulations of the Ministry of Transport, Communications and High Technologies of the Republic of Azerbaijan* (Jan. 12, 2018), https://www.e-gov.az/en/news/read/630

of Internal Affairs, Ministry of Justice, and Special State Protection Service are also involved in the implementation of the data protection legal regime.[568]

Azerbaijan is party to Convention 108 or Convention for the Protection of Individuals with regard to Automatic Processing of Personal Data.[569] Convention 108 reaffirms the fundamental values of respect for privacy. Article 5 defines the need for personal data to be "obtained and processed fairly and lawfully." Convention 108 requires Azerbaijan to establish an independent data protection authority.

The Council of Europe (CoE) Action Plan for Azerbaijan 2022–2025 has a specific chapter on data protection and artificial intelligence. Azerbaijan is committed to:

- Move closer to signing and ratifying the Convention for the Protection of Individuals with regard to Automatic Processing of Personal Data (Convention 108+)
- Adopt legislation compliant with the Convention for the Protection of Individuals with regard to Automatic Processing of Personal Data (ETS No. 108)
- Establish a dedicated independent authority for personal data protection
- Disseminate and implement Council of Europe guidelines on the use of artificial intelligence
- Conduct an awareness-raising campaign on the importance of data protection for the benefit of the local population

The Chief of Special Communication and Information Security State Service of Azerbaijan, Tural Mammadov, said during the Cyber-Secure Economy: Reforms, Innovative Approaches and Solutions event in 2022 that a draft law on ensuring personal data protection has been prepared in Azerbaijan based on European standards, the GDPR in particular.[570]

The Electronic Security Service under Azerbaijan's Ministry of Digital Development and Transport has drafted a new version of the law on the Protection of Personal Data. Reportedly, the amendments seek to "balance security and [...] ensure the reliable protection of citizens' rights."[571]

Azerbaijan is not an accredited member of the Global Privacy Assembly and has not sponsored the 2018 GPA Resolution on AI and Ethics,[572] the 2020 Resolution

---

[568] Council of Europe, *Data Protection Systems in the Republic of Azerbaijan* (Dec. 2016), https://rm.coe.int/16806ee927

[569] Council of Europe Portal, *Data Protection: Azerbaijan* (May 3, 2010), https://www.coe.int/en/web/data-protection/azerbaijan

[570] Azernews, *Azerbaijan Applying Effective Measures to Ensure Population's Protection from Cyber Attacks* (Sept. 21, 2022), https://www.azernews.az/business/199707.html

[571] Report News Agency, *Azerbaijan updates law on protection of personal data* (Oct. 9, 2025), https://report.az/en/ict/azerbaijan-updates-law-on-protection-of-personal-data

[572] International Conference of Data Protection & Privacy Commissioners, *Declaration on Ethics and Data Protection in Artificial Intelligence* (Oct. 23, 2018), https://globalprivacyassembly.org/wp-content/uploads/2018/10/20180922_ICDPPC-40th_AI-Declaration_ADOPTED.pdf

on AI and Accountability,[573] the 2022 Resolution on Facial Recognition Technology, or the 2023 GPA Resolution on Generative AI Systems.[574]

Algorithmic Transparency

Azerbaijan has neither signed nor ratified the protocol modernizing Convention 108, which provides for algorithmic transparency.[575] National legislation does not provide for algorithmic transparency.

AI Readiness and Digitization

The Azerbaijan government has taken concrete steps to promote technology innovation. In June 2019, Presidential Decree 718 created a Centralized Government Cloud[576] for the effective organization of the formation, storage, maintenance and integration of state information systems and reserves. The Decree also supports innovative solutions based on Artificial Intelligence and robotics and committed Azerbaijan to use "machine learning" (M2M), "artificial intelligence" (EU), "big data," and "internet of things" (IoT) as bases for providing public services in the country.

The e-Gov Development Center of the State Agency for Public Service and Social Innovations organized the International Conference on Artificial Intelligence in Digital Governance. The first panel discussions were dedicated to the transformation of healthcare with AI, the second panel focused on AI and society, and the last session addressed the future of artificial intelligence and how it will change human life.[577]

The Center for Analysis and Coordination of the Fourth Industrial Revolution (C4IR) held several events dedicated to AI and machine learning. The events focused on exchanging views and experience on the development of AI and machine learning

---

[573] Global Privacy Assembly, *Adopted Resolution on Accountability in the Development and Use of Artificial Intelligence* (Oct 2020), https://globalprivacyassembly.org/wp-content/uploads/2020/11/GPA-Resolution-on-Accountability-in-the-Development-and-Use-of-AI-EN.pdf

[574] Global Privacy Assembly, *Resolution on Generative Artificial Intelligence Systems* (Oct. 2023), https://globalprivacyassembly.org/wp-content/uploads/2023/10/5.-Resolution-on-Generative-AI-Systems-101023.pdf

[575] Council of Europe Treaty Office, *Charter of Signatures and Ratifications of Treaty 223* (Jan. 26, 2025), https://www.coe.int/en/web/conventions/full-list?module=signatures-by-treaty&treatynum=223

[576] President of the Republic of Azerbaijan, *Strategic Roadmap for Development of Telecommunications and Information Technologies in Azerbaijan Republic* (Dec. 6, 2016), https://monitoring.az/assets/upload/files/6683729684f8895c1668803607932190.pdf

[577] State Agency for Public Service and Social Innovations under the President of the Republic of Azerbaijan, *The International Conference on Artificial Intelligence in Digital Governance Has Ended* (Apr. 28, 2021), https://www.digital.gov.az/en/media/press/the-international-conference-on-artificial-intelligence-in-digital-governance-has-ended

with the involvement of government agencies, private companies, scientific and educational institutions, and civil society.[578]

In October 2022, a mission of the Center for Analysis and Coordinator of the Fourth Industrial Revolution (C4IR) participated in a "Global Dialogue: The role of artificial intelligence in the new global development—innovation and inclusiveness," presenting the measures taken by the government on the digital economy in the country.[579]

AI and the Judiciary

The Head of the Innovation and Digital Development Agency of Azerbaijan, Inara Valiyeva, announced the introduction of the Unified Court System based on AI for 2023. The Unified Court System is currently operating in test mode in coordination with the Supreme Court with about 50,000 documents used in the creation of the system.[580] Azerbaijan plans to focus on improving the quality of decision-making and management by using Big Data and AI technologies.[581] The project, conducted with the cooperation of the Council of Europe for the Efficiency of Justice (CEPEJ) and the World Bank, includes an automated random allocation of cases based on criteria regarding subject matter, weight of the case, and the court's or judge's caseload; a "fast-track procedure for uncontested small clams; an "e-Order" automated system"; and a unified judicial portal.[582]

The Ministry of Digital Development and Transport and the Ministry of Justice jointly launched the E-qanun.ai, an AI-driven search platform for Azerbaijan's legal documents and legislative database.[583] The platform was developed as part of

---

[578] Ministry of Economy of Azerbaijan, *Another Artificial Intelligence Event Organized* (Nov. 27, 2021), https://old.economy.gov.az/en/article/suni-intellekt-uzre-novbeti-tedbir-teshkil-edilib/32162; The Ministry of Economy of Azerbaijan , *Sessions Held within "Trends of the Fourth Industrial Revolution" Event* (Feb. 24, 2022), https://www.economy.gov.az/en/article/dorduncu-senaye-i-nqilabinda-trendler-movzusunda-tedbir-cherchivesinde-sessiyalar-kechirilib/32341.

[579] Azertag, *Azerbaijan's Center for Analysis and Coordination of Fourth Industrial Revolution Attends International Event in Geneva* (Oct. 12, 2022), https://azertag.az/en/xeber/Azerbaijans_Center_for_Analysis_and_Coordination_of_Fourth_Industrial_Revolution_attends_international_event_in_Geneva-2331065

[580] Report News Agency, *Azerbaijan Creates Unified Judicial System Based on Artificial Intelligence* (Dec. 16, 2022), https://report.az/en/ict/azerbaijan-creates-unified-judicial-system-based-on-artificial-intelligence/

[581] Ibid

[582] Judge Dr. Ramin Gurbanov, Project Coordinator, President of the European Commission for the Efficiency of Justice (CEPEJ), *Azerbaijan: e-Courts and the Joint Achievements in Digitalization of Justice* (Jun. 2021), https://thedocs.worldbank.org/en/doc/3ecf7262788a3ec69c8a45bbd3342a28-0080022021/related/29-06-21-Presentation-e-Court-WB-Dr-Ramin-Gurbanov.pdf; European Commission for the Efficiency of Justice, *Strengthening the Efficiency and Quality of the Judicial System in Azerbaijan* (Mar. 2019–Feb. 2023), https://www.coe.int/en/web/cepej/strengthening-the-efficiency-and-quality-of-the-judicial-system-in-azerbaijan.

[583] Ministry of Digital Development and Transport, *E-qanun.ai Platform Officially Presented* (Sept. 25, 2025), https://www.mincom.gov.az/en/media-en/news/e-qanunai-platform-officially-presented

the implementation of the national AI Strategy to support natural language processing priorities outlined in subsection 8.2.2 of the Strategy.

Biometrics

The Law on Biometric Data adopted in 2008 determines the formation and requirements for biometric information resources, the organization and purpose of the biometric identification system, and the application of biometric technologies. The law also regulates the relations arising in this area. Azerbaijan began issuing biometric-based electronic identity cards in September 2018. The cards contain information on place of residence and marital status as well as photos and fingerprints if the citizen is over age 15.[584]

AI Surveillance

A Freedom House 2022 report indicated that "state surveillance is pervasive, though the exact extent to which security agencies monitor ICT activity or track users remains unclear."[585] Usage of surveillance tools by the Azerbaijani Government was also reported by Amnesty International.[586] The OCCRP (Organized Crime and Corruption Reporting Project) named Azerbaijan's surveillance a Digital Autocracy.[587] A report by Open Tech found that Azerbaijan is one of several countries where censorship, facial recognition, and access to encrypted communication have been prevalent as forms of information control.[588]

The Law on Operative-Search Activity authorizes agencies to conduct surveillance without a court order in cases where it is regarded as necessary to prevent serious crimes against individuals or especially dangerous crimes against the state. The vaguely written provision leaves the law open to abuse. It has long been believed that the State Security Service and the Ministry of Internal Affairs monitor the communications of individuals, especially foreigners, prominent political activists, and business figures.[589]

---

[584] Azerbaijan Milli Majlis [National Assembly], *Law No. 651–IIIQ: Biometric Data Law (Ədliyyə Nazirliyi Hüquqi aktların vahid elektron bazası, Biometrik informasiya haqqında)* (Jun. 13, 2008), https://e-qanun.az/framework/15144

[585] Freedom House, *Freedom of the Net 2022* (2022), https://freedomhouse.org/country/azerbaijan/freedom-net/2022

[586] Amnesty International, *Demand an End to the Targeted Surveillance of Human Rights Defenders* (Aug. 16, 2016), https://www.amnesty.org/en/petition/targeted-surveillance-human-rights-defenders/

[587] Miranda Patrucic and Kelly Bloss, *Life in Azerbaijan's Digital Autocracy: 'They Want to be in Control of Everything* (Jul. 18, 2021), https://www.occrp.org/en/the-pegasus-project/life-in-azerbaijans-digital-autocracy-they-want-to-be-in-control-of-everything

[588] Valentin Weber, *The Worldwide Web of Chinese and Russian Information Controls*, p. 20, https://public.opentech.fund/documents/English_Weber_WWW_of_Information_Controls_Final.pdf

[589] Freedom House, *Freedom on the Net 2021* (2021), https://freedomhouse.org/country/azerbaijan/freedom-net/2021

In 2015, leaked documents showed that the Azerbaijani government was a client of the Italian surveillance company Hacking Team.[590] In the previous year, Citizen Lab reported that the government was using RCS (Remote Control System) spyware sold by Hacking Team. The RCS endpoint was active in Azerbaijan between June and November 2013. Azerbaijan hit international headlines in 2013 when the results of the October presidential elections were accidentally released before voting began.[591] RCS allows anyone with access to activate a targeted device's camera and microphone and to steal videos, photos, documents, contact lists, and emails.[592]

In 2017, Amnesty International reported that Azerbaijani human rights activists, journalists, and political dissidents have been the targets of a sustained spear phishing campaign using emails and Facebook chat, apparently aimed at gaining access to their personal information and private communications.[593] In the same year, malware targeted Azerbaijani dissidents.[594]

An April 2018 report by Qurium revealed that Azerbaijan had purchased specialized security equipment, in particular Deep Packet Inspection (DPI) technology, from the Israeli company Allot Communications for some $3 million.[595]

In October 2018, Israeli newspaper Haaretz reported that Israel's Verint Systems had sold surveillance equipment and software to the Azerbaijani government. Local police later used the equipment to identify the sexual orientation of users on Facebook.[596] Haaretz reported that a few years after Verint's systems began being used Azerbaijani police arrested and tortured 45 gay men and transgender women.[597]

In July 2021, an investigative initiative led by Forbidden Stories concluded that the Pegasus software produced by the Israeli cybersurveillance company NSO

---

[590] Cora Currier, *A Detailed Look at Hacking Team's Emails About Its Respective Clients* (Jul. 7, 2015), https://theintercept.com/2015/07/07/leaked-documents-confirm-hacking-team-sells-spyware-repressive-countries/, also see Amnesty International, *New EU Dual Use Regulation Agreement "A Missed Opportunity" to Stop Exports of Surveillance Tools to Repressive Regimes* (Mar. 25, 2021), https://www.amnesty.org/en/latest/news/2021/03/new-eu-dual-use-regulation-agreement-a-missed-opportunity-to-stop-exports-of-surveillance-tools-to-repressive-regimes/

[591] Citizen Lab, *Mapping Hacking Team's "Untraceable" Spyware* (Feb. 17, 2014), https://citizenlab.ca/2014/02/mapping-hacking-teams-untraceable-spyware/

[592] Freedom House, *Freedom on the Net 2021* (2021), https://freedomhouse.org/country/azerbaijan/freedom-net/2021

[593] Amnesty International, *Azerbaijan: Activists Targeted by "Government-Sponsored" Cyberattack* (Mar. 10 2017), https://www.amnesty.org/en/latest/news/2017/03/azerbaijan-activists-targeted-by-government-sponsored-cyber-attack/

[594] Amnesty International, *False Friends: How Fake Accounts and Crude Malware Targeted Dissidents in Azerbaijan* (Mar. 10, 2017), https://www.amnesty.org/en/latest/research/2017/03/false-friends-spearphishing-of-dissidents-in-azerbaijan/

[595] Qurium, *Corruption, Censorship and a Deep Packet Inspection Vendor* (Apr. 10, 2018), https://www.qurium.org/alerts/azerbaijan/corruption_censorship_and_a_dpi_vendor/

[596] Haaretz, *Revealed: Israel's Cyber-Spy Industry Helps World Dictators Hunt Dissidents and Gays* (Oct. 20, 2018), https://www.haaretz.com/israel-news/.premium.MAGAZINE-israel-s-cyber-spy-industry-aids-dictators-hunt-dissidents-and-gays-1.6573027

[597] Ibid.

Group was used in Azerbaijan to target more than 40 Azerbaijani journalists.[598] Reporters with the Organized Crime and Corruption Reporting Project (OCCRP), which was among the groups working on the project, found some 250 potential targets in Azerbaijan, the majority of whom were "dissidents, activists, journalists, and opposition politicians." The report added that "journalists came under particular pressure, with dozens of prominent names, including OCCRP's Khadija Ismayilova, appearing on the list."[599]

The day full-scale war erupted between Azerbaijani and Armenian forces in Nagorno-Karabakh, the Ministry of Transport, Communications and High Technologies throttled mobile and fixed-line broadband internet across Azerbaijan and blocked a number of social media platforms and websites, including Facebook, WhatsApp, and Skype. The action lasted 46 days, Azerbaijan's longest internet disruption to date.[600]

Facial Recognition

A 2019 report by Qurium indicated that Azerbaijan's government may be using facial-recognition technology. Qurium identified an AzerTelecom server running the software.[601]

The Ministry of Transport, Communications and High Technologies (now Ministry of Digital Development and Transport) developed a Public-Private Partnership with SVORT and Sinam to establish AzInTelecom LLC. AzInTelecom is developing a new generation cloud digital signature called SIMA.[602] SIMA's initial purpose is to facilitate access to e-government services but will expand to the use of services through banks, mobile operators, internet providers, and household appliance stores. SIMA combines face recognition, public key, and cloud technologies and is based on biometric authentication.

Based on SIMA signatures, the Information Computing Centre is developing a payment system with face recognition technology.[603] Users scan their identity card and verify their identity via face recognition through the front camera of the mobile

---

[598] Amnesty International, *Massive Data Leak Reveals Israeli NSO Group's Spyware Used to Target Activists, Journalists, and Political Leaders Globally* (Jul. 19, 2021), https://www.amnesty.org/en/latest/news/2021/07/the-pegasus-project-2/

[599] Freedom House, *Freedom on the Net 2021* (2021), https://freedomhouse.org/country/azerbaijan/freedom-net/2021

[600] Ibid

[601] Qurium, *Find Face and Internet Blocking in Azerbaijan* (Dec. 3, 2019), https://www.qurium.org/alerts/azerbaijan/find-face-and-internet-blocking-in-azerbaijan/

[602] Ministry of Digital Development and Transport of Azerbaijan, *New Generation Digital Signature Presented in Azerbaijan – SIMA* (Feb. 11, 2022), https://mincom.gov.az/az/media/xeberler/azerbaycanda-yeni-nesil-reqemsal-imza-teqdim-edilib-sima1431

[603] Xeberler, *A Payment System Will Be Introduced in Azerbaijan* (Nov. 5, 2021), https://xeberler.az/new/details/azerbaycanda-uztanima-ile-odenis-sistemi-tetbiq-edilecek--27254.htm

devices.[604] E-signatures are regulated by the Law of Azerbaijan on Electronic Signature and Electronic Document of March 9, 2004.[605] No amendments were introduced concerning SIMA, biometric authentication, or facial recognition.

Smart Cities

As part of its plan to repopulate the Karabakh area following the war with Armenia, Azerbaijan plans to build Smart Cities/Villages in Zangilan. After the first pilot project of the Aghali village, three additional smart villages were planned for 2023.[606]

The President of Azerbaijan, Ilham Aliyev, announced in January 2021 that "settlements recently liberated from Armenian occupation will be re-established based on the concept of smart city/village."[607] The Concept of Smart Cities and Smart Villages was later approved by presidential order.[608] The Smart City/Village concept is to be implemented through working groups representing various ministries.[609] A publication of the Baku Research Institute provides insights about the Smart Cities concept as a means to promote sustainable development of territories and attract population back.[610] The World Bank published a framework and Smart Villages Readiness Index for Azerbaijan to support the country in ensuring that economic opportunities, access of service, and governance are applied in the process.[611]

---

[604] Sadraddin Aghjayev, *Azerbaijan Eyes Creating New Payment Service Using e-Signature Capabilities (Interview)* (Mar. 14, 2022), https://en.trend.az/business/3568769.html

[605] Digital Trade Hub of Azerbaijan, *Law of the Republic of Azerbaijan on Electronic Signature and Electronic Document* (Mar. 9, 2004), https://dth.az/legal-framework

[606] Asif Mehman, *Azerbaijani State to Fund Implementation of "Smart Village" Project in Aghdam* (Sept. 6, 2023), https://en.trend.az/azerbaijan/politics/3793294.html; Dilara Aslan Özer, Daily Sabah, *Azerbaijan to Build 3 More "Smart Villages" in Zangilan by End of 2023* (Sept. 22, 2022), https://www.dailysabah.com/politics/azerbaijan-to-build-3-more-smart-villages-in-zangilan-by-end-of-2023/news.

[607] Ruslan Rehimov, AA (Anadolu Agency), *Azerbaijan to Build "Smart Cities" in Liberated Regions* (Jan. 26, 2021), https://www.aa.com.tr/en/azerbaijan-front-line/azerbaijan-to-build-smart-cities-in-liberated-regions/2123643

[608] President of the Republic of Azerbaijan, *Order No. 2584 on Developing the Concept of Smart City and Smart Village* (Apr. 19, 2021), https://minenergy.gov.az/en/prezidentin-ferman-ve-serencamlari/page/3

[609] Ruslan Rehimov, AA (Anadolu Agency), *Azerbaijan to Build 'Smart Cities' in Liberated Regions* (Jan. 26, 2021), https://www.aa.com.tr/en/azerbaijan-front-line/azerbaijan-to-build-smart-cities-in-liberated-regions/2123643

[610] Baku Research Institute, *Building Smart Cities and Villages in Azerbaijan: Challenges and Opportunities* (Aug, 6, 2021), https://bakuresearchinstitute.org/en/building-smart-cities-and-villages-in-azerbaijan-challenges-and-opportunities/

[611] World Bank, *A Framework for Developing Smart Villages in Azerbaijan* (Oct. 19, 2021), https://www.worldbank.org/en/country/azerbaijan/publication/a-framework-for-developing-smart-villages-in-azerbaijan

Environmental Impact of AI

The Minister of Digital Development and Transport emphasized the country's attention the principles of sustainable development in its technology and innovation agenda in his announcement of the national strategy for AI development.[612] However, no available policies demonstrate how officials plan to implement the principles.

Lethal Autonomous Weapons

During his swearing-in ceremony in February 2024, Azerbaijan's President, Ilham Aliyev stated, "Today, in the field of industrial development, in military industrial complex and, in general, in the military field, technological development and superiority are of great importance. Everyone saw this during the Second Karabakh War and the anti-terrorist operation. […] Technological development, digitization, cybersecurity, application of artificial intelligence—all these should become part of our daily life. Government agencies, the private sector, and all other segments of society must be ready for this. […] I do not see any dangers or threats to Azerbaijan but I know why—because they know that our response will be merciless, it will be crushing, and it will be given regardless of anything. Therefore, if we do not achieve technological development, if there are just a few thousand not tens of thousands of Azerbaijanis in this business, we may be overtaken. Therefore, I think that all institutions and, at the same time, society should know and see this as a duty."[613]

Azerbaijan is not a party to the Convention on Conventional Weapons (CCW),[614] although the country participated as an observer on CCW meetings on killer robots in November 2019.[615] Azerbaijan is not among the countries that support a ban on lethal autonomous weapons.[616] At the Plenary meeting on December 22, 2023, Azerbaijan abstained from voting on General Assembly Resolution 78/241, which asks UN Secretary-General António Guterres to seek the views of countries

---

[612] Nazrin Abdul, Azernews, *Azerbaijan Announces Draft Strategy for Artificial Intelligence Development* (Oct. 10, 2024), https://www.azernews.az/business/232395.html

[613] Report News Agency, *Inauguration Ceremony of President Ilhan Aliyev Held at Milli Majlis* (Feb. 14, 2024), https://report.az/en/domestic-politics/inauguration-ceremony-of-president-ilham-aliyev-held-at-milli-majlis/

[614] United Nations Office for Disarmament Affairs, *High Contracting Parties and Signatories CCW* (Jun. 25, 2024), https://www.un.org/disarmament/the-convention-on-certain-conventional-weapons/high-contracting-parties-and-signatories-ccw/

[615] United Nations Office for Disarmament Affairs, *Final Report of the Meeting of High Contracting Parties to the Convention on Conventional Weapons*, CCW/MSP/2019/9 (Nov. 13–15, 2019), https://documents-dds-ny.un.org/doc/UNDOC/GEN/G19/343/64/PDF/G1934364.pdf?OpenElement

[616] Human Rights Watch, *Stopping Killer Robots: Country Positions on Banning Fully Autonomous Weapons and Retaining Human Control* (Aug. 10, 2020), https://www.hrw.org/report/2020/08/10/stopping-killer-robots/country-positions-banning-fully-autonomous-weapons-and

and other stakeholders on addressing the challenges of autonomous weapons systems from various perspectives and report back in the second half of 2024.[617]

Over the last decade, Azerbaijan has steadily built up its armed forces and purchased weapons from Russia, Israel, and Türkiye.[618] Azerbaijan also developed drone arsenal composed of Turkish and Israeli UAVs.[619]

In 2016, during the Four-Day War, also known as the Nagorno-Karabakh conflict in which the Azerbaijani Armed Forces faced the Armenian-backed Artsakh Defense Army, Azerbaijan used the Harop UAV to hit many targets, including artillery, air defense systems, and a busload of Armenian troops.[620] As Forbes reported,[621] the Harops were supplied by Israel. In 2017 a team from Aeronautics was in Azerbaijan to finalize a contract for Orbiter 1K kamikaze drones and were asked to attack enemy positions. Apparently when the Israeli drone operators refused, "senior representatives of the company took control and operated the craft themselves, ultimately missing their targets." Israeli authorities imposed a two-year ban on Aeronautics for this stunt. But when the ban expired in 2019, the company promptly announced a $13m deal to sell drones to Azerbaijan. In the longer term, Azerbaijan plans to produce a licensed copy of the Orbiter known as Zarba themselves.

An Azerbaijani company announced in 2018 that it was working on three different sizes of kamikaze drones,[622] including one with an 11-pound warhead able to cruise for three hours looking for targets, while another company announced manufacturing kamikaze-drone called Bat.[623]

Algorithm Watch reported that Azerbaijani forces used at least three different models of drones capable of identifying and destroying a target automatically[624] and

---

[617] Human Rights Watch, *Resounding Support for Killer Robots Treaty* (May 2, 2024), https://www.hrw.org/news/2024/05/02/resounding-support-killer-robots-treaty

[618] Human Rights Watch, *Stopping Killer Robots: Country Positions on Banning Fully Autonomous Weapons and Retaining Human Control* (Aug. 10, 2020), https://www.hrw.org/report/2020/08/10/stopping-killer-robots/country-positions-banning-fully-autonomous-weapons-and

[619] Centre for Strategic and International Studies, *The Air and Missile War in Nagorno-Karabakh: Lessons for the Future of Strike and Defence* (Dec. 8, 2020), https://www.csis.org/analysis/air-and-missile-war-nagorno-karabakh-lessons-future-strike-and-defense

[620] Forbes, *The Weird and Worrying Drone War in the Caucasus* (Jun. 22, 2020), https://www.forbes.com/sites/davidhambling/2020/06/22/the-weird-and-worrying-drone-war-in-the-caucasus/?sh=5a3148f145da

[621] Ibid

[622] Azeri Defence, *Azerbaijani Company Has Made Prototype of Three Kinds of Loitering Munitions* (Mar. 15, 2018), https://web.archive.org/web/20180528041925/http:/en.azeridefence.com/azerbaijani-company-has-made-prototype-of-three-kinds-of-loitering-munitions/

[623] Azernews, *Azerbaijan Academy of Sciences Produces Kamikaze Drone* (Apr. 26, 2018), https://www.azernews.az/nation/138175.html

[624] Algorithm Watch, *The Year Algorithms Escaped Quarantine: 2020 in Review* (Dec. 28, 2020), https://algorithmwatch.org/en/review-2020/

several autonomous weapons in the Second Nagorno-Karabakh conflict, known as the 2020 War. [625]

The 2020 War raised serious questions concerning the legality of UAVs and autonomous weapons. Ulrike Franke, autonomous weapons expert from the European Council on Foreign Relations, saw a watershed in warfare and commented: "The really important aspect of the conflict in Nagorno-Karabakh, in my view, was the use of these loitering munitions, so-called "kamikaze drones"—these pretty autonomous systems."[626] Human Rights Watch, in a statement to support the International Committee of the Red Cross initiative to ban Killer Robots, indicated: "the increased use of weapons systems with autonomy in today's armed conflicts underscores the importance of creating a new international legal standard now, before it is too late." [627]

The Republic of Nagorno-Karabakh was dissolved in 2024.[628] In 2023, with Russia distracted and bogged down by its war in Ukraine, Azerbaijan launched a blockade of Nagorno Karabakh, eventually seizing the region and causing the majority of its Armenian population to flee.[629]

In the preliminary report on the draft of the UNESCO Recommendation on the Ethics of AI, Azerbaijan was referenced as a case of the use of AI technologies in armed conflicts. The report concluded that: "from the point of view of international humanitarian law, it is highly recommended to have certain regulations concerning the use of AI in military technology, so-called lethal autonomous weapons."[630]

AI Literacy

There is growing emphasis on digital skills and AI readiness in Azerbaijan's AI strategy and educational policy. UNESCO reported that Azerbaijan's education

---

[625] Robyn Dixon, Washington Post, *Azerbaijan's Drones Owned the Battlefield in Nagorno-Karabakh — And Showed Future of Warfare* (Nov. 11, 2020), https://www.washingtonpost.com/world/europe/nagorno-karabkah-drones-azerbaijan-aremenia/2020/11/11/441bcbd2-193d-11eb-8bda-814ca56e138b_story.html

[626] Deutsche Welle, *DW Report on Cyber and Autonomous Weapons: Future Wars — And How to Prevent Them* (Jun. 7, 2021), https://www.dw.com/en/dw-report-on-cyber-and-autonomous-weapons-future-wars-and-how-to-prevent-them/a-57801575

[627] Human Rights Watch, *International Committee of the Red Cross Backs Killer Robot Ban* (May 13, 2021), https://www.hrw.org/news/2021/05/13/international-committee-red-cross-backs-killer-robot-ban

[628] CNN, *Nagorno-Karabakh to Officially Dissolve, Marking an End to Its Decades-Long Struggle* (Sept. 28, 2023), https://edition.cnn.com/2023/09/28/europe/nagorno-karabakh-officially-dissolve-intl/index.html

[629] Simon Anglim, *Azerbaijan's Victory: Initial Thoughts, Observations (and Caveats for the 'Innovative')*, Military Strategy Magazine 7, no. 3, pp. 10–17 (2021), https://www.militarystrategymagazine.com/article/azerbaijans-victory-initial-thoughts-and-observations-and-caveats-for-the-innovative/

[630] UNESCO, *Compilation of Comments Received from Member States on the First Draft of the Recommendation* (2021), https://unesdoc.unesco.org/ark:/48223/pf0000376747

authorities define digital literacy as a basic competency for all learners and data classes will be integrated throughout the educational path.[631] However, there is no documented national program for the public or for civil servants nor a publicly available AI or digital literacy curriculum or standards.

Human Rights

Azerbaijan is a Member State of the United Nations and of the Council of Europe. Azerbaijan endorsed the UN Universal Declaration of Human Rights[632] and ratified the European Convention on Human Rights. Azerbaijan also has human rights obligations at the regional[633] and universal levels.

The 2025 Freedom House report on Azerbaijan ranks the country as "Not Free" with a score of 7 out of 100.[634] Among key aspects, the report cited the concentration of power in the authoritarian regime headed by President Ilham Aliyev since 2003 and his family. According to Freedom House, "Corruption is rampant, and the formal political opposition has been weakened by years of persecution. The authorities have carried out an extensive crackdown on civil liberties in recent years, leaving little room for independent expression or activism." The fleeing of nearly all ethnic Armenians from the Nagorno-Karabakh region following Azerbaijani control in 2023 also contributed to the country's "not free" status.

Azerbaijan scores 34/100 on internet freedom.[635] Internet freedom continues to be restricted with blocked access to several social media sites and self-censorship through a media registry. "The government also launched a media registry, required by the new media law adopted in 2022, and rejected the applications of several independent news outlets to join the registry. Prosecution of activists for their online criticism of the government continued during the coverage period. Additionally, activists faced online harassment, doxing, and blackmail." [636]

The Council of Europe (CoE) Action Plan for Azerbaijan 2022–2025 includes strategic programming to align legislation, institutions, and practices further with CoE standards in human rights, the rule of law, and democracy.[637] Among the objectives of the Action Plan are: the implementation of the UN 2030 Agenda for Sustainable

---

[631] UNESCO, *Global Education Monitoring Report: Azerbaijan, Technology* (Aug. 5, 2024), https://education-profiles.org/northern-africa-and-western-asia/azerbaijan/~technology

[632] International Justice Resource Center, *Azerbaijan Factsheet* (Jul. 2019), https://ijrcenter.org/country-factsheets/country-factsheets-europe/azerbaijan-human-rights-factsheet/

[633] Council of Europe, *46 Member States* (2023), https://www.coe.int/en/web/portal/46-members-states

[634] Freedom House, *Freedom in the World 2025:Azerbaijan* (2025), https://freedomhouse.org/country/azerbaijan/freedom-world/2025

[635] Freedom House, *Freedom on the Net 2024: Azerbaijan* (2024), https://freedomhouse.org/country/azerbaijan/freedom-net/2024

[636] Freedom House, *Freedom on the Net 2023: Azerbaijan* (2023), https://freedomhouse.org/country/azerbaijan/freedom-net/2023

[637] Council of Europe, *Council of Europe Action Plan for Azerbaijan 2022–2025* (Feb. 1, 2022), https://rm.coe.int/action-plan-azerbaijan-2022-2025-eng/1680a59aa3

Development,[638] and the identification of areas for cooperation in artificial intelligence.

Azerbaijan is also part of Partnership for Good Governance, through which the European Union and the Council of Europe cooperate to strengthen governance in the Eastern Partnership region[639] and advance localized efforts to improve "stability, prosperity, and resilience." [640]

OECD / G20 AI Principles

Azerbaijan is a member of neither the OECD nor the Global Partnership on Artificial Intelligence (GPAI)[641] and has not endorsed the OECD AI Principles.[642] The OECD Development Centre has been supporting Azerbaijan in the promotion of enterprise digitalization. A 2022 study highlighted the significance of digitalization to allow Small Medium Enterprises access to strategic resources and integration into global markets.[643]

Council of Europe AI Treaty

Azerbaijan contributed as a Council of Europe Member State in the negotiations of the Council of Europe Framework Convention on AI and Human Rights, Democracy, and the Rule of Law.[644] However, Azerbaijan has not signed the treaty.[645]

---

[638] United Nations, *Sustainable Development Goals* (2023), https://www.un.org/sustainabledevelopment/

[639] Diplomatic Service of the European Union, *Eastern Partnership)* (Mar. 17, 2022), https://www.eeas.europa.eu/eeas/eastern-partnership_en

[640] European Commission for the Efficiency of Justice, *Partnership for Good Governance in Azerbaijan* (2026), https://www.coe.int/en/web/cepej/cooperation-programmes/partnership-for-good-governance-azerbaijan

[641] OECD AI Policy Observatory, *About the Global Partnership on Artificial Intelligence (GPAI)* (2026), https://oecd.ai/en/about/about-gpai

[642] OECD Legal Instruments, *Recommendation of the Council on Artificial Intelligence, Adherents* (May 3, 2024), https://legalinstruments.oecd.org/en/instruments/oecd-legal-0449#adherents

[643] OECD, *Promoting Enterprise Digitalisation in Azerbaijan* (2022), https://doi.org/10.1787/6a612a2a-en

[644] Council of Europe, *Framework Convention on Artificial Intelligence* (2026), https://www.coe.int/en/web/artificial-intelligence/the-framework-convention-on-artificial-intelligence

[645] Council of Europe Treaty Office, *Chart of Signatures and Ratifications of Treaty 225* (Mar. 12, 2026), https://www.coe.int/en/web/conventions/full-list?module=signatures-by-treaty&treatynum=225

UNESCO Recommendation on AI Ethics

Azerbaijan is a UNESCO member since 1992 and is one of the 193 countries that endorsed the UNESCO Recommendation on the Ethics of AI. [646] Azerbaijan has not initiated the Readiness Assessment Methodology (RAM),[647] a foundational tool for countries to implement the UNESCO Recommendation.

Evaluation

Azerbaijan has set an ambitious goal of embracing new technologies to propel the economic growth of the country. The digitalization process in Azerbaijan has been accelerating and has stimulated several initiatives, including the release of a National AI Strategy and AI standards supporting its implementation. The country has not yet signed the protocol modernizing Convention 108 nor created an independent agency or mechanism for data (and AI) oversight to modernize its data protection law regime. Concerns exist that, with the use of AI, Azerbaijan is turning into a "digital autocracy." The country has not endorsed any declaration against the use of LAWS and has been a laboratory for their use.

---

[646] UNESCO, *UNESCO Member States Adopt the First Ever Global Agreement on the Ethics of Artificial Intelligence* (Apr. 21, 2022), https://www.unesco.org/en/articles/unesco-member-states-adopt-first-ever-global-agreement-ethics-artificial-intelligence

[647] UNESCO Global AI Ethics and Governance Observatory, *Global Hub* (Oct. 2025), https://www.unesco.org/ethics-ai/en/global-hub

Bahrain

In 2025, Bahrain launched a National AI Policy to guide the use of AI in the public sector. The country also published an AI readiness assessment report built on the UNESCO RAM.

National AI Strategy

The Bahraini government and legislature have initiated and operationalized a multi-layered framework including a national policy, a regionally aligned ethical manual to ensure cultural and legal interoperability, and a binding standalone law to regulate high-risk applications and establish legal certainty. The Kingdom of Bahrain envisions AI development and application as an opportunity to establish the country's leadership in digital transformation and a tool to improve services to citizens and streamline government processes more broadly.[648] Fundamental principles for the National AI Policy[649] (iGA National Policy) include human decision-making and control; safety and prevention of harm; fairness, equity, and non-discrimination; transparency and explainability; responsibility, accountability, and awareness; integrity and non-fabrication; privacy and data protection; reliability and safety; investment; promotion innovation; and protection of intellectual property rights. The Information and eGovernment Authority (iGA) structured the National AI Policy structured around four key pillars: Policies and Legislation, AI Use and Adoption, Awareness and Education, and Enhancing Local and International Cooperation.

Bahrain adopted the Gulf Cooperation Council (GCC) Guiding Manual on the Ethics of Artificial Intelligence to promote the AI use consistent with human values, local laws, and the UNESCO Recommendation on the Ethics of AI.[650] The guide focuses on four main values: promoting human dignity and autonomy; respecting Islamic Sharia, the Constitution, and cooperation among GCC countries; protecting the environment and promoting sustainability; and achieving human well-being. The principles center on four areas: human autonomy and decision-making; system safety and harm prevention; justice, fairness, and non-discrimination; and privacy and data protection.

---

[648] Government of Bahrain, *Artificial Intelligence: Leveraging Artificial Intelligence (AI) in Bahrain* (Feb. 11, 2026), https://www.bahrain.bh/wps/portal/en/BNP/ExploreBahrain/ArtificialIntelligence
[649] Information & eGovernment Authority, *General Policy for the Use of Artificial Intelligence*, pp. 5–6 (May 20, 2025), https://www.iga.gov.bh/Media/Publications/National Digital Policies/General Policy for the Use of AI - Final 30 Jul 2025.pdf
[650] Gulf Cooperation Council, *Guiding Manual on the Ethics of Artificial Intelligence Use in Member States of the Gulf Cooperation Council (GCC), Version 1*, pp. 3–4 (Nov. 2023), https://www.bahrain.bh/wps/wcm/connect/f863157d-2753-4ef7-a4dd-21f8f7c01cdb/The+Guiding+Manual+on+the+Ethics+of+Artificial+Intelligenc+for+GCC+Member+States.pdf?MOD=AJPERES&CVID=psWZnfZ

A 2023 Telecommunications Plan[651] and iGA National Policy are part of government plans to transition Bahrain to a knowledge-based economy that keeps pace with the global trends in science and technology.[652] In the opening of the 5th legislative cycle of 2019, for example, H.M. King Hamada bin Isa Al Khalifa directed the government to undertake a national plan to increase the readiness for the digital economy "by adopting and employing artificial intelligence technology in the production and service sectors, through the establishment of the necessary systems and technical frameworks, as well as through encouraging quality investments, in order to guarantee the maximum benefit to our national economy."[653]

The broader strategic digital and economic initiatives include the Kingdom's Economic Vision 2030,[654] which addresses AI policy and development in the country.[655] The Digital Transformation Journey of the Kingdom focuses on fostering an innovation ecosystem for the use of emerging technologies for the benefit of the Bahraini society. The Kingdom "encourages those involved in designing, developing, and deploying new technologies to ensure that they are consistent with the Kingdom's values and adhere to international norms and standards."[656]

AI Legislation

Bahrain's Shura Council unanimously approved a new standalone law to regulate AI in April 2024.[657] The law identifies offenses for manipulating content without consent, punishable by fines and imprisonment.[658] The AI Regulation Law

---

[651] Ministry of Transportation and Telecommunications, *Sixth National Telecommunication Plan* (Nov. 2023), https://www.mtt.gov.bh/projects/sixth-national-telecommunications-plan

[652] Oxford Business Group, *Viewpoint: King Hamad bin Isa Al Khalifa* (2020), https://oxfordbusinessgroup.com/articles-interviews/eye-on-the-future-king-hamad-bin-isa-al-khalifa-on-modernisation-and-the-ongoing-transition-to-a-knowledge-based-economy-viewpoint

[653] Bahrain News Agency, *HM King Opens Second Session of 5th Legislative Term* (Oct. 13, 2019), https://www.bna.bh/en/HMKingopenssecondsessionof5thLegislativeTerm.aspx?cms=q8FmFJgiscL2fwIzON1+DnPyGy%2FzC+rQlxqVtgB48bs=

[654] Bahrain's National Portal, *The Economic Vision 2030 Bahrain* (2016), https://www.mofne.gov.bh/en/project-initiatives/bahrain-economic-vision-2030/

[655] Information & eGovernment Authority, *Emerging Technologies* (2022), https://www.iga.gov.bh/en/category/emerging-technologies

[656] Bahrain's National Portal, *Government of Bahrain's Digital Transformation Journey* (Jan. 16, 2022), https://www.bahrain.bh/wps/portal/en/BNP/BahrainAtAGlance/BahrainDigitalGovernmentJourney/

[657] Shura Council, *Proposed Law Regulating Artificial Intelligence Technologies and Their Uses* (Apr. 29, 2024), https://perma.cc/PBY7-CD94

[658] Gulf Daily News Online, *New Law to Regulate Use of AI Approved in Bahrain* (Apr. 29, 2024), https://www.zawya.com/en/world/middle-east/new-law-to-regulate-use-of-ai-approved-in-bahrain-djpvmm4r

also requires ministers to establish sector-specific regulations.[659] The Council of Representatives is reviewing the law.

Public Participation

Since 2016, Bahrain has been very active in promoting and developing public participation in relation to its digital services and technology. This has consistently involved a theme of strong encouragement of citizens and users to engage with the government via primarily digital means and social media. Bahrain's eGov Strategy 2016 included a digitized National Suggestions and Complaints System, an open data platform, and continued encouragement of uptake of eGovernment services. In 2018, the strategy directly prioritized objectives including "Nurturing active participation and engagement with constituents."[660]

The iGA National Policy specifies where updates will be posted and provides an email address connected to an invitation that "All comments and suggestions related to the General Policy for the Use of Artificial Intelligence in the Kingdom of Bahrain are welcomed and appreciated."[661] The invitation notes that feedback will help improve the document and ensure it is comprehensive and aligned with the user's needs.

The Public consultation on the implementation of at least six draft regulations to complement the Personal Data Protection Law (PDPL)[662] resulted in the enactment of ten ministerial resolutions in March 2022 to align the PDPL more closely to international standards and the GDPR.[663]

In the Digital Government Strategy 2022, the government developed the "Leave No One Behind" digital policy to focus on citizen needs, underpinned by the Digital First principle.[664] This principle prioritizes the design of public services

---

[659] Inter-Parliamentary Union, *Parliamentary Actions on AI Policy*, Bahrain (Feb. 25, 2026), https://www.ipu.org/impact/democracy-and-strong-parliaments/artificial-intelligence/parliamentary-actions-ai-policy

[660] Government of Bahrain, *The Information & eGovernment Authority's Strategy, eGov Strategy 2018*, https://www.bahrain.bh/wps/portal/en/!ut/p/z0/fY69TsNQDEZfxUtmX4r6wwhUImWgSAyUu1ROY27cJnZ661Tk7QlMTIyfdM7RhxF3GJWuksjFlNppf8TF_mYzD-XqMYTy5XUdFttyt ZwtH263dzN8Y8VnjP9DU0WO53O8x3gwdf5y3FXa71mLQJUNDt4wnERTbV0RKmoyiUItSZxaS HblrB2rw9GGrDwWIXNLzjW49XK4FOEnIPppufu9DqQ18B-RBm8si49w8TyZacT-9PT-DVRXnPM!/

[661] Information & eGovernment Authority, *General Policy for the Use of Artificial Intelligence*, p. 13 (May 20, 2025), https://www.iga.gov.bh/Media/Publications/National Digital Policies/General Policy for the Use of AI - Final 30 Jul 2025.pdf

[662] Bahrain News Agency, *Consultation on Implementing Bahrain's Personal Data Protection Law Launched Online* (Jun. 14, 2021), http://bna.bh/en/HMKingissuesdecree46/ConsultationonimplementingBahrainsPersonalDataProtectionLawlaunchedonline.aspx?cms=q8FmFJgiscL2fwIzON1+DhLTzyak+XqL0ckS1SQmlDM=

[663] DLA Piper, *Data Protection Laws of the World: Bahrain* (Jan. 17, 2024), https://www.dlapiperdataprotection.com/index.html?t=law&c=BH

[664] Kingdom of Bahrain, Digital-First Policies*: Bahrain's Digital Path* (Jul. 30, 2024), https://www.bahrain.bh/wps/portal/en/BNP/HomeNationalPortal/ContentDetailsPage/!ut/p/z0/lY67bs

digitally first to encourage citizens voicing their concerns and opinions via portals such as Tawasul. The design should require citizens and corporations to supply data only once to a government entity and the pooling of government data for greater public accessibility and civic engagement, which supports Bahrain's Digital Identities initiative.

Bahrain has established a system to receive public feedback on how to run services more efficiently and effectively. However, outcomes and findings from participatory action are not always publicly available.

Data Protection

Bahrain has led the region in the enactment of legislation aligned to the highest global standards. The Kingdom was the second state in the Gulf Region, following Qatar, to address personal data protection as a right, with the enactment of Law No. (30) of 2018 with Respect to Personal Data Protection Law (PDPL).[665] The Royal Decree No (78) of 2019 established the Personal Data Protection Authority (PDPA),[666] and granted the Minister of Justice, Islamic Affairs and Waqf the chair of the Board of Directors of the PDPA.[667] The PDPA has the responsibility to enforce the law and investigate potential violations.

The PDPL, inspired by the EU GDPR, entered into effect in August 2019. The Law protects the rights of individuals regarding the collection, processing, and storage of their personal data (whether by automatic or manual means), and establishes obligations to data controllers or data processors in their relationship with data owners or subjects.[668]

The PDPL shares similarities with the GDPR[669] in the rights to notification of data collection and processing, right to erasure, and right of rectification, but differs in some specifics related to the right to be informed (Article 18) and the right to object to processing of personal data (Articles 19, 20, and 21). Exceptions to the right not to be subject to decision-making based on solely automated means (Article 22), the

---

JAEEV_ZVO4RDPYxEAZgoQTiYSAxGMbtLZH9oTNrLEXlM_HUPGo6O69Oro6oGENWsyRC-PZibFt3-h42_14xWTwjjgb90L8WQ7n8WjaxekkggUJflK-hpKv2Rjj72TQD_ujCCM8v_Dvfq_fQGdOPP17WKdSbUkCNKk7eOVLUjuWInd_AaamrA0LK0sFN_bi0gRYUxspV95VnLW9IVNnpWK54e5tHpWftMm5YG-sqpzljKl5XKDaTVYdvXk5AeziS-M!/

[665] Information Guide, *Personal Data Protection, Law No. 30 of 2018 with Respect to Personal Data Protection* (2018), https://www.bahrain.bh/wps/wcm/connect/ab8b334e-8c6f-4ff9-90b6-94135da559ca/Law+No.+%2830%29+of+2018+DPL.pdf?MOD=AJPERES&CVID=oFapPNI

[666] Personal Data Protection Authority, *Royal Decree No. 78*, Lexis Middle East (2022), https://www.lexismiddleeast.com/law/Bahrain/Decree_78_2019/en

[667] Akin Gump, *Bahrain Ministry of Justice to Act as Data Protection Authority* (Oct. 15, 2019), https://www.mondaq.com/data-protection/857510/bahrain-ministry-of-justice-to-act-as-data-protection- authority

[668] DLA Piper, *Bahrain Publishes Personal Data Protection Law* (Sept. 17, 2018), https://www.dlapiper.com/en/qatar/insights/publications/2018/09/bahrain-publishes-personal-data-protection-law/

[669] Securiti, *Bahrain's PDPL vs. GDPR* (2023), https://securiti.ai/bahrains-pdpl-vs-gdpr/

absence of the right to data portability, and the shorter period of response to data subjects' requests (15 days vs one month in the GDPR) are important differences.

Unlike the GDPR, the PDPL mandates data managers to recognize the right of Bahraini data owners to object to personal data that causes harm or distress to the data owner or any persons. Prior written approval from the Personal Data Protection Authority is mandatory before processing certain personal data. Article 58 defines criminal and civil penalties for violations including prison, in specified cases, in contrast with GDPR where only monetary penalties are defined.

Following public consultations, the Minister of Justice issued ten decisions to guide implementation of the PDPL. These decisions specified technical requirements to protect personal data, rules and procedures for professing, the transfer of data outside of Bahrain, conditions for creating publicly accessible personal data records, and administrative fees and policies.[670] Terms and language such as transparency and accountability are mentioned briefly, as well as human rights and freedoms.

Other laws that provide general rights to privacy, including digital privacy,[671] are: Article 26 of the Kingdom's constitution, which safeguards confidentiality of postal, telegraphic, telephonic or electronic communication;[672] Law No. 16 on Electronic Transactions; Law No. 48 on Telecommunications,; Law No. 60 on Information Technology Crimes; and Law No. 16 on Protection of State Information and Documents.[673]

Previous laws related to data protection that complement PDPL also include the Central Bank of Bahrain and Financial Institutions Law 2006 and Labour Law 2012, which regulates data protection between employees and employers[674]

In January 2017, Bahrain ratified the Arab Treaty on Combating Cybercrime to establish new rules on retaining user data and real-time monitoring of activities

Bahrain is not a member of the Global Privacy Assembly (GPA).[675] Bahrain's Personal Data Protection Authority gained Observer status for the 2025 GPA session

---

[670] Ministry of Justice, Islamic Affairs and Waqf, *The Minister of Justice Issues Ten Decisions to Implement the Personal Data Protection Law* (Mar. 20, 2022), https://www.moj.gov.bh/index.php/en/news-archived-291

[671] Government of Bahrain, Information Guide, *Personal Data Protection* (Mar. 25, 2025), https://www.bahrain.bh/wps/portal/en/!ut/p/z1/tZLNcoIwFIVfhY1L5l6CELe2TovtjFotItk4AYKmShI xtT9PX3DXhWIXzSp35pxJzvkuMFgBU_wkN9xKrfi-mVMWrr1xgNHgHjGazEYYTqMBJ fTOn858SH4LvNcY8WU5Hs6nDyMP-x6wW_x44QzxNv9FAenyL4EBM7ksIPUDUpQlJS6h NHT7WZ-6WVkSl_OQC-oFNCiwVefKGruFNFNmLVQPeabfrWO3wtlJtSl01c OMb2sulXT2YiOP-3Obxx7WormKwrHayLyZjaiPbc1OwS13TK2tyFtpVyr_nOpKaUn7zytcnr FD0ILreoRd75ZA2qSgFwUBgeQkxQfEStdVs2uLP6KIEJ668Db7K98OBzZssGllxaeF1T9yWwgFp orjuBr4X-5uHn3fTdzHxP0By3kwjw!!/

[672] Constitute Project, *Bahrain's Constitution of 2002 with Amendments through 2017* (Aug. 26, 2021), https://www.constituteproject.org/constitution/Bahrain_2017.pdf?lang=en

[673] Council of Europe, *Bahrain* (May 28, 2020), https://www.coe.int/en/web/octopus/-/bahrain

[674] OneTrust DataGuidance, *Bahrain Data Protection Overview* (Aug. 2021), https://www.dataguidance.com/notes/bahrain-data-protection-overview

[675] Global Privacy Assembly, *List of Accredited Members* (2026), https://globalprivacyassembly.com/participation-in-the-assembly/list-of-accredited-members/

and will hold the status through the 2028 closed session.[676] Bahrain neither sponsored nor endorsed the 2018 GPA Resolution on AI and Ethics,[677] the 2020 GPA Resolution on AI and Accountability,[678] the 2022 GPA Resolution on AI and Accountability,[679] and the 2023 GPA Resolution on Generative AI Systems.[680]

Algorithmic Transparency

Bahrain has not enacted specific algorithmic transparency legislation. However, the te Information & eGovernment Authority (iGA) General Policy for the Use of Artificial Intelligence[681] establishes a binding framework for government agency practices. The iGA National Policy codifies transparency and explainability; fairness, equity, and non-discrimination, and privacy and data protection. The PDPL in Article 15, requires permission from the Authority before the automated processing of personal data of sensitive nature, biometric data, genetic data, linking data to two or more data controllers for different purposes. Article 22 includes the right to object to decisions based solely on automated processing for purposes of assessing performance at work, financial standing, credit scoring, reliability or conduct. Yet the PDPL does not include provisions to ensure transparency in the methods of processing.

According to Bahrain's National Digital Portal, the country introduced legal frameworks to protect individual privacy and reduce algorithmic biases in 2024. Bahrain has enacted a new AI regulatory framework which consists of 38 articles, imposing fines of up to BD1,000 for using AI in decision-making without human oversight. More severe penalties, including fines up to BD2,000, apply to AI misuse that infringes privacy, promotes discrimination, or deviates from intended purposes.[682]

[676] Global Privacy Assembly, *List of Observers* (2026), https://globalprivacyassembly.com/participation-in-the-assembly/list-of-observers/

[677] International Conference of Data Protection & Privacy Commissioners, *Declaration on Ethics and Data Protection in Artificial Intelligence* (Oct. 23, 2018), https://globalprivacyassembly.org/wp-content/uploads/2018/10/20180922_ICDPPC-40th_AI-Declaration_ADOPTED.pdf

[678] Global Privacy Assembly (GPA), *Adopted Resolution on Accountability in the Development and Use of Artificial Intelligence* (Oct. 2020), https://globalprivacyassembly.org/wp-content/uploads/2020/11/GPA-Resolution-on-Accountability-in-the-Development-and-Use-of-AI-EN.pdf

[679] Global Privacy Assembly (GPA), *Resolution on Principles and Expectations for the Appropriate Use of Personal Information in Facial Recognition Technology* (Oct. 2022) https://globalprivacyassembly.org/wp-content/uploads/2022/11/15.1.c.Resolution-on-Principles-and-Expectations-for-the-Appropriate-Use-of-Personal-Information-in-Facial-Recognition-Technolog.pdf

[680] Global Privacy Assembly, *Resolution on Generative Artificial Intelligence Systems* (Oct. 2023), https://globalprivacyassembly.org/wp-content/uploads/2023/10/5.-Resolution-on-Generative-AI-Systems-101023.pdf

[681] Information & eGovernment Authority, *General Policy for the Use of Artificial Intelligence* (May 20, 2025), https://www.iga.gov.bh/Media/Publications/National Digital Policies/General Policy for the Use of AI - Final 30 Jul 2025.pdf

[682] Hyscaler, *Bahrain Pioneers AI Regulation with New Standalone Law* (Apr. 29, 2024), https://hyscaler.com/insights/bahrain-pioneers-ai-regulation/

Previously, Bahrain joined other member states of the Digital Cooperation Organization (DCO) to adopt the Riyadh AI Call for Action Declaration (RAICA) during the Global AI Summit of 2022.[683] The Declaration marks a commitment to develop AI technology that benefits people, communities, and nations. Item 4 of the Call for Action addresses safeguards to prevent unfairness of algorithms against individuals due to their orientation, culture, gender, or race.

Mass Surveillance

Amnesty International reported about an ongoing investigation on the use of Pegasus Spyware to infect the devices of three political activists in Bahrain in 2021.[684] The misuse of the spyware provided by Israeli NSO Group to trace mobile phones targeted at least 50,000 devices of journalists and human rights activists. According to Citizen Lab, Bahraini authorities used cyber surveillance since 2013, including phone forensics technology sold by Israeli company Cellebrite to extract private data from the devices of arrested activists.[685]

The Carnegie Endowment for Peace report on AI surveillance, found that Bahrain is using facial recognition for smart policing, mainly from Chinese technology providers such as Dahua.[686] Reports of use of surveillance cameras in public spaces also surfaced.

The OHCHR reiterated concerns about surveillance of individuals and groups cooperating with the United Nations.[687] The Ministry of Interior has issued statements warning about legal procedures against activists and their followers who participate in cybercrime on social media.[688]

The Cybercrimes Directorate and the Cyber Safety Directorate monitor social media and websites to identify and prevent crimes, including forbidden speech, false news, and threats to national security. "Department officials receive training to

[683] WIRED, *The Key Wins of Saudi Arabia's Global AI Summit* (Sept. 27, 2021), https://wired.me/technology/saudi-arabia-global-ai-summit/

[684] Amnesty International, *Bahrain: Devices of Three Activists Hacked with Pegasus Spyware* (Feb. 18, 2022), https://www.amnesty.org/en/latest/news/2022/02/bahrain-devices-of-three-activists-hacked-with-pegasus-spyware/

[685] The Citizen Lab, *From Pearl to Pegasus. Bahraini Government Hacks Activists with NSO Group Zero-Click iPhone Exploits* (Aug. 24, 2021), https://citizenlab.ca/2021/08/bahrain-hacks-activists-with-nso-group-zero-click-iphone-exploits/

[686] Carnegie Endowment for International Peace, *The Global Expansion of AI Surveillance* (Sept. 2019), https://carnegieendowment.org/files/WP-Feldstein-AISurveillance_final1.pdf

[687] OHCHR, *Report Details Disturbing Trends as Reprisals Continue against People Cooperating with the UN* (Sept. 29 2022), https://www.ohchr.org/en/press-releases/2022/09/report-details-disturbing-trends-reprisals-continue-against-people

[688] Ministry of Interior of Bahrain, *Anti-Cyber Crime Message on Twitter* (May 21, 2019), https://twitter.com/moi_bahrain/status/1130802153663213568

identify "permitted speech and forbidden speech," whether it is text, video, or audio.[689]

Bahrain's eGovernment authority released the "BeAwareBahrain" app to "ensure the safety of all its citizens and residents."[690] The app alerted individuals in case of close contact with a confirmed case of COVID-19 and was able to confirm the location of people who had moved more than 15 meters from their phone. A 2020 report by Amnesty International on 11 countries using data tracking apps for COVID-19 concluded that the BeAwareBahrain app was among "the most alarming tools [...] carrying out live or near-live tracking of users' locations by frequently uploading GPS coordinates to a central server."[691] The surveillance activities included the real-time broadcast of users' locations to a government database, the publication of online sensitive information linked to their national ID, and the enforced use of a Bluetooth bracelet during quarantine.

Biometric Identification

Bahrain has integrated biometrics and digital IDs as part of the Digital First Policy. The use of a nationwide digital identity scheme to include Sijilat (the Commercial Registration Portal), the National Taxation System, and Sehati (the National Social Health Insurance Program), aims to facilitate users access to government services through one-time input into the system.[692] Users can visit a Government Service Center if they lack internet access. The iGA provided the ID card & Birth Certificate eServices through the unified MyGov app in 2025.[693] In an interview, the iGA Chief Executive stated that the new card functions as an ICAO-compliant digital travel document.[694] The card also uses advanced fingerprint matching technology to securely verify the holder's identity.

Bahrain introduced biometric identification for automated border control in Bahrain International Airport (BIA) in 2019, as part of the Airport Modernization

---

689 BCHR, *Bahrain: The "Cyber Safety Directorate" Monitors Internet Activity in Style Similar to Big Brother* (Nov. 25, 2013), http://bahrainrights.net/?p=5973

690 Ministry of Health, *BeAwareBahrain* (2023), https://healthalert.gov.bh/en/category/beaware-bahrain-app

691 Amnesty International, *Bahrain Kuwait and Norway Contact Tracing Apps among Most Dangerous for Privacy* (Jun. 16, 2020), https://www.amnesty.org/en/latest/news/2020/06/bahrain-kuwait-norway-contact-tracing-apps-danger-for-privacy/

692 Kingdom of Bahrain, Digital-First *Policies: Bahrain's Digital Path* (Jul. 30, 2024), https://bh.bh/new/en/digitalfirst_en.html

693 Bahrain News Agency, *iGA facilitates Access to ID Card, Birth Certificate Services via MyGov App* (Mar. 17, 2025), https://www.bna.bh/en/news?cms=q8FmFJgiscL2fwIzON1+DsGlD77r3jhz7FB%2FBbCEsEM=

694 Daily Tribune– News of Bahrain, *Bahrain's Next-Gen ID Card: A Game-Changer for Travel and Technology* (Mar. 2, 2025), https://www.newsofbahrain.com/bahrain/109426.html

Programme (AMP).[695] In July 2022, Bahrain joined the Global Entry Partnership, the U.S. Customs and Border Protection (CBP) program for expedited entry into the U.S. for pre-approved, low-risk travelers. The program makes use of facial recognition and travelers can make use of biometric kiosks at airports to process them.[696]

Tamkeen (Labour Fund), Bahrain's government agency promoting private-sector business and individual development, launched digital onboarding using biometric identification to register customers. Tamkeen uses B2B payment integrated with BENEFIT Pay.[697] BenefitPay is a financial network established in 1997 under license of the Central Bank of Bahrain.[698] Benefit uses facial recognition for validation and authentication of users, recognizing all GCC IDs and passports of all nationalities. The service is available for FinTech, insurance, and investment companies to curtail cybercrime.[699]

With the growing trend towards digitization, only around 5% of transactions require physical contact, mainly for issues related to issuing residence permits and biometric updates. New initiatives include a national birth system, digital archiving of records, and a Central Population Registration System.[700]

Environmental Impact of AI

Bahrain revised its strategy roadmap in the wake of the COVID-19 pandemic to reprioritize digital projects and ensure alignment with the UN Sustainable Development Goals.[701] The government continues to drive its National Digital

---

[695] Security World Market, *Bahrain Enhances Border Control with Biometric Identification* (Jun. 6, 2019), https://www.securityworldmarket.com/me/Newsarchive/bahrain-enhances-border-control-with-biometric-identification1

[696] Frank Hersey, *Bahrain Latest Addition to US Biometric Global Entry Partnership,* BiometricUpdate (Jul. 18, 2022), https://www.biometricupdate.com/202207/bahrain-latest-addition-to-us-biometric-global-entry-partnership

[697] Government of Bahrain, *National Portal of the Kingdom of Bahrain*, https://www.bahrain.bh/wps/portal/en/!ut/p/z1/rVLLbsIwEPwVesgx8SYmr95SRGkRjwoKJb4gJziPktghGGj_vka0BypelerbWjOzszuLCJohwuk2T6nMBaeFqkPizJ-G4JiWZ3U9PMEQDJ3WyG-D2QEbvR0DAOP2HuC-uFPfcjxA5BY-WP2W-dxU_H4fIPAeRr3XxxZAB1_jTxFBJOaykhkKWSq2laglLeaMa5CJkmmw2uTxslHkfLnWgEZiIxt7HKt5ybjUgH_P2jgw93pVnC9QuFgktu0noPs4wnoTfKpTz3R126VmQpPEik33x_-ZF9w4_wUAuSw_ZlxpkOM2J1I46FxY87VGoTLqnjWqFMZ_3Fz32mrU7eXvqxUJVMCCS_Yh0exfElad00JEh_MOeIS9FJGaJaxmtbGp1XcmZbW-10CD3W5npEKkBTNiYWyWGpwiZWKtzP3GoqqcTEoPf-Z6t50MBjqJ7GLbC-6-AKLSLMY!/dz/d5/L3dHQSEvUUt3SS9nQSEh/

[698] Benefit, *The Pulse of Bahrain* (2023), https://benefit.bh/

[699] News of Bahrain, *BENEFIT Holds Identification and Verification Service Workshops* (Jul. 25, 2022), https://www.newsofbahrain.com/business/82774.html

[700] Biometric Update, *Bahrain Sees High Uptake of Digital ID, CRVS Services through Online Platforms* (Feb. 5, 2024), https://www.biometricupdate.com/202402/bahrain-sees-high-uptake-of-digital-id-crvs-services-through-online-platforms

[701] Kingdom of Bahrain, *The Information & eGovernment Authority's Strategy*, https://www.bahrain.bh/wps/portal/en/!ut/p/z0/fY69TsNQDEZfxUtmX4r6wwhUImWgSAyUu1ROY

Economy Strategy (NDES) for 2022–2026[702] to build a strong and sustainable digital economy. This effort is aligned with the Kingdom of Bahrain Economic Vision 2030, which seeks to ensure "a safe, secure, sustainable and attractive environment and world class infrastructure."[703]

Bahrain has developed ethics to guide the responsible development and adoption of AI in the country, including a principle that "AI systems should be developed and used in a way that is sustainable and minimizes their environmental impact."[704]

AI is also a driver of sustainable development in Bahrain. The University of Bahrain and the Benefit Company signed a Memorandum of Understanding to use AI and advanced computing for research and innovation into sustainable energy, climate change, and sea-level prediction.

Lethal Autonomous Weapons

Bahrain is a High Contracting party to the Convention on Certain Conventional Weapons' (CCW)[705] and a member of Non-Aligned Movement (NAM).[706] In a statement presented during the Group of Governmental Experts on Emerging Technologies (GGW) on LAWS Meeting in Geneva in 2022, the NAM called for the negotiation of a "legally binding international instrument stipulating prohibitions and regulations on lethal autonomous weapons systems."[707]

27cJnZ661Tk7QlMTIyfdM7RhxF3GJWuksjFlNppf8TF_mYzD-XqMYTy5XUdFttytZwtH263dzN8Y8VnjP9DU0WO53O8x3gwdf5y3FXa71mLQJUNDt4wnERTbV0RKmoyiUItSZxaSHblrB2rw9GGrDwWIXNLzjW49XK4FOEnIPppufu9DqQ18B-RBm8si49w8TyZacT-9PT-DVRXnPM!/

[702] Kingdom of Bahrain, *National Digital Economy Strategy* (Mar. 2025), https://www.bahrain.bh/wps/wcm/connect/49ed95ce-58db-4908-8085-c82656d6af77/National+Digital+Economy+Strategy.pdf?MOD=AJPERES&CVID=pwiSJDs

[703] Kingdom of Bahrain, *The Economic Vision 2030 for Bahrain*, https://www.moh.gov.bh/Content/Files/Vision_2030.pdf

[704] Kingdom of Bahrain, *Artificial Intelligence* (Sept. 24, 2024), https://www.bahrain.bh/wps/portal/en/BNP/HomeNationalPortal/ContentDetailsPage/!ut/p/z1/vZJLb4JAFIX_Ci5ckrk8hSXW-kqNFUWFjRlgwFGcQRy1_fcdu2hSGovddHaTnHNzz_kuitAaRQxfaI4F5QwX8h9G9kYbWTB0ngBee6YOs6Xr292JBpOBgVbfBdoiACkYef6039PA1FD0iB_uPA8e838J3IVrg63rnqnPBwZMG_1LFKGoTGiKQmxkhq1ZppolrquaSeqo2AJDjTOcpbGbkljTb-qEiVJsURizckNYG3DMz0IRW6LsKctTfmhDSnMqcKHszidBE6IkvCikrPqstQ0VKbAgqSJ4SZOTnFAJmtGESgdlghQFzQmTNsxSpeIxlzNO9aQ_o0S_F1nzD2eBI5t6GU9tvWtMn2ukfha5umVvYN20Q22CGwR9sJ1B3zRtx-9YHRTKJTt3Q44k7wslVxQwXh3kec7_SG8IaNx0EfLk6e54jDxJmksabwKt_xn1nDBUHoIgODjGO1X3vnNdZNviknut1geZeYBX/

[705] UN Office of Disarmament Affairs, *High Contracting Parties and Signatories CCW* (Mar. 18, 2025), https://disarmament.unoda.org/en/our-work/conventional-arms/convention-certain-conventional-weapons/high-contracting-parties-and-signatories-ccw

[706] UN Office of Disarmament Affairs, *Regional Groups—CCW* (Dec. 2022), https://front.un-arm.org/wp-content/uploads/2022/12/Regional-Groups-Dec-2022.pdf

[707] Non-Aligned Movement, *Working Paper by the Bolivarian Republic of Venezuela on Behalf of the Non-Aligned Movement (NAM) and Other State parties to the Convention on Certain Conventional*

Bahrain voted for the UN General Assembly Resolution 78/241[708] on lethal autonomous weapons systems, which directed the UN Secretary-General to prepare a report reflecting the views of member and observer states. Bahrain also supported the more recent UN General Assembly Resolution L.77. The resolution raised concerns about the impact of autonomous weapons on global security and regional and international stability and provides for open consultations to consider the Secretary-General's report.

Bahrain also endorsed the Political Declaration resulting from the Responsible AI in the Military Domain Summit (REAIM 2023) co-hosted by the Netherlands and the Republic of Korea in November 2023.[709] The Declaration followed a joint call to action on the responsible development, deployment, and use of AI in the military domain issued during the summit.[710]

AI Literacy

The Third Pillar of the iGA National AI Policy is dedicated to Awareness and Education.[711] Objectives center on raising awareness among government employees "about the latest AI technologies and how to benefit from them in service development" and ensuring they have the technical capabilities use the technology safely and effectively. Rules and requirements instruct government entities to provide training programs but no specific initiatives are proposed.

The National Digital Economy Strategy 2022–2026[712] envisions a future-ready workforce where graduates "are equipped with advanced digital skills and are prepared for future job markets." Strategic initiatives include incorporating ICT curriculum in primary and secondary education as well as empowering the Skills Bahrain program and incentivizing other upskilling and reskilling programs.[713]

Weapons (CCW) (Jul. 2022), https://documents.unoda.org/wp-content/uploads/2022/08/WP-NAM.pdf

[708] Automated Decision Research, *State Positions: Bahrain*, https://automatedresearch.org/news/state_position/bahrain/

[709] US Department of State, *Political Declaration on Responsible Military Use of Artificial Intelligence and Autonomy*, Endorsing States (Nov. 27, 2024), https://www.state.gov/political-declaration-on-responsible-military-use-of-artificial-intelligence-and-autonomy/

[710] Government of Netherlands, *Call to Action on Responsible Use of AI in the Military Domain* (Feb. 16, 2023), https://www.government.nl/documents/publications/2023/02/16/reaim-2023-call-to-action

[711] Information & eGovernment Authority, *General Policy for the Use of Artificial Intelligence*, pp. 10–11 (May 20, 2025), https://www.iga.gov.bh/Media/Publications/National DigitalPolicies/General Policy for the Use of AI - Final 30 Jul 2025.pdf

[712] Government of Bahrain, *National Digital Economy Strategy*, p. 6 (Mar. 2025), https://www.bahrain.bh/wps/wcm/connect/49ed95ce-58db-4908-8085-c82656d6af77/National+Digital+Economy+Strategy.pdf?MOD=AJPERES&CVID=pwiSJDs

[713] Ibid, p. 10

Human Rights

Bahrain supports the Universal Declaration of Human Rights.[714] The Prime Minister has expressed the importance of the UDHR in public statements.[715] The country has completed four cycles of Universal Periodic Reviews, the most recent in November 2022.[716] The UPR of Bahrain noted the recommendations of various states for Bahrain to ratify international human rights covenants. The country made a voluntary pledge and commitment to implement projects of the national human rights plan (102 projects), for 2022–2026 and to submit voluntary reports every two years on the ongoing efforts to enhance the country's human right system.[717]

Bahrain is one of the 16 State Parties to the Arab Charter on Human Rights and has ratified the International Covenant on Civil and Political Rights (ICCPR), the International Covenant on Economic, Social, and Cultural Rights (ICESCR), and the Convention on the Rights of the Child (CRC).[718]

Freedom House rated Bahrain as "Not Free" (12/100) in 2025.[719] Restrictions to freedom of expression and beliefs and the use of spyware and surveillance technology to target activists and dissidents are areas of concern. Internet freedom in Bahrain remained restricted in 2024 with content critical of the government blocked and removed. Freedom House notes that despite the continued use of social media for activism, self-censorship prevailed due to fear of state surveillance.[720]

Transparency International's 2023 Corruption Perception Index ranks Bahrain 76th of 180 countries with a score of 42/100.[721] Bahrain showed significant progress in 2024 with a move up eleven places to 53.[722]

---

[714] UN Bahrain, *Universal Declaration of Human Rights at 75: Our Shared Values and Path to Solutions* (Dec. 12, 2023), https://bahrain.un.org/en/255528-universal-declaration-human-rights-75-our-shared-values-and-path-solutions

[715] Bahrain News Agency, *HRH Prime Minister Highlights Universal Declaration of Human Rights* (Dec. 8, 2018), https://www.bna.bh/en/ConstitutionalCourttoconsiderConstitutionalCase1/HRHPrimeMinisterhighlightsUniversalDeclarationofHumanRights.aspx?cms=q8FmFJgiscL2fwIzON1%2BDktmz32o3rvQb7QWjpEbQ0M%3D

[716] United Nations Human Rights Council, *Universal Periodic Review: Bahrain* (2023), https://www.ohchr.org/en/hr-bodies/upr/bh-index

[717] United Nations Human Rights Council, Report of the Working Group on the Universal Periodic Review: Bahrain, (Jan. 11, 2023), https://docs.un.org/en/A/HRC/52/4

[718] United Nations Human Rights Treaty Bodies, *Ratification Status for Bahrain*, https://tbinternet.ohchr.org/_layouts/15/TreatyBodyExternal/Treaty.aspx?CountryID=13&Lang=en

[719] Freedom House, Freedom in the World 2025: Bahrain (2025), https://freedomhouse.org/country/bahrain/freedom-world/2025

[720] Freedom House, *Freedom on the Net 2025: Bahrain* (2025), https://freedomhouse.org/country/bahrain/freedom-net/2025

[721] Transparency International, *Corruption Perceptions Index* (2023) https://www.transparency.org/en/cpi/2023/index/bhr

[722] Transparency International, *Corruption Perceptions Index* (2024), https://www.transparency.org/en/cpi/2024/index/bhr

Although Bahrain granted amnesty to over 2,500 prisoners in 2024,[723] Human Rights Watch reported on the ongoing imprisonment of human rights defenders, some for more than a decade, and pro-democracy activists and abysmal conditions and denial of adequate healthcare in prisons. Limits on freedom of expression and the return of the death penalty, especially given unfair trials, remain concerns.[724]

OECD / G20 AI Principles

Bahrain is not a member of the OECD and has not endorsed the OECD AI Principles.[725] Bahrain has not submitted reports to the OECD AI Policy Observatory in relation to ongoing policies, strategies, or activities associated with AI.[726] Bahrain has launched initiatives such introducing AI and cybersecurity into education that align with the OECD AI principles. Bahrain is a member of the MENA-OECD Initiative on Governance and Competitiveness created in 2021 to implement reforms to improve governance structures and cooperation.[727]

Council of Europe AI Treaty

Bahrain has not endorsed the Council of Europe Framework Convention on Artificial Intelligence and Human Rights, Democracy, and the Rule of Law, the first legally binding international AI treaty.[728]

UNESCO Recommendation on AI Ethics

Bahrain is a UNESCO member since 1972[729] and is one of the member States which endorsed the Recommendation on the Ethics of AI.[730] Bahrain has published a readiness assessment report[731] based on the UNESCO Readiness Assessment

---

723 Human Rights Watch, *World Report 2025, Bahrain: Events of 2024*, https://www.hrw.org/world-report/2025/country-chapters/bahrain

724 Human Rights Watch, *World Report 2024, Bahrain: Events of 2023*, https://www.hrw.org/world-report/2024/country-chapters/bahrain?gad_source=1&gclid=Cj0KCQiAst67BhCEARIsAKKdWOlMPuCV_m45obpoTCUyKVfF0hCineKJ6Mg5Q3oWaPUPyf4LWbQsvTgaAgCiEALw_wcB

725 OECD, *OECD AI Principles Overview: Countries Adhering to the AI Principles* (May 2024), https://oecd.ai/en/ai-principles

726 OECD.AI, *National AI Policies & Strategies*, https://oecd.ai/en/dashboards/overview

727 OECD, *MENA-OECD Initiative on Governance & Competitiveness for Development, 2021 Ministerial Conference* (2021), https://www.oecd.org/en/regions/middle-east-and-north-africa.html

728 Council of Europe Treaty Office, *Chart of Signatures and Ratifications of Treaty 225* (Apr. 5, 2026), https://www.coe.int/en/web/conventions/full-list?module=signatures-by-treaty&treatynum=225

729 UNESCO, *Member States, Bahrain*, https://www.unesco.org/en/countries/bh

730 UNESCO, *UNESCO Member States Adopt the First Ever Global Agreement on the Ethics of Artificial Intelligence* (Nov. 25, 2021), https://en.unesco.org/news/unesco-member-states-adopt-first-ever-global-agreement-ethics-artificial-intelligence

731 Government of Bahrain, *UNESCO's Artificial Intelligence Readiness Assessment Methodology Report – RAM* (Nov. 12, 2025), https://www.iga.gov.bh/Media/Agencies/Kingdom of Bahrain AI RAM Report.pdf

Methodology (RAM). However, UNESCO does not yet feature Bahrain with a country profile.[732] Key recommendations in the report include developing data infrastructure; inclusive STEM education focusing on women, youth, and people with disabilities; national campaigns and outreach to promote AI literacy and public trust; and research focused on the social, economic, and environmental impacts of AI as well as system development.[733]

Evaluation

Bahrain's National AI Policy incorporated principles from the GCC AI Ethics Guide and UNESCO Recommendation to drive AI development aligned to ethical and human-centered principles. Combined with the Economic Vision 2030, Digital Government Strategy 2022, and Digital Economy Strategy 2022–2026, the Kingdom of Bahrain has provided a roadmap and KPIs for the governance of AI. The participation of multiple governmental agencies advancing the digitalization and AI plans is a strength in the pathway to governing AI.

In terms of data protection and independent oversight, the alignment of the Bahrain's Personal Data Protection Law (PDPL) to GDPR, and the draft regulations for Data Protection Impact Assessments (DPIA) are positive steps, which can be enhanced by the adoption of legislation to enforce algorithmic transparency and the creation of an independent data and AI supervisory authority.

Although Bahrain has signed the Universal Declaration of Human Rights and the Arab Charter on Human Rights, the country stands as "not free" in relation to human rights protection. Concerns exist regarding the use of AI for mass surveillance purposes.

---

[732] UNESCO Global AI Ethics and Governance Observatory, *Global Hub* (Oct. 2025), https://www.unesco.org/ethics-ai/en/global-hub

[733] Government of Bahrain, *UNESCO's Artificial Intelligence Readiness Assessment Methodology Report – RAM*, pp. 10–11 (Nov. 12, 2025), https://www.iga.gov.bh/Media/Agencies/Kingdom of Bahrain AI RAM Report.pdf

Bangladesh

In 2025, Bangladesh submitted a draft National AI Policy for public consultation. The country also completed the UNESCO RAM process and published the report.

National AI Strategy

Bangladesh published its National Strategy on Artificial Intelligence in March 2020.[734] The goal is to make Bangladesh a "technologically advanced nation by the next decade." The National AI Strategy is driven by the slogan "AI for Innovative Bangladesh." To create a "sustainable AI Ecosystem," the report proposes six strategic pillars:

- research and development
- skilling and reskilling of AI workforce
- data and digital infrastructure
- ethics, data privacy, security & regulations
- funding and accelerating AI startups
- industrialization for AI technologies

Each pillar consists of a strategic brief, a roadmap, action plan, related stakeholders and lead ministries. Finally, a summary roadmap in the report includes steps for the development of AI over the next five years. Under the Ethics, Data Privacy, Security & Regulations pillar, the Bangladeshi government aims to create a new set of AI ethics guidelines to address issues such as fairness, safety, cybersecurity, and transparency. The ICT Division and Ministry of Law, Justice and Parliamentary Affairs intended to formulate Right to Explanation (RTE) Guideline for AI Algorithm.[735]

The national Parliament passed the Agency to Innovate (a2i) Bill 2023, creating the a2i as Bangladesh's national innovation agency, a statutory body. The a2i is a flagship program of the 2019 Smart Bangladesh's Vision 2041 and its 20-year Perspective Plan,[736] which aim to utilize technology for societal advancement by 2041. The Vision and Perspective Plan focus on four institutional pillars: governance, democratization, decentralization, and capacity building. As part of the three-prong strategy for Bangladesh's Innovation Economy, the Plan aims to leverage the fourth industrial revolution, including AI and smart machines, for competitiveness and a

---

[734] Information and Communication Technology Division, *National Strategy for Artificial Intelligence Bangladesh* (Mar. 2020), https://file-rangpur.portal.gov.bd/files/pbs2.dinajpur.gov.bd/files/1885c0a0_28a4_4fcc_8d4d_dcd7ce23fd8b/7b684f19a15dfcd0f542382764572486.pdf

[735] Ibid, pp. 40-41

[736] General Economics Division (GED) Bangladesh Planning Commission Ministry of Planning, *Making Vision 2041 a Reality: Perspective Plan of Bangladesh 2021–2041* (Mar. 2020), https://oldweb.lged.gov.bd/uploadeddocument/unitpublication/1/1049/vision%202021-2041.pdf

low-carbon economy. To this end, Bangladesh aims to develop a robust legal framework and governance structure for e-government.[737]

Minister of Law, Justice, and Parliamentary Affairs Mr. Anisul Huq announced the government's initiative to draft a law on AI in February 2024. The Minister emphasized plans to consult with international stakeholders before finalizing the act.[738] Minister Huq led a discussion with the Ministry of Posts on the draft framework for the AI law, stressing the importance of addressing critical aspects related to human welfare and the ethical use of AI. He highlighted the commencement of discussions on AI's legal framework, with a focus on safeguarding human rights and promoting the beneficial application of AI across various sectors.[739] A draft policy[740] was released in 2024 but is no longer available on the Ministry website.

A formal AI policy initiative under a Steering Committee in July 2025.[741] The initiative contributes to a vision to develop a "responsible, inclusive, and innovation-driven national framework for artificial intelligence that empowers every citizen and industry." The committee held three meetings in 2025 and released a draft AI policy for public consultation in early 2026. Bangladesh released a revised National AI policy in February 2026.[742] The draft policy charges the Ministry of Law, Justice, and Parliamentary Affairs with drafting "comprehensive AI liability legislation" by 2028.

Public Participation

The National AI Strategy of Bangladesh identified engagement with media and civil societies for creating a "robust ethics, data privacy, security and regulations guideline" for emerging technologies.[743] As part of its National Internet of Things (IoT) Strategy, the Bangladesh government proposed to establish "an Advisory

---

[737] Bangladesh Parliament, *Legislation* (Mar. 2024), https://www.parliament.gov.bd/acts-of-parliament

[738] Dhaka Tribune, *Govt to Make Law on Artificial Intelligence* (Feb. 13, 2024), https://www.dhakatribune.com/bangladesh/339317/anisul-govt-to-make-law-on-artificial

[739] Dhaka Tribune, *Law Minister: Draft of Law on AI and Its Use to Be Formulated by September* (Mar. 21, 2024), https://www.dhakatribune.com/bangladesh/laws-rights/342388/draft-of-law-on-ai-and-its-use-to-be-formulated-by

[740] Regulations.ai, *Bangladesh—National AI Policy (2024)* (Apr. 2, 2024), https://regulations.ai/regulations/RAI-BD-NA-NAI2DXX-2024

[741] AI Policy National Initiative, *AI Policy for Bangladesh* (2026), https://aipolicy.gov.bd/

[742] AI Policy National Initiative, *Bangladesh National AI Policy 2026–2030, Draft v2* (Feb. 9, 2026), https://aipolicy.gov.bd/docs/national-ai-policy-bangladesh-2026-2030-draft-v2.0.pdf

[743] Information and Communication Technology Division, *National Strategy for Artificial Intelligence Bangladesh,* pp. 40-41 (Mar. 2020), https://file-rangpur.portal.gov.bd/files/pbs2.dinajpur.gov.bd/files/1885c0a0_28a4_4fcc_8d4d_dcd7ce23fd8b/7b684f19a15dfcd0f542382764572486.pdf

Committee including representatives from Government, industry, academia, and community" to provide ongoing guidance in the emerging areas of IoT.[744]

The Bangladeshi government sought civil society input on the draft of the Cyber Security Act in 2023, receiving around 900 recommendations in the two weeks of the comment period.[745] Bangladesh's first National AI-driven design competition, National AI Art-A-Thon, organized by the United Nations Development Programme (UNDP) Bangladesh and the Bangladesh Computer Council (BCC) under the Bangladesh's Information and Communication Technology (ICT) Ministerial Division, brought together a diverse group of people —professional and student artists, creators, architects, designers, content makers, and technologists—for a creative challenge that used artificial intelligence to reinterpret and showcase Bangladesh's rich cultural heritage.[746]

The Steering Committee overseeing the AI Policy national initiative held a public consultation on the first draft of the National AI policy during 2 weeks in early 2026.[747] The website for the national initiative includes all meeting notes and recordings for the Steering Committee; however, the comments received from the consultation are not available

Data Protection

Although the Constitution of Bangladesh does not explicitly grant the fundamental right to privacy, Article 43 of the Constitution recognizes this right under certain restrictions and states that, "every citizen shall have the right, subject to any reasonable restrictions imposed by law in the interests of the security of the State, public order, public morality or public health – (a) to be secured in his home against entry, search and seizure; and b) to the privacy of his correspondence and other means of communication."[748]

Bangladesh proposed a draft Data Protection Act in 2022.[749] One critic noted that "The proposed Data Protection Act 2022 and the Bangladesh Telecommunication

---

744 Information and Communication Technology Division, *National Internet of Things Strategy,* p. 11 (Mar. 2020), https://ictd.gov.bd/pages/legislative-informations/national-internet-of-things-strategy-bangladesh-4e9ee4-69414b22c4774958d7b55699

745 Amnesty International, *Bangladesh: Government Must Remove Draconian Provisions from the Draft Cyber Security Act* (Sept. 5, 2023), www.amnesty.org/en/latest/news/2023/08/bangladesh-government-must-remove-draconian-provisions-from-the-draft-cyber-security-act/

746 UNDP Bangladesh, *First-Ever AI Art-A-Thon to Reimagine Heritage through Human-AI Synergy* (Apr. 20, 2025), https://www.undp.org/bangladesh/press-releases/first-ever-ai-art-thon-reimagine-heritage-through-human-ai-synergy

747 AI Policy National Initiative, *AI Policy for Bangladesh, Public Consultation* (2026), https://aipolicy.gov.bd/feedback

748 Sadiya S. Silvee and Sabrina Hasan, *The Right to Privacy in Bangladesh in the Context of Technological Advancement*, International and Comparative Law Journal 1(2) (Dec. 8, 2018), https://ssrn.com/abstract=3298069 or http://dx.doi.org/10.2139/ssrn.3298069

749 Information and Communication Technology Division, *The Proposed Data Protection Act 2022* (Jul. 16, 2022), https://www.cirt.gov.bd/acts/data-protection-act; [unofficial translation] https://dpo-

Regulatory Commission Regulation for Digital, social media and OTT Platforms, 2021 attempt to protect citizen information against American tech companies, but both draft regulations grant Bangladesh authorities powers to control everything on the Bangladeshi internet."[750] The purpose of this law is to provide security for personal data. The law does not provide a definition of personal data.[751] The proposed data protection law mandates storage of citizen data within Bangladesh. "The localization of the data within Bangladesh gives authorities broad powers to access people's personal data without judicial oversight and accountability for any violation of people's right to privacy,"[752] wrote Amnesty International in feedback to the proposed bill.

In November 2023, the Cabinet of Bangladesh gave its approval in principle to the Draft Data Protection Act 2023. The Act shall now be enacted by the Parliament.[753] The revised Act addressed stakeholders' feedback on the transition period and data localization requirements. However, the definition of personal data was not addressed.[754] The Act provides for the establishment of a Board in charge of overseeing data protection in the country. The Board shall be constituted by the Government and consisting of a chairman and four members.

The Telecommunications Act (2000) is a law "for the purpose of development and efficient regulation of telecommunication systems and telecommunication services in Bangladesh."[755] Under Section 67 (b) of the Act no person can "intercept any radio communication or telecommunication nor shall [utilize] or divulge the intercepted communication, unless the originator of the communication or the person to whom the originator intends to send it has consented to or approved the interception or divulgence." Under Section 97 of the Act, the government may ask the telecommunication operator to maintain records relating to the communications of a specific user under the broad definition of National Security and Public Interest.

---

india.com/Resources/Privacy_Regulations_in_Asia_Pacific_Countries/The-Data-Protection-Act-2022-Bangladesh.pdf

[750] Nilesh Christopher, *Bangladesh's New Data Protection Law Grants More Power to the State than Its People, Rest of World* (Aug. 24, 2022), https://restofworld.org/2022/newsletter-south-asia-bangladeshs-data-protection-law/

[751] Harisur Rohoman, *Data Protection Act 2022: More Questions than Answers*, Dhaka Tribune (Oct.16, 2022), https://www.dhakatribune.com/op-ed/2022/10/17/data-protection-act-2022-more-questions-than-answers

[752] Amnesty International, *Bangladesh: New Data Protection Bill Threatens People's Right to Privacy* (Apr. 27, 2022), https://www.amnesty.org/en/latest/news/2022/04/bangladesh-new-data-protection-bill-threatens-peoples-right-to-privacy/

[753] Dhaka Tribune, *Cabinet Gives In-Principle Approval to Draft Personal Data Protection Act* (Nov. 27, 2023), https://www.dhakatribune.com/bangladesh/government-affairs/332341/cabinet-gives-in-principle-approval-to-draft

[754] Atlantic Council South Asia Center, *Bangladesh Draft Data Protection Act 2023: Potential and Pitfalls* (May 8, 2023), https://www.atlanticcouncil.org/wp-content/uploads/2023/05/Bangladesh-Draft-Data-Protection-Act-2023-Potential-and-Pitfalls.pdf

[755] Bangladesh Telecommunication Regulatory Commission, *The Bangladesh Telecommunications Act 2001* (Apr. 16, 2001), https://btrc.gov.bd/site/view/law/-

The Information Communication Technology Act (2006) imposes responsibility on any individual or body corporate handling personal or sensitive data and requires them to maintain and implement reasonable security practices.[756] Section 46 of the Information Communication Technology Act states that the state can intercept, monitor, or decrypt data in the interest of:

- state sovereignty, integrity, or security
- relations with foreign states
- public order
- preventing the commission of or incitement to commit any offense relating to the above
- investigation of any offense

The Digital Security Act came into force in October 2018 and pertained to "offences committed through digital devices."[757] Section 26 of the Act provides for "punishment for unauthorized collection, use etc. of identity information." The Digital Security Act 2018 repealed 5 provisions of the Information Communication Technology Act 2006 including section 57, which was deemed to be in violation of rights protected under the Bangladesh Constitution.[758]

Bangladesh's parliament passed the Cyber Security Act (CSA), replacing the Digital Security Act of 2018, to combat cybercrime and disinformation. The CSA outlines measures for detecting, preventing, and prosecuting crimes committed through digital or electronic means. Additionally, Section 5 of the Act establishes a National Cyber Security Agency, which will operate under the ICT Division to fulfill the Act's objectives, including supervising and coordinating with the National Computer Emergency Response Team, Computer Emergency Teams, or the Computer Incident Response Team.

The Cyber Security Act has largely been critiqued as a replication of the DSA. As Amnesty International reported, "The only changes the CSA makes are related to sentencing, which can be summarized as follows: lowering the maximum applicable prison sentence for eight [offenses], removing a sentence of imprisonment for two [offenses], increasing the maximum applicable fine for three [offenses] and removing the higher applicable penalty for all repeat [offenses]."[759]

---

[756] Bangladesh Computer Council, *ICT Act 2006* (Jan. 30. 2023), https://bcc.portal.gov.bd/site/page/8a843dba-4055-49af-83f5-58b5669c770d/-

[757] Official Gazette, *Digital Security Act 2018*, UN Office on Drugs and Crime (Oct. 8, 2018), https://www.unodc.org/cld/uploads/res/digital-security-act-2018_html/Digital_Security_Act_2018.pdf

[758] Clooney Foundation for Justice and Center for Governance Studies (CGS), *The Information and Communication Technology Act of 2006: Bangladesh's Zombie Cyber Law* (Nov. 2024), https://cfj.org/wp-content/uploads/2024/11/Bangladesh-ICT-Act-Report_November-2024-1.pdf

[759] Amnesty International, *Bangladesh: Open Letter to the Government: Feedback on Proposed Cyber Security Act* (Aug. 22, 2023), www.amnesty.org/en/documents/asa13/7125/2023/en/

Bangladesh's interim government enacted the Cyber Security Ordinance 2025,[760] replacing the CSA. The new ordinance introduces provisions criminalizing the unlawful use of artificial intelligence (AI)—such as unauthorized access using AI tools, identity theft, and improper use of personal data—making the ordinance the country's first binding legal measures to directly address AI risks.[761] The ordinance represents a strong step towards enforceable law that recognizes AI could accelerate cybersecurity threats. Additionally, the ordinance signals Bangladesh's government concerns about AI threats (e.g., deepfakes, impersonation, and automated attacks). Still, the Cyber Security Ordinance faces international criticism for the lack of oversight structures, algorithmic accountability frameworks, or safeguards for civil liberties. The Cyber Security Ordinance's ambiguous terms could pose serious risk of government overreach and violation of human rights.

Algorithmic Transparency

While there is no explicit right to algorithmic transparency in Bangladeshi law, the government acknowledged the lack of transparency of machine learning. The national AI strategy identified a Right to Explanation Guideline for AI Algorithms as one of the outcomes and noted GDPR as a potential model for integrating this right into the national legal framework.[762]

Under the principle of oversight and accountability, the second draft of the National AI Policy 2026–2030 stresses an individual's right to "clear accessible explanation of outcomes, contestability, and human review"[763] when affected by high-impact decisions.

Biometric Identification

Since 2008, the Election Commission of Bangladesh has issued a National Identity Card (NID) which is compulsory for every Bangladeshi citizen over the age of 18 for voting and for availing 22 types of services, including banking, taxpayer identity number (TIN), driving license, and passport. In 2016, the government started issuing a machine readable "smart NID card" with a chip that can store encrypted data such as biometric and identification data for enhancing security and reducing

[760] *Cyber Security Ordinance, 2025(Ordinance No 25 of 2025)*, https://www.researchgate.net/publication/398270759_The_Cyber_Security_Ordinance_2025_An_Unofficial_English_Translation_1

[761] Human Rights Watch, *Joint Letter to Bangladesh Chief Adviser Yunus, Regarding Follow-up on Human Rights CSO Meeting during UNGA 2025* (Oct. 2025), https://www.hrw.org/news/2025/10/19/joint-letter-to-bangladesh-chief-adviser-yunus

[762] Information and Communication Technology Division, *National Strategy for Artificial Intelligence Bangladesh*, pp. 41, 47 (Mar. 2020), https://file-rangpur.portal.gov.bd/files/pbs2.dinajpur.gov.bd/files/1885c0a0_28a4_4fcc_8d4d_dcd7ce23fd8b/7b684f19a15dfcd0f542382764572486.pdf

[763] AI Policy National Initiative, *Bangladesh National AI Policy 2026–2030, Draft v2*, p. 14 (Feb. 9, 2026), https://aipolicy.gov.bd/docs/national-ai-policy-bangladesh-2026-2030-draft-v2.0.pdf

forgery.[764] In June 2023, authority for issuing NID cards was transferred from the Election Commission to the Bangladesh Home Ministry.[765] This change, part of the National Identity Registration Act, aims to streamline citizen identification processes.

Environmental Impact of AI

The draft National AI Policy 2024 emphasizes Sustainability as a Key Principle, aiming to develop AI technologies that minimize their carbon footprint and promote environmentally friendly practices.[766]

Lethal Autonomous Weapons

Bangladesh is a High Contracting Party to the Convention on Certain Conventional Weapons (CCW).[767] Bangladesh engages actively in discussions surrounding lethal autonomous weapons systems (LAWS). At a 2016 UN General Assembly, Bangladesh's representative noted Bangladesh is "of the view that further attention of the international community is required in this area so that consensus can be reached in near future so that use of such weapons can be regulated by international laws and regulations."[768] In October 2023, at the 78th UN General Assembly, Bangladesh voted in favor of resolution L.56, emphasizing the international community's need to address concerns regarding autonomous weapons.[769] Bangladesh advocates for a comprehensive assessment of these systems' compliance with international law, including humanitarian and human rights laws. Bangladesh is part of the Non-Aligned Movement (NAM) in the CCW,[770] which supports negotiating a legally binding instrument on autonomous weapons. The NAM emphasizes the

---

[764] Mizan Rahman, *Bangladesh Launches Smart National ID Cards*, Gulf Times (Oct. 16, 2016), https://web.archive.org/web/20180517001739/http://www.gulf-times.com/story/515953/Bangladesh-launches-smart-national-ID-cards

[765] BiometricUpdate, *Home Affairs Dept to Become National ID-Issuing Authority in Bangladesh* (Jun. 15, 2023), https://www.biometricupdate.com/202306/home-affairs-dept-to-become-national-id-issuing-authority-in-bangladesh

[766] Regulations.ai, *Bangladesh—National AI Policy (2024)* (Apr. 2, 2024), https://regulations.ai/regulations/RAI-BD-NA-NAI2DXX-2024

[767] UN Office of Disarmament Affairs, *High Contracting Parties and Signatories CCW* (Mar. 18, 2025), https://disarmament.unoda.org/en/our-work/conventional-arms/convention-certain-conventional-weapons/high-contracting-parties-and-signatories-ccw

[768] Permanent Mission of Bangladesh to the United Nations, New York, *Intervention [...] in the Thematic Discussion on "Conventional Weapons" in the First Committee* (Oct. 19/20, 2016), https://nypm.mofa.gov.bd/pages/static-pages/6952667c35ce18e1c05a97aa

[769] UN Office of Disarmament Affairs, *Statement by Mr. Toufiq Islam Shatil, Deputy Permanent Representative of Bangladesh to the UN Thematic Debate: "Conventional Weapons" First Committee 78th Session United Nations General Assembly* (Oct. 2023), https://docs-library.unoda.org/General_Assembly_First_Committee_-Seventy-Eighth_session_(2023)/Bangladesh.pdf

[770] UN Office of Disarmament Affairs, *Regional Groups—CCW* (Dec. 2022), https://front.un-arm.org/wp-content/uploads/2022/12/Regional-Groups-Dec-2022.pdf

urgency of regulating emerging technologies related to autonomous weapons systems through a legally binding instrument under the Convention.

AI Literacy

The Bangladesh United Nations Development Programme recognized the importance of equipping young professionals, one-third of Bangladesh's population, with AI Literacy skills to thrive in a technology-driven world. Bangladesh's UNDP made a commitment to empower youth providing AI skills to strengthen the national innovation and entrepreneurship ecosystem. Bangladesh's UNDP commitment to empower youth with AI skills alongside public and private sector partnerships resulted in programs to support youth-led start-ups, to expand access to software programming certifications and to connect youth with job opportunities.[771]

The draft National AI Policy frames public awareness campaigns as foundational to and charges the government with creating education programs to build public trust, digital literacy, and acceptance of AI systems.[772] More specific skills-based and digital literacy initiatives emphasize objectives to build an AI-ready workforce and AI use and development. For example, the policy proposes introducing AI and computational thinking education beginning in grades 8 or 9 to build foundational skills in math, programming, and machine learning.[773] The goal is to "cultivate a generation capable of innovating, not just consuming, within the global AI ecosystem." Under Education, Research, and Development, the National AI Policy acknowledges AI ethics and over-reliance on AI under the Risk Management and Accountability Measures.[774]

Human Rights

Freedom House ranked Bangladesh as "Partly Free" with a score of 44/100 for political and civil rights in the 2026 Freedom in the World Report.[775] The report highlights reforms to pen political and civic space since the ruling Awami League (AL) was overthrown in 2024 after 15 years in power. Extralegal violence and political persecution remain problems, however. The country's first general election

---

[771] Bangladesh United Nations Development Programme (UNDP), *Empowering Youth with Digital Skills and AI* (Jul. 2025), https://www.undp.org/bangladesh/news/empowering-youth-digital-skills-and-ai

[772] AI Policy National Initiative, *Bangladesh National AI Policy 2026–2030, Draft v2*, p. 23 (Feb. 9, 2026), https://aipolicy.gov.bd/docs/national-ai-policy-bangladesh-2026-2030-draft-v2.0.pdf

[773] Ibid, p. 38

[774] Ibid, p. 28

[775] Freedom House, *Freedom in the World 2026: Bangladesh* (2026), https://freedomhouse.org/country/bangladesh/freedom-world/2026

in more than a decade were "conducted in a free, fair and peaceful manner in line with international standards."[776]

The UN Office of the High Commissioner for Human Rights (OHCHR) noted concerns over excessive use of force by security forces against protesters in n August 2024 before the AL was overthrown. The OHCHR received multiple reports of extrajudicial killings, arbitrary arrests, and suppression of political dissent. The report called for independent investigations into these incidents to ensure accountability.[777]

OECD / G20 AI Principles

Bangladesh is not a member of the OECD or the G20 and has not endorsed AI principles.[778] Bangladesh does not participate in the Global Partnership on AI.[779] The draft National AI Policy[780] commits to endorsing the principles as a non-member state under the International Alignment and Cooperation foundation. The country is still early in the process of implementing policies to ensure the human-centered development of trustworthy AI.

Council of Europe AI Treaty

Bangladesh has not endorsed the Council of Europe Framework Convention on AI and Human Rights, Democratic Values, and Rule of Law.[781]

UNESCO Recommendation on AI Ethics

Bangladesh officially endorsed the UNESCO Recommendation on the Ethics of Artificial Intelligence in 2021 and has since taken steps to integrate principles including transparency, fairness, and human-centered AI development into national frameworks.[782]

UNESCO collaborated with Bangladesh's ICT Ministry and Aspire to Innovate (a2i) to conduct the Readiness Assessment Methodology (RAM) released in

---

[776] South Asia Newslog, *BNP Wins Over Two-Thirds Majority in 13th Bangladesh General Eletions* (Feb. 13, 2026), https://southasia.news.blog/2026/02/13/bnp-wins-over-two-thirds-majority-in-13th-bangladesh-general-elections/

[777] UN Human Rights Office of the High Commission, *Preliminary Analysis of Recent Protest and Unrest in Bangladesh,* (Aug 16 , 2024), https://www.ohchr.org/sites/default/files/2024-08/OHCHR-Preliminary-Analysis-of-Recent-Protests-and-Unrest-in-Bangladesh-16082024_2.pdf

[778] OECD Legal Instruments, *Recommendation of the Council on Artificial Intelligence, Adherents* (May 3, 2024), https://legalinstruments.oecd.org/en/instruments/OECD-LEGAL-0449#adherents

[779] OECD AI Policy Observatory, *About the Global Partnership on Artificial Intelligence (GPAI)* (2026), https://oecd.ai/en/about/about-gpai

[780] AI Policy National Initiative, *Bangladesh National AI Policy 2026–2030, Draft v2*, p. 25 (Feb. 9, 2026), https://aipolicy.gov.bd/docs/national-ai-policy-bangladesh-2026-2030-draft-v2.0.pdf

[781] Council of Europe Treaty Office, *Chart of Signatures and Ratifications of Treaty 225* (Mar. 28, 2026), https://www.coe.int/en/web/Conventions/full-list/?module=signatures-by-treaty&treatynum=225

April 2025. UNESCO highlights the country's advances in AI research, AI startups, the training of models on Bengali, and the draft National AI Strategy as positive directions for AI development.[783] Recommendations center on ensuring the AI strategy and policy are inclusive and rights-protecting, strengthening the data protection and cybersecurity regimes, and developing a more structured AI and digital literacy curriculum.

Evaluation

Bangladesh set out a national strategy for AI that recognizes the importance of AI ethics and endorsed the UNESCO Recommendation on the Ethics of AI. The data protection ordinance in 2025 marked an improvement and there is support in the national AI strategy for a GDPR-style law and for an explicit right of algorithmic transparency in the draft National AI Policy 2024. Bangladesh's extensive biometric identification program through the national ID system remains concerning, especially with ongoing violence even after the change in government in late 2024.

---

[783] UNESCO Global AI Ethics and Governance Observatory, *Bangladesh* (2026), https://www.unesco.org/ethics-ai/en/bangladesh

Belgium

In 2025, Belgium released a Charter for the Responsible Use of AI in Public Services and designated authorities for the protection of fundamental rights in line with Article 77 of the EU AI Act.

National AI Strategy

In October 2022, the Council of Ministers of the Belgian Federal Government approved a National Convergence Plan for the Development of Artificial Intelligence, reducing the fragmentation of regional AI strategies to "exploit the opportunities offered by AI to the full" in a Smart AI Nation.[784] The Plan centers on nine objectives: "(1) Promoting trustworthy AI; (2) Guaranteeing cybersecurity; (3) Strengthening Belgium's competitiveness and attractiveness through AI; (4) Developing a data-driven economy and a high-performance infrastructure; (5) AI at the heart of healthcare; (6) Driving more sustainable mobility; (7) Protecting the environment; (8) Better lifelong training; (9) Providing citizens with better services and protection." A steering committee created by FPS Policy and Support (BOSA) and the FPS Economy executes and coordinates the Convergence Plan.

The Convergence Plan envisions a "Smart AI Nation in which truly every citizen shares in the benefits AI provides."[785] To achieve a Smart AI Nation "we must take particular care to protect fundamental rights such as privacy and non-discrimination, and ensure that new technologies are developed within an appropriate ethical and legal framework." Under the first objective, the Convergence Plan references steps toward an EU AI Act and acknowledges "new rights and obligations may be needed to ensure that the framework for AI innovation provides the necessary protection [...] and that it is accessible and transparent to all users of AI services." Dialogue with citizens, raising awareness about AI uses and harms, and studying impacts on individuals and societies are other action items.[786]

At the beginning of 2024, Belgium launched a call for candidates to set up a Data and AI Ethics Advisory Committee for the federal administration. Its members were officially appointed by ministerial decree on May 8, 2024.[787] The creation of this Committee has several objectives, including raising awareness among civil servants

---

[784] FPS Policy and Support (BOSA), *National Convergence Plan for the Development of Artificial Intelligence*, https://bosa.belgium.be/en/themes/digital-administration/digital-strategy-and-policy/national-convergence-plan-development

[785] FSP Policy and Support BOSA, *National Convergence Plan for the Development of Artificial Intelligence: Becoming an AI Smart Nation*, p. 3 (Oct. 2022), https://bosa.belgium.be/sites/default/files/content/images/DigitaleOverheid/AI/Plan_AI_EN.pdf

[786] Ibid, pp. 5–8

[787] FPS BOSA, *Appointment of the Data and AI Ethics Advisory Committee for the Federal Administration* (May 8, 2024), https://bosa.belgium.be/en/news/appointment-data-and-ai-ethics-advisory-committee-federal-administration

in relation to the use of data and AI and its ethical considerations; preserving human oversight; and respect for values such as human rights, democracy and the rule of law.

In 2019, the Information Report on the necessary cooperation between the Federal State and the federated entities regarding the impact, opportunities, possibilities and risks of the digital "smart society" was released by a working group created by the Belgian Senate that has been meeting since 2018.[788] Their findings and recommendations are grouped in six chapters: governance, ethics and human rights, and legislation; economy, labor market and taxation; education and training; attention economy: impact on people; privacy and cybersecurity; research and development. The report states that "The development and use of artificial intelligence shall be based on the following guiding principles: prudence, vigilance, loyalty, reliability, justification and transparency, accountability, limited autonomy, humanity, human integrity, and balancing of individual and collective interests." "Fundamental rights, in particular human dignity and freedom, and privacy, must be the basis and starting point for all actions and legislation in the field of artificial intelligence."[789]

AI Regulations in the Federated States

The national Convergence Plan centers on working "in synergy with the federated states that have been pursuing a very dynamic policy for the introduction of AI for several years."[790] Belgium is a federal state with three regions as well as three communities, all of which have their own government, and many of which have also developed strategies and initiatives on digitalization or AI. "The Flemish Community, the French Community and the German-speaking Community are divided according to language and culture. They are responsible for language, culture, education, audiovisual media, and individual assistance such as specific parts of health policy and social welfare. The regions (the Flemish Region, the Brussels Capital Region and the Walloon Region) are divided based on territory."[791] They are responsible for the economy, employment, housing, public works, energy transportation, environmental and spatial planning and have some things to say concerning international affairs. The

---

[788] Sénat de Belgique, *Rapport d'information relatif à la nécessaire collaboration entre l'État fédéral et les entités fédérées en ce qui concerne les retombées, les opportunités, les potentialités et les risques de la « société intelligente » numérique* (Mar. 2019), https://www.senate.be/www/webdriver?MItabObj=pdf&MIcolObj=pdf&MInamObj=pdfid&MItypeObj=application/pdf&MIvalObj=100664119

[789] Sénat de Belgique, *Rapport d'information relatif à la nécessaire collaboration entre l'État fédéral et les entités fédérées en ce qui concerne les retombées, les opportunités, les potentialités et les risques de la « société intelligente » numérique*, (Mar. 2019), https://www.senate.be/www/webdriver?MItabObj=pdf&MIcolObj=pdf&MInamObj=pdfid&MItypeObj=application/pdf&MIvalObj=100664119

[790] Ibid, p. 3

[791] Belgium.be, Belgian Federal Government, *Belgium, A Federal State*, https://www.belgium.be/en/about_belgium/government/federale_staat

Federal Government is responsible for foreign affairs, defense, justice, finance, social security, healthcare and internal affairs.

The Flemish Region (*Vlaanderen*) released the Flanders Radically Digital program in 2019, which was renewed as Flanders Radically Digital II (VRD2) in 2021. This regional digital strategy aims to foster digital public services, data processing, and automation in the public sector.[792] The strategy focuses specifically on algorithmic transparency, privacy, explainability, and human-centered, safe, sustainable, and trustworthy AI.[793] The plan includes the allocation of 5 million euros to initiatives related to AI ethics and education.[794] Another AI Action plan,[795] covering 2024–2028, continued these efforts following a positive external evaluation of progress.

In 2022, the Flemish government adopted a new strategy for data governance, which also comprises initiatives for ethical data use, privacy, transparency, and AI analytics for the public sector.[796] In September 2024, in continuation of the Flemish Government's commitment to AI, a project to make Flemish Public Administration AI-ready was launched.[797]

The Walloon government published the Digital Wallonia 2019–2024 strategy "based on values including a cross-disciplinary approach, transparency, coherence, openness and flexibility."[798] In 2019, the Walloon government also launched the regional strategy DigitalWallonia4.ai. The strategy aims to accelerate the development of the Walloon AI ecosystem and "sustainably include Wallonia in

---

[792] Digitaal Vlaanderen, *Flanders Radically Digital II [Vlaanderen Radicaal Digitaal II]*, https://www.vlaanderen.be/uw-overheid/werking-en-structuur/hoe-werkt-de-vlaamse-overheid/informatie-en-communicatie/vlaanderen-radicaal-digitaal-ii

[793] De Vlaamse minister van Werk, Economie, Innovatie en Sport, *QUATERNOTA AAN DE VLAAMSE REGERING. Betreft: Vlaams Beleidsplan Artificiële Intelligentie* (2019), https://www.ewi-vlaanderen.be/sites/default/files/quaternota_aan_de_vlaamse_regering_-_vlaams_beleidsplan_artificiele_intelligentie.pdf

[794] Flanders Department for Economy, Science and Innovation, *Vlaams actieplan Artificiële Intelligentie gelanceerd* (Mar. 22, 2019), https://www.ewi-vlaanderen.be/nieuws/vlaams-actieplan-artificiele-intelligentie-gelanceerd

[795] Flanders Minister of Economy, Innovation, Labor, Social Economy and Agriculture, *Memo to the Flemish Government regarding Flemish Artificial Intelligence Policy Plan AI Research Program—Working Year 2024 [Betreft: Vlaams Beleidsplan Artificiële Intelligentie 2024–2028, Onderzoeksprogramma AI - werkjaar 2024]* (2024), https://live-ewi-ai.rcaonline.be/sites/default/files/2024-04/Vlaams%20Beleidsplan%20AI%202024-2028.pdf

[796] Digitaal Vlaanderen, *Actieplan 2022: Vlaamse Datastrategie* (Feb. 2022), https://assets.vlaanderen.be/image/upload/v1647858968/Vlaamse_datastrategie_kacrph.pdf

[797] UNESCO, *AI-Ready Flemish Public Administration* (Sept. 2024), https://www.unesco.org/en/artificial-intelligence/recommendation-ethics/flemish-government

[798] Digitalwallonia.be, *Digital Wallonia 2019–2024* (Jun. 2018), https://www.digitalwallonia.be/en/posts/digital-wallonia-2019-2024

national and European AI initiatives in order to build a foundation of trust around transparent, ethical and responsible AI."[799]

In 2022, the Brussels Capital Region started a pilot project with the FARI Institute, a non-profit initiative led by Vrije Universiteit Brussel and Université Libre de Bruxelles, to develop an AI strategy for the region.[800] In January 2023, FARI published outcomes from the 2022 Conference on AI, Data and Robotics in Cities, including recommendations from the Observer Committee for the Brussels Capital Region. The recommendations included raising AI awareness and literacy; governing with various stakeholders, including citizens; determining AI-related responsibilities; collaborating on multi- and interdisciplinary levels; (responsibly) enabling data flows; reusing and readapting the existing AI guidelines, principles, and regulations. [801] The adoption of the 2022 Convergence Plan was also preceded by several initiatives at national and European level.

EU Digital Services Act

The EU Digital Services Act (DSA) applies in Belgium.[802] The DSA regulates online intermediaries and platforms. Its main objective is to prevent illegal and harmful activities online and the spread of disinformation.

As required by DSA, Belgium designated competent authorities to implement provisions of the DSA. As numerous competences are involved in the implementation of the DSA, at federal and community level, four competent authorities have been designated in Belgium.[803]

A Bill to implement the DSA was introduced to the Belgian Chamber of Representatives on January 31, 2024. The Bill aims to harmonize Belgian legislation with the new requirements for digital services under the DSA. Key provisions include the designation of the Belgian Institute for Postal Services and Telecommunications (BIPT) as the competent authority to supervise the provisions of the DSA at the federal level, the issuance of decrees designating the competent authorities for federated

---

[799] Digitalwallonia.be, *DigitalWallonia4.ai : Artificial Intelligence at the Service of Citizens and Companies in Wallonia* (Apr. 2019), https://www.digitalwallonia.be/en/posts/digitalwallonia4-ai-artificial-intelligence-at-the-service-of-citizens-and-companies-in-wallonia/

[800] FARI, *AI Strategy for the Brussels Region* (Sept. 2022), https://www.fari.brussels/research-and-innovation/project/ai-strategy-for-the-brussels-region

[801] FARI, Brussels Conference 2022: Summary and Recommendations, *What Can the Brussel Capital Region Do to Foster Responsible AI?* (Jan. 2023), https://issuu.com/faribrussels/docs/en.fari_conference_summary_2022

[802] *Regulation (EU) 2022/2065 of the European Parliament and of the Council of 19 October 2022 on a Single Market for Digital Services and amending Directive 2000/31/EC (Digital Services Act)*, (Oct. 21, 2022), https://eur-lex.europa.eu/legal-content/EN/TXT/?uri=celex%3A32022R2065

[803] Belgian Institute for Postal Services and Telecommunications, *Digital Services Act: Belgium Has Designated Its Four Competent Authorities to Enforce the Provisions of the European Regulation Brussels* (May 24, 2024), https://www.bipt.be/consumers/publication/digital-services-act-belgium-has-designated-its-four-competent-authorities-to-enforce-the-provisions-of-the-european-regulation-brussels

entities, and the establishment of a cooperation agreement between the federal state and the communities. The Bill introduces additional obligations for providers of very large online platforms and search engines. These obligations include the management, assessment, and limitation of systemic risks and the implementation of crisis response mechanisms. They apply to platforms and search engines with an average monthly number of users in the EU of 45 million or more. The Bill also proposes changes to other aspects of Belgian law to align with the DSA and introduces sanctions for non-compliance with orders issued by the competent authorities.[804]

EU AI Act

As an EU member State, Belgium is bound by the EU AI Act.[805] Belgium's FSA Policy and Support (BOSA) identifies the EU AI Act as an important aid in implementing the National Convergence Plan. A Royal Decree established the Steering Committee as the primary federal contact on matters related to AI policy, alongside the Federal Government Data Ethics and AI Advisory Committee established in 2023.[806]

Belgium has not designated an official single point of contact for the EU AI Act but has identified a number of authorities to protect fundamental rights in line with Article 77.[807] Belgium has designated 18 authorities at the federal level as well as authorities for each of the regions. Some of these authorities, such as Unia, an institute for equality, overlap. The data protection authority is also listed, along with bodies dedicated to cybersecurity, policing, and migration.

---

[804] La Chambre.Be, *Draft law implementing Regulation (EU) 2022/2065 [...] on a single market for digital services and amendinding [...] reslating to the status of the regulator of the Belgian postal and telecommunications sectors [Document parlementaire 55K3799: Project de loi mettant en oeuvre le règlement (UE) 2022/2065 [...] à un marché unique des services numériques et modifiant [...] au statut du régulateur des secteurs des postes et des télécommunictions belges]* (Apr. 18, 2024), https://www.dekamer.be/kvvcr/showpage.cfm?section=/flwb&language=fr&cfm=/site/wwwcfm/flwb/flwbn.cfm?lang=N&legislat=55&dossierID=3799

[805] European Parliament, *Artificial Intelligence Act, European Parliament legislative resolution of 13 March 2024 on the proposal for a regulation of the European Parliament and of the Council on laying down harmonised rules on Artificial Intelligence (Artificial Intelligence Act) and amending certain Union Legislative Acts (COM(2021)0206 – C9-0146/2021 – 2021/0106(COD))*, P9_TA(2024)0138, https://www.europarl.europa.eu/RegData/seance_pleniere/textes_adoptes/definitif/2024/03-13/0138/P9_TA(2024)0138_EN.pdf

[806] Service Public Federal Economie, PME, Classes Moyennes et Energie, *Royal Decree Instituting a Steering Committee on Artificial Intelligence [Arrêté royal instituant un Comité d'orientation en matière d'intelligence artificielle*, Moniteur Belge n.121, 6 June 2024 (May 25, 2024), https://www.ejustice.just.fgov.be/mopdf/2024/06/06_1.pdf#Page201

[807] SPF Economie, PME Classes Moyennes et Energie, *List of Belgian Authorities Charged with the Protection of Fundamental Rights [Liste des autorités belges en charge de la protection des droits fondamentaux (Article 77 AI Act)]* (Mar. 2025), https://economie.fgov.be/sites/default/files/Files/Online/Liste-des-autorites-belges-en-charge-de-la-protection-des-droits-fondamentaux.pdf

Belgium voted against EU AI Code of Conduct, or the General-Purpose AI Code of Practice, over copyright issues. Commenting on the vote, the federal minister for Digitalisation, Vanessa Matz, told the press, "Belgium and other European countries have emphasised that this is not the end of the process. The Commission has acknowledged that the code may need to be evaluated and amended sooner than legally required. I will continue to work with the relevant services to ensure that AI models take the interests of journalists, publishers, producers and creators into account."[808] The General-Purpose AI Code of Practice is a voluntary tool to help industry comply with the AI Act's obligations for providers of general-purpose AI models.[809]

Public Opinion

A 2019 opinion survey by AI4Belgium, a committee under the FSA BOSA, examined the public perception of AI, its perceived impact, and the role the government should play in AI implementation.[810] According to the survey, 76% of the respondents hold a positive attitude towards technological developments, while only 6% hold a negative attitude. Most respondents were worried about the loss of privacy, security and integrity of their personal information (85%), less use of human common sense (85%), less human interaction (83%) and the loss of trust and control over robots and artificial intelligence (77%).

When asked which activity to prioritize, the highest priority concerned "the management of ethical risks around AI. For example, discrimination, privacy, etc." (74%). This was followed by "supporting employees and employers in the transition to AI in the workplace" (65%); "improving public service through AI" (58%); "supporting research and development (R&D) and innovation in the field of AI" (52%); "facilitating and supporting enterprise access to AI technologies" (48%); and "supporting start-ups engaged in AI" (45%). The majority of citizens suspect that AI will increase inequality between highly educated and low- or unskilled people (66%) and between persons with a privileged background and persons without one (60%).

Public Participation

As part of its Presidency of the Council of the European Union in the first half of 2024, Belgium launched a pioneering democratic initiative: a citizens' panel on

---

[808] Belga News Agency, *Belgium Votes against EU AI Code of Conduct amid Copyright Concerns* (Aug. 1, 2025), https://www.belganewsagency.eu/belgium-votes-against-eu-ai-code-of-conduct-amid-copyright-concerns

[809] European Commission, *The General-Purpose AI Code of Practice* (Jul. 10, 2025), https://digital-strategy.ec.europa.eu/en/policies/contents-code-gpai

[810] AI4Belgium, *Artificial Intelligence Perception [Perception intelligence artificielle]* (Feb. 2019), https://bosa.belgium.be/sites/default/files/content/documents/DTdocs/AI/Etude-Perception-intelligence-artificielle-Ipsos.pdf

artificial intelligence (AI).[811] This initiative underlines the country's commitment to an inclusive and participatory approach towards the formulation of European policies in the field. The citizens' panel, which is made up of 60 people selected at random from over 16,000 invitations sent out across Belgium, brought together all strata of the population in terms of age, gender, levels of education, and other demographic criteria. This diversity ensured that the discussions and recommendations reflect a wide range of perspectives and experiences rooted in people's lived experiences.[812]

Belgian Minister of Foreign and European Affairs described the initiative as their determination "to put the Belgian people at the heart of the decision-making process. [813] Civil society must be heard on issues as important as artificial intelligence and contribute towards ambitious policies that meet their expectations." The panel's conclusions were compiled into a report, which featured 9 key messages aimed at contributing to Belgian and EU policy on AI. These messages reflected citizens' aspirations for a responsible, ambitious, and beneficial approach to AI to ensure that AI serves the interests of all and leaves no one behind.[814] On May 25, 2024, the report was presented and debated in Brussels, in the presence of Belgian and European politicians, representatives of civil society and tech companies.[815]

Data Protection

Since Belgium is an EU Member State, the General Data Protection Regulation (GDPR) is directly applicable in Belgium. Taking stock of the GDPR, the Belgium Privacy Commission was reformed in 2018. It is now called the Belgian Data Protection Authority (DPA) and aims to ensure compliance with the GDPR.[816] The DPA has direct sanctioning powers as well as extended enforcement capabilities. The

---

[811] Belgian Federal Government, *Belgium Launched AI Presidency Citizen Panel* (Feb. 27, 2024), https://belgian-presidency.consilium.europa.eu/en/news/launch-of-citizens-panel-on-artificial-intelligence/

[812] Belgium Federal Government, *Belgian Citizens Meet to Reflect on AI* (May 29, 2024), https://diplomatie.belgium.be/en/policy/policy-areas/highlighted/belgian-citizens-meet-reflect-ai

[813] Belgian Federal Government, *Belgium Launched AI Presidency Citizen Panel* (Feb. 27, 2024), https://belgian-presidency.consilium.europa.eu/en/news/launch-of-citizens-panel-on-artificial-intelligence/

[814] Belgium EU Presidency, *The Citizen's Panel on AI Issues Its Report* (May 22, 2024), https://wayback.archive-it.org/12710/20241018172535/https://belgian-presidency.consilium.europa.eu/en/news/the-citizens-panel-on-ai-issues-its-report/

[815] Belgium EU Presidency, *beEU Citizen Panel Closing Event* (May 25, 2024), https://wayback.archive-it.org/12710/20240718214829/https://belgian-presidency.consilium.europa.eu/en/events/beeu-citizen-panel-closing-event/

[816] PWC Legal, *The New Belgian Data Protection Authority: Who's Who and How Will It Work* (Jan. 23, 2019), https://www.pwclegal.be/en/news/the-new-belgian-data-protection-authority---whos-who-and-how-wil.html

Belgian DPA published guidance on the interplay between the GDPR and the AI Act on September 19, 2024.[817]

In March 2022, the DPA expressed concern over the Belgian draft legislation on amending the Act of December 3, 2017, establishing the Data Protection Authority. In particular, the DPA mentioned that the draft law jeopardizes its efficiency and independence.[818] The draft legislation introduces parliamentary interference regarding the DPA internal organization and setting its priorities. The European Data Protection Board expressed similar concerns in its letter of April 2022 in support of the Belgian DPA.[819] The reform proposal of the DPA Act amending the law of December 3, 2017, was approved and entered into force on June 1, 2024.[820]

Belgium also established a Federal Data and AI Ethics Advisory Committee to promote ethical AI use in public administration, with objectives that include raising awareness among civil servants on data and AI ethics.[821]

Regarding the activities of law enforcement authorities, Belgium transposed[822] the EU Data Protection Law Enforcement Directive (LED).[823] The LED provides for the prohibition of any decision based solely on automated processing, unless it is provided by law, and of profiling that results in discrimination,[824] and "will in particular ensure that the personal data of victims, witnesses, and suspects of crime

---

[817] Belgium DPA, *Artificial Intelligence Systems and the GDPR A Data Protection Perspective*, (Sep 19, 2024), https://www.gegevensbeschermingsautoriteit.be/publications/information-brochure-on-artificial-intelligence-systems-and-the-gdpr.pdf

[818] Belgian Data Protection Authority, *A New Draft Law Threatens the Independence and Functioning of the BE DPA* (Mar. 8, 2022), https://www.dataprotectionauthority.be/citizen/a-new-draft-law-threatens-the-independence-and-functioning-of-the-be-dpa

[819] European Data Protection Board letter (Apr. 6, 2022). https://edpb.europa.eu/system/files/2022-04/edpb_letter_out_2022-0022_belgian_draft_legislation_en.pdf

[820] Data Protection Authority, *A New Chapter for the APD – Amendment of the Law and New ROI*, (May 31, 2024), https://www.autoriteprotectiondonnees.be/citoyen/actualites/2024/05/31/un-nouveau-chapitre-pour-lapd-modification-de-la-loi-et-nouveau-roi

[821] FPS BOSA, *Appointment of the Data and AI Ethics Advisory Committee for the Federal Administration* (May 8, 2024), https://bosa.belgium.be/en/news/appointment-data-and-ai-ethics-advisory-committee-federal-administration

[822] Belgian Official Journal, *Act on the protection of natural persons with regard to the processing of personal data* (Jul. 30, 2018), https://www.dataprotectionauthority.be/publications/act-of-30-july-2018.pdf

[823] *Directive (EU) 2016/680 of the European Parliament and of the Council of 27 April 2016 on the protection of natural persons with regard to the processing of personal data by competent authorities for the purposes of the prevention, investigation, detection or prosecution of criminal offences or the execution of criminal penalties, and on the free movement of such data, and repealing Council Framework Decision 2008/977/JHA*, https://eur-lex.europa.eu/legal-content/EN/TXT/?uri=celex%3A02016L0680-20160504

[824] Article 11 (1) and (2) of the LED, https://eur-lex.europa.eu/legal-content/EN/TXT/?uri=celex%3A02016L0680-20160504

are duly protected and will facilitate cross-border cooperation in the fight against crime and terrorism."[825]

Following the transposition of the EU Law Enforcement Directive in Belgian law, the Supervisory Body for Police Information, "the oversight body which looks at how the police use information (COC) was reformed to function as an independent data protection body."[826] It is intended to oversee how the police use data.[827]

Both the DPA and the COC are members of the Global Privacy Assembly (GPA). The DPA co-sponsored the 2018 GPA Declaration on Ethics and Data Protection in Artificial Intelligence[828] but did not endorse the 2020 Resolution on Accountability in the Development and Use of Artificial Intelligence,[829] 2022 GPA Resolution on Facial Recognition Technology,[830] or 2023 GPA Resolution on Generative AI Systems.[831]

Belgium is also a member of the Council of Europe and ratified the Council of Europe's Convention 108+ for the protection of individuals with regard to the processing of personal data.[832]

In 2019, a national human rights institution, the Federal Institute for the Protection and Promotion of Human Rights, was established. Its main goal is to facilitate cooperation between the existing human rights oversight mechanisms and fill the gaps in the existing landscape.[833] More specifically, it "ensures that the use of

---

[825] European Commission, *Legal Framework of Data Protection in the EU*, https://commission.europa.eu/law/law-topic/data-protection/legal-framework-eu-data-protection_en

[826] Algorithm Watch, *Automating Society Report 2020: Belgium* (2020), https://automatingsociety.algorithmwatch.org/report2020/belgium/

[827] Supervisory Body for Police Information, https://www.controleorgaan.be/en/

[828] Global Privacy Assembly, *Declaration on Ethics and Data Protection in Artificial Intelligence* (Oct. 23, 2018), https://globalprivacyassembly.org/wp-content/uploads/2018/10/20180922_ICDPPC-40th_AI-Declaration_ADOPTED.pdf

[829] Global Privacy Assembly, *Resolution on Accountability in the Development and Use of Artificial Intelligence* (Oct. 2020), https://globalprivacyassembly.org/wp-content/uploads/2020/10/FINAL-GPA-Resolution-on-Accountability-in-the-Development-and-Use-of-AI-EN-1.pdf

[830] Global Privacy Assembly, *Resolution on Principles and Expectations for the Appropriate Use of Personal Information in Facial Recognition Technology* (Oct. 2022), https://globalprivacyassembly.org/wp-content/uploads/2022/11/15.1.c.Resolution-on-Principles-and-Expectations-for-the-Appropriate-Use-of-Personal-Information-in-Facial-Recognition-Technolog.pdf

[831] Global Privacy Assembly, *Resolution on Generative Artificial Intelligence Systems* (Oct. 2023), https://globalprivacyassembly.org/wp-content/uploads/2023/10/5.-Resolution-on-Generative-AI-Systems-101023.pdf

[832] Council of Europe, *Convention 108 and Protocols* (2026), https://www.coe.int/en/web/data-protection/convention108-and-protocol

[833] European Networks of National Human Rights Institutions, *ENNHRI Welcomes New Law Adopted on National Human Rights Institution in Belgium* (May 9, 2019), http://ennhri.org/news-and-blog/ennhri-welcomes-new-law-adopted-on-national-human-rights-institution-in-belgium/

new technologies and the digitalisation of society contribute to strengthening our rights rather than limiting them."[834]

Algorithmic Transparency

Belgium is subject to the GDPR and Convention 108+. Belgians have a general right to obtain access to information about automated decision-making and to the factors and logic of an algorithm.[835] They also have the right to meaningful contestation.[836] Belgium has signed but not yet ratified Convention 108+.[837] Belgium's ratification is required for the Convention to enter into force.

In 2019, the Federal Institute for the Protection and Promotion of Human Rights issued its first opinion on the use of algorithms and artificial intelligence by the administration and argued in favor of more transparency from public authorities.[838]

Algorithm Registers

Brussels, in collaboration with eight other cities across Europe and with the help of Eurocities' Digital Forum, adopted an algorithm register, the Algorithmic Transparency Standard. The data officer for the Brussels Capital region will oversee the register, which allows residents to access information about the use of algorithms by municipalities and their impact. The register includes information such as the type and purpose of an algorithm, the department using the algorithm, the geographical area and domain it relates to, and a risk category. The register also references the data source and training data, any bias and mitigation, and human oversight. This initiative builds on similar algorithm registers launched in Amsterdam and Helsinki.

According to André Sobczak, Secretary General, Eurocities, "The efforts undertaken by these cities aim to set a standard for the transparent and ethical use of algorithms while their use is still in its relative infancy across city administrations in Europe. In this way, they seek to offer both a safeguard for people whose data may be

---

[834] Federal Institute for the Protection and Promotion of Human Rights, *Technology, Digitalisation and Human Rights, What Does FIRM/IFDH Do?*, https://www.federalinstitutehumanrights.be/en/uw-rechten/themes/technology-digitalisation-and-human-rights

[835] See Recital 63 and Article 22 of the GDPR. Article 9 c) of the Convention 108+ as well as Recital 77, *Explanatory Report, Convention 108+*, p. 24, https://rm.coe.int/convention-108-convention-for-the-protection-of-individuals-with-regar/16808b36f1

[836] *Recommendation CM/Rec(2020)1 of the Committee of Ministers to Member States on the Human Rights Impacts of Algorithmic Sstems* (Apr. 8, 2020), https://search.coe.int/cm/pages/result_details.aspx?objectid=09000016809e1154

[837] Council of Europe Treaty Office, *Chart of Signatures and Ratifications of Treaty 223* (Feb. 8, 2026), https://www.coe.int/en/web/conventions/full-list;?module=signatures-by-treaty&treatynum=223

[838] Federal Institute for the Protection and Promotion of Human Rights, *L'IFDH demande plus de transparence sur l'utilisation des algorithmes par les autorités* (Oct. 7, 2021), https://www.federalinstitutehumanrights.be/en/nieuws/lifdh-demande-plus-de-transparence-sur-lutilisation-des-algorithmes-par-les-autorites

used by algorithms, and have created a validated model that other cities can use straight away." [839]

AI in Public Administration

The Council of Ministers proposed the establishment of an Advisory Committee on Ethics on Data and Artificial Intelligence for the federal administration[840] to a) empower civil servants in the use of data and AI, b) raise awareness among civil servants about the ethical aspects of data use, and c) show citizens that the Federal Administration is setting an example and dealing with digital technology in an ethical and innovative way. The committee was established in 2024.[841]

In January 2024, Wallonia announced its aim to integrate AI into the public service sector.[842] To this end, Wallonia launched a public call for the development of a Proof of Concept (PoC) with public entities.[843] The Flemish region also initiated a project in 2024 to ensure the Flemish Public Administration AI-ready.[844] This project is funded by the European Commission's Directorate-General for Structural Reform Support. UNESCO's Ethics of AI Unit, Social and Human Sciences Sector, is tasked with implementing the project, which will leverage UNESCO's Readiness Assessment Methodology (RAM).

The Council of Europe Technical Support Project launched an initiative to strengthen the administrative capacity of equality bodies in Belgium (as well as Finland and Portugal) over 2024–2026, increasing the Belgian Interfederal Center for Equal Opportunities (Unia) readiness to uncover discrimination in AI deployment in public administration and support those facing discrimination by AI, in line with Council of Europe standards.[845]

---

[839] Eurocities, *Nine Cities Set Standards for the Transparent Use of Artificial Intelligence* (Jan. 23, 2023), https://eurocities.eu/latest/nine-cities-set-standards-for-the-transparent-use-of-artificial-intelligence/

[840] News.belgium, *Création d'un comité consultatif d'éthique des données et de l'intelligence artificielle* (Jul. 7, 2023), https://news.belgium.be/fr/creation-dun-comite-consultatif-dethique-des-donnees-et-de-lintelligence-artificielle

[841] FPS Justice, *Ministerial Degree Appointing the Members of the Advisory Committee on Data Ethics and Artificial Intelligence of the Federal Administration [Arrêté ministériel portant désignation des membres du Comité Consultatif d'Ethique des Données et d'Intelligence Artificielle de l'administration fédérale]* (May 8, 2024), https://www.ejustice.just.fgov.be/cgi/article.pl?language=fr&sum_date=2024-05-29&lg_txt=F&numac_search=2024004813

[842] Wallonia Regional Government of Belgium, *Wallonia to Integrate AI into Public Services* (Jan 7, 2024), https://www.digitalwallonia.be/fr/publications/ia-services-publics/

[843] Ibid

[844] UNESCO, *AI-Ready Flemish Public Administration* (Sept. 2024), https://www.unesco.org/en/artificial-intelligence/recommendation-ethics/flemish-government

[845] Council of Europe, *Upholding Equality and Non-Discrimination by Equality Bodies regarding the Use of Artificial Intelligence (AI) in Public Administrations* (Nov. 2024), https://www.coe.int/en/web/inclusion-and-antidiscrimination/upholding-equality-and-non-

As part of Objective 9 of the National AI Convergence Plan to "provide citizens with better services and better protection,"[846] the AI4Belgium Ethics and Law Committee developed the Charter for the Responsible Use of Artificial Intelligence in Public Services[847] in collaboration with experts and government agencies.[848] Fundamental principles and commitments underlying the charter include: the option to choose Human interaction over AI, Transparency about when AI is used, Human control or supervision for high-risk AI systems, the right to challenge an AI-supported decision, Data quality and governance requirements, Risk assessment and mitigation before deployment, Sustainability and environmental-impact-aware applications, and Continuing education about AI for public officials. Forty public service entities across the Federal Public Services and Programming, Public Interest Organizations, and Public Social Security Institutes have agreed to adhere to the Charter.

Facial Recognition

According to AlgorithmWatch, facial recognition has been used at the Brussels Airport, football matches, for school registration and for healthcare. A "smart" video surveillance system is also in use to locate criminals, solve theft cases and collect statistical information.[849] AlgorithmWatch stressed that there is no legal framework governing this activity by the police.

Additionally, Belgium's involvement in the Prüm Convention facilitates the sharing of biometric data, including facial recognition images, across European borders. In March 2024, Belgium signed the Prüm II expansion, which includes provisions for cross-border facial recognition technology. This expansion underscores the need for Belgium to implement robust data protection measures and clear guidelines to ensure that cross-border use of Facial Recognition Technology (FRT) aligns with privacy standards.[850]

---

discrimination-by-equality-bodies-regarding-the-use-of-artificial-intelligence-ai-in-public-administrations1

[846] FPS Policy and Support (BOSA), *National Convergence Plan for the Development of Artificial Intelligence*, https://bosa.belgium.be/en/themes/digital-administration/digital-strategy-and-policy/national-convergence-plan-development

[847] AI4Belgium Ethics and Law Committee, *Charter for the Responsible Use of Artificial Intelligence in Public Services [Charte pour l'utilisation responsable de l'intelligence artificielle dans les services publics]* (Jul. 11, 2025), https://bosa.belgium.be/sites/default/files/publications/documents/Charte%20pour%20l%27utilisation%20responsable%20de%20l%27IA%20dans%20les%20services%20publics.pdf

[848] FPS BOSA, *Artificial Intelligence in Public Services: A Charter for Responsible Use [L'intelligence artificielle dans les services publics: Une charte pour l'utilisation responsible]*, https://bosa.belgium.be/fr/themes/administration-numerique/intelligence-artificielle-et-nouvelles-technologies/lintelligence

[849] AlgorithmWatch, *Automating Society 2020* (Oct. 2020), https://automatingsociety.algorithmwatch.org/report2020/belgium/

[850] EUCRIM, *Prüm II Regulation Enters into Force* (2024), https://eucrim.eu/news/prum-ii-regulation-enters-into-force/

The COC has criticized the use of facial recognition at the Brussels airport, stating that there is "too little information about the implementation and risks of the technology as there was no clear policy or data protection impact assessment conducted to come to a conclusion or offer advice." In a 2022 report, the COC reiterated "the lack of legal basis for the use of facial recognition technology by the Brussels airport police" and noted that it took "corrective action" in order to halt the pilot project.[851] The COC also reported that Belgian police had used the controversial facial recognition system Clearview AI without a legal basis and advised Belgian police to cease this activity."[852]

Belgium's 2025–2029 federal coalition agreement commits to deploying AI tools in law enforcement, including facial recognition "for the detection of convicts and suspects."[853] The coalition agreement text says that the government will work on "a concrete policy" regarding the use of these technologies "within a strict and precisely defined legal framework." The text also mentions an expansion of the camera legislation to allow broader (smart) surveillance.

Regulatory Sandboxes

In Belgium, the Flemish and Walloon regions have both developed regulatory sandboxes aimed at creating self-contained, low-regulation environments where software developers can experiment with innovative solutions without taking on significant risks. In Flanders, the Sandbox Flanders project is coordinated through the Flemish government agency for digitalization, Digital Flanders. Sandbox Flanders is designed to develop innovative solutions for the public sector by "matching" start-ups with government agencies and departments. The project also aims to be a "permanent marketplace" for ideas without a profit motive.[854] Many of the projects in question use AI systems to solve practical issues in particular "challenges," such as predicting

---

[851] Controleorgaan.be, *Advies betreffende een voorstel van resolutie over een driejarig moratorium op het gebruik van gezichtsherkenningssoftware en – algoritmen in vaste of mobiele beveiligingscamera's in openbare en privéplaatsen* (Jan. 2022), https://www.controleorgaan.be/files/DA210029_Advies_N.pdf

[852] Controleorgaan.be, *toezichtrapport van het controleorgaan op de politionele informatie met betrekking tot het gebruik van clearview ai door de geïntegreerde politie* (Feb. 2022), https://www.controleorgaan.be/files/DIO21006_Toezichtrapport_Clearview_N_00050443.pdf

[853] Gouvernement Fédéral Belge, *Federal Coalition Agreement [Accord de coalition fédérale 2025-2029]* (Jan. 31, 2025), https://www.belgium.be/sites/default/files/resources/publication/files/Accord_gouvernemental-Bart_De_Wever_fr.pdf; Knowledge Centre Data & Society, *Belgium—Federal Government Agreement 2025–2029* (Feb. 18, 2025), https://data-en-maatschappij.ai/en/publications/belgi%C3%AB-federaal-regeerakkoord-2025-2029

[854] Digitaal Vlaanderen, *Sandbox Vlaanderen: Ruimte voor innovatie en experiment*, https://www.vlaanderen.be/digitaal-vlaanderen/onze-oplossingen/sandbox-vlaanderen-ruimte-voor-innovatie-en-experiment

exam enrollments and automated policy impact assessments.[855] In the Walloon case, a private entity (LUDEBO) coordinates the sandbox ecosystem together with Digital Wallonia and the blockchain startup-collective Walchain, but the goal of this regulatory sandbox is only to foster blockchain technology development.[856]

Starting in July 2023, regional authorities commenced sectorial AI testing under the Digital Europe Program.[857] The Testing and Experimentation Facilities (TEFs) are poised to aid in the implementation of the EU AI Act, particularly by supporting regulatory sandboxes in collaboration with national authorities. This initiative encompasses a facility spearheaded by Digitaal Vlaanderen, focused on smart cities and communities, with the objective of ensuring AI technologies align with European values prior to market introduction. The Flemish Government published a reflection paper outlining models and implementation pathways for an AI regulatory sandbox targeting SMEs and high-risk AI systems within Belgium's governance structure.[858] The paper provides a foundational framework and practical options for an AI regulatory sandbox within Belgium, including support for SMEs and high-risk AI innovators, alignment with the strategic objectives set out by the EU AI Act, and specific operational structures and legal considerations for implementation.

In 2024, the Brussels-Capital Region launched a mobility data and AI regulatory sandbox as part of the European AI Testing and Experimentation Facility for Smart and Sustainable Cities and Communities. Coordinated by Digitaal Vlaanderen, this initiative supports Belgian startups in developing AI solutions by providing access to real data and controlled testing environments. The sandbox specifically focuses on AI applications in the mobility sector, offering a structured environment for experimentation that ensures compliance with European standards. This project underscores Belgium's commitment to fostering AI innovation while aligning with EU regulatory standards for safe and ethical AI development.[859]

---

[855] Digitaal Vlaanderen, *Challenges Sandbox Vlaanderen,* https://www.vlaanderen.be/digitaal-vlaanderen/onze-diensten-en-platformen/sandbox-vlaanderen-ruimte-voor-innovatie-en-experiment/experimenten-sandbox-vlaanderen

[856] Digitalwallonia.be, *Sandbox Wallonia (SBW)*, https://www.digitalwallonia.be/en/cartography/sandbox-wallonia/

[857] Agentschap Digitaal Vlaanderen, *Digitaal Vlaanderen bouwt mee aan Europese testfaciliteiten voor AI-oplossingen* (Jul. 17, 2023), https://www.vlaanderen.be/digitaal-vlaanderen/nieuwsberichten/digitaal-vlaanderen-bouwt-mee-aan-europese-testfaciliteiten-voor-ai-oplossingen

[858] Vlaamse Overheid, Departement Economie, Wetenschap en Innovatie (EWI), *Reflectiepaper AI Regulatory Sandbox in Vlaanderen* (Oct. 11, 2023), https://publicaties.vlaanderen.be/view-file/60622

[859] Gegevensbeschermingsautoriteit, *Presentatie van Karl-Filip Coenegrachts – CITCOM.AI: A Brussels Mobility Data Project* (2024), https://www.gegevensbeschermingsautoriteit.be/publications/presentatie-van-karl-filip-coenegrachts---citcom.ai---a-brussels-mobility-data-project.pdf

AI in Healthcare

The use of AI in healthcare is one of the cornerstones of Belgium's AI strategy. An entire chapter of its national convergence plan is devoted to "AI at the heart of healthcare"[860] and the plan states that "the creation of a healthcare data agency reflects the importance and the necessity of facilitating the re-use of healthcare for research and innovation." The national strategy also emphasizes the "perspective of the user" in that it aims to stimulate human-centric use of AI in healthcare, while acknowledging the necessity of clear accountability rules, attention to data bias and privacy, and giving patients agency over their own data.[861]

In 2024, the Belgian EBCP (Evidence-Based Care Policies) mirror group released a policy brief focused on enhancing AI applications in healthcare, particularly in cancer care. The brief identifies critical areas for improvement, including data access, IT infrastructure, and legal and ethical frameworks, while emphasizing the importance of building public trust in AI applications. The brief also highlights Belgium's active role in EU initiatives, such as the European Health Data Space (EHDS) and the EU Cancer Imaging Infrastructure (EUCAIM), advocating for continued investments to align with European data-sharing and healthcare innovation standards.

Belgium's Health Data Agency funds projects "to facilitate the development and management of data in the healthcare sector, improving the quality, affordability, prevention and targeting of healthcare for every citizen and every patient."[862] Three of the four projects approved for the first round of funding center on Ai in the healthcare space. They include a project using LLMs to convert unstructured medical text, like notes, into more structured data and medical letters in Dutch to make the data more useful for analysis, another applying generative AI to match children to clinical trials, and develop an AI tool to determine whether patient data can be truly anonymized for export.

Environmental Impact of AI

Belgium's Convergence Plan for developing AI includes protecting the environment among the main objectives.[863] However, there are no specific action

[860] FPS Policy and Support (BOSA), *National Convergence Plan for the Development of Artificial Intelligence*, https://bosa.belgium.be/en/themes/digital-administration/digital-strategy-and-policy/national-convergence-plan-development
[861] Ibid
[862] Health Data Agency, *Innovation Project Funding* (2026), https://www.hda.belgium.be/en/services/innovation-project-funding#atab_932_panel_2
[863] FPS Policy and Support (BOSA), *National Convergence Plan for the Development of Artificial Intelligence*, https://bosa.belgium.be/en/themes/digital-administration/digital-strategy-and-policy/national-convergence-plan-development

steps to ensure that AI development is sustainable. The EU AI Act, for example, has been criticized for emphasizing AI *for* sustainability over AI that *is* sustainable.[864]

Lethal Autonomous Weapons

In 2018, the Belgian Parliament passed a Resolution to prohibit use, by the Belgian Defense, of killer robots and armed drones.[865] In this resolution, the Parliament states that Belgium should:

- Participate in international working groups within the framework of the United Nations and the Convention on Certain Conventional Weapons (CCW) in particular to work towards an internationally recognized definition of killer robots and to determine which types of weapons will fall into this category in the future
- Advocate in international fora, together with like-minded countries, for a global ban on the use of killer robots and fully automated armed drones
- Ensure that the Belgian Defense never deploys killer robots in military operations
- Support the development and use of robotic technology for civilian purposes.

The culmination of Belgium's efforts, which began in 2018 to ban lethal autonomous weapons, was marked by the Belgian Defense Committee's approval of a bill to this effect.[866] Consequently, in January 2023, Belgium became one of the first countries in the world to implement an outright ban on lethal autonomous weapons.

Belgium has taken a central role in putting Lethal Autonomous Weapons Systems (LAWS) on the international agenda. During its chairmanship of the Group of Governmental Experts (GGE) on Lethal Autonomous Weapons Systems, Belgium aimed to "find consensus in relation to the clarification, consideration, and development of aspects of the normative and operational framework on emerging technologies in the area of LAWS."[867]

Belgium was one of the 70 countries that endorsed a joint statement on autonomous weapons systems at the 2022 United Nations General Assembly. The joint statement urged "the international community to further their understanding and address these risks and challenges by adopting appropriate rules and measures, such

---

[864] Zuzanna Warso and Kris Shrishak, *Hope: The AI Act's Approach to Address the Environmental Impact of AI*, Tech Policy Press (May 21, 2024), https://www.techpolicy.press/hope-the-ai-acts-approach-to-address-the-environmental-impact-of-ai/

[865] Chambre des représentants, *Proposal for a Resolution Regarding the Creation of an Inclusive and Sustainable Robo-Digital Agenda [Proposition de résolution relative à la création d'un agenda robonumérique inclusif et durable]* (Jul. 27, 2017), https://www.lachambre.be/doc/flwb/pdf/54/2643/54k2643001.pdf

[866] Nick Amies, *Belgium upholds decision to ban 'killer robots'*, https://www.brusselstimes.com/350980/belgium-upholds-decision-to-ban-killer-robots

[867] Delegation of the European Union to the UN and other international organisations in Geneva, *EU lines to take: Group of Governmental Experts on emerging technologies in the area of Lethal Autonomous Weapons Systems* (Mar. 2022), https://www.eeas.europa.eu/eeas/eu-lines-take-group-governmental-experts-emerging-technologies-area-lethal-autonomous-weapons_en?s=62; Dig.watch, *GGE on lethal autonomous weapons systems*, https://dig.watch/processes/gge-laws

as principles, good practices, limitations and constraints. We are committed to upholding and strengthening compliance with International Law, in particular International Humanitarian Law, including through maintaining human responsibility and accountability in the use of force."[868]

At the 78th UN General Assembly First Committee in 2023, Belgium voted in favor[869] of resolution L.56[870] on autonomous weapons systems, emphasizing the "urgent need for the international community to address the challenges and concerns raised by autonomous weapons systems." Belgium co-sponsored and supported Draft Resolution L.77 in 2024.[871] Belgium stated, "scientific development is outpacing regulations in the field of autonomous weapon systems. Innovative systems and artificial intelligence are being introduced on the battlefield. Now is the time for a normative framework, if we wish to regulate the development and use of these systems before being confronted with a fait accompli."[872]

Belgium was also part of a cohort of 42 states that delivered a vital joint statement in the 2025 GGE asserting that the rolling text, which serves as a starting point for the rules of a treaty, "is a sufficient basis to fulfil the mandate of this GGE in its current form" and a "sufficient basis for negotiations on an instrument on lethal autonomous weapons systems."[873]

Belgium has also engaged in international discussions on LAWS outside the UN, including the REAIM Summits. Belgium endorsed the joint call for action on the responsible development, deployment and use of artificial intelligence in the military domain during the 2023 REAIM Summit in the Netherlands.[874] Belgium also

---

[868] United Nations (UN) General Assembly, First Committee, *Joint Statement on Lethal Autonomous Weapons Systems First Committee, 77th United Nations General Assembly Thematic Debate – Conventional Weapons* (Oct. 21, 2022), https://estatements.unmeetings.org/estatements/11.0010/20221021/A1jJ8bNfWGlL/KLw9WYcSnnAm_en.pdf

[869] Stop Killer Robots, *164 states vote against the machine at the UN General Assembly*, https://www.stopkillerrobots.org/news/164-states-vote-against-the-machine/

[870] General Assembly, *Lethal Autonomous Weapons*, Resolution L56 (Oct.12, 2023), https://reachingcriticalwill.org/images/documents/Disarmament-fora/1com/1com23/resolutions/L56.pdf

[871] UN General Assembly, *General and Complete Disarmament: Lethal Autonomous Weapons Systems* (Oct. 18, 2024), https://documents.un.org/doc/undoc/ltd/n24/305/45/pdf/n2430545.pdf

[872] UN General Assembly, *Belgium Statement for the General Debate UN General Assembly 79 First Committee* (Oct. 14, 2024), https://newyorkun.diplomatie.belgium.be/sites/default/files/1COMstatementBELGIUM_2024.pdf

[873] UN Office of Disarmament Affairs, *Joint Statement to the September 2025 Session of the CCW GGE LAWS* (Sept. 2025), https://docs-library.unoda.org/Convention_on_Certain_Conventional_Weapons_-Group_of_Governmental_Experts_on_Lethal_Autonomous_Weapons_Systems_(2025)/Joint_statement_to_the_September_2025_session_of_the_CCW_GGE_LAWS_-_As_delivered.pdf

[874] Government of Netherlands, *Call to Action on Responsible Use of AI in the Military Domain*, Press Release (Feb.16, 2023), https://www.government.nl/latest/news/2023/02/16/reaim-2023-call-to-action

endorsed the resulting Political Declaration on Responsible Military Use of AI and Autonomy issued in November 2023.[875] At the second REAIM summit, hosted by the Republic of Korea in 2024,[876] Belgium endorsed the outcome document, Blueprint for Action, which translated the principles agreed upon at REAIM 2023 into action steps[877].

Belgium endorsed the Chair's Summary of the Vienna conference Humanity at the Crossroads: Autonomous Weapons Systems and the Challenge of Regulation, in April 2024, affirming its commitment to work urgently towards an international legal instrument to regulate autonomous weapons systems.[878]

AI Literacy

Belgium expanded national policy actions for AI literacy in line with the EU AI Act, focusing on both public engagement with digital policymaking and capacity building for AI oversight entities. Through initiatives like the Action plan DigiJump and the Flemish AI Academy, the government invested in curriculum reform, teacher training, and digital skills programs to empower citizens and ensure that oversight professionals receive specialized training for regulatory compliance and responsible AI adoption, and workers to meet AI and STEM workforce demands, with new programs tailored for compliance with the EU AI Act and greater adoption across public and industrial sectors.[879]

Human Rights

Belgium is a signatory to many international human rights treaties and conventions. In 2025, Belgium received a rating of "Free" with a score of 96/100 in the Freedom House Freedom in the World report. Freedom House reported that

---

[875] US Department of State, *Political Declaration on Responsible Military Use of Artificial Intelligence and Autonomy*, Endorsing States (Feb. 12, 2024), https://www.state.gov/political-declaration-on-responsible-military-use-of-artificial-intelligence-and-autonomy/

[876] Government of the Netherlands, *Speech by Minister Hanke Bruins Slot at the High Level Segment of the Conference on Disarmament* (Feb. 27, 2024), https://www.government.nl/documents/speeches/2024/02/27/speech-by-minister-hanke-bruins-slot-at-the-conference-on-disarmament

[877] Republic of Korea Ministry of Foreign Affairs, *Outcome of Responsible AI in Military Domain (REAIM) Summit 2024* (Sept.12, 2024), https://www.mofa.go.kr/eng/brd/m_5674/view.do?seq=321057

[878] Federal Ministry for European and International Affairs of Austria, *2024 Vienna Conference on Autonomous Weapons Systems* (Apr. 30, 2024), https://www.bmeia.gv.at/en/european-foreign-policy/disarmament/conventional-arms/autonomous-weapons-systems/2024-vienna-conference-on-autonomous-weapons-systems

[879] OECD, *Progress in Implementing the European Union Coordinated Plan on Artificial Intelligence (Volume 1), Belgium Country Notes* (Nov. 10, 2025), https://www.oecd.org/content/dam/oecd/en/publications/reports/2025/10/progress-in-implementing-the-european-union-coordinated-plan-on-artificial-intelligence-volume-1-country-notes_b0385317/belgium_61b627d9/a82b426c-en.pdf

"Belgium is a stable electoral democracy with a long record of peaceful transfers of power. Political rights and civil liberties are legally guaranteed and largely respected. Major concerns in recent years have included the threat of terrorism, corruption scandals, and rising xenophobia."[880]

Per the country chapter in the European Commission's 2025 Rule of Law report,[881] the Belgian government strengthened legal protections for journalists by decriminalizing defamation and introducing tougher penalties for crimes against them, acknowledging their public interest role. Belgium is also advancing the transposition of the EU anti-SLAPP Directive with a draft law offering broad protections and developed in consultation with journalist associations. However, journalist groups report rising legal intimidation, political interference, and harassment, especially online and during protests.

In a 2020 Recommendation to member States on the human rights impacts of algorithmic systems, the Council of Europe Committee of Ministers clarified that Member States' commitment to "ensuring the rights and freedoms enshrined in the Convention for the Protection of Human Rights and Fundamental Freedoms to everyone within their jurisdiction" required them to "ensure that any design, development and ongoing deployment of algorithmic systems occur in compliance with human rights and fundamental freedoms, which are universal, indivisible, inter-dependent and interrelated, with a view to amplifying positive effects and preventing or minimising possible adverse effects."[882]

OECD / G20 AI Principles

Belgium has endorsed the OECD/G20 AI Principles. In its 2021 survey, the OECD noted several examples of implementation of the AI Principles by Belgium, including the establishment of an AI Observatory, providing financial and non-financial support to retrain and attract top AI talent, development of an AI self-assessment tool, and the resolution to prohibit the use of lethal autonomous weapons by local armed forces.[883]

Belgium is a member of the Global Partnership for AI, a multi-stakeholder initiative that aims to foster international cooperation on AI research and applied

---

[880] Freedom House, *Freedom in the World 2025: Belgium* (2025), https://freedomhouse.org/country/belgium/freedom-world/2025

[881] European Commission, *2025 Rule of Law Report: Country Chapter on the Rule of Law Situation in Belgium*, pp. 12–15 (Jul. 8, 2025), https://commission.europa.eu/document/download/5fe5bb71-2898-4079-ad5a-19a3ff595a62_en?filename=5_1_63936_coun_chap_belgium_en.pdf

[882] *Recommendation CM/Rec(2020)1 of the Committee of Ministers to Member States on the Human Rights Impacts of Algorithmic Systems* (Apr. 8, 2020), https://search.coe.int/cm/pages/result_details.aspx?objectid=09000016809e1154

[883] OECD, *State of Implementation of the OECD AI Principles: Insights from National AI Policies*, pp. 10, 14, 29, 30 (Jun. 18, 2021), https://www.oecd.org/digital/state-of-implementation-of-the-oecd-ai-principles-1cd40c44-en.htm

activities and is "built around a shared commitment to the OECD Recommendation on Artificial Intelligence."[884]

Belgium's 2022 national AI strategy covers all the OECD AI principles: inclusive growth, sustainable development and well-being; human-centered values and fairness, transparency and explainability; robustness, security and safety; and accountability.[885] However, the strategy does not explicitly mention the OECD as the basis for its common objectives.[886]

Council of Europe AI Treaty

Belgium contributed as a Council of Europe and EU Member State in the negotiations of the Council of Europe Framework Convention on AI, Human Rights, Democracy and the Rule of Law.[887] Belgium is party to the Convention through the European Commission signature on September 5, 2024. However, Belgium has not taken individual action to endorse or ratify the treaty.[888]

UNESCO Recommendation on AI Ethics

Belgium is a signatory to the UNESCO Recommendation on the Ethics of Artificial Intelligence. The 2022 National Convergence Plan does not explicitly refer to the UNESCO Recommendation.[889] Still, actions by the AI4Belgium Ethics and Law Committee reflect alignment with UNESCO Policy Recommendations. However, the Flemish region is collaborating with UNESCO on project to ready the public administration for AI systems.[890] UNESCO's Ethics of AI Unit, Social and Human Sciences Sector, will leverage the Readiness Assessment Methodology (RAM) underway in Flanders as part of the implementation.

---

[884] OECD AI Policy Observatory, *About the Global Partnership on Artificial Intelligence (GPAI)* (2026), https://oecd.ai/en/about/about-gpai

[885] OECD, *OECD AI Principles Overview* (2026), https://oecd.ai/en/ai-principles

[886] FSP Policy and Support (BOSA), *National Convergence Plan for the Development of Artificial Intelligence: Becoming an AI Smart Nation* (Oct. 2022), https://bosa.belgium.be/sites/default/files/content/images/DigitaleOverheid/AI/Plan_AI_EN.pdf

[887] Council of Europe, *Framework Convention on Artificial Intelligence* (2026), https://www.coe.int/en/web/artificial-intelligence/the-framework-convention-on-artificial-intelligence

[888] Council of Europe Treaty Office, *Chart of Signatures and Ratifications of Treaty 225* (Feb. 8, 2026), https://www.coe.int/en/web/conventions/full-list?module=signatures-by-treaty&treatynum=225

[889] FSP Policy and Support (BOSA), *National Convergence Plan for the Development of Artificial Intelligence: Becoming an AI Smart Nation* (Oct. 2022), https://bosa.belgium.be/sites/default/files/content/images/DigitaleOverheid/AI/Plan_AI_EN.pdf

[890] UNESCO, *AI-Ready Flemish Public Administration* (Sept. 2024), https://www.unesco.org/en/artificial-intelligence/recommendation-ethics/flemish-government

Evaluation

Belgium has a national AI strategy with a strong focus on AI ethics that should ensure coherence among the regional AI strategies and align toward common objectives. Regional AI policy innovation has been strengthened to align with the UNESCO Recommendation on AI Ethics. Belgium also established a Federal Data and AI Ethics Advisory Committee within the Federal Public Service (FPS) to enhance AI oversight and educate civil servants on data ethics, human oversight, and fundamental rights.

However, some crucial points are still awaiting action, such as national action on the Council of Europe treaty and the establishment of an independent national supervisory agency under the EU AI Act.

Benin

In 2025, Benin hosted the Regional Summit on Digital Transformation in Western and Central Africa, where states affirmed a commitment to creating governance structures to foster ethical AI development. The country progressed on implementing the National AI Strategy by partnering with UNESCO to develop national AI competency frameworks for teachers and students.

National AI Strategy

The Ministerial Council of Benin approved and adopted the National Artificial Intelligence and Big Data Strategy (*Stratégie Nationale d'Intelligence Artificielle et des Mégadonnées*, SNIAM 2023–2027)[891] on January 18, 2023. The SNIAM 2023–2027 is deeply rooted in the National Development Plan (PND) 2018–2025[892] and the Government Action Program (PAG),[893] positioning digital technology as a lever for accelerating structural economic transformation.

The SNIAM 2023–2027 identifies six major problems addressed through 4 Strategic Programs "that can be rolled out in three phases spread over five years, with a portfolio containing one hundred and twenty-three (123) actions impacting both the public and the private sectors."[894] Program 1 focuses on AI development and use in high priority sectors (agriculture, health, education, and living environment). Programs 2 and 3 aim to strengthen human capacity and skills for working in AI professions and provide support for training, research, and innovation in the private sector while fostering cooperation. Program 4 implements institutional and governance frameworks to regulate the ethical and liability issues related to AI and big data.

A robust legal framework, specifically the Digital Code[895] and data protection law,[896] provides a basis for Program 4. Although the SNIAM 2023–2027 does not

---

[891] Ministère du Numérique et de la Digitalisation, *National Artificial Intelligence and Big Data Strategy 2023–2027* (2023), https://numerique.gouv.bj/assets/documents/national-artificial-intelligence-and-big-data-strategy-1682673348.pdf

[892] Gouvernement de la République, *Plan National de Développement 2018–2025* (2018), https://www.gouv.bj/download/2/mpd_plan-national-developpement_2018-2025_final_14_janv.pdf

[893] Présidence de la République, *Programme d'Actions du Gouvernement 2021–2026* (2021), https://beninrevele.bj/pag-2021-2026/

[894] Ministère du Numérique et de la Digitalisation, *National Artificial Intelligence and Big Data Strategy 2023–2027*, p. 2 (2023), https://numerique.gouv.bj/assets/documents/national-artificial-intelligence-and-big-data-strategy-1682673348.pdf

[895] Assemblée Nationale, *Digital Code in the Republic of Benin [Loi n° 2017-20 portant code du numérique en République du Bénin]* (Jun. 13, 2017), https://apiprod.apdp.bj/storage/c46ab3f55e93d40c8545b70add0ea8c7/CODE-DU-NUMERIQUE-DU-BENIN_2018-version-APDP.pdf

[896] Présidence de la République, *Law No. 2009-09, Dealing with the Protection of Personally Identifiable Information* (May 22, 2009), https://cyrilla.org/api/files/1599165019443wr78rbbpf9m.pdf

explicitly align with international governance frameworks, it functionally aligns with the core principles of global standards such as the OECD AI Principles, Universal Guidelines for Artificial Intelligence,[897] UNESCO Recommendation on the Ethics of AI, Council of Europe AI Treaty, or African Union Continental AI Strategy. Key areas of alignment include Accountability as guiding principle: "all stakeholders will need to be involved in the design, implementation, and ongoing monitoring and evaluation of the strategy."[898] The SNIAM Program 4 on the implementation of a governance framework for AI outlines specific actions to establish monitoring as liability mechanism to ensure that "AI systems are designed with respect for people, the planet, prosperity, peace, transparency, justice and fairness, accountability, non-maleficence, privacy, benevolence, freedom and autonomy, sustainability, dignity, and solidarity."[899]

Additionally, under Action 3.3: Bolstering institutional cooperation and collaboration, the strategy mandates the strengthening of "win-win" international and regional cooperation to support the national AI ecosystem. The SNIAM explicitly identifies UNESCO and the African Union (AU)—operating through bodies such as Smart Africa and ECOWAS (Economic Community of West African States)—as key institutional partners.[900]

Regional and AU Collaboration

Benin's SNIAM 2023–2027[901] functionally aligns with the five core focus areas in the African Union Digital Transformation Strategy,[902] effectively serving as a national-level implementation of the continental vision.

The Cotonou Declaration serves as the explicit bridge, where Benin and regional peers formally committed "to building, by 2030, a Single African Digital Market based on connectivity, trust, innovation, sustainability, and ethical AI, so that no country and no citizen is left behind in the digital economy."[903] To this end, the Cotonou Declaration reiterates commitments to "Ethical artificial intelligence and inclusive innovation" by embedding "transparency, fairness and inclusivity into

---

[897] CAIDP, *Universal Guidelines for Artificial Intelligence*, https://www.caidp.org/universal-guidelines-for-ai/
[898] Ministère du Numérique et de la Digitalisation, *National Artificial Intelligence and Big Data Strategy 2023–2027*, p. 12 (2023), https://numerique.gouv.bj/assets/documents/national-artificial-intelligence-and-big-data-strategy-1682673348.pdf
[899] Ibid, p. 35
[900] Ibid, p. 33
[901] Ibid
[902] African Union, *The Digital Transformation Strategy for Africa (2020–2030*) (2020), https://au.int/sites/default/files/documents/38507-doc-dts-english.pdf
[903] World Bank, *Regional Summit on Digital Transformation in Western and Central Africa—Cotonou Declaration,* Preamble (Nov. 18, 2025), https://www.worldbank.org/en/news/statement/2025/11/18/regional-summit-on-digital-transformation-in-western-and-central-africa-cotonou-declaration

artificial intelligence governance to ensure that technology benefits all communities while mitigating risks such as bias, misuse and disinformation."[904]

Benin's SNIAM 2023–2027 aligns with the African Union Continental Artificial Intelligence Strategy[905] on five priority key areas, particularly ensuring ethical and secure AI that minimizes risks, protects data privacy, and implements ethical, human-centric AI aligning with AU principles.

Public Participation

According to the official SNIAM 2023–2027, "The process of developing Benin's National Artificial Intelligence and Big Data Strategy (SNIAM) was headed up by the Ministry of Digital Affairs and Digitalization, in an approach involving all the stakeholders."[906]

SNIAM established Accountability as one of the key principles driving the implementation of the strategy. Accountability, that goes "hand in hand with the principle of the effective participation of the various stakeholders (government, private sector, and the end users) in the design, development, implementation, and ongoing monitoring and evaluation of the strategy."[907] Although the SNIAM 2023–2027 described an approach intended to involve all stakeholders, no open public participation mechanisms were found to have been established during the development phase.

Data Protection

The Digital Code,[908]as amended by law no. 2020-35 of 06 January 2021,[909] and the Law dealing with the Protection of Personally Identifiable Information.[910] The Digital Code deals with the collection, treatment, transmission, storage, and use of personal data by a person, the state, local authorities, and legal persons, as well as automated processing and non-automated processing of personal data contained in

---

[904] Ibid, IV, 9(b)

[905] African Union, *Continental Artificial Intelligence Strategy* (Aug. 9, 2024), https://au.int/en/documents/20240809/continental-artificial-intelligence-strategy

[906] Ministère du Numérique et de la Digitalisation, *National Artificial Intelligence and Big Data Strategy 2023–2027* (2023), https://numerique.gouv.bj/assets/documents/national-artificial-intelligence-and-big-data-strategy-1682673348.pdf

[907] Ibid, p. 13

[908] Assemblée Nationale, *Digital Code in the Republic of Benin [Loi n° 2017-20 portant code du numérique en République du Bénin]* (Jun. 13, 2017), https://apiprod.apdp.bj/storage/c46ab3f55e93d40c8545b70add0ea8c7/CODE-DU-NUMERIQUE-DU-BENIN_2018-version-APDP.pdf

[909] Assemblée Nationale, *Loi No. 2020-35 du 06 janvier 2021 modifiant la loi no 2017-20 du 20 avril 2018 portant code du numérique en République du Bénin* (Jan. 6, 2021), https://documentation-anbenin.org/s/textes-de-lois/item/2170

[910] Présidence de la République, *Law No. 2009-09, Dealing with the Protection of Personally Identifiable Information* (May 22, 2009), https://cyrilla.org/api/files/1599165019443wr78rbbpf9m.pdf

files, or any processing of data for public security, defense, research, prosecution of criminal offenses, or the security and essential interests of the state.[911] The 2020 amendments have been criticized as overly broad, leaving journalists and activists vulnerable to accusations of libel and disinformation.[912]

The Personal Data Protection Authority (*Autorité de Protection des Données à caractère Personnel*, APDP) serves as the independent regulatory body established to ensure protection of personal data and privacy and enforcement of the Digital Code.[913] The personal data protection law originally named the National Commission of Information (Freedom) and Privacy (*Commision Nationale de l'Informatique et des Libertés*) but the body transformed in 2018 under the Digital Code.

The APDP serves a critical function as the government moves to implement the data-intensive initiatives outlined in the SNIAM 2023–2027.[914] The APDP is an active member of the Network of African Data Protection Authorities (RAPDP/NADPA) and the Francophone Association of Personal Data Protection Authorities (AFAPDP).

Benin's approach to data privacy is not isolated; it is deeply entrenched in broader regional goals, established in the African Union´s Digital Transformation Strategy for Africa (2020-2030)—along with the recent commitments acquired in the Cotonou 2025 Declaration.[915] Benin also aligns with the African Union (AU) Continental Artificial Intelligence (AI) Strategy, which prioritizes creating national and regional data pools, facilitating cross-border data flows, and implementing robust data protection regulations based on the Malabo Convention.[916]

To fortify its national data governance, Benin is developing a specific National Data Strategy[917] that explicitly operationalizes the African Union's Continental Data

---

[911] Assemblée Nationale, *Digital Code in the Republic of Benin [Loi n° 2017-20 portant code du numérique en République du Bénin]* (Jun. 13, 2027), https://apiprod.apdp.bj/storage/c46ab3f55e93d40c8545b70add0ea8c7/CODE-DU-NUMERIQUE-DU-BENIN_2018-version-APDP.pdf

[912] Amesty International, *Benin: New Laws, New Human Rights Restrictions*, pp. 2, 4 (Jan. 2022), https://www.amnesty.org/fr/wp-content/uploads/2022/07/AFR1457362022ENGLISH.pdf

[913] Assemblée Nationale, *Digital Code in the Republic of Benin [Loi n° 2017-20 portant code du numérique en République du Bénin]* (Jun. 13, 2027), https://apiprod.apdp.bj/storage/c46ab3f55e93d40c8545b70add0ea8c7/CODE-DU-NUMERIQUE-DU-BENIN_2018-version-APDP.pdf

[914] Ministère du Numérique et de la Digitalisation, *National Artificial Intelligence and Big Data Strategy 2023–2027* (2023), https://numerique.gouv.bj/assets/documents/national-artificial-intelligence-and-big-data-strategy-1682673348.pdf

[915] UN Economic Commission for Africa, *Cotonou Declaration on Accelerating the Digital Transformation of Africa* (May 14–16, 2025), https://uneca.org/eca-events/sites/default/files/resources/documents/TCND/africa-wsis-annual-review/2025/2500615e.pdf

[916] African Union, *African Union Convention on Cybersecurity (Malabo Convention)* (Jun. 27, 2014), https://au.int/en/treaties/african-union-convention-cyber-security-and-personal-data-protection

[917] Smart Africa Secretariat, *Terms of Reference (ToR) Recruitment of a Consultancy Firm or Consortium for the Development of Benin's National Data Strategy* (2025), https://smartafrica.org/wp-content/uploads/2025/11/ToR-Benin-National-Data-Stratetegy_EN-1.pdf

Policy Framework[918] and the binding principles of the Malabo Convention on Cybersecurity and Personal Data Protection. Backed by the technical and financial support of the Smart Africa Alliance and German Cooperation (GIZ) under the AU-EU Data Governance in Africa project, this initiative prioritizes the creation of a regulated, secure, and rights-based digital environment.

The Global Privacy Assembly (GPA) lists the National Commission for Information (Freedom) and Liberties as an accredited member since 2015 and does not list the APDP. Benin's data protection authority did not endorse the 2018 Declaration on Ethics and Data Protection in Artificial Intelligence,[919] 2020 Resolution on Accountability in the Development and Use of Artificial Intelligence,[920] 2022 Resolution on Principles and Expectations for the Appropriate Use of Personal Information in Facial Recognition Technology,[921] or 2023 Resolution on Generative Artificial Intelligence Systems.[922]

Algorithmic Transparency

Benin's Digital Code provides basic provisions related to algorithmic transparency within Book V on Data Protection.[923] The Code prohibits automated decisions and profiling in judicial cases and requires notice of automated processing as well as access to the logic of decisions. The Code does not mention redress or the right to contest a decision. This Code is enforced by the Personal Data Protection Authority (APDP).

Article 401 prohibits court decisions and decisions producing legal effects that may significantly affect an individual based solely on automated processing of

---

[918] African Union, *African Union Data Policy Framework* (Jul. 28, 2022), https://au.int/sites/default/files/documents/42078-doc-DATA-POLICY-FRAMEWORKS-2024-ENG-V2.pdf

[919] International Conference of Data Protection & Privacy Commissioners, *Declaration on Ethics and Data Protection in Artificial Intelligence* (Oct. 23, 2018), https://globalprivacyassembly.com/wp-content/uploads/2018/10/20180922_ICDPPC-40th_AI-Declaration_ADOPTED.pdf

[920] Global Privacy Assembly (GPA), *Adopted Resolution on Accountability in the Development and Use of Artificial Intelligence* (Oct. 2020), https://globalprivacyassembly.com/wp-content/uploads/2020/11/GPA-Resolution-on-Accountability-in-the-Development-and-Use-of-AI-EN.pdf

[921] Global Privacy Assembly (GPA), *Resolution on Principles and Expectations for the Appropriate Use of Personal Information in Facial Recognition Technology* (Oct. 2022), https://globalprivacyassembly.com/wp-content/uploads/2022/11/15.1.c.Resolution-on-Principles-and-Expectations-for-the-Appropriate-Use-of-Personal-Information-in-Facial-Recognition-Technolog.pdf

[922] Global Privacy Assembly (GPA), *Resolution on Generative Artificial Intelligence Systems* (Oct. 2023), https://globalprivacyassembly.com/wp-content/uploads/2023/10/5.-Resolution-on-Generative-AI-Systems-101023.pdf

[923] Assemblée Nationale, *Digital Code in the Republic of Benin [Loi n° 2017-20 portant code du numérique en République du Bénin]*, pp. 339–447 (Jun. 13, 2027), https://apiprod.apdp.bj/storage/c46ab3f55e93d40c8545b70add0ea8c7/CODE-DU-NUMERIQUE-DU-BENIN_2018-version-APDP.pdf

personal data, including profiling intended to evaluate aspects of personality, though exceptions apply when such decisions occur within a contractual framework or legal provision that includes appropriate safeguards, including the data subject's right to express their point of view.[924] Article 405 of the Code specifies that "automated or non-automated data processing operations carried out by public or private organizations and involving personal data must, prior to their implementation, be subject to a prior declaration to the Authority or be registered in a register maintained by the person designated for this purpose by the data controller."[925]

Under Articles 415 and 416, data controllers are obliged to inform subjects of the data about "the existence of automated decision-making, including profiling, as referred to in Article 401 and, at least in such cases, meaningful information about the logic involved, as well as the significance and the envisaged consequences of such processing for the data subject."[926] However, the Code does not define what constitutes "meaningful information about the logic, nor does it establish specific standards for algorithmic explainability, technical documentation requirements, or the level of detail required when explaining automated systems. The Code lacks provisions for algorithmic impact assessments, mandatory auditing of automated decision-making systems, or specific oversight mechanisms beyond general data protection compliance.

Benin has a national Open Data Portal designed, developed, and deployed to increase transparency in government operations.

Environmental Impact of AI

Benin currently does not have national laws or policies that explicitly address the environmental impact of artificial intelligence.

Lethal Autonomous Weapons

Benin is a High Contracting Party to the Convention on Certain Conventional Weapons (CCW)[927] and has supported a ban on lethal autonomous weapons systems (LAWS) since at least 2018 as a member of the African Group[928] and Non-Aligned Movement (NAM) in the CCW.[929] The NAM advocated for the CCW Group of

---

[924] Ibid, pp. 371–372

[925] Ibid, pp. 374–375

[926] Ibid, pp. 389–393

[927] UN Office for Disarmament Affairs, *High Contracting Parties and Signatories CCW* (Mar. 18, 2025), https://disarmament.unoda.org/en/our-work/conventional-arms/convention-certain-conventional-weapons/high-contracting-parties-and-signatories-ccw

[928] Permanent Observer Mission of the African Union to the UN, *African Group* (2022), https://www.africanunion-un.org/africangroup

[929] UN Office for Disarmament Affairs, *Regional Groups—CCW* (Dec. 2022), https://front.un-arm.org/wp-content/uploads/2022/12/Regional-Groups-Dec-2022.pdf

Governmental Experts (GGE) on LAWS to implement a treaty.[930] The African Group reiterated support for a legally binding instrument within the CCW in 2023,[931] with the view that "the world no longer needs more weapons, particularly those with unpredictable functionality, which operate on data and algorithm biases that could self-learn and take away meaningful human control to potentially cause unnecessary human suffering and damage to property." A 2018 African Group statement to the GGE on LAWS rationalized the need for human control: "The African Group finds it inhumane, abhorrent, repugnant, and against public conscience for humans to give up control to machines, allowing machines to decide who lives or dies, how many lives and whose life is acceptable as collateral damage when force is used."[932]

AI Literacy

The National Artificial Intelligence and Big Data Strategy (SNIAM 2023–2027) highlights capacity building and stakeholder sensitization as core objectives. The SNAIAM proposes programs to strengthen human capital, deliver training, and raise awareness among public-sector actors and beneficiaries.[933]

Benin is partnering with UNESCO and the KIX Africa 21 Hub to develop national AI competency frameworks for teachers and students.[934] This collaboration aims to integrate AI and digital skills into education systems, strengthen foundational literacy through local languages, and ensure that AI supports inclusive, equitable learning aligned with national strategies like Benin's AI and Big Data Strategy.

Human Rights

Benin's Constitution, ratified in 1991, establishes commitments to the rule of law, fundamental rights, and human dignity. Benin is also a signatory to key

---

[930] UN Office for Disarmament Affairs, *Working Paper by the Bolivarian Republic of Venezuela on Behalf of the Non-Aligned Movement (NAM) and Other Parties to the Convention on Certain Conventional Weapons (CCW)* (Mar. 7–11, Jul. 25–29, 2022), https://documents.unoda.org/wp-content/uploads/2022/08/WP-NAM.pdf

[931] UN Office for Disarmament Affairs, *Statement by the Federal Republic of Nigeria to the United Nations and Other International Organizations in Geneva on Behalf of the African Group Member States* (Nov. 15–17, 2023), https://docs-library.unoda.org/Convention_on_Certain_Conventional_Weapons_-Meeting_of_High_Contracting_Parties_(2023)/Nigeria_-_African_Group_-_2023_MHCP.pdf

[932] UN Office for Disarmament Affairs, *Statement by the African Group* (Apr. 9, 2018), https://reachingcriticalwill.org/images/documents/Disarmament-fora/ccw/2018/gge/statements/9April_African-Group.pdf

[933] Ministère du Numérique et de la Digitalisation, *National Artificial Intelligence and Big Data Strategy 2023–2027* (2023), https://numerique.gouv.bj/assets/documents/national-artificial-intelligence-and-big-data-strategy-1682673348.pdf

[934] UNESCO, *UNESCO and Partners Pioneer National AI Competency Frameworks and Local Language Learning in Africa* (Oct. 17, 2025), https://www.unesco.org/en/articles/unesco-and-partners-pioneer-national-ai-competency-frameworks-and-local-language-learning-africa

international human rights instruments.[935] Once considered a stable democracy in Sub-Saharan Africa, Benin has seen deterioration in the protection of human rights since 2016. In the 2025 Freedom House Freedom in the World Report, Benin scores 60/100 and ranks as "Partly Free" for political rights and civil liberties.[936]

Attacks on independent media outlets and arbitrary arrests of dissidents have led to a shrinkage of civic space. Specifically, the criminalization of "publication of false information" and "electronic harassment" in the Digital Code has led to suspensions of media and self-censorship amongst journalists.[937]

Benin's revised Code of Ethics[938] in the media sector aims to address digital & AI challenges in journalism and other media. The Code intends to safeguard professional integrity and ethical standards with a strong emphasis on preventing disinformation.

OECD / G20 AI Principles

Benin is not a member of the OECD and has not officially endorsed the OECD AI principles.[939] The National Development Plan 2018–2025 base for the AI strategy aims for "strong, inclusive and sustainable economic growth within the framework of more effective national and local governance, achievable by focusing on the development of human capital and infrastructure."[940] These align with the OECD Principles. Benin is not a member of the Global Partnership on Artificial Intelligence (GPAI).[941]

---

[935] UN Human Rights, *Ratification of 18 International Human Rights Treaties, Benin* (2014), https://indicators.ohchr.org/

[936] Freedom House, *Freedom in the World 2025: Benin* (2025), https://freedomhouse.org/country/benin/freedom-world/2025

[937] Amnesty International, *Benin: Election Candidates Must Commit to Protecting Human Rights amid Shrinking Civic Space* (Jan. 8, 2026), https://www.amnesty.org/en/latest/news/2026/01/benin-candidates-human-rights/

[938] Media Foundation for West Africa, *Benin Updates Media Ethics Code to Address Digital and AI Challenges* (Mar. 3, 2025), https://mfwa.org/country-highlights/benin-updates-media-ethics-code-to-address-digital-and-ai-challenges/

[939] OECD Legal Instruments, *Recommendation of the Council on Artificial Intelligence, Adherents* (May 3, 2024), https://legalinstruments.oecd.org/en/instruments/oecd-legal-0449#adherents

[940] Ministère du Numérique et de la Digitalisation, *National Artificial Intelligence and Big Data Strategy 2023–2027*, p. 11 (2023), https://numerique.gouv.bj/assets/documents/national-artificial-intelligence-and-big-data-strategy-1682673348.pdf

[941] OECD AI Policy Observatory, *About the Global Partnership on Artificial Intelligence (GPAI)* (2026), https://oecd.ai/en/about/about-gpai

Council of Europe AI Treaty

Benin has not signed the Council of Europe Framework Convention on Artificial Intelligence and Human Rights, Democracy, and the Rule of Law.[942]

UNESCO Recommendation on AI Ethics

Benin adopted the UNESCO Recommendation on the Ethics of Artificial Intelligence together with 192 other countries.[943] Benin's partnership with UNESCO to develop national AI literacy competency frameworks for teachers and students aligns with the recommendation on Literacy and Awareness and inclusive AI development.[944]

Benin has not initiated the UNESCO Readiness Assessment Methodology (RAM),[945] a tool to facilitate implementation of the UNESCO Recommendation.

Evaluation

Benin set out a national strategy for AI that recognizes digital technology as a lever for accelerating structural economic transformation. The country protects personal data and the right to privacy through a comprehensive legal framework. However, it does not explicitly protect a right to algorithmic transparency by law.

Benin´s domestic activities and ethical guidelines are functionally aligned with global standards. Significantly, the country aligns with regional initiatives, anchoring its digital agenda in the Digital Transformation Strategy for Africa (2020–2030)[946] and the Cotonou 2025 Declaration to foster a sovereign and interconnected African digital market. Benin is also a signatory to key international human rights instruments; however, the criminalization of "publication of false information" and "electronic harassment" in the Digital Code[947] has led to suspensions of media and self-censorship among journalists, raising concerns over freedom of expression and the press.

---

[942] Council of Europe Treaty Office, *Chart of Signatures and Ratifications of Treaty 225* (Feb. 16, 2026), https://www.coe.int/en/web/conventions/full-list;?module=signatures-by-treaty&treatynum=225

[943] UNESCO, *Recommendation on the Ethics of Artificial Intelligence* (2021), https://unesdoc.unesco.org/ark:/48223/pf0000380455

[944] UNESCO, *UNESCO and Partners Pioneer National AI Competency Frameworks and Local Language Learning in Africa* (Oct. 17, 2025), https://www.unesco.org/en/articles/unesco-and-partners-pioneer-national-ai-competency-frameworks-and-local-language-learning-africa

[945] UNESCO Global AI Ethics and Governance Observatory, *Global Hub* (Oct. 2025), https://www.unesco.org/ethics-ai/en/global-hub

[946] African Union, *The Digital Transformation Strategy for Africa (2020–2030)* (2020), https://au.int/sites/default/files/documents/38507-doc-dts-english.pdf

[947] Assemblée Nationale, *Digital Code in the Republic of Benin [Loi n° 2017-20 portant code du numérique en République du Bénin]* (Jun. 13, 2017), https://apiprod.apdp.bj/storage/c46ab3f55e93d40c8545b70add0ea8c7/CODE-DU-NUMERIQUE-DU-BENIN_2018-version-APDP.pdf

Brazil

In 2025, Brazil finalized an updated national AI plan (PBIA) while legislators continued work on a legal framework for AI. The country also saw publication of the UNESCO RAM report.

National AI Strategy

Then-President Jair Bolsonaro declared before the United Nations General Assembly in September 2020 that Brazil is "open for the development of state-of-the-art technology and innovation efforts, such as 4.0 Industry, artificial intelligence, nanotechnology, and 5G technology, with all partners who respect our sovereignty and cherish freedom and data protection."[948]

The Brazilian government adopted a national AI strategy, *Estratégia Brasileira de Inteligência Artificial* (EBIA) in 2021,[949] following on the Digital Transformation Strategy (E-Digital) released in 2018.[950] Release of the EBIA followed restructuring of the Ministry of Science, Technology, Information, and Communications, which was divided into the Ministry of Science, Technology, and Information (*Ministério da Ciência, Tecnologia e Inovação*, MCTI) and the Ministry of Communication (MCom). A broad Directorate on Science and Digital Innovation under the Secretary of Entrepreneurship and Innovation of MCTI oversees AI policy.[951]

The EBIA included developing ethical principles to guide responsible use of AI, removing barriers to innovation, and improving collaboration between government, the private sector, and researchers among the key objectives. A governance committee composed of representatives from the public sector, companies, civil society, and academia identified specific actions to reinforce these

---

[948] President Jair Bolsonaro, *Remarks at the General Debate of the 75th Session of the United Nations General Assembly* (Sept. 22, 2020), https://www.gov.br/mre/en/content-centers/speeches-articles-and-interviews/president-of-the-federative-republic-of-brazil/speeches/remarks-by-president-jair-bolsonaro-at-the-general-debate-of-the-75th-session-of-the-united-nations-general-assembly-september-22-2020

[949] Government of Brazil, *Brazilian Strategy for Digital Transformation* (Jul. 13, 2021), https://www.gov.br/governodigital/pt-br/estrategias-e-governanca-digital/estrategias-e-politicas-digitais/estrategia-brasileira-de-inteligencia-artificial

[950] Brazil Government, *Brazilian Strategy for Digital Transformation* (2018), https://www.gov.br/mcti/pt-br/centrais-de-conteudo/comunicados-mcti/estrategia-digital-brasileira/estrategiadigital.pdf

[951] Ministério da Ciência, Tecnologia e Inovação (MCTI), *Organization Chart* (2021), https://www.gov.br/mcti/pt-br/imagens/organograma/sempi.pdf

objectives in a 2022 Working Plan.[952] Updates to the EBIA website in 2026 emphasized the basis of the EBIA's objective in the OECD AI Principles.[953]

The federal government finalized the Brazilian Artificial Intelligence Plan 2024–2028 (*Plano Brasileiro de Inteligência Artificial*, PBIA) 2025–2028.[954] The new plan, developed through 2024–2025, reassesses objectives and actions in the country's AI strategy and seeks to better align them with national interests and priorities.[955] The plan outlines Brazil's vision for a human-centered AI, promoting national autonomy, international cooperation, and alignment with the Sustainable Development Goals (SDGs). The PBIA emphasizes ethical standards, bias prevention, data sovereignty, and inclusivity in AI applications, aiming to position Brazil as a pioneer in socially responsible AI development. The PBIA includes foundational projects meant to develop the knowledge base and regulatory environment needed for sustainable and trustworthy AI. Pillars include Training and Workforce Development, Improvement of Public Services, and Support for AI Regulation and Governance.

Under the guidance of the earlier PBIA, the country launched the Brazilian AI Observatory (OBIA) in 2024[956] to monitor AI implementation nationwide and promote transparency and fairness in key sectors like healthcare. The initiative complements Senate guidance on ethical AI policies.

AI Legislation

Both of Brazil's congressional bodies, the Federal Senate and house of representatives (*Câmara dos Deputados*) have approved versions of bills to establish a legal framework for AI in Brazil and to provide ethical frameworks and guidelines for the development and use of AI in the country.[957] After academics and NGOs warned that a draft bill (21/2020) "may help perpetuate recent cases of algorithmic

---

[952] Ministério da Ciência, Tecnologia e Inovação (MCTI), *EBIA—Working Plan 2022* (May 12, 2022), https://www.gov.br/mcti/pt-br/acompanhe-o-mcti/transformacaodigital/arquivosestrategiadigital/copy_of_RelatoriofinalEBIA_Eixo6V2.4.pdf

[953] Ministério da Ciência, Tecnologia e Inovação (MCTI), *Estratégia Brasileira de Inteligência Artificial* (Jan. 13, 2026), https://www.gov.br/mcti/pt-br/acompanhe-o-mcti/transformacaodigital/estrategia-brasileira-de-inteligencia-artificial

[954] Ministério da Ciência, Tecnologia e Inovação (MCTI), *Plano Brasileiro de Inteligência Artificial* (Jun. 12, 2025), https://www.gov.br/mcti/pt-br/centrais-de-conteudo/publicacoes-mcti/plano-brasileiro-de-inteligencia-artificial/pbia_mcti_2025.pdf

[955] Ministério da Ciência, Tecnologia e Inovação (MCTI), *Plano Brasileiro de Inteligência Artificial (PBIA), 2024–2028* (Jul. 29, 2024), https://www.gov.br/mcti/pt-br/acompanhe-o-mcti/noticias/2024/07/plano-brasileiro-de-ia-tera-supercomputador-e-investimento-de-r-23-bilhoes-em-quatro-anos/ia_para_o_bem_de_todos.pdf/view

[956] Network Information Center, *Brazilian Artificial Intelligence Observatory Will Be Launched This Tuesday* (Sept. 2, 2024), https://www.nic.br/noticia/na-midia/observatorio-brasileiro-de-inteligencia-artificial-sera-lancado-nesta-terca-feira/

[957] Brazilian House of Representatives, *House Approves Project That Regulates the Use of Artificial Intelligence* (Sept. 29, 2021), https://www.camara.leg.br/noticias/811702-camara-aprova-projeto-que-regulamenta-uso-da-inteligencia-artificial/

discrimination through provisions that hinder accountability for AI-induced errors and restrict the scope of the rights established in the LGPD and in the Brazilian Constitution,"[958] the Federal Senate appointed a temporary Commission of Jurists (*Comissão de Juristas*, "CJSUBIA") to analyze the three draft AI bills and consolidate them into a proposal for a new Brazilian AI Act. The process included public consultations and international seminars.[959]

The Commission of Jurists submitted a draft legal framework to the Federal Senate in 2022.[960] The explanatory statement of the report stresses "Therefore, this substitute bill is based on the premise that there is no trade-off—a mutually exclusive choice—between protection of fundamental rights and freedoms, appreciation of work and dignity of the human person in the face of the economic order and the creation of new value chains. [...] Its normative objective is to conciliate an approach based on risks with a regulatory model based on rights [...] weight of regulation is dynamically calibrated according to the potential risks of the technology application context."

The proposed framework is a risk-based approach, imposing different obligations depending on the level of risk of the AI system, while recognizing for all individuals affected by AI systems, regardless of the system risk levels. The rights for all affected individuals include right to prior information before interacting with AI systems, right to explanation, right to challenge decisions or predictions made by AI, right to human participation in decisions in AI systems, right to non-discrimination, and right to privacy and protection of personal data. The draft bill addresses liability, defining that in the case of high risk or excessive risk of the AI systems, the supplier or operator is objectively liable for the damage caused. Otherwise, the liability will be presumed and there will be a reversal of the burden of proof in favor of the victim. A competent authority to enforce the legislation would be a body or entity of the Federal Government, still to be defined.

---

[958] José Renato Laranjeira de Pereira and Thiago Guimarães Moraes, *Promoting Irresponsible AI: Lessons from a Brazilian Bill*, Heinrich Böll Stiftung (Feb. 14, 2022, https://eu.boell.org/en/2022/02/14/promoting-irresponsible-ai-lessons-brazilian-bill; Change.org, *Open Letter from Jurists to the Federal Senate against Article 6, Item VI of PL 21–A/2020* (Oct. 26, 2021), https://www.change.org/p/senado-federal-carta-aberta-de-juristas-ao-senado-federal-contra-o-artigo-6o-inciso-vi-do-pl-21-a-2020

[959] Federal Senate, *CJSUBIA—Commission of Jurists Responsible for Elaboration of Substitute on Artificial Intelligence in Brazil* (2022), https://legis.senado.leg.br/comissoes/comissao?codcol=2504

[960] Federal Senate, *Jurists Commission Concludes Text of the Legal Framework for Artificial Intelligence* (Dec. 6, 2022), https://www12.senado.leg.br/noticias/materias/2022/12/06/comissao-conclui-texto-sobre-regulacao-da-inteligencia-artificial-no-brasil

After amendments and a supplemental vote document[961] to update a June 2024 report that recommended the approval of Bill 2338/2023,[962] the Senate approved Bill 2338 in December 2024.[963] The bill advanced to the Chamber of Deputies for additional voting.[964]

The State of Ceará adopted a law in 2021 establishing obligations and guidelines for the implementation and use of AI systems within the state.[965] This law requires AI systems to be designed in a safe way, based on ethics and in accordance with this law and Brazilian laws more generally. The law defines an AI system as computer science technologies that enable computers to interact with humans, through technological mechanisms that enable the simulation of human reasoning. The law also states that companies headquartered in the State of Ceará, or that have their artificial intelligence systems in use and operation in the State of Ceará, must be responsible for how their systems operate, being liable for any damages in accordance with the law.

Public Participation

The (then) Ministry of Science, Technology, Innovations, and Communications (MCTIC) organized an online public consultation between December 2019 and February 2020 to gather inputs for "a National Artificial Intelligence Strategy that allows to enhance the benefits of AI for the country, mitigating any negative impacts."[966] According to the terms of the public consultation, "the objective of the strategy is to solve concrete problems in the country, identifying priority areas in the development and use of AI-related technologies in which there is greater potential for obtaining benefits. It is envisaged that AI can bring gains in promoting competitiveness and increasing Brazilian productivity, in providing public

---

[961] Senate Agency, *Rapporteur Presents Updated Report on AI Regulation* (Jul. 4, 2024). https://www12.senado.leg.br/noticias/materias/2024/07/04/relator-apresenta-relatorio-atualizado-sobre-regulamentacao-da-ia

[962] Senate Agency, *Report on AI Presented this Tuesday Includes Protection of Work and Identity* (Jun. 18, 2024), https://www12.senado.leg.br/noticias/materias/2024/06/18/relatorio-sobre-ia-apresentado-nesta-terca-inclui-protecao-ao-trabalho-e-a-identidade

[963] Senate Agency, *Senate Approves Regulation of Artificial Intelligence; Text Goes to the House* (Dec. 10, 2024), https://www12.senado.leg.br/noticias/materias/2024/12/10/senado-aprova-regulamentacao-da-inteligencia-artificial-texto-vai-a-camara

[964] Federal Senate, *Bill No. 2338, of 2023: Bill Providing for the Use of Artificial Intelligence* (Mar. 17, 2025), https://www25.senado.leg.br/web/atividade/materias/-/materia/157233

[965] Legislative Assembly of the State of Ceará, *Ordinary Law No. 17.611, Establishes Responsibilities and Guidelines for Artificial Intelligence Systems within the State of Ceará* (Aug. 11, 2021), https://www2.al.ce.gov.br/legislativo/legislacao5/leis2021/17611.htm

[966] Participate Brazil, Ministério da Ciência, Tecnologia, Inovações e Comunicações, *Brazilian Artificial Intelligence Strategy, Public Consultation Presentation and Instructions* (2020), http://participa.br/estrategia-brasileira-de-inteligencia-artificial/blog/apresentacao-e-instrucoes

services, in improving people's quality of life and in reducing social inequalities, among others."[967]

The consultation addressed thematic areas related to AI and focused on the government's role regarding the impact of AI technologies on society. Relevant documents to artificial intelligence were made available on the consultation website. The consultation collected about 1,000 contributions in total, which were considered for the development of the strategy proposal.[968]

On the other hand, Academics and NGOs have stated that the debate on Bill No. 21/2020 lacked public participation. According to José Renato Laranjeira de Pereira and Thiago Guimarães Moraes, "The debate on the bill ignored the claims of experts and civil society organisations to address the high risks of the technology regarding fundamental rights. In contrast, Members of Congress delivered favorable speeches on the positive impacts of AI in society, especially as a tool for efficiency and innovation."[969]

As indicated above, the Federal Senate responded to these criticisms by establishing a Commission of Jurists to prepare a new consolidated proposal for a Brazilian AI Act, the preparation of which involved the establishment of public participation mechanisms. The Commission of Jurists initiated the organization of meetings, seminars, and public hearings divided into thematic axes, with the participation of experts and national and international representatives. 12 thematic panels were created and the Commission of Jurists received 102 statements from civil society entities, consolidated in the report submitted to the Federal Senate.[970]

Between May and September 2025, the Chamber of Deputies' Special Commission on AI held twelve public hearings on the legal framework for Artificial Intelligence bill, inviting academia, civil society, and industry. Separately, the ANPD conducted a nationwide consultation (*Tomada de Subsídios)* on AI and automated decisions. The ANPD published results of the nationwide inquiry in May 2025.[971]

---

[967] Ibid

[968] OECD AI Policy Observatory, *Policy Initiatives for Brazil, Brazilian AI Strategy* (Jan. 11, 2026), https://oecd.ai/en/dashboards/policy-initiatives/brazilian-ai-strategy-2027

[969] José Renato Laranjeira de Pereira and Thiago Guimarães Moraes, *Promoting Irresponsible AI: Lessons from a Brazilian Bill*, Heinrich Böll Stiftung (Feb. 14, 2022), https://eu.boell.org/en/2022/02/14/promoting-irresponsible-ai-lessons-brazilian-bill

[970] Federal Senate, *Jurists Commission Concludes Text of the Legal Framework for Artificial Intelligence* (Dec. 6, 2022), https://www12.senado.leg.br/noticias/materias/2022/12/06/comissao-conclui-texto-sobre-regulacao-da-inteligencia-artificial-no-brasil

[971] Câmara dos Deputados, *Comissão Especial sobre Inteligência Artificial realiza audiências públicas* (2025), https://www.camara.leg.br/noticias/1196564-comissao-debate-impacto-da-inteligencia-artificial

Data Protection

Brazil's President signed the data protection law, *Lei Geral de Proteção de Dados Pessoais* (LGPD), in September 2020.[972] The LGPD is the first comprehensive data protection law in Brazil and mirrors the European Union's GDPR.[973] The LGPD is relevant to the processing of personal data in relation to AI applications.[974]

The LGPD established the National Data Protection Authority (*Autoridade Nacional de Proteção de Dados*, ANDP) to oversee data protection and privacy in the country. A new law in 2022 modified the LGPD to transform the ANDP into an independent agency, separate from the executive with its own budget.[975] The ANPD is guaranteed technical and decision-making autonomy[976] and is given important attributions related to the interpretation, application and enforcement of the LGPD.[977] Following public consultation on automatic decisions, the ANPD added the issue to the 2025–2026 agenda.[978]

The Brazilian Congress enacted Constitutional Amendment (EC) 115, which establishes personal data protection as a fundamental right in February 2022.[979] EC 115 also gives the Federal Government exclusive jurisdiction to legislate on personal data protection and processing. The amendment establishes the competence of the highest court within Brazil's judiciary, the Federal Supreme Court (*Supremo Tribunal Federal*, STF), to adjudicate legal issues involving personal data. Previously, the STF

---

[972] Presidency of the Republic, Sub-General Secretariat for Legal Affairs, *General Law on Protection of Personal Data (LGPD)* (Aug. 14, 2020), http://www.planalto.gov.br/ccivil_03/_ato2015-2018/2018/Lei/L13709.htm; Katitza Rodriguez, Veridiana Alimonti, *A Look-Back and Ahead on Data Protection in Latin America and Spain* (Sept. 21, 2020), https://www.eff.org/deeplinks/2020/09/look-back-and-ahead-data-protection-latin-america-and-spain

[973] Hogan Lovells Engage, *Brazil Creates a Data Protection Authority* (Jan. 11, 2019), https://www.engage.hoganlovells.com/knowledgeservices/news/brazil-creates-a-data-protection-authority

[974] Lexology, *An Interview with Demarest Advogados Discussing Artificial Intelligence in Brazil* (Nov. 27, 2020), https://www.lexology.com/library/detail.aspx?g=70705701-b4c6-4aa7-8a8a-344dd757f578

[975] Presidency of the Republic, *Law No. 14.460, Transforms the National Data Protection Authority into a Sovereign Entity* (Oct. 25, 2022), https://www.planalto.gov.br/ccivil_03/_ato2019-2022/2022/lei/L14460.htm

[976] Presidency of the Republic, *General Law on Protection of Personal Data* (LGPD), Article 55-B (Aug. 14, 2020), http://www.planalto.gov.br/ccivil_03/_ato2015-2018/2018/Lei/L13709.htm

[977] Ibid, Art. 55-J LGPD; Centre for Information Policy Leadership (CIPL) and Centro de Direito, Internet e Sociedade of Instituto Brasiliense de Direito Público (CEDIS-IDP), *The Role of the Brazilian Data Protection Authority (ANPD) under Brazil's New Data Protection Law (LGPD)* (Apr. 17, 2020), https://www.informationpolicycentre.com/uploads/5/7/1/0/57104281/[en]_cipl-idp_paper_on_the_role_of_the_anpd_under_the_lgpd__04.16.2020__3_.pdf

[978] Autoridade Nacional de Proteção de Dados (ANPD), *Nota Técnica 12/2025, Decisões Automatizadas e IA* (May 2025), https://www.gov.br/anpd/pt-br/acesso-a-informacao/participacao-social/outras-acoes/documentos/ts-06-2024-nt-12-2025-consolidacao-das-contribuicoes.pdf/view

[979] Federal Senate, *Constitutional Amendment on Protection of Personal Data Adopted* (Feb. 10, 2022), https://www12.senado.leg.br/noticias/materias/2022/02/10/promulgada-emenda-constitucional-de-protecao-de-dados

had examined related cases based on other constitutional guarantees, such as the inviolability of intimacy and individuals' private life.

The ANPD approved regulation on international data transfers and the content of standard contractual clauses in August 2024.[980] The regulation implements the international data transfer framework under the LGPD.[981] Publication of the regulation came amid discussions of creating a "sovereign cloud" to ensure government data is stored within national borders as part of Brazil's Artificial Intelligence Plan.[982]

The National Council for the Rights of Children and Adolescents (CONANDA) developed a resolution to protect children and adolescents' rights in virtual environments. The resolution establishes specific data protection requirements such as the prohibition of processing minors' personal data for commercial purposes including targeted advertisement and profiling in connection with consumption, behavior, and market segmentation.[983]

The ANPD became a member of the Global Privacy Assembly (GPA) in 2023.[984] The ANPD has not endorsed the 2018 GPA Declaration on Ethics and Data Protection in Artificial Intelligence,[985] the 2020 GPA Resolution on AI Accountability,[986] the 2022 GPA Resolution on Facial Recognition Technology,[987] or the 2023 GPA Resolution on Generative AI Systems.[988]

---

980 Official Journal of the Union, *Resolution CD/ANDP No. 19, of August 23, 2024* (Aug. 23, 2024), https://www.in.gov.br/en/web/dou/-/resolucao-cd/anpd-n-19-de-23-de-agosto-de-2024-580095396

981 ANPD, *ANPD Approves Regulation on International Data Transfers* (Aug. 23, 2024), https://www.gov.br/anpd/pt-br/assuntos/noticias/resolucao-normatiza-transferencia-internacional-de-dados

982 Ministério da Ciência, Tecnologia e Inovação (MCTI), *Plano Brasileiro de Inteligência Artificial (PBIA), 2024–2028* (Jul. 29, 2024), https://www.gov.br/mcti/pt-br/acompanhe-o-mcti/noticias/2024/07/plano-brasileiro-de-ia-tera-supercomputador-e-investimento-de-r-23-bilhoes-em-quatro-anos/ia_para_o_bem_de_todos.pdf/view

983 Official Diary of the Union, *Resolution No. 245 of April 5, 2024, Providing for the Rights of Children and Adolescents in the Digital Environment* (Apr. 5, 2024), https://www.gov.br/participamaisbrasil/https-wwwgovbr-participamaisbrasil-blob-baixar-7359

984 Global Privacy Assembly, *List of Accredited Members* (2026), https://globalprivacyassembly.com/participation-in-the-assembly/list-of-accredited-members/

985 Global Privacy Assembly, *Declaration on Ethics and Data Protection in Artificial Intelligence* (Oct. 23, 2018), https://globalprivacyassembly.org/wp-content/uploads/2018/10/20180922_ICDPPC-40th_AI-Declaration_ADOPTED.pdf

986 Global Privacy Assembly, *Resolution on Accountability in the Development and Use of Artificial Intelligence* (Oct. 2020), https://globalprivacyassembly.org/wp-content/uploads/2020/10/FINAL-GPA-Resolution-on-Accountability-in-the-Development-and-Use-of-AI-EN-1.pdf

987 Global Privacy Assembly, *Resolution on Principles and Expectations for the Appropriate Use of Personal Information in Facial Recognition Technology* (Oct. 2022), https://globalprivacyassembly.org/wp-content/uploads/2022/11/15.1.c.Resolution-on-Principles-and-Expectations-for-the-Appropriate-Use-of-Personal-Information-in-Facial-Recognition-Technolog.pdf

988 Global Privacy Assembly, *Resolution on Generative Artificial Intelligence Systems* (Oct. 2023), https://globalprivacyassembly.org/wp-content/uploads/2023/10/5.-Resolution-on-Generative-AI-Systems-101023.pdf

AI Oversight

The ANPD adopted a 2023 resolution emphasizing the need to strengthen data protection and oversight concerning AI applications.[989]

As a member of the Ibero-American Network for the Protection of Personal Data (*Red Ibero-Americano de Protección de Datos,* RIPD), the ANPD endorsed the General Recommendations for the Processing of Personal Data in Artificial Intelligence[990] and the accompanying Specific Guidelines for Compliance with the Principles and Rights that Govern the Protection of Personal Data in Artificial Intelligence Projects.[991] The RIPD data protection authorities initiated a coordinated action on ChatGPT, investigating risks for the rights and freedoms of users in relation to the processing of their personal data. Concerns centered on the risk of misinformation: "ChatGPT does not have knowledge and/or experience in a specific domain, so the precision and depth of the response may vary in each case, and/or generate responses with cultural, racial or gender biases, as well as false ones."[992]

Algorithmic Transparency

Article 20 of the LGPD establishes the right of any individual "to request the review of decisions taken solely on the basis of automated processing of personal data that affect his interests, including decisions designed to define his personal, professional, consumer and credit profile or aspects of his personality."[993]

As a result, "the controller shall provide, whenever requested, clear and adequate information regarding the criteria and procedures used for the automated decision, observing the commercial and industrial secrets."[994] Where the information is not provided due to the observance of commercial and industrial secrecy, the

---

[989] ANPD, *Resolution CD/ANPD No 10, of December 5, 2023*, https://www.gov.br/anpd/pt-br/documentos-e-publicacoes/documentos-de-publicacoes/nota-tecnica-no-19-2023-fis-cgf-anpd.pdf

[990] Red Ibero-Americano de Protección de Datos (RIPD), *General Recommendations for the Processing of Personal Data in Artificial Intelligence* (Jun. 2019), https://www.redipd.org/sites/default/files/2020-02/guide-general-recommendations-processing-personal-data-ai.pdf

[991] Red Iberoamericano de Protección de Datos (RIPD), *Specific Guidelines for Compliance with the Principles and Rights that Govern the Protection of Personal Data in Artificial Intelligence Projects* (Jun. 2019), https://www.redipd.org/en/documents/guide-general-recommendations-processing-personal-data-ai

[992] Red Iberoamericano de Protección de Datos (RIPD), *Authorities of the Ibero-American Network for the Protection of Personal Data Initiate a Coordinated Action in Relation to the Service ChatGPT* (May 8, 2023), https://www.redipd.org/noticias/autoridades-red-iberoamericana-de-proteccion-de-datos-personales-inician-accion-chatgpt

[993] Presidency of the Republic, Sub-General Secretariat for Legal Affairs, *General Law on Protection of Personal Data (LGPD)* (Aug. 14, 2020), https://www.planalto.gov.br/ccivil_03/_ato2015-2018/2018/lei/l13709.htm; Katitza Rodriguez, Veridiana Alimonti, *A Look-Back and Ahead on Data Protection in Latin America and Spain* (Sept. 21, 2020), https://www.eff.org/deeplinks/2020/09/look-back-and-ahead-data-protection-latin-america-and-spain

[994] Article 20.1 LGPD

national data protection authority "may perform audits to verify discriminatory aspects in automated processing of personal data."[995]

Some Brazilian researchers and judicial advocates, including Prof. Renato Leite Monteiro and entities such as the Brazilian Institute of Consumer Protection (IDEC), assert that a comprehensive interpretation of the LGPD, in conjunction with the Constitution, consumer law, and other legal provisions, guarantees the existence of a right to explanation in Brazil. However, this position demands greater judicial consolidation.[996]

The ANPD public consultation on automated decisions and artificial intelligence and related report offer steps toward providing this consolidation.[997] The AI Regulatory Sandbox initiative can provide further clarity with tests of algorithmic transparency and data protection mechanisms in a controlled environment.[998]

The national AI strategy outlined transparency as a critical element of AI governance both regarding explainability of decisions taken by autonomous systems and the transparency of methodologies used in the development of AI systems, including data sources and project procedures.[999]

With regard to the transparency principle, the RIPD Specific Guidelines for Compliance with the Principles and Rights that Govern the Protection of Personal Data in Artificial Intelligence Projects provide, "The information provided regarding the logic of the AI model must include at least basic aspects of its operation, as well as the weighting and correlation of the data, written in a clear, simple and easily understood language, it will not be necessary to provide a complete explanation of the algorithms used or even to include them. The above always looking not to affect the user experience."[1000]

The Brazilian Artificial Intelligence Plan 2024–2028 established and funded the National Center for Algorithmic Transparency and Trustworthy AI, a national

---

[995] Article 20.2 LGPD

[996] Institute for Research on Internet and Society, *Automated Decisions and Algorithmic Transparency* (Nov. 16, 2019), https://irisbh.com.br/en/automated-decisions-and-algorithmic-transparency/

[997] Autoridade Nacional de Proteção de Dados (ANPD), *ANPD apresenta resultados da Tomada de Subsídios sobre Inteligência Artificial e Revisão de Decisões Automatizadas* (May 20, 2025), https://www.gov.br/anpd/pt-br/assuntos/noticias/anpd-apresenta-resultados-da-tomada-de-subsidios-sobre-tratamento-automatizado-de-dados-pessoais

[998] Autoridade Nacional de Proteção de Dados (ANPD), *Após recursos, ANPD divulga resultado definitivo do projeto Sandbox Regulatório* (Oct. 14, 2025), https://www.gov.br/anpd/pt-br/assuntos/noticias/apos-recursos-anpd-divulga-resultado-definitivo-do-projeto-sandbox-regulatorio

[999] Ministério da Ciência, Tecnologia e Inovação (MCTI), *Estratégia Brasileira de Inteligência Artificial* (Jan. 13, 2026), https://www.gov.br/mcti/pt-br/acompanhe-o-mcti/transformacaodigital/estrategia-brasileira-de-inteligencia-artificial

[1000] Ibero-American Network for the Protection of Personal Data (RIPD), *Specific Guidelines for Compliance with the Principles and Rights that Govern the Protection of Personal Data in Artificial Intelligence Projects*, p. 17 (Jun. 2019), https://www.redipd.org/sites/default/files/2020-02/guide-specific-guidelines-ai-projects.pdf

center to develop research and studies on AI risks, security, transparency, and reliability.[1001]

Use of AI in Public Administration

As a member of the Latin American Centre for Development Administration (CLAD), Brazil approved the Ibero American Charter on Artificial Intelligence in Civil Service.[1002] The Charter aims to provide a roadmap and common framework for CLAD member states to learn about the challenges and opportunities involved in the implementation of AI in public administration and adapt their AI policy strategies and laws accordingly. The guiding principles are human autonomy; transparency, traceability and explainability; accountability, liability and auditability; security and technical robustness; reliability, accuracy, and reproducibility; confidence, proportionality, and prevention of damage; privacy and protection of personal data; data quality and safety; fairness, inclusiveness, and non-discrimination; human-centering, public value, and social responsibility; and sustainability and environmental protection.

The third axis of the proposal for the Brazilian Artificial Intelligence Plan 2024–2028 focuses on the use of AI to improve public services. The plan includes 19 suggested actions to foster the use of AI in the public sector to improve the delivery of public services and enhance impact on development and social inclusion. Some of the areas targeted for proposed AI solutions to improve public services include procurement, management of human resources, management of educational resources, cybersecurity, management of assets, and prevention of extreme climate events.[1003] The updated PBIA 2024–2028 introduces "AI for Public Power" actions to deploy AI in health, agriculture, and social protection. Thirty-one Immediate Actions prioritize public sector digitalization and risk assessment frameworks.[1004]

Algorithm-Based Advertising

The Brazilian Advertising Self-Regulation Council (CONAR), a nongovernmental organization that acts as a tribunal in the advertising industry to

---

[1001] Ministério da Ciência, Tecnologia e Inovação (MCTI), *Plano Brasileiro de Inteligência Artificial (PBIA), 2024–2028* (Jul. 29, 2024), https://www.gov.br/lncc/pt-br/assuntos/noticias/ultimas-noticias-1/plano-brasileiro-de-inteligencia-artificial-pbia-2024-2028

[1002] Latin American Centre for Development Administration (CLAD), *Ibero American Charter on Artificial Intelligence in Civil Service* (Nov. 2023), https://web-api-backend.clad.org/uploads/ciia-en-03-2024.pdf

[1003] Ministério da Ciência, Tecnologia e Inovação (MCTI), *Plano Brasileiro de Inteligência Artificial (PBIA), 2024-2028* (Jul. 29, 2024), https://www.gov.br/lncc/pt-br/assuntos/noticias/ultimas-noticias-1/plano-brasileiro-de-inteligencia-artificial-pbia-2024-2028

[1004] Ministério da Ciência, Tecnologia e Inovação (MCTI), *PBIA 2025–2028*, pp. 22–28 (Jun. 12, 2025), www.gov.br/mcti/pt-br/acompanhe-o-mcti/noticias/2025/06/plano-brasileiro-de-inteligencia-artificial-pbia-_vf.pdf

ensure advertising practices are lawful and ethical, [1005] warned advertisers to be cautious when using algorithm-based advertising after deciding a case involving algorithm-based advertising in December 2021.[1006] The ruling responded to a consumer complaint against a streaming platform, alleging that an ad did not reflect the actual content of the platform. Although CONAR found the streaming platform had committed no irregularities, the council cautioned advertisers that algorithmic-based ads designed to increase engagement rather than inform may hinder consumers' ability to find solutions they seek.

A Brazilian court fined Meta Platforms up to $3.62 million after finding the company had accepted paid advertisements that fraudulently used a retailer's name to mislead consumers. Ads revealed that artificial intelligence had been used to mimic a businessman's voice.[1007]

The Brazilian government sanctioned the Digital Statute for Children and Adolescents in 2025.[1008] This law creates a significant new restriction on algorithm-based advertising by explicitly prohibiting profiling to direct commercial advertising to children and adolescents. The law defines profiling in Article 2 (Section V).

EdTech

The EdTech apps Descomplica and Stoodi used in Sao Paulo were part of a 2022 global study conducted by Human Rights Watch on education technology endorsed by 49 governments, including Brazil, during the pandemic.[1009] Based on technical and policy analysis of these EdTech products, Human Rights Watch found that Brazil's endorsement of these online learning platforms put at risk or directly violated children's rights. Both Descomplica and Stoodi have "the ability to collect their users' advertising IDs. This allowed these apps to tag children and identify their devices for the sole purpose of advertising to them," in violation of child data protection principles as well as corporations' human rights responsibilities outlined in the United Nations Guiding Principles on Business and Human Rights.

In March 2024, Human Rights Watch submitted recommendations to the UN Committee on the Rights of the Child urging the ANPD to seek actionable steps against the EdTech companies to protect children and adolescents' data privacy.

[1005] CONAR, *About CONAR*, http://www.conar.org.br/

[1006] CONAR, *Complaint No. 203/21, Ruling Issued on 7 Dec. 2021*, Conar Gazette: Ethics in Practice, Issue 222, pp. 6–7 (Jan. 2022), http://www.conar.org.br/pdf/conar222.pdf

[1007] Reuters, *Meta faces $3.6 Millon Fine in Brazil for Allowing Bogus Havan Ads* (Aug. 28, 2024) https://www.reuters.com/technology/meta-faces-36-million-fine-brazil-allowing-bogus-havan-ads-2024-08-28/

[1008] Presidência da República, *Lei Nº 15.211, de 17 de Setembro de 2025 (Estatuto Digital da Criança e do Adolescente)* (Sept. 17, 2025), https://www.planalto.gov.br/ccivil_03/_ato2023-2026/2025/lei/L15211.htm

[1009] Human Rights Watch, *How Dare They Peep into My Private Life* (May 25, 2022), https://www.hrw.org/report/2022/05/25/how-dare-they-peep-my-private-life/childrens-rights-violations-governments

Human Rights Watch also recommended that ANPD build on the current "case law and Supreme court rulings protecting children's right in educational settings" when addressing online education.[1010] Brazil's legislature took direct action to address these concerns. The government sanctioned Digital Statute for Children and Adolescents in 2025,[1011] establishing new binding rules for all platforms accessed by minors.

A federal law introduced in February 2025 would also limit smartphone use in public and private elementary and high schools across Brazil.[1012]

Use of AI in Courts

With a current backlog of 78 million lawsuits, the Brazilian judicial system operates with substantial challenges in case flow management. The lack of resources to meet this demand[1013] has led to numerous initiatives involving Artificial Intelligence.

In this context, the President of the National Council of Justice (*Conselho Nacional de Justiça*, CNJ), a judicial agency responsible for the administrative and financial control of the judiciary and the supervision of judges,[1014] published in August 2020 a Resolution on ethics, transparency and governance in the production and use of artificial Intelligence in the Judiciary. [1015] The NCJ Resolution addresses AI-related requirements such as respect for human rights; preservation of equality, non-discrimination, plurality, and solidarity; transparency (from disclosure to explainability); data security; user control; and accountability.

The NCJ passed a new resolution updating the previous guidelines to create a new norm around AI in the judiciary with guidelines, requirements, and a governance structure for the development, use, and auditability of artificial intelligence tools.[1016]

[1010] Human Rights Watch, *Brazil: Submission to the UN Committee on the Rights of the Child, 98th Pre-Sessional Working Group* (Mar. 27, 2024), https://www.hrw.org/news/2024/03/27/brazil-submission-un-committee-rights-child

[1011] Presidência da República, *Lei Nº 15.211, de 17 de setembro de 2025 (Estatuto Digital da Criança e do Adolescente)* (Sept. 17, 2025), https://www.planalto.gov.br/ccivil_03/_ato2023-2026/2025/lei/L15211.htm

[1012] Federal Official Gazette, *Lei Nº 15.100, de 13 de janeiro de 2025* (Jan. 14, 2025), https://www.in.gov.br/en/web/dou/-/lei-n-15.100-de-13-de-janeiro-de-2025-606772935

[1013] SIPA, *The Future of AI in the Brazilian Judicial System: AI Mapping, Integration, and Governance* (Jun. 2020), https://itsrio.org/wp-content/uploads/2020/06/SIPA-Capstone-The-Future-of-AI-in-the-Brazilian-Judicial-System-1.pdf

[1014] US Law Library of Congress, *Brazil, Legal Research Guide: The Judicial Branch* (2011), https://www.loc.gov/item/2019671039/

[1015] Conselho Nacional de Justiça (CNJ), *Resolution No. 332 Provides for Ethics, Transparency and Governance in the Production and Use of Artificial Intelligence in the Judiciary and Provides Other Measures* (Aug. 21, 2020), https://atos.cnj.jus.br/atos/detalhar/3429

[1016] Conselho Nacional de Justiça (CNJ), *CNJ Approves Resolution Regulating the Use of AI in the Judiciary* (Feb. 18, 2025) https://www.cnj.jus.br/cnj-aprova-resolucao-regulamentando-o-uso-da-ia-no-poder-judiciario/

Another Resolution established guidelines for generative AI in Brazilian courts, mandating human oversight, risk management plans, and auditable records.[1017]

The Attorney General of the Union (AGU) began using artificial intelligence tools in the production of legal documents in 2024. Federal lawyers and prosecutors may use generative AI tools through the institutional document management system known as Super Sapiens to make summaries of extensive texts, classify and sort documents, extract data from a process, or obtain suggestions for petition models.[1018]

The Public Prosecutor's Office of the State of Rio de Janeiro reportedly invested in data science and AI to expedite investigations and prevent crimes.[1019] The system has allowed information from different sources and bodies to be collected as well as real-time data to be collected from suspected criminals. Likewise, Brazil's federal and state police are using AI applications such as military drones[1020] and crime prediction software.[1021]

Computational Propaganda

According to Freedom House's Freedom on the Net report, Brazil is only "partly free" because of the prevalence of manipulated content online and the "notable proliferation of disinformation during the 2018 and 2022 election campaigns."[1022] Several studies[1023] have pointed out the use of bots on online gatekeepers in the

---

[1017] Conselho Nacional de Justiça (CNJ), *Resolução nº 615, de 11 de março de 2025*, https://atos.cnj.jus.br/atos/detalhar/6001

[1018] Attorney General of the Union, *AGU Starts Using Artificial Intelligence Tools in the Production of Legal Documents* (Sept. 24, 2024), https://www.gov.br/agu/pt-br/comunicacao/noticias/agu-passa-a-utilizar-ferramentas-de-inteligencia-artificial-na-producao-de-documentos-juridicos

[1019] Felipe Grandin and Marco Antônio Martins, *MPRJ Bets on Artificial Intelligence to Speed Up Investigations in Rio*, O Globo, G1 (Oct. 1, 2018), https://g1.globo.com/rj/rio-de-janeiro/noticia/2018/10/01/mp-aposta-em-inteligencia-artificial-para-agilizar-investigacoes-no-rj.ghtml

[1020] ISTOE, *Against Organized Crime, PF Puts Unmanned Aerial Vehicle in the Amazon* (Aug. 20, 2016), https://istoe.com.br/contra-o-crime-organizado-pf-poe-veiculo-aereo-nao-tripulado-na-amazonia/

[1021] Sarah Griffiths, *CrimeRadar Is Using Machine Learning to Predict Crime in Rio*, Wired UK (Aug. 18, 2016), https://www.wired.co.uk/article/crimeradar-rio-app-predict-crime; United for Smart Sustainable Cities, *Crime Prediction for More Agile Policing in Cities, Rio de Janeiro, Brazil: Case Study of the U4SSC City Science Application Framework* (Oct. 2019), https://igarape.org.br/wp-content/uploads/2019/10/460154_Case-study-Crime-prediction-for-more-agile-policing-in-cities.pdf

[1022] Freedom House, *Freedom on the Net 2022: Brazil* (2022), https://freedomhouse.org/country/brazil/freedom-net/2022

[1023] Samantha Bradshaw, Hannah Bailey, Philip N. Howard, *Industrialized Disinformation: 2020 Global Inventory of Organized Social Media Manipulation*, Oxford Internet Institute (2021), https://demtech.oii.ox.ac.uk/wp-content/uploads/sites/12/2021/02/CyberTroop-Report20-Draft9.pdf; Joao Guilherme Bastos dos Santos, Arthur Ituassu, Sergio Lifschitz, Thayane Guimaraes, Diego Cerqueira, Debora Albu, Redson Fernando, Julia Hellen Ferreira, Maria Luiza Mondelli, *From Digital Militias to Coordinated Behavior: Interdisciplinary Methods of Analysis and Analysis and Identification of Bots in Brazilian Elections*, Sociedade Brasileira de Computação, Annals of the X

electoral context. Misinformation campaigns in Brazil have consisted in "pro-government and pro-party propaganda"; "attacking the opposition or mounting smear campaigns"; "suppressing participation through trolling and harassment"; and social media campaigns that "drive division and polarize citizens."[1024] Brazil's Supreme Court opened a probe into so-called "digital militias" in 2021 and received a report from the Federal Police detailing their structure for coordinating attacks against rival politicians and democratic institutions and the dissemination of "false news."[1025]

The Brazilian Judiciary responded to these challenges—primarily through the Superior Electoral Court and the Federal Supreme Court—by implementing innovative measures to mitigate the harmful effects of disinformation and ensure the integrity of the electoral process.[1026] In February 2024, the Superior Electoral Court approved a resolution[1027] that addressed several threats related to artificial intelligence. The resolution included prohibition of deep fakes; obligation to warning about the use of AI in electoral propaganda; restriction of the use of "robots" to mediate contact with the voter (the campaign cannot simulate dialogue with a candidate or anyone else); and accountability of "big techs" that do not immediately remove content that violates these guidelines from their platforms. The violation of this norm carries strict punishment for candidates. Article 9c highlights that a candidate can have their registration or mandate canceled for infringing on the norm and Article 9-E establishes the joint and severe liability of the providers, in a civil and administrative way, if they do not immediately remove certain content and/or accounts that violate these guidelines during the electoral period.

Facial Recognition

Facial recognition is implemented in both the public and private sectors in Brazil. According to Instituto Igarapé, a Brazilian think tank, there were at least 48

Brazilian Workshop on Social Network Analysis and Mining, pp. 187–192 (2021), https://doi.org/10.5753/brasnam.2021.16138

[1024] Samantha Bradshaw, Hannah Bailey, and Philip N. Howard, *Industrialized Disinformation: 2020 Global Inventory of Organized Social Media Manipulation*, Oxford Internet Institute, p. 13 (2021), https://demtech.oii.ox.ac.uk/wp-content/uploads/sites/12/2021/02/CyberTroop-Report20-Draft9.pdf

[1025] Lais Martins, *Under Bolsonaro, Political Attacks Gain Institutional Legitimacy in Brazil*, Global Voices (Oct. 2, 2022), https://advox.globalvoices.org/2022/10/02/under-bolsonaro-political-attacks-gain-institutional-legitimacy-in-brazil/

[1026] Gustavo Borges, Wilson Center, *Combating Disinformation by the Brazilian Judiciary: Initiatives for the 2024 Municipal Elections*, Brazil Builds blog (Oct. 25, 2024), https://www.wilsoncenter.org/blog-post/combating-disinformation-brazilian-judiciary-initiatives-2024-municipal-elections

[1027] Superior Electoral Court (Tribunal Superior Eleitoral), *TSE Prohibits the Use of Artificial Intelligence to Create and Spread Fake Content in the Elections* (Feb. 28, 2024), https://www.tse.jus.br/comunicacao/noticias/2024/Fevereiro/tse-proibe-uso-de-inteligencia-artificial-para-criar-e-propagar-conteudos-falsos-nas-eleicoes

facial recognition applications throughout 16 Federal States between 2011 to 2019.[1028] The main use sectors are public security, border control, transportation, and education.[1029]

In August 2018, the Brazilian Institute for Consumer Protection (IDEC) brought a civil public class action[1030] before the Court of Justice of Sao Paulo for breach of privacy and consumer legislation against the São Paulo Metro operator, regarding an AI crowd analytics system that claimed to predict the emotion, age, and gender of metro passengers without processing personal data.[1031] In May 2021, the São Paulo 37th Civil Court ordered the company to pay compensation for passengers' data collection without their consent and prohibited the company from continuing the implementation and use of cameras that recognized human presence or identified emotion, gender, and age groups.[1032]

Another monitoring system with built-in facial recognition installed in the São Paulo subway network was challenged in Court. In early 2020, the operating company was requested to provide clarification on the risks and impact assessment expected from the implementation of the new technology, on how personal data will be processed, on technical databases and security systems issues, and on actions to mitigate the potential risk of a data breach.[1033] In March 2022, the 6th Public Treasury Court of São Paulo delivered a preliminary injunction ordering the operating company to suspend the deployment and use of the facial recognition system in the São Paulo metro stations.[1034] In April 2022, the preliminary injunction was confirmed in appeal

---

[1028] Instituto Igarapé, *Facial Recognition in Brazil*, https://igarape.org.br/infografico-reconhecimento-facial-no-brasil/

[1029] Thiago Moraes, *Facial Recognition in Brazil*, Wired (Nov. 20, 2019), https://medium.com/@lapinbr/face-recognition-in-brazil-f2a23217f5f7

[1030] Instituto Brasileiro de Defesa do Consumidor, *ViaQuatro* (Aug. 30, 2018), https://idec.org.br/sites/default/files/acp_viaquatro.pdf

[1031] AccessNow, *Facial Recognition on Trial: Emotion and Gender "Detection" under Scrutiny in a Court Case in Brazil* (Jun. 29, 2020), https://www.accessnow.org/facial-recognition-on-trial-emotion-and-gender-detection-under-scrutiny-in-a-court-case-in-brazil/

[1032] Tribunal de Justicia do Estado de São Paulo, 37th Civil Court, *IDEC v. ViaQuatro* (May 7, 2021), https://idec.org.br/sites/default/files/75432prot_sentenca-viaquatro.pdf

[1033] Tribunal de Justicia do Estado de São Paulo, 1st Public Treasury Court of São Paulo, *Public Defenders Office of the State of Sao Paulo and Others v. Company of the Metropolitan of Sao Paulo, METRO* (Feb. 12, 2020), https://idec.org.br/sites/default/files/doc_76725029_1.pdf; Tozzini Freire, *Facial Recognition Is Disputed in Court* (Feb. 14, 2020), https://tozzinifreire.com.br/en/boletins/facial-recognition-is-disputed-in-court

[1034] Court of Justice of the State of São Paulo, 6th Public Treasury Court of São Paulo, *Public Defenders Office of the State of Sao Paulo and Others v. Company of the Metropolitan of Sao Paulo, METRO* (Mar. 22, 2022), https://images.jota.info/wp-content/uploads/2022/03/liminar-metro-reconhecimento-facial.pdf
Angelica Mari, *São Paulo Subway Ordered to Suspend Use of Facial Recognition* (Mar. 25, 2022), https://www.zdnet.com/article/sao-paulo-subway-ordered-to-suspend-use-of-facial-recognition/

by the São Paulo 5th Public Law Chamber.[1035] However, in October 2022, the same Chamber ruled that the operating company was entitled to deploy and use facial recognition technology on the grounds that the LGPD allows such use for public security purposes. Governor Rodrigo Garcia officially inaugurated the facial recognition system the following month. The aim is for 5,000 biometric cameras to be deployed in transit stations over the next 30 months.[1036]

In July 2023, the government of São Paulo introduced the Muralha Paulista system, a security network that interconnects cameras and radars in different cities to the Metropolitan Region of São José do Rio Preto.[1037] The Muralha Paulista system has since been used by football clubs to identify perpetrators of violence or harassment at large sporting events, specifically to flag people with open warrants or judicial restrictions.[1038]

In November 2023, Brazil's ANDP issued a technical note that sheds light on the legality of facial recognition technology in sports stadiums. Amid controversies about striking a balance between enhancing public safety and safeguarding individual privacy, the note lays down concrete suggestions to ensure compliance with the General Data Protection Law. Key among these suggestions is the call for comprehensive Data Protection Impact Assessments that specifically address the processing of biometric data. The note outlines critical guidelines for avoiding legal pitfalls and details the consequences of failing to adhere to these standards, marking a step forward in the ongoing discussions around ethical deployment of AI technologies in public spaces.[1039]

The ANPD announced its investigation into irregularities in the use of facial recognition systems for the sale of tickets and entry to stadiums by 23 football clubs. The irregularities pertain to the compliance with transparency requirements and the handling of data concerning children and adolescents under the data protection law. The inspector issued a preventive measure requiring football clubs to publish

---

[1035] Court of Justice of the State of São Paulo, 5th Public Law Chamber (Appeal), *Public Defenders Office of the State of Sao Paulo and Others v. Company of the Metropolitan of Sao Paulo, METRO* (Apr. 12, 2022), https://internetlab.org.br/wp-content/uploads/2022/04/doc_429164790.pdf

[1036] Biometric Update, *Brazil Deploys ISS Facial Recognition to Secure São Paulo Metro* (Dec. 9, 2022), https://www.biometricupdate.com/202212/brazil-deploys-iss-facial-recognition-to-secure-sao-paulo-metro

[1037] O Globo, g1, *What Is the Electric Wall, a Security System that Will Be Implemented in 17 Cities of the Valley* (Jul. 21, 2023), https://g1.globo.com/sp/vale-do-paraiba-regiao/noticia/2023/07/21/o-que-e-a-muralha-eletronica-sistema-de-seguranca-que-sera-implantado-em-17-cidades-do-vale-do-paraiba.ghtml

[1038] Reuters, *Palmeiras' Facial Recognition on Match Tickets Helps Police Arrest Criminals* (Sept. 23, 2023), https://www.reuters.com/sports/soccer/palmeiras-facial-recognition-match-tickets-helps-police-arrest-criminals-2023-09-23/

[1039] Autoridade Nacional de Proteção de Dados (ANDP) Coordenação-Geral de Fiscalização, *Technical Note No. 175/2023/CGF/ANPD* (Nov. 2023), https://www.gov.br/anpd/pt-br/documentos-e-publicacoes/documentos-de-publicacoes/nota-tecnica-no-175-2023-cgf-anpd-acordo-de-cooperacao-mjsp-e-cbf.pdf

information on the registration and biometric identification procedures for fans on ticket selling platforms within 20 business days as the agency investigates.[1040]

The Brazilian police have also been using live facial recognition for Carnival and has plans to use the technology in events involving crowds to find wanted criminals. In 2020, police forces rolled out facial recognition in six capitals across the country. When announcing the use of live facial recognition, the São Paulo police said they would create a "situation room," where they would monitor the images from the cameras, which would then be compared with a database managed by a biometrics lab. According to the police, the aim is to reduce the likelihood of mistakes, such as wrongful arrests.[1041]

In 2021, Brazil rolled out end-to-end biometric identification technologies by IDEMIA, a technology solution provider, for use in passenger identification at several airports, including domestic airports in São Paulo and Rio de Janeiro.[1042] Many have voiced concern at the government's embrace of facial recognition technology, especially surrounding issues of racial bias, given that the LGPD does not address these technologies.[1043]

In the first quarter of 2022, several campaigns started in Brazil opposing the widespread use of facial recognition technologies in public spaces.

One of them is called *Sai da minha cara* (Get off my face) and is led by civil society organizations, including the IDEC.[1044] A localized campaign with several states and cities, *Sai da minha cara* advocates for laws to ban facial recognition. Including bills to restrict the use of facial recognition by public authorities before the State Legislative Assembly of São Paulo[1045] and before the City Council in Porto

[1040] Autoridade Nacional de Proteção de Dados (ANDP), *ANDP Monitors the Use of Facial Recognition Systems in Ticket Sales and Stadium Entrances by 23 Football Clubs* (Feb. 18, 2025), https://www.gov.br/anpd/pt-br/assuntos/noticias/anpd-fiscaliza-uso-de-sistema-de-reconhecimento-facial-na-venda-de-ingressos-e-na-entrada-de-estadios-por-23-clubes-de-futebol

[1041] Angelica Mari, *Brazilian Police Introduces Live Facial Recognition for Carnival,* Brazil Tech (Feb. 25, 2020), https://www.zdnet.com/article/brazilian-police-introduces-live-facial-recognition-for-carnival/

[1042] Angelica Mari, *Brazilian Airports Expand Facial Recognition Trials*, ZDNet (Nov. 22, 2021), https://www.zdnet.com/article/brazilian-airports-expand-facial-recognition-trials/; Chris Burt, *Brazil's Pilot of IDEMIA Face Biometrics Advances to Simultaneous Operation at Capital Airports,* Biometric Update (Jun. 16, 2021), https://www.biometricupdate.com/202106/brazils-pilot-of-idemia-face-biometrics-advances-to-simultaneous-operation-at-capital-airports

[1043] Charlotte Peet, *Brazil's Embrace of Facial Recognition Worries Black Communities*, Rest of World (Oct. 22, 2021), https://restofworld.org/2021/brazil-facial-recognition-surveillance-black-communities/; Leaders League, *The Controversial Use of Facial Recognition in Brazil and Europe* (Aug. 12, 2021), https://www.leadersleague.com/en/news/the-controversial-use-of-facial-recognition-in-brazil-and-europe

[1044] IDEC, *Lawmakers from All Regions of Brazil Present Bills to Ban Facial Recognition in Public Spaces* (Jun. 20, 2022), https://idec.org.br/release/parlamentares-de-todas-regioes-do-brasil-apresentam-projetos-de-lei-pelo-banimento-do

[1045] São Paulo Legislative Assembly, *State Law Proposal No. 385/2022* (Jun. 23, 2022), https://www.al.sp.gov.br/propositura/?id=1000448817

Alegre, the capital of the state of Rio Grande do Sul.[1046] A bill before the City Council in Curitiba, capital city of the state of Paraná, would restrict the use of facial recognition for mass surveillance by public authorities.[1047]

Another campaign started in May 2022 under the name "Tire o Meu Rosto da Sua Mira" (Get My Face Out of Your Sight).[1048] Civil society organizations demand a total ban on the use of facial recognition technologies for public security purposes in Brazil. According to the campaign's open letter, in Brazil, a country with the third largest incarcerated population in the world, the use of facial recognition technologies for public security purposes would lead to the worsening of racist practices.

A Federal Law proposal seeking to regulate the use of facial recognition technologies in Brazil both for the private and public sectors was submitted to the Brazilian House of Representatives in August 2022.[1049] The legislative process began that October and the proposed legislation was forwarded to the responsible commissions.

A new study published by the Center for Security and Citizenship Studies (CESeC) in March 2024 showed that 67.4 million Brazilians, close to a third of the population, are potentially under surveillance by facial recognition cameras. In updates in January 2025, the CESeC cited 337 facial recognition projects in the country and more than 81 million people surveilled.[1050]

The banning of the use of facial recognition systems for public security purposes was one of the key points that emerged during the public hearings promoted by the Commission of Jurists that elaborated the proposal for the regulation of artificial intelligence in Brazil.[1051] Several Brazilian rights organizations called for a ban on facial recognition by law enforcement during discussions around the proposed AI regulation bill.[1052] For example, the Coalition Rights on the Net (*Coalizão Direitos*

[1046] Porto Alegre City Council, *Process No. 00499/22* (Jun. 21, 2022), https://www.camarapoa.rs.gov.br/processos/137992

[1047] Curitiba City Council, *Law Proposal No. 005.00138.2022* (Jul. 1, 2022), https://www.cmc.pr.gov.br/wspl/sistema/ProposicaoDetalhesForm.do?select_action=&pro_id=459750

[1048] Tire Meu Rostro da Sua Mira, *A Call for a General Ban on the Use of Facial Recognition Technologies in Public Security*, https://tiremeurostodasuamira.org.br

[1049] Brazilian House of Representatives, *Federal Law Proposal No. 2392/2022* (Aug, 31, 2022), https://www.camara.leg.br/proposicoesWeb/fichadetramitacao?idProposicao=2334803

[1050] Centro de Estudos de Segurança e Cidadania (CESeC), *The Panopticon: Monitor of New Technology in the Public Security of Brazil* (Jan. 27, 2025), https://www.opanoptico.com.br/

[1051] Federal Senate, *Debates Suggest the End of Facial Recognition in Public Security* (May 18, 2022), https://www12.senado.leg.br/noticias/materias/2022/05/18/debates-apontam-para-fim-do-reconhecimento-facial-na-seguranca-publica

[1052] Masha Borak, *As Brazil Debates AI Bill, Calls for Racial Recognition Bans Emerge* (Jul. 8, 2024), https://www.biometricupdate.com/202407/as-brazil-debates-ai-bill-calls-for-facial-recognition-bans-emerge

na Rede) cited several examples of arrests after mistaken identity in their open letter advocating for the Brazilian AI regulation to ban facial recognition technologies.[1053]

On November 10th, 2025, Resolution No. 17 was published by Brazil's National Human Rights Council. It establishes guidelines for the use of facial recognition technologies in public spaces.[1054]

Environmental Impact of AI

Brazil's AI Plan (PBIA) recognized the environmental implications of AI technologies, focusing on integrating sustainability within AI development. The strategy includes the monitoring of AI's carbon footprint, advocating for energy-efficient AI systems, and promoting best practices to reduce environmental damage. The policy also encourages the creation of regulatory sandboxes for AI, which allow for controlled testing of AI technologies with an emphasis on minimizing their environmental impact, particularly in terms of energy consumption and e-waste. Brazil is also aligning its AI policies with global climate goals, supporting AI's potential to address challenges related to climate change and sustainable development.[1055] In addition, the Brazilian government announced a "new global initiative focused on information integrity, with particular attention to climate change." This is in line with the G20 and the 30th UN Framework Convention on Climate Change (COP 30).[1056]

Lethal Autonomous Weapons

Brazil has consistently called for the negotiation of a legally binding instrument to ensure meaningful human control over critical functions in lethal autonomous weapons systems (LAWS). Brazil issued a joint statement with Austria and Chile at a 2018 meeting of the Convention on Certain Conventional Weapons (CCW) Group of Government Experts (GGE) on LAWS that proposed to establish an open-ended GGE to negotiate a legally binding instrument to ensure meaningful

---

[1053] Coalizão Direitos Na Rede (Coalition Rights on the Net), *Open Letter: Advocating for Brazilian AI Regulation that Protects Human Rights* (Jul. 8, 2024), https://direitosnarede.org.br/2024/07/08/open-letter-advocating-for-brazilian-ai-regulation-that-protects-human-rights/

[1054] Presidência da República, *Resolução nº 17, de 10 de novembro de 2025 - Dispõe sobre o uso de tecnologias de reconhecimento facial em espaços públicos dos estados e do Distrito Federal*, https://www.gov.br/participamaisbrasil/resolucao-dispoe-sobre-uso-de-tecnologias-de-reconhecimento-facial-em-espacos-publicos-dos-estados-e-df

[1055] Ministério da Ciência, Tecnologia e Inovação (MCTI), *Plano Brasileiro de Inteligência Artificial (PBIA), 2024–2028* (Jul. 29, 2024), https://www.gov.br/mcti/pt-br/acompanhe-o-mcti/noticias/2024/07/plano-brasileiro-de-inteligencia-artificial-ia_para_o_bem_de_todos.pdf/view

[1056] G20 Brasil, *UNESCO Offers Recommendations for Regulation and National Policies on AI* (Sept. 12, 2024), https://www.gov.br/g20/en/news/unesco-offers-recommendations-for-regulation-and-national-policies-on-ai

human control over critical functions in LAWS.[1057] Brazil's representative at the 77th meeting of the UN General Assembly First Committee reiterated the need for a regulation that recognizes the centrality of human control in the development and use of autonomous systems, in line with international humanitarian law;[1058] however, Brazil was not among the 70 countries that endorsed a joint statement on autonomous weapons systems at the meeting.[1059] Brazil voted in favor of the adoption of the 2024 UN General Assembly Resolution on lethal autonomous weapons.[1060] Brazil delivered a statement on behalf of 39 High Contracting Parties to the CCW at the 2025 GGE meeting[1061] stressing the "need for urgent action" on negotiating a legal instrument in line with the GGE mandate.

Brazil joined more than 30 other Latin American and Caribbean states to endorse the Belén Communiqué in 2023.[1062] The Communiqué calls for "urgent negotiation" of a binding international treaty to regulate and prohibit the use of autonomous weapons to address the grave concerns raised by removing human control from the use of force.

AI Literacy

The PBIA 2025–2028 lists Dissemination, Training, and Capacity-Building as a strategic axis, promoting digital skills training and AI literacy for citizens and civil servants.[1063] The axis aims to reach 30% of the adult population with the national

---

[1057] GGE LAWS, *Proposal for a Mandate to Negotiate a Legally-Binding Instrument that Addresses the Legal, Humanitarian and Ethical Concerns Posed by Emerging Technologies in the Area of Lethal Autonomous Weapons Systems (LAWS)*, U.N. Doc. CCW/ GGE.2/2018/WP.7 (Aug. 30, 2018), https://documents-dds-ny.un.org/doc/UNDOC/GEN/G18/264/05/PDF/G1826405.pdf?OpenElement

[1058] UN press release, *Enough Bullets Made Each Year to Kill 'Twice the Number of Planet's Inhabitants,' First Committee Hears during Debate on Conventional Weapons*, 77th Session (Oct. 21, 2022), https://press.un.org/en/2022/gadis3695.doc.htm

[1059] UN General Assembly, First Committee, *Joint Statement on Lethal Autonomous Weapons Systems First Committee, 77th United Nations General Assembly Thematic Debate – Conventional Weapons* (Oct. 21, 2022), https://estatements.unmeetings.org/estatements/11.0010/20221021/A1jJ8bNfWGlL/KLw9WYcSnnAm_en.pdf

[1060] UN Digital Library, *Lethal Autonomous Weapons: Resolution/adopted by the General Assembly*, A/RES/79/62 (Dec. 2, 2024), https://digitallibrary.un.org/record/4068497?ln=en

[1061] UN Office of Disarmament Affairs, *Joint Statement to the September 2025 Session of the CCW GGE LAWS* (Sept. 2025), https://docs-library.unoda.org/Convention_on_Certain_Conventional_Weapons_-Group_of_Governmental_Experts_on_Lethal_Autonomous_Weapons_Systems_(2025)/Joint_statement_to_the_September_2025_session_of_the_CCW_GGE_LAWS_-_As_delivered.pdf

[1062] Latin American and the Caribbean Conference of Social and Humanitarian Impact of Autonomous Weapons, *Communiqué* (Feb. 24, 2023), https://www.rree.go.cr/files/includes/files.php?id=2261&tipo=documentos

[1063] Ministério da Ciência, Tecnologia e Inovação (MCTI), *Plano Brasileiro de Inteligência Artificial*, p. 33 (Jun. 12, 2025), https://www.gov.br/mcti/pt-br/centrais-de-conteudo/publicacoes-mcti/plano-brasileiro-de-inteligencia-artificial/pbia_mcti_2025.pdf

awareness campaign within 3 years. Within 5 years, the plan intends for 85% of adults "have a basic understanding of AI, its benefits and risks."[1064] The Plan proposes a program for AI learning, including about the impacts of AI targeting students and teachers in basic education, though details are few. Implementation is shared by MCTI and the Ministry of Education.

Human Rights

Brazil is a signatory to the Universal Declaration of Human Rights and other international human rights treaties and conventions.[1065] Freedom House rated Brazil as "Free" in the 2025 Freedom in the World Report with a score of 72/100 for political rights and civil liberties. According to Freedom House, Brazil is "a democracy that holds competitive elections. Its political arena, though polarized, is characterized by vibrant public debate. However, independent journalists and civil society activists risk harassment and violent attack and political violence is high. Minority groups suffer from crime, disproportionate violence, and economic exclusion, issues the government struggles to address. Corruption is endemic at top levels, contributing to widespread disillusionment among the public. Societal discrimination and violence against LGBT+ people remain serious problems."

Freedom House rated Brazil as "Partly Free" with a score of 65/100 in the Freedom on the Net report.[1066] The report cited obstacles to internet access, limits on digital content, and violation of user rights as concerns.

AI Safety Summit

Brazil participated in the first AI Safety Summit in 2023 and endorsed the Bletchley Declaration. [1067] Brazil thus committed to participate in international cooperation on AI "to promote inclusive economic growth, sustainable development and innovation, to protect human rights and fundamental freedoms, and to foster public trust and confidence in AI systems to fully realise their potential." Brazilian representatives did not attend the 2024 Summit in Seoul, in part because of the country's schedule conflict with a G20 event hosted at the same time.[1068]

[1064] Ibid, p. 36–37

[1065] UN Treaty Body Database, *Ratification Status by Country of by Treaty: Brazil*, https://tbinternet.ohchr.org/_layouts/15/TreatyBodyExternal/treaty.aspx

[1066] Freedom House, *Freedom on the Net: Brazil* (2024), https://freedomhouse.org/country/brazil

[1067] UK Department for Science, Innovation & Technology, Foreign, Commonwealth &Development Office, Prime Minister's Office, *The Bletchley Declaration by Countries Attending the AI Safety Summit* (Nov. 2023), https://www.gov.uk/government/publications/ai-safety-summit-2023-the-bletchley-declaration/the-bletchley-declaration-by-countries-attending-the-ai-safety-summit-1-2-november-2023

[1068] Martin Coulter, *Second Global AI Safety Summit Faces Tough Questions, Lower Turnout*, Reuters (Apr. 29, 2024), https://www.reuters.com/technology/second-global-ai-safety-summit-faces-tough-questions-lower-turnout-2024-04-29/

At the AI Action Summit in Paris, Brazil signed the Statement on Inclusive and Sustainable Artificial Intelligence for People and the Planet, affirming priorities that include AI access, sustainability, safety, and transparency.[1069]

OECD / G20 AI Principles

Brazil has endorsed the OECD AI Principles and the G20 AI Guidelines.[1070] Brazil's most recent AI strategy explicitly integrates the OECD AI Principles as the its base.[1071] Brazil is also a member of the Global Partnership on AI, a multi-stakeholder initiative that aims to foster international cooperation on AI research and applied activities.[1072] The GPAI is "built around a shared commitment to the OECD Recommendation on Artificial Intelligence."[1073]

With the G20 Presidency in 2024, Brazil led progress on the realization of the G20 AI Guidelines. The G20 Digital Economy ministers, led by Brazil, endorsed principles for digital inclusion and AI for inclusive sustainable development, inequality reduction, and combating disinformation.[1074]

Brazil and the European Union convened for the 13th Edition of the bilateral Digital Dialogue[1075] in Brussels in 2025 and published a Joint Communique on their commitment to digital development and global technology governance. Both parties evaluated their advancements in key areas of cooperation from the 2024 Dialogue,[1076] including AI regulation, and reaffirmed their commitment to foster digital development, facilitate digital innovation and inclusion, and preserve democratic principles. Importantly, while approving a joint work plan to guide the collaboration

---

[1069] Élysée, *Statement on Inclusive and Sustainable Artificial Intelligence for People and the Planet* (Feb. 11, 2025), https://www.elysee.fr/en/emmanuel-macron/2025/02/11/statement-on-inclusive-and-sustainable-artificial-intelligence-for-people-and-the-planet

[1070] OECD AI Policy Observatory, *OECD AI Principles Overview, Countries Adhering to the AI Principles*, https://oecd.ai/en/ai-principles

[1071] Ministério da Ciência, Tecnologia e Inovação, *Estratégia Brasileira de Inteligência Artificial* (Jan. 13, 2026), https://www.gov.br/mcti/pt-br/acompanhe-o-mcti/transformacaodigital/estrategia-brasileira-de-inteligencia-artificial

[1072] OECD AI Policy Observatory, *About the Global Partnership on Artificial Intelligence (GPAI)* (2026), https://oecd.ai/en/about/about-gpai

[1073] Government of Canada, *Canada Concludes Inaugural Plenary of the Global Partnership on Artificial Intelligence with International Counterparts in Montreal* (Dec. 4, 2020), https://www.canada.ca/en/innovation-science-economic-development/news/2020/12/canada-concludes-inaugural-plenary-of-the-global-partnership-on-artificial-intelligence-with-international-counterparts-in-montreal.html

[1074] G20 Digital Economy Ministers, *G20 DEWG Maceio Ministerial Declaration* (Sept. 13, 2024), https://g7g20-documents.org/database/document/2024-g20-brazil-sherpa-track-digital-economy-ministers-ministers-language-g20-dewg-maceio-ministerial-declaration

[1075] European Union, *13th EU-Brazil Digital Dialogue Reinforces Digital Cooperation* (Feb. 12, 2025), https://digital-strategy.ec.europa.eu/en/news/13th-eu-brazil-digital-dialogue-reinforces-digital-cooperation

[1076] European Union, *The EU and Brazil Strengthen their Digital Cooperation,* (Mar. 21, 2024), https://digital-strategy.ec.europa.eu/en/news/eu-and-brazil-strengthen-their-digital-cooperation

over the next two years, the parties agreed to continue cooperating in multilateral fora as decided in the G20 Maceió Ministerial Declaration on Digital Inclusion.

Council of Europe AI Treaty

Brazil did not participate in the negotiations for the Council of Europe Framework Convention on Artificial Intelligence and Human Rights, Democracy, and the Rule of Law and has not signed the treaty.[1077]

UNESCO Recommendation on AI Ethics

Brazil has endorsed the UNESCO Recommendation on the Ethics of AI and recognizes the obligation to implement the framework. UNESCO partnered with the Regional Centre for Information Society Development Studies (CETIC.br) to host a launch event for the Portuguese version of the UNESCO AI Recommendation, which aimed to "discuss and promote this subject in Brazilian society."[1078] At the launch, Director and UNESCO representative in Brazil Marlova Noleto reiterated Brazil's active engagement in preparing the document and the country's preemptive alignment with the recommendations. Brazilian Ambassador to UNESCO and president of UNESCO's 41st General Conference, Santiago Mourão, emphasized that the Recommendations fully align with the Brazilian government's guidelines and actions on AI themes.[1079]

CAF, the development bank of Latin America, and UNESCO signed a letter of intent to collaborate on the implementation of the Recommendation on AI Ethics in Latin America and the Caribbean.[1080] They established a Regional Council comprising national and local governments from Latin America and the Caribbean, including Brazil, and convened for the inaugural meeting in October 2023.[1081]

Brazil also signed the resulting 2023 Santiago Declaration to Promote Ethical Artificial Intelligence.[1082] The Santiago Declaration reflects UNESCO's

---

[1077] Council of Europe Treaty Office, *Chart of Signatures and Ratifications of Treaty 225* (Jan. 13, 2026), https://www.coe.int/en/web/Conventions/full-list/?module=signatures-by-treaty&treatynum=225

[1078] UNESCO, *UNESCO Launches Portuguese Version of Publication of Artificial Intelligence*, launch video (May 11, 2022), https://brasil.un.org/pt-br/181308-unesco-lanca-versao-em-portugues-de-publicacao-sobre-inteligencia-artificial

[1079] Ibid

[1080] Ángel Melguizo and Gabriela Ramos, *Ethical and Responsible Artificial Intelligence: From Words to Actions and Rights*, Somos Ibero-America (Feb. 1, 2023), https://www.somosiberoamerica.org/pt-br/tribunas/inteligencia-artificial-etica-e-responsavel-das-palavras-aos-fatos-e-direitos/

[1081] UNESCO, *Chile Will Host the First Latin American and Caribbean Ministerial and High Level Summit on the Ethics of Artificial Intelligence* (Sept. 25, 2023), https://www.unesco.org/en/articles/chile-will-host-first-latin-american-and-caribbean-ministerial-and-high-level-summit-ethics

[1082] Ministerial and High Level Summit of Latin America and the Caribbean, *Santiago Declaration "To Promote Ethical Artificial Intelligence in Latin America and the Caribbean"* (Oct. 2023),

Recommendation on the Ethics of AI and establishes fundamental principles that should guide public policy on AI. These include proportionality, security, fairness, non-discrimination, gender equality, accessibility, sustainability, privacy and data protection.

Brazil was one of the first countries to complete the UNESCO Readiness Assessment Methodology (RAM), a tool to support the effective implementation of the Recommendation.[1083] The RAM helps countries and UNESCO identify and address any institutional and regulatory gaps.[1084] The process of implementing the RAM "highlighted progress, but also that significant challenges still need to be addressed. [1085] Issues such as the significant inclusion of marginalized groups and transparency in the processes of development and use of artificial intelligence (AI) demand constant attention and improvement. On the other hand, important advances were noted, demonstrating a maturation in the country's regulatory, policy and infrastructure approaches. UNESCO published the Brazil RAM Report in 2025.[1086]

In 2024, initiatives in Brazil sought to highlight how marginalized groups such as the indigenous community are excluded from AI debates under UNESCO's Recommendation on the Ethics of AI.[1087]

Evaluation

Brazil has developed a robust national strategy for AI and has established a comprehensive law for data protection that includes algorithmic transparency. The federal government's Brazilian Artificial Intelligence Plan 2024–2028 (PBIA), now updated for 2025 to better align objectives with national interests and priorities, marks significant progress. ANPD has also progressed on data protection in the context of the proposed AI plan. While AI legislation is still under review, the introduction of a new bill in the Chamber of Deputies, while the Senate bill is under committee review, demonstrates a commitment to establishing principles, guidelines and standards for

---

https://minciencia.gob.cl/uploads/filer_public/40/2a/402a35a0-1222-4dab-b090-5c81bbf34237/declaracion_de_santiago.pdf

[1083] UNESCO Global AI Ethics and Governance Observatory, *Brazil* (Nov. 2025), https://www.unesco.org/ethics-ai/en/brazil

[1084] UNESCO Global AI Ethics and Governance Observatory, *Readiness Assessment Methodology* (2025), https://www.unesco.org/ethics-ai/en/ram

[1085] UNESCO Global AI Ethics and Governance Observatory, *Brazil* (Nov. 2025), https://www.unesco.org/ethics-ai/en/brazil

[1086] UNESCO, *Brazil: Readiness Assessment Report on Artificial Intelligence* (2025), https://unesdoc.unesco.org/ark:/48223/pf0000393091

[1087] Edson Prestes, Lutiana Valadares Fernandes Barbosa, Viviane Ceolin Dallasta Del Grossi, Cynthia Picolo Gonzaga de Azevedo, Gustavo Macedo, Renan Maffei, *AI and Brazil's Indigenous Populations: A Call for Participation* (Jul. 31, 2024), https://www.unesco.org/en/articles/ai-and-brazils-indigenous-populations-call-participation

the development, implementation and responsible use of AI systems in Brazil.[1088] Initiatives to tackle disinformation and abuses from facial recognition have been inadequately addressed, however, leaving human rights groups continuing to call for bans.

Brazil has emerged as an international leader in championing inclusive and more ethical AI. Brazil led on implementation of the UNESCO Recommendation through completion of the RAM and participation in regional discussions of best practices. Brazil's leadership in furthering digital inclusion in AI and beyond in the G20 is also noteworthy. Addressing national concerns in areas such as facial recognition could make Brazil a model for implementing international frameworks.

[1088] Câmara dos Deputados, *PL 526/2025 – provides for the regulation of use of AI in Brazil and provides for other measures* (Feb. 18, 2025), https://www.camara.leg.br/proposicoesWeb/fichadetramitacao?idProposicao=2484379

Canada

In 2025, Canada launched the AI Strategy for the Federal Service and continued to work on responsible AI in government. The country also endorsed the Council of Europe AI Treaty.

National AI Strategy

Canada's national AI Strategy has focused on the use of AI in government and public services. The Canadian government stated "Artificial intelligence (AI) technologies offer promise for improving how the Government of Canada serves Canadians. As we explore the use of AI in government programs and services, we are ensuring it is governed by clear values, ethics, and laws."[1089] Canada has set out Guiding Principles[1090] to ensure the "effective and ethical use of AI" in government. The government has committed to "understand and measure" impacts, be transparent about use, "provide meaningful explanations" for AI decision-making, "be as open as we can be," and provide sufficient training."

The AI Strategy for Federal Public Service 2025–2027 prioritizes responsible use of AI, expertise, security, and public trust by providing training and talent development, and building public trust through transparency.[1091] The strategy aims to accelerate responsible AI adoption throughout the federal public service, leveraging AI within the Government of Canada to enhance productivity, enhance capacity in science and research, and deliver improved digital services. The Advisory Council on Artificial Intelligence[1092] formed in 2019 and expanded in 2025 advises the Government on opportunities for AI applications to ensure economic growth benefits all Canadians and reflections national values. The expansion of the mandate in 2025 includes advice on budget initiatives and the development of the Safe and Secure Artificial Intelligence Advisory Group to guide the government on emerging risks from AI systems across the economy. An AI Strategy Taskforce[1093] charged with

---

[1089] Government of Canada, *Responsible Use of Artificial Intelligence in Government* (Mar. 4, 2025), https://www.canada.ca/en/government/system/digital-government/digital-government-innovations/responsible-use-ai.html

[1090] Government of Canada, *Guiding Principles for the Use of AI in Government* (Mar. 4, 2025), https://www.canada.ca/en/government/system/digital-government/digital-government-innovations/responsible-use-ai/principles.html

[1091] Treasury Board of Canada Secretariat, *Canada Launches First-Ever Artificial Intelligence Strategy for the Federal Public Service* (Mar. 4, 2025), https://www.canada.ca/en/treasury-board-secretariat/news/2025/03/canada-launches-first-ever-artificial-intelligence-strategy-for-the-federal-public-service.html

[1092] Government of Canada, *Advisory Council on Artificial Intelligence* (Oct. 31, 2025), https://ised-isde.canada.ca/site/advisory-council-artificial-intelligence/en

[1093] Government of Canada, *AI Strategy Taskforce* (Oct. 20, 2025), https://ised-isde.canada.ca/site/advisory-council-artificial-intelligence/en/ai-strategy-taskforce

providing recommendations for a new national AI Strategy also falls under the Advisory Council.

The federal strategy builds on collaborations with provinces, such as the collaboration with the government of Quebec to establish the Centre of Expertise in Montréal for the Advancement of Artificial Intelligence (CEIMIA).[1094] The CEIMIA, formerly known as ICEMAI, continues to facilitate international collaboration on responsible AI as an International Centre of Expertise under the Global Partnership on Artificial Intelligence (GPAI). As a meeting place for academia, civil society, and government, CEIMIA "enables Quebec to highlight the important role of its AI ecosystem, specifically in the area of responsible development of AI, and to take its place internationally as an essential partner and subject-matter expert."

Directive on Automated Decision-Making

The Government of Canada's Treasury Board Secretariat (TBS) established a Directive on Automated Decision-Making to ensure that administrative decisions are "compatible with core principles of administrative law such as transparency, accountability, legality, and procedural fairness."[1095] Canada has developed a questionnaire for an Algorithmic Impact Assessment to "assess and mitigate the risks associated with deploying an automated decision system" and to comply with the Directive.[1096] The Directive took effect on April 1, 2019, with compliance required by no later than April 1, 2020. The Directive is reviewed and updated every two years with feedback from the public. Existing systems have one year to comply with updated requirements.

The TBS issued the Guide on the Scope of the Directive on Automated Decision-Making in 2024 to clarify the conditions for exception and applicability of the Directive.[1097] The guide itemized five key elements for applicability: used by a department; developed after April 1, 2020; used in an administrative decision-making process; replaces or assists judgment; and used in a production environment. The Guide recommends compliance with the Directive in National Security Systems (NSSs) while nothing that NSSs are out of scope.

[1094] Government of Canada, *The Governments of Canada and Quebec and the International Community Join Forces to Advance the Responsible Development of Artificial Intelligence* (Jun. 15, 2020), https://www.canada.ca/en/innovation-science-economic-development/news/2020/06/the-governments-of-canada-and-quebec-and-the-international-community-join-forces-to-advance-the-responsible-development-of-artificial-intelligence.html

[1095] Government of Canada, *Directive on Automated Decision-Making* (Jun. 24, 2025), https://www.tbs-sct.gc.ca/pol/doc-eng.aspx?id=32592

[1096] Government of Canada, *Algorithmic Impact Assessment Tool* (Mar. 19, 2025), https://www.canada.ca/en/government/system/digital-government/digital-government-innovations/responsible-use-ai/algorithmic-impact-assessment.html

[1097] Treasury Board of Canada Secretariat, *Guide on the Scope of the Directive on Automated Decision-Making* (Jun. 27, 2024), https://www.canada.ca/en/government/system/digital-government/digital-government-innovations/responsible-use-ai/guide-scope-directive-automated-decision-making.html#toc2

In a parallel effort to support the Directive, the TBS worked with Public Services and Procurement Canada (PSPC) to establish a pre-qualified AI Vendor procurement program to streamline the procurement of AI solutions and services in the government.[1098] This new public AI procurement program was used to help government departments and agencies build awareness of the solutions offered by AI. It also provided small and medium AI companies with an opportunity to provide their services to the government. The Government of Canda posted a tender for companies to qualify for the AI source list in October 2024. The procurement aims to "establish a pre-qualified list of suppliers […] who meet all the mandatory criteria to provide Canada with responsible and effective AI services, solutions and products."[1099]

Digital Charter Implementation Act

Bill C-27, the Digital Charter Implementation Act (DCIA), represented Canada's first comprehensive Ai legislation, introduced in June 2022.[1100] The Bill, which comprised three separate pieces of legislation: the Consumer Privacy Protection Act (CPPA), Personal Information and Data Protection Tribunal Act (PIDPTA), and the Artificial Intelligence and Data Act (AIDA),[1101] faced public criticism because of what was perceived as insufficient protections and inadequate public consultation. Amendments incorporating some of the public feedback in November 2023 failed to earn public trust in AIDA as an effective, adaptable regulatory framework for AI. Civil society organizations[1102] called for the withdrawal of AIDA, noting "consultation with Indigenous rights-holders, civil society, the private sector, and other stakeholders that should have taken place prior to AIDA's introduction" before reintroducing it in revised form.

AIDA expired in a parliamentary committee, leaving the future of AI legislation in Canada uncertain.[1103] The Digital Charter Implementation Act also expired in Committee.[1104] The government has since initiated renewed public consultations to inform future legislation on AI regulation and data governance.

[1098] Government of Canada, *List of Interested Artificial Intelligence (AI) Suppliers* (Nov. 24, 2025), https://www.canada.ca/en/government/system/digital-government/digital-government-innovations/responsible-use-ai/list-interested-artificial-intelligence-ai-suppliers.html

[1099] Government of Canada, *Invitation to Qualify to Artificial Intelligence Source List* (Oct. 8, 2024), https://canadabuys.canada.ca/en/tender-opportunities/tender-notice/ws4286933967-doc4822970058

[1100] Parliament of Canada, 44th Parliament, 1st Session, *Digital Charter Implementation Act, 2022* (Jan. 6, 2025), https://www.parl.ca/legisinfo/en/bill/44-1/c-27

[1101] Ibid

[1102] Canadian Civil Liberties Association, *CCLA Joins Call from Civil Society to Withdraw AIDA from Bill C-27* (Apr. 25, 2024), https://openmedia.org/assets/AIDA_joint_letter.pdf

[1103] Blair Attard-Frost, *The Death of Canada's Artificial Intelligence and Data Act: What Happened, and What's Next for AI Regulation in Canada*, Op-Ed, Montreal AI Ethics Institute (Jan. 17, 2025), https://montrealethics.ai/the-death-of-canadas-artificial-intelligence-and-data-act-what-happened-and-whats-next-for-ai-regulation-in-canada/

[1104] Parliament of Canada, 44th Parliament, 1st Session, *C-27 An Act to Enact the Consumer Privacy Protection Act, the Personal Information and Data Protection Tribunal Act and the Artificial*

Public Participation

In 2019, Canada established an Advisory Council on Artificial Intelligence to "inform the long-term vision for Canada on AI both domestically and internationally."[1105] Composed of researchers, academics, and business leaders, the Council advises the Government of Canada on how to build on Canada's AI strengths to support entrepreneurship, drive economic growth and job creation and build public trust in AI. The Council created two working groups to date, one on Commercialization and another one on Public Awareness. Public awareness is a key area for the Council that emphasized that policy design, including sectoral priorities, require the trust and support of the public to succeed.[1106]

Canada's AI Advisory Council created its public engagement and consultation processes using both consultation and deliberation. The national survey elicited an array of citizens' input on AI use in different sectors. Online workshops aimed to find ways to address ethical concerns raised by citizens via the survey. Among the goals of the deliberative process is to shape a new set of guidelines and recommendations for the development of AI.[1107]

The government of Canada published a public consultation in 2023 to gather insights on the negotiations on the Council of Europe Convention on AI, Human Rights, Democracy and the Rule of Law. Officially, the role of Canada in the negotiations is to help shape "the treaty to reflect Canadian values and interests, while promoting Canadian objectives on AI in the context of potential risks to human rights, democracy and the rule of law."[1108]

The Minister of Innovation, Science and Industry launched a public consultation to help inform the design and implementation of a new AI Compute Access Fund and a Canadian AI Sovereign Compute Strategy proposed in Budget 2024. The consultation was aimed at researchers, innovators, civil societies, Indigenous groups, and businesses through various means including online.[1109]

Intelligence and Data Act and to Make Consequential and Related Amendments to Other Acts [Digital Charter Implementation Act, 2022] (Jan. 6, 2025), https://www.parl.ca/legisinfo/en/bill/44-1/c-27

[1105] Government of Canada, *Advisory Council on Artificial Intelligence* (Oct. 31, 2025), https://ised-isde.canada.ca/site/advisory-council-artificial-intelligence/en

[1106] Innovation Science and Economic Development Canada, *Public Awareness Working Group*, https://ised-isde.canada.ca/site/advisory-council-artificial-intelligence/en/public-awareness-working-group

[1107] OECD (2021), *State of Implementation of the OECD AI Principles: Insights from national AI policies* (Jun 18, 2021), https://doi.org/10.1787/1cd40c44-en; OECD AI Policy Observatory, *AI Initiatives from Canada* (2026), https://oecd.ai/en/dashboards/national/canada

[1108] Global Affairs Canada, *Share Your Thoughts: Canada's Participation in Treaty Negotiations on Artificial Intelligence at the Council of Europe,* (Nov. 27 2023), https://international.canada.ca/en/global-affairs/consultations/foreign-affairs/2023-11-27-artificial-intelligence

[1109] Innovation, Science and Economic Development Canada, *Government of Canada Launches Public Consultation on Artificial Intelligence Computing Infrastructure* (Jun. 26, 2024),

The Government of Canada established an AI Strategy Task Force and launched a public consultation to gather stakeholder input on the next national AI strategy and AI computing infrastructure.[1110] This engagement targeted researchers, civil society groups, industry representatives, and Indigenous communities, marking a renewed federal effort to ensure meaningful participation in shaping Canada's future AI policy.[1111]

Data Protection

The Office of the Privacy Commissioner (OPC) of Canada provides advice and information for individuals about protecting personal information.[1112] The agency also enforces two federal privacy laws that set out the rules for how federal government institutions and certain businesses must handle personal information. The Privacy Act regulates the collection and use of personal data by the federal government.[1113] The Personal Information Protection and Electronic Documents Act (PIPEDA) applies to personal data collected by private companies.[1114]

First in November 2020, and then in November 2022, after an extensive period of consultation, the Privacy Commissioner issued proposals on regulating artificial intelligence.[1115] The recommendations "aim to allow for responsible AI innovation and socially beneficial uses while protecting human rights." The Commissioner recommended amending PIPEDA to:

- allow personal information to be used for new purposes towards responsible AI innovation and for societal benefits
- authorize these uses within a rights-based framework that would entrench privacy as a human right and a necessary element for the exercise of other fundamental rights

---

https://www.canada.ca/en/innovation-science-economic-development/news/2024/06/government-of-canada-launches-public-consultation-on-artificial-intelligence-computing-infrastructure.html

[1110] Government of Canada, *AI Strategy for the Federal Public Service 2025–2027: Engagement and Consultation* (Feb. 25, 2026), https://www.canada.ca/en/government/system/digital-government/digital-government-innovations/responsible-use-ai/gc-ai-strategy-engagement-consultation.html

[1111] Government of Canada, *Canada Launches AI Strategy Task Force and Public Consultation* (Sept, 2025), https://www.canada.ca/en/innovation-science-economic-development/news/2025/09/government-of-canada-launches-ai-strategy-task-force-and-public-engagement-on-the-development-of-the-next-ai-strategy.html

[1112] Office of the Privacy Commissioner of Canada, *The Privacy Act in Brief* (Aug. 2019), https://www.priv.gc.ca/en/privacy-topics/privacy-laws-in-canada/the-privacy-act/pa_brief/

[1113] Ibid

[1114] Office of the Privacy Commissioner of Canada, *PIPEDA in Brief* (May 2019), https://www.priv.gc.ca/en/privacy-topics/privacy-laws-in-canada/the-personal-information-protection-and-electronic-documents-act-pipeda/pipeda_brief/

[1115] Office of the Privacy Commissioner of Canada, *Commissioner Issues Proposals on Regulating Artificial Intelligence* (Nov. 2020), https://www.priv.gc.ca/en/opc-news/news-and-announcements/2020/nr-c_201112/

- create a right to meaningful explanation for automated decisions and a right to contest those decisions to ensure they are made fairly and accurately
- strengthen accountability by requiring a demonstration of privacy compliance upon request by the regulator
- empower the OPC to issue binding orders and proportional financial penalties to incentivize compliance with the law
- require organizations to design AI systems from their conception in a way that protects privacy and human rights.

The Commissioner also highlighted a public consultation initiated by the OPC that received 86 comments from industry, academia, civil society, and the legal community, among others. Those inputs were incorporated in a separate report, which informs the recommendations for law reform.[1116]

Federal, provincial and territorial privacy commissioners published the principles for responsible, trustworthy and privacy-protective generative AI technologies in December 2023. The objective is to help organizations that are developing, providing or using generative AI to be compliant with fundamental rights and privacy legislation.[1117]

The Office of the Privacy Commissioner of Canada published the Strategic Plan considering its role in the development of AI in Canada. The OPC considered governing and developing AI among its three priorities for 2024–2027. The plan outlines how the OPC will maximize its impact in fully and effectively promoting and protecting the fundamental right to privacy, addressing and advocating for privacy in this time of technological change, and championing children's privacy rights.[1118]

The Privacy Commissioner of Canada is an accredited member of the Global Privacy Assembly (GPA) since 2002.[1119] The Privacy Commissioner has endorsed the 2018 GPA Resolution on AI and Ethics,[1120] the 2020 GPA Resolution on Facial

---

[1116] Office of the Privacy Commissioner of Canada, *Policy Proposals for PIPEDA Reform to Address Artificial Intelligence Report* (Nov. 2020), https://www.priv.gc.ca/en/about-the-opc/what-we-do/consultations/completed-consultations/consultation-ai/pol-ai_202011/.

[1117] Office of the Privacy Commissioner of Canada, *Principles for Responsible, Trustworthy and Privacy-Protective Generative AI Technologies* (Dec. 7, 2023), https://www.priv.gc.ca/en/privacy-topics/technology/artificial-intelligence/gd_principles_ai/

[1118] Office of the Privacy Commissioner of Canada, *Office of the Privacy Commissioner of Canada Strategic Plan 2024-27: A Roadmap for Trust, Innovation and Protecting the Fundamental Right to Privacy in the Digital Age* (Jan. 22, 2024), https://www.priv.gc.ca/media/6112/strategic-plan-2024-27.pdf

[1119] Global Privacy Assembly, *List of Accredited Members* (2024), https://globalprivacyassembly.org/participation-in-the-assembly/list-of-accredited-members/

[1120] International Conference on Data, *Declaration on Ethics and Data Protection in Artificial Intelligence* (Oct. 2018), https://globalprivacyassembly.org/wp-content/uploads/2018/10/20180922_ICDPPC-40th_AI-Declaration_ADOPTED.pdf

Recognition,[1121] the 2022 GPA Resolution on AI and Accountability,[1122] and the 2023 GPA Resolution on Generative AI Systems.[1123]

Algorithmic Transparency

The PIPEDA includes strong rights for individual access concerning automated decisions.[1124] The PIPEDA Reform Report for AI builds on public consultations and proposes to "Provide individuals with a right to explanation and increased transparency when they interact with, or are subject to, automated processing."[1125] The Cofone Report on the reform also explains that "the right to explanation is connected to the principles of privacy, accountability, fairness, non-discrimination, safety, security, and transparency. The effort to guarantee these rights supports the need for a right to explanation." The amended Directive on Automated Decision-Making strengthens transparency, accountability, and testing/monitoring, with updates reflected in the Algorithmic Impact Assessment tool.[1126]

In the last Open Algorithms Network meeting that was co-chaired by the Government of Canada, with participation from the Governments of Estonia, Norway, the United Kingdom, and Scotland as well as civil society respondents, the participants started considering issues of "equality, bias, and discrimination in their algorithmic commitments in Open Government Partnership (OGP) action plans and across government AI strategies."[1127] "Many of these commitments are grounded in the idea that opening data and design of algorithms is an avenue to reduce bias and discrimination, and that the process of collecting data or design is as important as the

---

[1121] Global Privacy Assembly, *Adopted Resolution on Accountability in the Development and Use of Artificial Intelligence* (Oct. 2022), https://globalprivacyassembly.org/wp-content/uploads/2020/10/FINAL-GPA-Resolution-on-Accountability-in-the-Development-and-Use-of-AI-EN-1.pdf

[1122] Global Privacy Assembly, *Resolution on Principles and Expectations for the Appropriate Use of Personal Information in Facial Recognition Technology* (Oct. 2022), https://globalprivacyassembly.org/wp-content/uploads/2022/11/15.1.c.Resolution-on-Principles-and-Expectations-for-the-Appropriate-Use-of-Personal-Information-in-Facial-Recognition-Technolog.pdf

[1123] Global Privacy Assembly, *Resolution on Generative Artificial Intelligence Systems* (Oct. 2023), https://globalprivacyassembly.org/wp-content/uploads/2023/10/5.-Resolution-on-Generative-AI-Systems-101023.pdf

[1124] Office of the Privacy Commissioner, *PIPEDA Fair Information Principle 9 – Individual Access* (Aug. 2020), https://www.priv.gc.ca/en/privacy-topics/privacy-laws-in-canada/the-personal-information-protection-and-electronic-documents-act-pipeda/p_principle/principles/p_access/

[1125] Professor Ignacio Cofone, *Policy Proposals for PIPEDA Reform to Address Artificial Intelligence Report* (Nov. 2020), https://www.priv.gc.ca/en/about-the-opc/what-we-do/consultations/completed-consultations/consultation-ai/pol-ai_202011/

[1126] Government of Canada, *Progress on AI in Government* (Nov. 28, 2025), https://www.canada.ca/en/government/system/digital-government/digital-government-innovations/responsible-use-ai/progress.html

[1127] Open Government Partnership, *Three Recommendations for More Inclusive and Equitable AI in the Public Sector* (Jan. 24, 2023), https://www.opengovpartnership.org/stories/three-recommendations-for-more-inclusive-and-equitable-ai-in-the-public-sector/

outcome."[1128] Algorithmic transparency[1129] is an emerging commitment[1130] area for OGP.

Canada's Federal AI Registry is a public database documenting all AI systems used by government agencies. The Registry includes information about systems' purpose, data sources, performance metrics, and human oversight mechanisms. This registry fulfills commitments made in Canada's 2024–2026 Open Government Partnership National Action Plan and establishes new standards for public sector algorithmic transparency.[1131]

Data Scraping

The Office of the Privacy Commissioner alongside the data protection authorities of 16 countries issued a statement in October 2024 calling for more protection of fundamental rights in mass data scraping and for AI model training.[1132] The 2024 statement followed up on a joint statement issued in August 2023, based on engagement with social media companies (SMCs) and other industry stakeholders. The statement provided additional guidance, emphasizing that contractual terms cannot render scraping lawful, and that SMCs and other organizations must ensure there is a legal basis, maintain transparency, and obtain consent when required by law. The statement also encouraged stakeholders to leverage AI to protect personal data from unlawful scraping.[1133]

Canada's Privacy Commissioner opened an investigation into social media platform X to ensure compliance for AI data use in accordance with the PIPEDA.[1134] The investigation will focus on the platform's compliance with federal privacy law with respect to data scraping of Canadians' personal information to train artificial intelligence models.

[1128] Ibid

[1129] Open Government Partnership, *Glossary: Transparency* (2026), https://www.opengovpartnership.org/glossary/transparency/

[1130] Open Government Partnership, *Glossary: Commitment* (2026), https://www.opengovpartnership.org/glossary/commitment/

[1131] Open Government Partnership, *Open Algorithms Network* (2026), https://www.opengovpartnership.org/about/partnerships-and-coalitions/open-algorithms-network/

[1132] Office of the Privacy Commissioner of Canada, *Concluding Joint Statement on Data Scraping and the Protection of Privacy* (Oct. 28, 2024), https://www.priv.gc.ca/en/opc-news/speeches-and-statements/2024/js-dc_20241028/

[1133] Global Privacy Assembly, *GPA's Working Group Published Its Concluding Joint Statement on Data Scraping and Privacy Protection* (Nov. 25, 2024), https://ai-regulation.com/gpa-statement-data-scraping-and-privacy-protection/

[1134] Office of the Privacy Commissioner of Canada, *Privacy Commissioner Investigation into Complaint about Social Media Platform X* (Feb. 27, 2025), https://www.priv.gc.ca/en/opc-news/news-and-announcements/2025/nr-c_250227/

Facial Recognition

Canada continues to ensure that facial recognition technologies do not threaten the rights of its citizens. Canada took decisive action against Clearview AI in 2021, after discovering the organization had scraped billions of images of people from across the Internet, in clear violation of Canadians' privacy rights.[1135] The Supreme Court of British Columbia upheld a ban on Clearview AI's collection of biometric data without consent in 2025.[1136] The court dismissed a Clearview petition to overturn an order prohibiting the company from offering its facial recognition services using images and biometric facial arrays collected from individuals in British Columbia.

Following the Clearview action, the privacy commissioners of Canada, Ontario, and Québec issued a joint statement recommending the creation of a legal regulatory framework for the use of facial recognition technology by police. The recommendation identifies four requirements:[1137]

- Clear and explicit definitions for the appropriate and prohibited use of facial recognition technologies by law enforcement including "no-go" zones and the prohibition of mass surveillance
- Restricting the use of such technologies based on strict necessity and proportionality
- The use of strong independent oversight that are based on proactive engagement, including program pre-authorization and advanced notice before initiating initiatives based on facial recognition technology
- Privacy rights protection that limits the risk to individuals including limitations on the duration of information retention and measures to ensure that data is accurate.

The Canada Border Services Agency (CBSA) secured Treasury Board approval for the Traveler Modernization initiative,[1138] which includes integrating facial recognition verification into the Advance Declaration system. The project emphasizes privacy, transparency, accessibility, and human oversight, and is being co-developed with Transport Canada and Immigration, Refugees and Citizenship Canada.

[1135] Government of Canada, Office of the Privacy Commissioner, *Clearview AI's Unlawful Practices Represented Mass Surveillance of Canadians, Commissioners Say* (Feb. 3, 2021), https://www.priv.gc.ca/en/opc-news/news-and-announcements/2021/nr-c_210203/?=february-2-2021

[1136] Biometric Update, *Canadian Court Upholds Clearview Biometric Data Ban: Supreme Court of British Columbia Nixes Petition* (Jan.10, 2025), https://www.biometricupdate.com/202501/canadian-court-upholds-clearview-biometric-data-ban

[1137] Office of Privacy Commissioner of Canada, *Recommended Legal Framework for Poilce Agencies' Use of Facial Recognition, Joint Statement by Federal, Provincial and Territorial Privacy Commissioners* (May 2, 2022), https://www.priv.gc.ca/en/opc-actions-and-decisions/advice-to-parliament/2022/s-d_prov_20220502/

[1138] Canada Border Services Agency, *Traveller Modernization: New Tools and Technologies for a Faster, Better and Safe Experience at the Border* (Dec. 23, 2024), https://www.cbsa-asfc.gc.ca/services/border-tech-frontiere/modern-eng.html

AI in Defense and Nuclear Policies

The Canadian Department of National Defence (DND) and Canadian Armed Forces (CAF) released their first-ever Artificial Intelligence Strategy in March 2024. The strategy focuses on five key areas: 1) Identifying and fielding capabilities required the institutions; 2) Creating a culture supportive of AI innovation; 3) Establishing principles, processes and practices to ensure that the use of AI is ethical, legal, inclusive, safe and trusted; 4) Managing talent and training to meet workforce needs; 5) Deepening strategic partnerships internally and externally with allies, industries and academia. The strategy "commits the Defence Team to becoming AI enabled by 2030."[1139]

The Canadian Nuclear Safety Commission joined the United Kingdom's Office for Nuclear Regulation and the United States Nuclear Regulatory Commission to establish high-level principles for the safe and effective integration of AI technologies in nuclear operations. The group jointly published the Trilateral Principles Paper on the Deployment of Artificial Intelligence in Nuclear Activities on September 5, 2024.[1140] Developed through a collaborative trilateral relationship formed in 2022, the paper addresses critical aspects such as safety and security engineering, human and organizational factors, AI system architecture, lifecycle management, and the documentation of safety measures. Highlighting the importance of data quality, the paper emphasizes that the performance of AI systems hinges on the quality of the data they process. The paper also identifies security challenges unique to AI, noting their potentially greater significance compared to conventional software issues. By providing this framework, the paper seeks to advance the responsible and secure use of AI within regulated nuclear activities.

Lethal Autonomous Weapons

Canadian academics urged Prime Minister Justin Trudeau to oppose Autonomous Weapon Systems in 2017 as part of the #BanKillerAI campaign.[1141] Trudeau included "Advance international efforts to ban the development and use of fully autonomous weapons systems," in the mandate he charged to Foreign Affairs Minister François Philippe Champagne in December 2019.[1142] However, this mandate has not been fulfilled. Canada CCW members delivered a 2022 proposal on Principles

---

[1139] Department of National Defence, *DM/CDS Message: Launch of the DND/CAF Artificial Intelligence Strategy* (Mar. 7, 2024), https://www.canada.ca/en/department-national-defence/maple-leaf/defence/2024/03/dm-cds-message-launch-dnd-caf-artificia-intelligence-strategy.html

[1140] Canadian Nuclear Safety Commission, *Trilateral Principles Paper on the Deployment of Artificial Intelligence in Nuclear Activities* (Sept. 27, 2024), https://www.cnsc-ccsn.gc.ca/eng/resources/research/technical-papers-and-articles/2024/canukus-ai-principles-paper/

[1141] Ian Kerr, *Weaponized AI Would Have Deadly, Catastrophic Consequences. Where Will Canada Side*? The Globe and Mail (Nov. 6, 2017), https://www.theglobeandmail.com/opinion/weaponized-ai-would-have-deadly-catastrophic-consequences-where-will-canada-side/article36841036/

[1142] Minister of Foreign Affairs, *Mandate Letter* (Dec. 13, 2019), https://walterdorn.net/home/281-mfa-mandate-letter-2019

and Good Practices on Emerging Technologies in the Area of Lethal Autonomous Weapons Systems.[1143] At the 78th UN General Assembly First Committee in 2023, Canada voted in favor[1144] of resolution L56[1145] on autonomous weapons systems, along with 163 other states. The Resolution emphasized the "urgent need for the international community to address the challenges and concerns raised by autonomous weapons systems," and mandated the UN Secretary-General to prepare a report reflecting the views of member and observer states on autonomous weapons systems. In the report to the UN Secretary-General,[1146] Canada advocated for a ban on LAWS that cannot be used in compliance with IHL, requirements that LAWS maintain "appropriate levels of human involvement," updating national policies to ensure technologies are used in compliance with international law, and open exchanges of information about emerging technologies.

Canada agreed to the joint call to action on the responsible development, deployment, and use of artificial intelligence (AI) in the military domain as part of the REAIM Summit in 2023 in the Netherlands.[1147] Canada also endorsed the resulting Political Declaration on Responsible Military Use of AI and Autonomy issued in November 2023.[1148] The second REAIM summit was held in South Korea. At the summit, 61 countries, including Canada,[1149] signed a Blueprint for Action for the ethical and human-centric use of AI in the military.[1150] The blueprint identifies the standards for AI in the military domain including the principles and framework for

---

[1143] Delegation of Japan to the Conference on Disarmament, CCW, *Principles and Good Practices on Emerging Technologies in the Area of Lethal Autonomous Weapons Systems* (Mar. 7, 2022), https://www.disarm.emb-japan.go.jp/Final proposal - laws principles and good practices - March 7 2022.pdf

[1144] Stop Killer Robots, *164 states vote against the machine at the UN General Assembly* (Nov. 1, 2023), https://www.stopkillerrobots.org/news/164-states-vote-against-the-machine/

[1145] General Assembly, *Lethal Autonomous Weapons, Resolution L56* (Oct. 12, 2023), https://reachingcriticalwill.org/images/documents/Disarmament-fora/1com/1com23/resolutions/L56.pdf

[1146] UN Office of Disarmament Affairs, *Canda's Views on Lethal Autonomous Weapons Systems as it Relates to the UN Resolution 78/241 "Lethal Autonomous Weapons Systems"* (2024), https://docs-library.unoda.org/General_Assembly_First_Committee_-Seventy-Ninth_session_(2024)/78-241-Canada-EN.pdf

[1147] Government of Netherlands, *Call to Action on Responsible Use of AI in the Military Domain* (Feb. 16, 2023), https://www.government.nl/latest/news/2023/02/16/reaim-2023-call-to-action

[1148] US Department of State, *Political Declaration on Responsible Military Use of Artificial Intelligence and Autonomy* (Nov. 9, 2023), Endorsing States (Nov. 27, 2024), https://www.state.gov/political-declaration-on-responsible-military-use-of-artificial-intelligence-and-autonomy/

[1149] Woojoo Hong, *U.S.-China Competition Looms Large at Seoul Summit on Use of AI in Military*, Asia Pacific Foundation of Canada (Oct. 9, 2024), https://www.asiapacific.ca/publication/us-china-competition-looms-large-seoul-summit-use-ai

[1150] REAIM 2024, *REAIM Blueprint for Action*, DigWatch (Sept. 11, 2024), https://dig.watch/resource/responsible-ai-in-the-military-domain-reaim-blueprint-for-action#REAIM_Blueprint_for_Action

future governance, compliance with international law, holding humans responsible, ensuring reliability and trustworthiness of AI, maintaining an appropriate human involvement and improving the ability to explain AI.

EdTech

In May 2022, Human Rights Watch published a global investigative report on the education technology (EdTech) endorsed by 49 governments, including Canada, for children's education during the pandemic. One of the case studies concerned "CBC Kids," offered by the Canadian Broadcasting Corporation and recommended by Canada's Quebec Education Ministry for pre-primary and primary school-aged children's learning. Based on technical and policy analysis of this EdTech product, Human Rights Watch found that the endorsement of this online learning platforms put at risk or directly violated children's rights, due to its tracking and profiling practices for advertising purposes. The report called on governments to develop, refine, and enforce modern child data protection laws and standards and ensure that children who want to learn are not compelled to give up their other rights to do so.[1151]

The Canadian Teachers Federation released a policy brief in September 2024, calling for urgent action to regulate the use of AI in Canadian K–12 public education.[1152] The Teachers' Federation requested that the Federal Government and the Council of Ministers of Education, Canada (CMEC) develop and implement policies that safeguard the rights of educators and students as AI systems proliferate in public education across the country.

AI Literacy

A recent global study by KPMG and the University of Melbourne reported that "Canada ranks among the least AI literate nations globally, holding the fourth-lowest position in AI training and literacy in a ranking of 47 countries."[1153] The Canadian Government has acknowledged the importance of a well-trained workforce to ensure AI is adopted responsibly. The government made "talent and training" a priority area in the 2025–2027 AI Strategy for the Federal Public Service.[1154] Key

---

[1151] Human Rights Watch, *How Dare They Peep into My Private Life* (May 25, 2022), https://www.hrw.org/report/2022/05/25/how-dare-they-peep-my-private-life/childrens-rights-violations-government

[1152] Canadian Teachers Federation, *Towards a Responsible Use of Artificial Intelligence in Canadian Public Education* (Sept. 30, 2024), https://www.ctf-fce.ca/blog-perspectives/towards-a-responsible-use-of-artificial-intelligence-in-canadian-public-education/

[1153] University of Melbourne and KPMG, *Trust, Attitudes, and Use of Artificial Intelligence: A Global Study 2025—Canadian Insights* (2025) https://assets.kpmg.com/content/dam/kpmg/ca/pdf/2025/07/trust-in-ai-en-report.pdf

[1154] Government of Canada, *AI Strategy for the Federal Public Service 2025-2027: Priority Areas* (Jun. 24, 2025), https://www.canada.ca/en/government/system/digital-government/digital-government-innovations/responsible-use-ai/gc-ai-strategy-priority-areas.html#sec2-toc-3

actions in the strategy include developing a training plan, benchmarking talent needs, and developing a talent plan.

Human Rights

Canada consistently ranks "Free" and among the top ten nations in the world for the protection of human rights and transparency on the Freedom House Freedom in the World Report. Canada's score dropped one point to 97 in 2025.[1155] Freedom House reported that, "Canada has a strong history of respect for political rights and civil liberties, though in recent years citizens have been concerned about laws related to the administration of elections, government transparency, the treatment of inmates in prisons, and restrictions on public sector employees wearing religious symbols. While members of minority groups, including Black and Indigenous Canadians, still face discrimination and economic, social, and political challenges, the federal government has acknowledged these problems and made some moves to address them."[1156]

The Law Commission of Ontario and Ontario Human Rights Commission, representing Canada's most populous province, jointly created an AI impact assessment tool in November 2024 to provide organizations a method to assess AI systems for compliance with human rights obligations.[1157] The human rights AI impact assessment (HRIA) assists developers and administrators of AI systems to identify, assess, minimize, or avoid discrimination and uphold human rights obligations throughout the lifecycle of an AI system by providing guidelines for when and how to conduct impact assessments. The HRIA is intended to: "Strengthen knowledge and understanding of human rights impacts; Provide practical guidance on specific human rights impacts, particularly in relation to non-discrimination and equality of treatment; and Identify practical mitigation strategies and remedies to address bias and discrimination from AI systems." [1158]

The Office of the Information and Privacy Commissioner of Alberta issued Comments Regarding Responsible AI Governance in Alberta, calling for provincial legislation to prohibit or restrict high-risk AI uses, such as training on sensitive data, and to embed human oversight, transparency, and fairness into AI decision-making.[1159] The report reflects emergent sub-national policy leadership on human

---

[1155] Freedom House, *Freedom in the World 2025: Canada* (2025), https://freedomhouse.org/country/canada/freedom-world/2025

[1156] Ibid

[1157] Law Commission of Ontario, Ontario Human Rights Commission, *Human Rights AI Impact Assessment* (Nov. 2024), https://www.lco-cdo.org/wp-content/uploads/2024/11/LCO-Human-Rights-AI-Impact-Assessment-EN.pdf

[1158] Ibid, p. 3

[1159] Office of the Information and Privacy Commissioner of Alberta, *Comments Regarding Responsible AI Governance in Alberta* (Jul 15, 2025), https://oipc.ab.ca/wp-content/uploads/2025/08/AI-Comments-from-the-OIPC-Regarding-Responsible-AI-Governance-in-Alberta-July-15-2025.pdf

rights-based AI governance in Canada, particularly focused on privacy, accountability, and proportionality in automated decision systems.

AI Safety Summit

Canada participated in the first AI Safety Summit and endorsed the Bletchley Declaration in 2023.[1160] Canada thus committed to participate in international cooperation efforts on AI "to promote inclusive economic growth, sustainable development and innovation, to protect human rights and fundamental freedoms, and to foster public trust and confidence in AI systems to fully realise their potential." Endorsing parties affirmed that for the good of all, AI should be designed, developed, deployed, and used, in a manner that is safe, in such a way as to be human-centric, trustworthy and responsible."

The formation of the International Network of AI Safety Institutes was announced at the AI Seoul Summit in May 2024. As a founding member of the Network, Canada will seek to actively coordinate efforts with international partners and collaborate on joint projects, including the development of guidance on AI safety.[1161] Canada launched the Canadian AI Safety Institute (CAISI) in November 2024. CAISI "is part of the Government of Canada's plan to support the safe and responsible development and deployment of artificial intelligence (AI). CAISI aims "to advance the science of AI safety, in collaboration with international partners, in order to ensure that governments are well-positioned to understand and act on the risks of advanced AI systems. These include risks posed by synthetic content, including impersonation and fraud, as well as risks posed by the development or deployment of systems that may be dangerous or hinder human oversight."[1162]

Canada participated in the 2025 AI Action Summit in Paris. The country signed the Statement on Inclusive and Sustainable Artificial Intelligence for People and the Planet.[1163]

[1160] UK Department for Science, Innovation & Technology, Foreign, Commonwealth & Development Office, *Prime Minister's Office, The Bletchley Declaration by Countries Attending the AI Safety Summit* (Nov. 2023), https://www.gov.uk/government/publications/ai-safety-summit-2023-the-bletchley-declaration/the-bletchley-declaration-by-countries-attending-the-ai-safety-summit-1-2-november-2023

[1161] Department for Science, Innovation & Technology, *Policy Paper: Seoul Statement of Intent toward International Cooperation on AI Safety Science, AI Seoul Summit 2024 (Annex)* (May 21, 2024), https://www.gov.uk/government/publications/seoul-declaration-for-safe-innovative-and-inclusive-ai-ai-seoul-summit-2024/seoul-statement-of-intent-toward-international-cooperation-on-ai-safety-science-ai-seoul-summit-2024-annex

[1162] Government of Canada, *Canadian Artificial Intelligence Safety Institute* (Nov. 2024), https://ised-isde.canada.ca/site/ised/en/canadian-artificial-intelligence-safety-institute

[1163] Élysée, *Statement on Inclusive and Sustainable Artificial Intelligence for People and the Planet*, AI Action Summit (Feb 11, 2025), https://www.elysee.fr/en/emmanuel-macron/2025/02/11/statement-on-inclusive-and-sustainable-artificial-intelligence-for-people-and-the-planet

OECD / G20 AI Principles

Canada endorsed the OECD and the G20 AI Principles, including the principles updated in May 2024.[1164]

In 2020, Canada and France, and a dozen other countries announced the Global Partnership on Artificial Intelligence (GPAI) to "support the responsible and human-centric development and use of AI in a manner consistent with human rights, fundamental freedoms, and our shared democratic values."[1165]

Canada and the European Union announced in 2020 that they were collaborating to leverage AI to help the international community respond to COVID-19. The initiative included the GPAI's group on AI and Pandemic Response and the annual EU-Canada Digital Dialogue.[1166]

The Safe and Secure AI Advisory Group initiated in 2025 incorporates the OECD AI Principles in the national strategy to promote AI while respecting human rights and democratic values.[1167]

Canada committed to a joint research partnership with the UK to "support pioneering work to keep advanced systems safe by maintaining transparency, predictability and responsiveness to human oversight" in line with OECD goals of accountability and explainability.[1168]

Council of Europe AI Treaty

Canada contributed as an Observer State in the negotiations of the Council of Europe Framework Convention on AI, Human Rights, Democracy, and the Rule of Law. The Committee of Ministers adopted the Framework during the 2024 ministerial

---

[1164] OECD AI Policy Observatory, *OECD AI Principles Overview, Countries Adhering to the Principles* (2026), https://oecd.ai/en/ai-principles

[1165] Government of Canada, *Joint Statement from Founding Members of the Global Partnership on Artificial Intelligence* (Jun. 15, 2020), https://www.canada.ca/en/innovation-science-economic-development/news/2020/06/joint-statement-from-founding-members-of-the-global-partnership-on-artificial-intelligence.html

[1166] European Union, *Joint Press Release Following the European Union-Canada Ministerial Meeting* (Sept. 9, 2020), https://eeas.europa.eu/headquarters/headquarters-homepage/84921/joint-press-release-following-european-union-canada-ministerial-meeting_en.

[1167] Government of Canada, *Canada Moves toward Safe and Responsible Artificial Intelligence* (Mar. 6, 2025), https://www.canada.ca/en/innovation-science-economic-development/news/2025/03/canada-moves-toward-safe-and-responsible-artificial-intelligence.html

[1168] Government of Canada, *Government of Canada Partners with United Kingdom to Invest in Groundbreaking AI Alignment Research* (Jul. 30, 2025), https://www.canada.ca/en/innovation-science-economic-development/news/2025/07/government-of-canada-partners-with-united-kingdom-to-invest-in-groundbreaking-ai-alignment-research.html

meeting[1169] and it opened for signature in September 2024. Canada signed this first international AI Treaty on February 11, 2025.[1170]

UNESCO Recommendation on AI Ethics

Canada is a signatory of the UNESCO Recommendation on the Ethics of Artificial Intelligence.[1171] Canada's principles for AI, directive on automated decision-making, and support of Impact Assessment tools reflect a commitment to the Recommendations in public deployment while guidelines for public procurement offer a means to extend those guidelines to the private sector.[1172] Canada's National AI Strategy for the Federal Public Service 2025–2027 based on international best practices represents further alignment with UNESCO ethical guidelines.[1173] Canada has not initiated the UNESCO Readiness Assessment Methodology (RAM), a tool to help countries ensure alignment with the Recommendation.

Evaluation

Canada is among the leaders in creating policies and guidelines for AI. In addition to endorsing the OECD/G20 AI Principles and establishing the GPAI with France, Canada has taken steps to establish model practices for the use of AI across government agencies. Canada has a solid record on human rights and is now working to update national privacy law to address the challenges of AI. The failure of the Digital Charter Implementation Act including the AIDA and data protection updates raises the possibility that Canada will remain without a comprehensive AI policy. However, the government is applying the lesson about the importance of public consultation in the formation of the next round of policies.

[1169] Council of Europe, *Framework Convention on Artificial Intelligence* (2025), https://www.coe.int/en/web/artificial-intelligence/the-framework-convention-on-artificial-intelligence

[1170] Council of Europe Treaty Office, *Chart of Signatures and Ratifications of Treaty 225* (Mar. 14, 2026), https://www.coe.int/en/web/Conventions/full-list/?module=signatures-by-treaty&treatynum=225

[1171] UNESCO, *Recommendation on the Ethics of Artificial Intelligence* (Nov. 23, 2021), https://unesdoc.unesco.org/ark:/48223/pf0000381137

[1172] Government of Canada, *Responsible Use of Artificial Intelligence in Government* (Mar. 4, 2025), https://www.canada.ca/en/government/system/digital-government/digital-government-innovations/responsible-use-ai.html#toc1

[1173] Government of Canada, *Canada launches New AI Strategy for AI in Public Service* (Mar. 6, 2025), https://www.canada.ca/en/government/system/digital-government/digital-government-innovations/responsible-use-ai/gc-ai-strategy-priority-areas.html

Chile

In 2025, Chile published the official update of the National AI Policy. The Ministerial Advisory Council offered recommendations to implement the new data protection law, which enters into force in December 2026.

National AI Strategy

Chile's Ministry of Science, Technology, Knowledge, and Innovation (*Ministerio de Ciencia, Tecnología, Conocimiento e Innovación*, MinCiencia) approved an updated National AI Policy focused on ethical considerations, responsible development, and the potential of AI to drive social and economic progress.[1174] The report followed Chile's milestone as the first country globally to apply and complete UNESCO's AI Readiness Assessment Methodology (RAM) and robust public consultations.[1175] The updated policy builds on the initial National AI Policy published in October 2021[1176] and aims to bring the Governance and Ethics axis in alignment with recent governance developments rather than formulating a new policy.[1177] The 3 policy axes on "enabling factors," "development and adoption," and "governance and ethics" address a broad range of AI-related issues from developing AI-ready talent at the school level to promoting investment in AI R&D, and adopting international standards for regulations. Strengthening international cooperation in the AI domain is a key focus.

The AI Action Plan[1178] complementing the original AI policy[1179] sets out 177 actions, aiming to complete 100 by the end of 2026. Priority tasks are in education, product development, talent management, and other areas. The AI Action Plan also deals with accountability and establishes timelines for priority tasks implementation.

[1174] Diario Oficial, Updates to the "National Artificial Intelligence Policy" Approved [Aprueba Actualización de la "Política Nacional de Inteligencia Artificial"] (Jan 28, 2025), https://www.diariooficial.interior.gob.cl/publicaciones/2025/01/28/44060/01/2600578.pdf

[1175] MinCiencia, *With More than 100 Actions Committed for 2026, Minister of Science Presents New Artificial Intelligence Policy* (May 2, 2024), https://minciencia.gob.cl/noticias/con-mas-de-100-acciones-comprometidas-para-2026-ministra-de-ciencia-presenta-nueva-politica-de-inteligencia-artificial/

[1176] Government of Chile, *Chile Presents the First National Policy on Artificial Intelligence* (Oct. 28, 2021), https://www.gob.cl/en/news/chile-presents-first-national-policy-artificial-intelligence/

[1177] MinCiencia, *Ministry of Science Opens Citizen Consultation to Update the National Artificial Intelligence Policy* (Jan. 19, 2024), https://www.minciencia.gob.cl/noticias/ministerio-de-ciencia-abre-consulta-ciudadana-para-actualizar-politica-nacional-de-inteligencia-artificial/

[1178] OECD Public Governance Reviews, *The Strategic and Responsible Use of Artificial Intelligence in the Public Sector of Latin America and the Caribbean*, Chapter 2, LAC Artificial Intelligence Strategies (Mar. 22, 2022), https://www.oecd-ilibrary.org/governance/the-strategic-and-responsible-use-of-artificial-intelligence-in-the-public-sector-of-latin-america-and-the-caribbean_1f334543-en

[1179] MinCiencia, *National AI Policy [Política Nacional de Inteligencia Artificial]*, p. 16 (Oct. 2021), https://www.minciencia.gob.cl/uploads/filer_public/bc/38/bc389daf-4514-4306-867c-760ae7686e2c/documento_politica_ia_digital_.pdf

The main AI Policy objective for Chile is "to place the country at the forefront of AI research, development, and innovation, with an ecosystem that creates new capacities in various sectors."[1180] Such an ecosystem should be designed according to transversal concepts of opportunity and responsibility, contribute to sustainable development, and improve the quality of life.[1181]

After discussing a draft Bill modeled on EU AI Act in 2023,[1182] the Chilean Parliament officially introduced a bill to promote the ethical development of AI, safeguard fundamental rights, and ensure consumer protection in the Chamber of Deputies (*Cámara de Diputadas y Diputados*).[1183] The Chamber of Deputies approved the draft Bill in 2025 and sent the bill back to the Commission on Future, Science, Technology Knowledge and Innovation for detailed amendments and revision. The draft Bill was sent to the Senate in October 2025, beginning the second constitutional stage.[1184]

Public Participation

MinCiencia opened a public consultation on the National Artificial Intelligence Policy 2024 in January,[1185] which remained accessible for three months. The consultation seeks to gather citizen feedback on proposed revisions to the 2021 National Artificial Intelligence Policy with a focus on ensuring ethical AI usage in the public sector and fostering international collaboration, particularly within Latin America. This consultation formed part of the ongoing update of the Ethics and Governance axis of the policy.

The drafting of the initial National AI Policy was marked by an effort to ensure wide and inclusive participation across the industry, academia, civil society and the

---

[1180] Ibid, p. 18

[1181] Ibid

[1182] DigWatch, *Chile Takes the First Steps toward AI Legislation* (Jun. 20, 2023), https://dig.watch/updates/chile-takes-the-first-steps-toward-ai-legislation; Cámara de Diputadas y Diputados, *Regulating AI Systems, Robotics, and Technological Connections in their Distinctive Application Environments [Regula los systemas de inteligencia artifical, la robotica y las tecnologias conexas, en sus distintos ambitos de aplicacion]* (Apr. 26, 2023), https://www.camara.cl/verDOC.aspx?prmID=72777&prmTipo=FICHAPARLAMENTARIA&prmFICHATIPO=DIP&prmLOCAL=0

[1183] Cámara de Diputadas y Diputados, *Legislature 372: Regulation of Artificial Intelligence Systems [Regula los sistemas de inteligencia artificial],* Proyectos de Ley (May 7, 2024), https://www.camara.cl/legislacion/ProyectosDeLey/tramitacion.aspx?prmID=17429&prmBOLETIN=16821-19; for an overview in English, see Guillermo Carey, José Ignacio Mercado, and Ricardo Alonso, *Bill to Regulate Artificial Intelligence Systems Is Introduced to the Chilean Chamber of Deputies*, Carey (May 13, 2024), https://www.carey.cl/en/bill-to-regulate-artificial-intelligence-systems-is-introduced-to-the-chilean-chamber-of-deputies/

[1184] Ibid

[1185] MinCiencia, *National Policy on Artificial Intelligence* (2024), https://minciencia.gob.cl/areas/inteligencia-artificial/politica-nacional-de-inteligencia-artificial/

public at large. The core team was composed of 12 experts in the field.[1186] Virtual and in-person AI seminars, discussion groups and workshops were held across the country, including the regions, where inputs for the AI Policy were collected.[1187] After the first draft of the AI Policy was produced, a citizens' consultation was held to collect their views. More than 9,000 people participated in the process.[1188] Civil society expressed criticism regarding the lack of transparency regarding the output of regional and online workshops and difficulties accessing meeting minutes.[1189] The inclusivity of the process was also called into question as only 21% of respondents to the second stage consultation were female and a consultation questionnaire was available only in Spanish.[1190]

The updated policies were also informed by a series of roundtables. A report published by the Library of the National Congress on one such roundtable on Public Policies on Artificial Intelligence in 2023 noted that the discussions allowed "the exchange of information and opinions from [participants] with diverse professional experiences and areas of focus" to inform the design of public policies in the country.[1191]

Chile's Digital Government requested public consultation on the 2030 Digital Government Strategy. This strategy outlines objectives for Chile to be above the OECD average in the Digital Transformation of the State and strengthen public trust.[1192] The strategy will evaluate emerging technologies such as AI and apply regulations within the public sector. A commission in charge of recommending how to implement the new data protection law also collected public comments over more than 1 month ahead of their final report in 2025.[1193]

[1186] MinCiencia, *Public Consultation on Artificial Intelligence: Report of Results* (Oct. 2021), https://minciencia.gob.cl/uploads/filer_public/6c/c1/6cc17cd7-ae58-48f0-ada1-d33a3e6e8958/informe_consulta_publica_ia_1.pdf

[1187] Government of Chile, *Ministry of Science Opens Participatory Process for the National Artificial Intelligence Policy* (Feb. 3, 2020), https://www.gob.cl/noticias/ministerio-de-ciencia-abre-proceso-participativo-para-la-politica-nacional-de-inteligencia-artificial/

[1188] OECD AI Policy Observatory, *AI in Chile: Chilean Participation Process on AI* (2020), https://oecd.ai/en/dashboards/countries/Chile

[1189] Velasco (Derechos Digitales), *The National Artificial Intelligence Policy of Chile and a Process for Citizen Participation* (Nov. 5, 2021), https://www.derechosdigitales.org/17010/la-politica-nacional-de-inteligencia-artificial-chilena-y-su-proceso-de-participacion-ciudadana/

[1190] Ibid

[1191] Biblioteca del Congreso Nacional, *Public Policies and Artificial Intelligence Roundtable Report* (Dec. 11, 2023), https://obtienearchivo.bcn.cl/obtienearchivo?id=repositorio/10221/35618/1/Informe_27_23_Informe_Submesa_Politicas_Publicas_e_Inteligencia_Artificial.pdf

[1192] Secretaría de Gobierno Digital, *Second Public Consultation: Digital Government Strategy 2030 [Segunda consulta pública: Estrategia de Gobierno Digital 2030]* (Feb. 6, 2025), https://participacion.digital.gob.cl/es-CL/projects/segunda-consulta-estrategia-de-gobierno-digital-2030

[1193] Secretaría de Gobierno Digital, Public Consultation: Convocation to Present Proposals to the Ministerial Advisory Commission for the Implementation of the Data Protection Law [Consulta

Data Protection

Article 19 of Chile's Constitution of 1980 protects the right to private life.[1194] In 2018, the constitutional guarantee was extended to explicitly protect personal data.[1195] The Law on Protection of Private Life (LPPL) of 1999 regulates the processing of personal data in public and private databases.[1196] A new Personal Data Protection and Processing Law[1197] replacing and modernizing the LPPL will enter into force in December 2026. The new law establishes the Personal Data Protection Agency (*Agencia de Protección de Datos Personales*) as an autonomous public-rights corporation to "ensure the effective protection of the rights that guarantee the private life of individuals and their personal data."[1198] The Agency will be technical and decentralized. The Agency will oversee impact assessments and determine requirements (art. 15) in addition to overseeing compliance with the law. In contrast to the Transparency Council (*Consejo para la Transparencia*) and consumer protection agency SERNAC (*Servicio Nacional del Consumidor*), which monitor data protection under the LPPL,[1199] the Agency will have the authority to sanction those who violate the law.[1200]

The processing of personal data in the financial and healthcare sectors is regulated by relevant sectoral laws.[1201] Legal actions regarding the violations of the constitutionally protected right to personal data can be brought before courts by the

---

Pública: Convocatoria a presentar propuestas a la Comisión Asesora Ministerial para la Implementación de la Ley de Protección de Datos] (Nov. 30, 2025), https://participacion.digital.gob.cl/es-CL/projects/leyprotecciondedatospersonales

[1194] Biblioteca del Congreso Nacional, *Constitution of the Republic of Chile, 24 October 1980*, Article 19 para 4 (Mar. 11, 2022), https://www.bcn.cl/leychile/navegar?idNorma=242302.

[1195] Biblioteca del Congreso Nacional, *Law No 21096 Concerning the Right to Protection of Personal Data* (Jun. 16, 2018), https://www.leychile.cl/Navegar?idLey=21096&tipoVersion=0.

[1196] Biblioteca del Congreso Nacional, *Law No 19628 on Protection of Private Life* (Feb. 28, 2020), https://www.bcn.cl/leychile/navegar?idNorma=141599

[1197] Biblioteca del Congreso Nacional, Ley 21719 Regulating the Protection and Treatment of Personal Data and Creating the Personal Data Protection Agency [Regula la protección y el tratamiento de los datos personales y crea la Agencia de Protección de Datos Personales] (Nov. 25, 2024), https://www.bcn.cl/leychile/navegar?idNorma=1209272

[1198] Ibid, Art. 30

[1199] DLA Piper, *Data Protection Laws of the World : Chile* (Jan. 24, 2022), https://www.dlapiperdataprotection.com/index.html?t=law&c=CL

[1200] Biblioteca del Congreso Nacional, Ley 21719 Regulating the Protection and Treatment of Personal Data and Creating the Personal Data Protection Agency [Regula la protección y el tratamiento de los datos personales y crea la Agencia de Protección de Datos Personales], Art. 30 bis., e. (Nov. 25, 2024), https://www.bcn.cl/leychile/navegar?idNorma=1209272

[1201] Biblioteca del Congreso Nacional, *Law No 19628 on Protection of Private Life* (Feb. 28, 2020), https://www.bcn.cl/leychile/navegar?idNorma=141599

Transparency Council or SERNAC.[1202] A Ministerial Advisory Committee[1203] tasked with advising the Executive on the regulatory and administrative measures necessary for implementing the new data protection framework urged the government to establish and fund the new Agency as soon as possible.[1204]

Chile's Transparency Council endorsed the Ibero-American Data Protection Network (*Red Iberoamericano de Protección de Datos*, RIPD) standards and guidelines and committed to integrated the principles of legitimation, lawfulness, loyalty, transparency, purpose, proportionality, quality, responsibility, safety and confidentiality in national policies.[1205] The 16 data protection authorities representing 12 countries initiated a coordinated action on ChatGPT, developed by OpenAI, in May 2023 on the basis that ChatGPT may entail risks for the rights and freedoms of users in the processing of their personal data because of the potential for "responses with cultural, racial or gender biases, as well as false ones."[1206]

The Transparency Council has been a member of the Global Privacy Assembly (GPA) since 2019.[1207] The Transparency Council has not endorsed any AI-related GPA Resolutions[1208] or the related ICDPPC Declaration on Ethics and Data Protection in Artificial Intelligence (2018).[1209]

1202 Ibid

1203 Diario Official, *Ministerio Secretaría General De La Presidencia Crea Comisión Asesora Ministerial para la Implementación de la Ley N° 21.719*, Art 1 (Jun. 17, 2025), https://www.diariooficial.interior.gob.cl/publicaciones/2025/06/17/44176/01/2660255.pdf

1204 Comisión Asesora Ministerial para la Implementación de la Ley No. 21.719, *Recomendaciones para la implementación de la nueva Ley de Protección de Datos Personales]* (2025–2026), https://www.linkedin.com/posts/rosarioleteliery_informe-comisi%C3%B3n-activity-7429540556025966593-r3Hu; from the Ministerio Secretaría General de la Presidencia, https://www.minsegpres.gob.cl/comision-asesora-ministerial

1205 Red Iberoamericano de Protección de Datos (RIPD), *Standards for Personal Data Protection for Ibero-American States* (2017), https://www.redipd.org/sites/default/files/2022-04/standars-for-personal-data.pdf

1206 Red Iberoamericano de Protección de Datos (RIPD), Authorities from the Ibero-American Network for the Protection of Personal Initiate a Coordinated Action against the ChatGPT Service (May 8, 2023), https://www.redipd.org/noticias/autoridades-red-iberoamericana-de-proteccion-de-datos-personales-inician-accion-chatgpt

1207 Global Privacy Assembly, *List of Accredited Members* (2023), https://globalprivacyassembly.org/participation-in-the-assembly/list-of-accredited-members/

1208 Global Privacy Assembly, *Resolution on Accountability in the Development and Use of Artificial Intelligence* (Oct. 2020), https://globalprivacyassembly.org/wp-content/uploads/2020/10/FINAL-GPA-Resolution-on-Accountability-in-the-Development-and-Use-of-AI-EN-1.pdf; Global Privacy Assembly, *Resolution on Principles and Expectations for the Appropriate Use of Personal Information in Facial Recognition Technology* (Oct. 2022), https://globalprivacyassembly.org/wp-content/uploads/2022/11/15.1.c.Resolution-on-Principles-and-Expectations-for-the-Appropriate-Use-of-Personal-Information-in-Facial-Recognition-Technolog.pdf; Global Privacy Assembly, *Resolution on Generative Artificial Intelligence Systems* (Oct. 2023), https://globalprivacyassembly.org/wp-content/uploads/2023/10/5.-Resolution-on-Generative-AI-Systems-101023.pdf

1209 40th International Conference of Data Protection and Privacy Commissioners, *Declaration on Ethics and Data Protection in Artificial Intelligence* (Oct. 23, 2018),

Algorithmic Transparency

Article 8 of the new data protection law that enters into force in December 2026 introduces a "right to object and not be subject to decisions based on the automated processing of their personal data, including profiling, which produces legal effects concerning them or significantly affects them" unless the decision is necessary for the conclusion or execution of a contract, based on data subject's consent, or provided for by law.[1210] In all decisions made based on automated processing, the data controller must take necessary measures to ensure the rights and freedoms of data subjects, including their right to information and transparency, to obtain explanation and human intervention, to express their point of view, and to request a review of the decision.

The new law aligns with the RIPD Specific Guidelines for Compliance with the Principles and Rights that Govern the Protection of Personal Data in Artificial Intelligence Projects, endorsed by the Transparency Council. The guidelines specify that "The information provided regarding the logic of the AI model must include at least basic aspects of its operation, as well as the weighting and correlation of the data, written in a clear, simple and easily understood language, it will not be necessary to provide a complete explanation of the algorithms used or even to include them. The above always looking not to affect the user experience."[1211]

In October 2022, the Transparency Council initiated efforts to draft a General Instruction on Algorithmic Transparency.[1212] The analysis of algorithmic transparency in the public sector carried out by the public innovation lab GobLab UAI and the Transparency Council shows that although information about processing activities and algorithmic logic is sometimes available, it is frequently fragmented and dispersed across different sources.[1213] The researchers pointed to the considerable effort required to collect, systematize and present the information in a manner understandable to the recipient.[1214]

---

http://globalprivacyassembly.org/wp-content/uploads/2018/10/20180922_ICDPPC-40th_AI-Declaration_ADOPTED.pdf

[1210] Biblioteca del Congreso Nacional, Ley 21719 Regulating the Protection and Treatment of Personal Data and Creating the Personal Data Protection Agency [Regula la protección y el tratamiento de los datos personales y crea la Agencia de Protección de Datos Personales], Art. 8 bis. (Nov. 25, 2024), https://www.bcn.cl/leychile/navegar?idNorma=1209272

[1211] Red Iberoamericano de Protección de Datos (RIPD), Specific Guidelines for Compliance with the Principles and Rights that Govern the Protection of Personal Data in Artificial Intelligence Projects, p. 17 (Jun. 2019), https://www.redipd.org/en/document/guide-specific-guidelines-ai-projects-en.pdf

[1212] María Paz Hermosilla and Ana María Muñoz, *Chile's Road to Algorithmic Transparency: Setting New Standards in Latin America,* Observatory of Public Sector Innovation (May 30, 2023), https://oecd-opsi.org/blog/chile-algorithmic-transparency/

[1213] GobLab UAI and Transparency Council, *Algorithmic Transparency in the Public Sector* (Oct. 2021), p. 19, https://directus.thegovlab.com/uploads/ai-ethics/originals/2e9f6394-7a3a-4034-a7cb-071923a32ae0.pdf

[1214] Ibid

The survey of 74 companies selling IoT devices carried out by SERNAC in February 2022 showed a widespread lack of knowledge regarding the use of AI and algorithmic technologies within the private sector. Many companies were not aware if manufacturers had employed such technologies in the products they were selling or distributing.[1215] This is aggravated by the lack of understanding about what data categories are collected by the IoT devices and the absence of privacy policies explaining processing to the consumers.

Use of AI in Public Administration

Chile's government published a circular entitled Guidelines for the use of AI tools in the public sector in December 2023. The circular addresses key themes such as human-centric AI, transparency and explainability, privacy, and data use. The guidelines came into effect on January 1, 2024, and the circular was distributed to all public services.[1216] The Guide to Ethical Formulation of Data Science Projects, a collaboration between the Digital government Secretariat and UAI and BID Lab,[1217] provides guidance to the public sector on ethical and legal risks associated with projects using personal data. The nation's public procurement body ChileCompra collaborated on recommendations for ethical and responsible data management as part of GobLab's Ethical Algorithms Project.[1218] This guide incorporated requirements related to transparency, privacy, non-discrimination, and explainability and helped Chile become amongst the first Latin American countries to establish data ethics requirements for AI procurement systems.

As a member of the Latin American Centre for Development Administration (CLAD), Chile approved the principles in the Ibero American Charter on Artificial Intelligence in Civil Service in late 2023.[1219] The guiding principles include: human autonomy; transparency, traceability and explainability; accountability, liability and auditability; security and technical robustness; reliability, accuracy, and

---

[1215] SERNAC, *Exploratory Study on the Risks of Use of IoT devices in Chile* (Feb. 25, 2022), https://www.sernac.cl/portal/619/w3-article-64912.html.

[1216] MinCiencia, *Government Publishes Circular for the Responsible Use of AI in Public Services* (Dec. 14, 2023), https://www.minciencia.gob.cl/noticias/gobierno-publica-circular-para-un-uso-responsable-de-la-ia-en-los-servicios-publicos/

[1217] Secretaría de Gobierno Digital, *Guide to Ethical Formulation of Data Science Projects [Guía Formulación Ética de Proyectos de Ciencia de Datos]* (Aug. 24, 2022), https://digital.gob.cl/biblioteca/estandares-y-guias/guia-formulacion-etica-de-proyectos-de-ciencia-de-datos/

[1218] ChileCompra, ChileCompra's New Directive Provides Recommendations for Managing Artificial Intelligence or Data Science Projects [Nueva Directiva de ChileCompra entrega recomendaciones para la gestión de proyectos de Inteligencia Artifical o Ciencia de Datos] (Dec 14, 2023), https://www.chilecompra.cl/2023/12/nueva-directiva-de-chilecompra-entrega-recomendaciones-para-la-gestion-de-proyectos-de-inteligencia-artificial-o-ciencia-de-datos/

[1219] Latin American Centre for Development Administration (CLAD), *Ibero American Charter on Artificial Intelligence in Civil Service* (Nov. 2023), https://web-api-backend.clad.org/uploads/ciia-en-03-2024.pdf

reproducibility; confidence, proportionality, and prevention of damage; privacy and protection of personal data; data quality and safety; fairness, inclusiveness, and non-discrimination; human-centering, public value, and social responsibility; and sustainability and environmental protection. The Charter also proposes prohibited uses, including facial recognition systems; systems aimed at behavior and cognitive manipulation of specific individuals or vulnerable groups (children or the elderly); and social ranking systems based on certain personality traits, individual behaviors, socio-demographic features or economic status.[1220]

Chile's AI Bill introduced in May 2024 aligns with the Charter's principles and outlines a comprehensive risk classification system for AI applications in public administration.[1221] This system mandates public agencies to conduct thorough assessments of the potential risks associated with the AI technologies they develop or procure.

The country introduced a public registry for AI systems,[1222] including high-risk AI applications, and implemented a mandatory algorithmic audit process.[1223] These updates are enforced by the newly created Data and AI Center,[1224] which will oversee the implementation and accountability of AI systems in public administration. These institutions implement specific provisions in the updated National AI Policy to mitigate algorithmic biases, with an emphasis on eliminating discrimination based on gender, ethnicity, and socio-economic status in public sector applications.[1225] The policy also includes requirements for continuous auditing of AI models to ensure transparency and the avoidance of "black box" effects.

[1220] Ibid, p. 21

[1221] Chamber of Deputies, *Legislature 372: Regulation of Artificial Intelligence Systems [Regula los Sistemas de Inteligencia Artificial]* (May 7, 2024), https://www.camara.cl/legislacion/ProyectosDeLey/tramitacion.aspx?prmID=17429&prmBOLETIN=16821-19; for an overview in English, see Guillermo Carey, José Ignacio Mercado, and Ricardo Alonso, *Bill to Regulate Artificial Intelligence Systems Is Introduced to the Chilean Chamber of Deputies*, Carey (Accessed Mar. 5, 2025), https://www.carey.cl/en/bill-to-regulate-artificial-intelligence-systems-is-introduced-to-the-chilean-chamber-of-deputies/

[1222] Algoritmos Públicos, *Repositorio* (2026), https://algoritmospublicos.cl/repositorio

[1223] Diario Oficial, Updates to the "National Artificial Intelligence Policy" Approved [Aprueba Actualización de la "Política Nacional de Inteligencia Artificial"] (Jan 28, 2025), https://www.diariooficial.interior.gob.cl/publicaciones/2025/01/28/44060/01/2600578.pdf

[1224] Contraloría General De La República, Resolution 2492 Approving the Organic Structure of the Data and AI Center and Defining the Scope of Its Functions [Resolución 2492 exenta Aprueba estructura orgánica del Centro de Datos e Inteligencia Artificial y Define Alcance de sus Funciones] (Jan. 18, 2025), https://www.bcn.cl/leychile/navegar?idNorma=1210446

[1225] Diario Oficial, Updates to the "National Artificial Intelligence Policy" Approved [Aprueba Actualización de la "Política Nacional de Inteligencia Artificial"] (Jan 28, 2025), https://www.diariooficial.interior.gob.cl/publicaciones/2025/01/28/44060/01/2600578.pdf

Facial Recognition

In 2019, a group of 28 civil society organizations and nearly 70 experts issued a public statement rejecting President Piñera's Mobile Surveillance System deployed the same year.[1226] The system was created under Chile's Safe Street (*Calle Segura*) plan and used drones equipped with high-definition cameras and facial recognition technology to monitor public areas to fight crime and improve coordination of security agencies. Despite the criticism, the surveillance system, which started with a fleet of 8 drones, was expanded and new public procurements were ongoing.[1227] Regulation DAN-151 issued by the General Directorate of Civil Aeronautics (DGAC) in 2024 defined appropriate conditions for drone use with an explicit reference to the requirement of respect to people's rights to privacy and intimacy in the operation of drones.[1228]

A number of cases recorded in Chile demonstrated errors, false positives, and a general lack of effectiveness in publicly funded and invasive facial recognition systems. The facial recognition system deployed in one of Santiago's shopping malls and used to cross-reference images with the Investigative Police database of wanted suspects, for example, was reported to result in a 90% rate of false positives.[1229] The Transparency Council criticized the system for being disproportionate and intrusive from a privacy perspective.[1230]

Inadequate results and widespread errors were also reported when using IDEMIA's real-time image analysis and facial recognition software in the municipality of Las Condes[1231] and the facial recognition system implemented by the Civil Registry.[1232] Likewise, a Ministry of Transport pilot payment system on buses

---

[1226] Civicus Monitor, *New Mass Surveillance System Concerns Chilean Civil Society* (Jul. 22, 2019), https://monitor.civicus.org/explore/new-mass-surveillance-system-chile/; Derechos Digitales, *Against Mass Surveillance in Public Spaces of the "Mobile Tele-Surveillance System"* (Apr. 2, 2019), https://www.derechosdigitales.org/12919/contra-la-vigilancia-masiva-en-los-espacios-publicos-del-sistema-de-televigilancia-movil/

[1227] Garcia, N., *Chile Will Lease Drone Flight Hours for the Surveillance of the Metropolitan Region* (Mar. 9, 2022), https://www.infodefensa.com/texto-diario/mostrar/3483902/chile-arrendara-horas-vuelo-drones-vigilancia-region-metropolitana.

[1228] Directorate General of Civil Aviation (DGAC), *DAN 151* (May 27, 2024), https://www.dgac.gob.cl/wp-content/uploads/2024/05/DAN-151-ED3-27MAY2024.pdf

[1229] Transparency Council, *CPLT Insists That Facial Recognition in the Capital's Mall Is "Disproportionate to the Purpose It Pursues"* (Mar. 14, 2019), https://www.consejotransparencia.cl/cplt-insiste-en-que-reconocimiento-facial-en-mall-capitalino-es-desproporcionado-para-el-fin-que-persigue/

[1230] Ibid

[1231] Derechos Digitales, *Which is Worse: A Facial Recognition System That Doesn't Work or One That Does?* (Sept. 10, 2021), https://www.derechosdigitales.org/16728/que-es-peor-un-sistema-de-reconocimiento-facial-que-no-funciona-o-uno-que-si-lo-hace/

[1232] University of Chile, *Pandemic and Technology: The Risks of Facial Recognition and Data Management* (Mar. 31, 2020), https://www.uchile.cl/noticias/162239/pandemia-y-tecnologia-los-riesgos-del-reconocimiento-facial

in Santiago that used facial recognition attracted considerable controversy due to concerns over data use, rights to privacy, and surveillance.[1233]

The data protection law due to come into force in December 2026 includes provisions strengthening the protection of biometric data, including facial recognition data, which is explicitly classified as sensitive personal data, and therefore subject to stricter safeguards and legal protection.[1234] The processing of such data requires the data subject's explicit consent, unless a specified exception applies (such as public health, legal obligations, or judicial purposes).[1235] Article 16(3) requires the data controller to provide the data subject with clear and comprehensive information regarding the purpose of the processing, legal basis, recipients of the data, retention period, and rights available to the data subject prior to processing.[1236]

EdTech

As part of a global investigative report on the education technology (EdTech) endorsed by 49 governments, including Chile, during the pandemic, Human Rights Watch found that Aprendo in Linea in Chile put at risk or directly violated children's rights. The EdTech product sent children's data to AdTech companies.[1237]

Key stakeholders, including policymakers, international organizations, private sector representatives, and academia, participated in discussions on the skills needed to meet the challenges and opportunities presented by artificial intelligence in education at a roundtable on Education and Skills in the Age of Artificial Intelligence hosted by the UNESCO Regional Office in Santiago in October 2023.[1238] Stakeholders highlighted the need for continuous training for teachers and school leaders regarding the ethical use of AI technologies. The discussion also emphasized the collaborative roles of various sectors in adapting education to current technological advancements. Valtencir Mendes, Head of Education at UNESCO

[1233] Fernanda Castillo, *Passage with Facial Recognition in RED: Experts War of Risks to User Privacy [Pasaje con reconocimiento facial en RED: Expertos advierten riesgos en la privacidad de los usuarios]*, BioBioChile (Jul. 11, 2025), https://www.biobiochile.cl/noticias/nacional/region-metropolitana/2025/07/11/pasaje-con-reconocimiento-facial-en-red-expertos-advierten-riesgos-en-la-privacidad-de-los-usuarios.shtml

[1234] Biblioteca del Congreso Nacional, *Law 21.719 Regulating the Protection and Treatment of Personal Data and Creating the Personal Data Protection Agency [Regula la protección y el tratamiento de los datos personales y crea la Agencia de Protección de Datos Personales]*, Art. 1-bis (5)(g) (Nov. 25, 2024), https://www.bcn.cl/leychile/navegar?idNorma=1209272

[1235] Ibid, Art. 16

[1236] Ibid, Art. 16-ter

[1237] Human Rights Watch, *How Dare They Peep into My Private Life* (May 25, 2022), https://www.hrw.org/report/2022/05/25/how-dare-they-peep-my-private-life/childrens-rights-violations-governments

[1238] UNESCO, *The Future of Education: UNESCO and Experts Debate the Role of Artificial Intelligence in Chile* (Oct. 11, 2024), https://www.unesco.org/en/articles/future-education-unesco-and-experts-debate-role-artificial-intelligence-chile

Santiago, concluded by stressing the need to protect students' rights while adapting the educational system to technological advancements.

Chile was among the 16 countries from Latin America and the Caribbean that committed to the promotion of ethical AI education as part of their endorsement of the Cartagena de Indias Declaration in August 2024.[1239] The declaration stresses the need for collaboration to foster ethical and inclusive AI development and commits the countries to enhancing digital education and sharing best practices while ensuring AI governance frameworks uphold human rights and promote innovation.

The Chilean Ministry of Education, in collaboration with UNESCO's Regional Office in Santiago, presented the first version of the National Plan for the Digital Transformation of Education in 2025.[1240] This strategic framework serves as a roadmap for the integrated and sustainable adoption of digital technologies across all levels of the Chilean education system, grounded in a participatory, ethically driven approach.

AI and Neurotechnology

Chile became the first country in the world to provide constitutional protection for neuro-rights in 2021. Neuro-rights are recognized as a subset of human rights aimed at balancing the potentially adverse effects of neurotechnology.[1241] Article 19 of the Chilean Constitution was amended by introducing the provision that "Scientific and technological development will be at the service of people and will be carried out with respect for life and physical and psychological integrity. The law will regulate the requirements, conditions and restrictions for its use on people, especially safeguarding brain activity, as well as the information derived from it."[1242] This guarantee is especially relevant in light of advances in the area of neuroscience and AI, such as brain-computer interfaces.[1243]

[1239] MinCiencia, *Undersecretary Gainza at AI Summit in Colombia* (Aug. 9, 2024), https://www.minciencia.gob.cl/noticias/subsecretaria-gainza-en-cumbre-de-ia-en-colombia-la-gobernanza-no-es-solo-regulacion-y-marcos-legales-sino-que-principalmente-tiene-que-ver-con-perspectivas-y-principios/

[1240] Ministerio de Educación, *Ministry of Education and UNESCO Present the First Version of the National Plan for the Digital Transformation of Education in Chile* (Aug. 26, 2025), https://www.mineduc.cl/mineduc-y-unesco-presentan-primera-version-plan-nacional-transformacion-digital/

[1241] A. McCay, *Neurorights: The Chilean Constitutional Change*, AI & Soc (2022), https://doi.org/10.1007/s00146-022-01396-0

[1242] Biblioteca del Congreso Nacional, *Law No 21383: Amendments to the Fundamental Rights Charter to Establish Scientific and Technological Development at the Service of People* (Oct. 25, 2021), https://www.bcn.cl/leychile/navegar?idNorma=1166983&idParte=10278855&idVersion=2021-10-25

[1243] Rafael Yuste et al., *Four Ethical Priorities for Neurotechnologies and* AI, Nature 551, pp. 159–163 (2017), https://doi.org/10.1038/551159a

Chile introduced a bill on the protection of neuro-rights and mental integrity and the development of research and neurotechnologies in October 2020. [1244] The bill, which was in the second constitutional procedure of the legislative process in 2024,[1245] aims to regulate neurotechnologies while preserving rights and data regarding brain activity and mental integrity.

Environmental Impact of AI

Chile's National AI Policy states that it will establish incentives for developers and deployers of AI systems to focus on energy efficiency and develop guidelines for environmentally responsible use in the public and private sectors.[1246] The updated policy emphasizes sustainable use and development of AI systems for social and environmental wellbeing as part of the human-centric principle and under the sustainable development principle.[1247] The Policy stresses Chile's renewable energy resources and legal norms around environmental impact assessments.[1248] Chile's endorsement of the Statement on Inclusive and Sustainable Artificial Intelligence for People and the Planet at the AI Action Summit reaffirms this commitment to developing energy-efficient, environmentally responsible, and socially inclusive AI ecosystems.[1249]

Lethal Autonomous Weapons

Chile has consistently advocated for an international treaty regulating and prohibiting the use of lethally autonomous weapons systems (LAWS) in the United Nations, Group of Governmental Experts on LAWS, and other international meetings. In 2018, Chile proposed along with Austria and Brazil an open-ended GGE to negotiate a legally binding instrument to ensure meaningful human control over

---

[1244] Chamber of Deputies, *No. 13828-19: Bill on the Protection of Neuro-Rights and Mental Integrity, and the Development of Research and Neurotechnologies* (Oct. 7, 2020), https://www.camara.cl/legislacion/ProyectosDeLey/tramitacion.aspx?prmID=14385&prmBOLETIN=13828-19

[1245] Chamber of Deputies, No. 13828-19: Bill on the Protection of Neuro-Rights and Mental Integrity, and the Development of Research and Neurotechnologies, Processing (Oct. 7, 2024), https://www.camara.cl/legislacion/ProyectosDeLey/tramitacion.aspx?prmID=14385&prmBOLETIN=13828-19

[1246] MinCiencia, *National Artificial Intelligence Policy* (May 2, 2024), https://www.minciencia.gob.cl/areas/inteligencia-artificial/politica-nacional-de-inteligencia-artificial/

[1247] Diario Oficial, *Updates to the "National Artificial Intelligence Policy" Approved [Aprueba Actualización de la "Política Nacional de Inteligencia Artificial"]* (Jan 28, 2025), https://www.diariooficial.interior.gob.cl/publicaciones/2025/01/28/44060/01/2600578.pdf

[1248] Ibid, pp. 11–12

[1249] Élysée, *Statement on Inclusive and Sustainable Artificial Intelligence for People and the Planet* (Feb. 11, 2025), https://www.elysee.fr/en/emmanuel-macron/2025/02/11/statement-on-inclusive-and-sustainable-artificial-intelligence-for-people-and-the-planet

critical functions in LAWS.[1250] In 2020, Chile also joined eight other CCW parties and reiterated the call for the development of a "normative and operational framework" for ensuring human control of LAWS.[1251]

Chile was one of 70 states that endorsed a joint statement on autonomous weapons systems in October 2022 calling for recognition of the dangers of autonomous weapons systems, acknowledged the need for human oversight and accountability, and emphasized the importance of an international framework of rules and constraints.[1252]At the 78th UN General Assembly First Committee in 2023, Chile voted in favor[1253] of resolution L.56[1254] on autonomous weapons systems, along with 163 other states, which mandated the UN Secretary-General to prepare a report reflecting the views of member and observer states on autonomous weapons systems. Chile reaffirmed its support for a legally binding international treaty prohibiting fully autonomous lethal weapons in 2025.[1255] In a joint statement, the Chilean delegation highlighted the need for meaningful human control over critical functions and reiterated a commitment to international humanitarian law and accountability in the use of AI in warfare.

Chile endorsed the Belén Communiqué in February 2023,[1256] which calls for "urgent negotiation" of a binding international treaty to regulate and prohibit the use

---

[1250] Austria, Brazil, and Chile, *Proposal for a Mandate to Negotiate a Legally-Binding Instrument that Addresses the Legal, Humanitarian and Ethical Concerns Posed by Emerging Technologies in the Area of Lethal Autonomous Weapons Systems (LAWS)*, U.N. Doc. CCW/ GGE.2/2018/WP.7 (Aug. 30, 2018), https://reachingcriticalwill.org/images/documents/Disarmament-fora/ccw/2018/gge/documents/29August_Proposal_Mandate_ABC.pdf

[1251] Austria, Belgium, Brazil, Chile, Ireland, Germany, Luxembourg, Mexico, and New Zealand, *Joint Commentary on Guiding Principles A, B, C and D (*Sept. 1, 2020), https://documents.unoda.org/wp-content/uploads/2020/09/GGE20200901-Austria-Belgium-Brazil-Chile-Ireland-Germany-Luxembourg-Mexico-and-New-Zealand.pdf

[1252] United Nations (UN) General Assembly, First Committee, *Joint Statement on Lethal Autonomous Weapons Systems First Committee*, 77th United Nations General Assembly Thematic Debate – Conventional Weapons (Oct. 21, 2022), https://estatements.unmeetings.org/estatements/11.0010/20221021/A1jJ8bNfWGlL/KLw9WYcSnnAm_en.pdf

[1253] Isabelle Jones, *164 States Vote against the Machine at the UN General Assembly*, Stop Killer Robots (Nov. 1, 2023), https://www.stopkillerrobots.org/news/164-states-vote-against-the-machine/

[1254] UN General Assembly, *Resolution L56: Lethal Autonomous Weapons* (Oct.12, 2023), https://reachingcriticalwill.org/images/documents/Disarmament-fora/1com/1com23/resolutions/L56.pdf

[1255] UN Office of Disarmament Affairs, *Joint Statement to the September 2025 Session of the CCW GGE LAWS* (Sept. 2025), https://docs-library.unoda.org/Convention_on_Certain_Conventional_Weapons_-Group_of_Governmental_Experts_on_Lethal_Autonomous_Weapons_Systems_(2025)/Joint_statement_to_the_September_2025_session_of_the_CCW_GGE_LAWS_-_As_delivered.pdf

[1256] Latin American and the Caribbean Conference on the Social and Humanitarian Impact of Autonomous Weapons, *Official Communique* (Feb. 24, 2023), https://conferenciaawscostarica2023.com/communique/?lang=en

of autonomous weapons to address the grave concerns raised by removing human control from the use of force.

Chile also participated in the international summit on the responsible application of artificial intelligence in the military domain hosted by the Netherlands in February 2023. At the end of the REAIM 2023, Chile, together with other countries, agreed on a joint call for action on the responsible development, deployment and use of artificial intelligence in the military domain.[1257]

AI Literacy

Chile's Digital Citizenship (*Ciudadanía Digital*) initiative launched in 2022 fosters AI and Digital Literacy in support of democracy and human rights.[1258] The initiative supports the "development of fundamental knowledge, skills, and attitudes to allow people to exercise their digital rights and strengthen democratic society through the safe, responsible, participatory, creative, critical, and reflective use of digital technology, understanding the influence of these technologies in their personal and public lives at local and global levels." Critical and Reflective Digital Literacy is the first of four dimensions of Digital Citizenship, along with Digital Care and Responsibility centering ethical use of technology, Participation in Digital Citizenship promoting caring for democracy through the use of digital technology, and Digital Creativity and Innovation. The Ministry of Education's Plan for Digital Citizenship and Literacy 2025–2026 incorporates AI content as part of the aim to "Raise awareness in the population of the challenges, opportunities, and risks of acting in digital environments so we can protect democracy and strengthen community life."[1259]

Human Rights

Chile ranked as "Free" with a score of 95/100 in the Freedom in the World report,[1260] with 38/40 for political rights and 57/60 for civil liberties in 2025.

The government of Chile has continued to demonstrate progress on improving the protection of human rights related to AI by updating the Ethics and Governance axis of the 2021 version of its National AI Policy[1261] in line with input from various

---

[1257] Government of Netherlands, *Call to Action on Responsible Use of AI in the Military Domain* (Feb. 16, 2023), https://www.government.nl/latest/news/2023/02/16/reaim-2023-call-to-action
[1258] Ministerio de Educación, *What Is Digital Citizenship? [¿Qué es la ciudadanía digital?]* (2025), https://ciudadaniadigital.mineduc.cl/que-es-la-ciudadania-digital/
[1259] Ministerio Secretaría General de Gobierno y Ministerio de Educación, *Digital Citizenship and Literacy Plan 2025–2026 [Plan de Ciudadanía y Alfabetización Digital 2025–2026]*, p. 6 (May 2025), https://ciudadaniadigital.mineduc.cl/wp-content/uploads/2025/05/Plan-Ciudadania-y-Alfabetizacion-Digital-2025.pdf
[1260] Freedom House, *Freedom in the World 2025: Chile* (2025), https://freedomhouse.org/country/chile/freedom-world/2025
[1261] MinCiencia, *National Policy on Artificial Intelligence* (May 2, 2024), https://minciencia.gob.cl/areas/inteligencia-artificial/politica-nacional-de-inteligencia-artificial/

stakeholder groups gathered during a 2-month public consultation period.[1262] The new AI bill[1263] aimed at regulating AI development in a way that safeguards and promotes democratic principles, the rule of law, and citizens' fundamental human rights will also ensure the ethical and responsible implementation of AI. At the Third World Summit of Committees of the Future in Santiago, organized by the Chilean Senate with the Inter-Parliamentary Union (IPU),[1264] parliamentary delegations from 17 countries produced an Outcome Document emphasizing that emerging technologies, including AI, must be governed in accordance with human rights, democracy, and the rule of law.[1265]

Despite Chile's ratification of several human rights instruments,[1266] some lingering issues remain regarding human rights protection. Chile still has problems with incarceration conditions in detention facilities and cases of torture and brutality by law enforcement.[1267] There are also reported cases of arbitrary or unlawful murders, as well as issues with the protection of the rights of women, indigenous populations, LGBTQI+, migrants, and refugees.[1268] In October 2020, following a series of protests and clashes with law enforcement authorities, more than 70% of citizens voted in favor of forming a constituent assembly to rewrite the constitution, which has remained unchanged since the dictatorship that officially ended in 1990.

[1262] MinCiencia, *Public Consultation: National Policy on Artificial Intelligence: Results Report 2024* (2024), https://www.minciencia.gob.cl/areas/inteligencia-artificial/politica-nacional-de-inteligencia-artificial/

[1263] Chamber of Deputies, *Legislature 372: Regulation of Artificial Intelligence Systems* [Spanish only] (May 7, 2024), https://www.camara.cl/legislacion/ProyectosDeLey/tramitacion.aspx?prmID=17429&prmBOLETIN=16821-19; for an overview in English, see Guillermo Carey, José Ignacio Mercado, and Ricardo Alonso, *Bill to Regulate Artificial Intelligence Systems Is Introduced to the Chilean Chamber of Deputies*, Carey (May 13, 2024), https://www.carey.cl/en/bill-to-regulate-artificial-intelligence-systems-is-introduced-to-the-chilean-chamber-of-deputies/

[1264] Senado, *Third World Summit of Parliamentary Commisions of the Future: Reaffirm Commitment to Anticipatory Governance [Tercera Cumbre Mundial de Comisiones Parlamentarias del Futuro: Reafirman el Compromiso con la Gobernanza Anticipatoria]* (Jan. 14, 2025), https://www.senado.cl/comunicaciones/noticias/tercera-cumbre-mundial-de-comisiones-parlamentarias-del-futuro-reafirman-el

[1265] Inter-Parliamentary Union (IPU), *For Innovative Anticipatory Governance that Incorporates the Future in Parliamentary Work—Outcome Document* (Jan. 14, 2025), https://www.ipu.org/file/20887/download

[1266] UN Treaty Body Database, *Ratification Status for Chile*, https://tbinternet.ohchr.org/_layouts/15/TreatyBodyExternal/Treaty.aspx?CountryID=35&Lang=EN

[1267] Human Rights Watch, *World Report 2022*, https://www.hrw.org/world-report/2022/country-chapters/chile

[1268] U.S. Department of State, Bureau of Democracy, Human Rights, and Labor, *2020 Country Reports on Human Rights Practices: Chile* (Mar. 20, 2021), https://www.state.gov/reports/2020-country-reports-on-human-rights-practices/chile/

AI Safety Summit

Chile participated in the first AI Safety Summit and endorsed the Bletchley Declaration in November 2023.[1269] Chile thus committed to participate in international cooperation efforts on AI "to promote inclusive economic growth, sustainable development and innovation, to protect human rights and fundamental freedoms, and to foster public trust and confidence in AI systems to fully realise their potential." Endorsing parties affirmed that for the good of all, AI should be designed, developed, deployed, and used, in a manner that is safe, in such a way as to be human-centric, trustworthy and responsible."

Chile did not participate in the AI Seoul Summit in May 2024.[1270]

Chile participated in the AI Action Summit 2025 in Paris and signed the Statement on Inclusive and Sustainable Artificial Intelligence for People and the Planet.[1271] Through this commitment, Chile pledged to promote transparent, ethical, and sustainable AI aligned with public-interest values and global democratic norms. At the Summit, Chile and Inria France jointly launched the Franco-Chilean Binational Center on AI, a strategic 2025–2026 initiative focused on AI safety evaluation, trustworthy generative AI, energy-efficient computing, open-source AI tools, and talent development.[1272] This Center strengthens Chile's institutional capacity for safe AI research and reinforces alignment with the OECD and UNESCO frameworks.

OECD / G20 AI Principles

As an OECD member, Chile endorsed the OECD Al Principles.[1273] Chile's Al-related documents, including in its National Al Policy,[1274] directly reference the OECD Principles. The May 2024 update to the National AI Policy strongly emphasizes responsible AI development, ethical AI deployment, and the importance of international collaboration on AI governance.

[1269] UK Department for Science, Innovation & Technology, Foreign, Commonwealth & Development Office, *Prime Minister's Office, The Bletchley Declaration by Countries Attending the AI Safety Summit*, (Nov. 2023), https://www.gov.uk/government/publications/ai-safety-summit-2023-the-bletchley-declaration/the-bletchley-declaration-by-countries-attending-the-ai-safety-summit-1-2-november-2023

[1270] UK Department for Science, Innovation & Technology, *AI Seoul Summit: Participants List (Governments and Organizations)* (Jun. 2024), https://www.gov.uk/government/publications/ai-seoul-summit-programme/ai-seoul-summit-participants-list-governments-and-organisations

[1271] Élysée, *Statement on Inclusive and Sustainable Artificial Intelligence for People and the Planet* (Feb. 11, 2025), https://www.elysee.fr/en/emmanuel-macron/2025/02/11/statement-on-inclusive-and-sustainable-artificial-intelligence-for-people-and-the-planet

[1272] Inria Chile & MinCiencia, *Launch of the Franco-Chilean Binational Center on AI at the AI Action Summit* (Feb. 14, 2025), https://inria.cl/en/paris-ai-action-summit-inria-and-ministry-science-chile-launch-binational-center-ai-and-announce-5

[1273] OECD Legal Instruments, *OECD/LEGAL/0449: Recommendation of the Council on Artificial Intelligence* (May 3, 2024), https://legalinstruments.oecd.org/en/instruments/oecd-legal-0449

[1274] MinCiencia, *National Artificial Intelligence Policy* (May 2, 2024), https://www.minciencia.gob.cl/areas/inteligencia-artificial/politica-nacional-de-inteligencia-artificial/

The OECD has noted Chile's AI policy efforts across various dimensions,[1275] including its promotion of societal dialogues on AI, its creation of partnerships with the private sector for human capacity development, its willingness for international cooperation on AI governance.

Chile's AI policy vision also echoes several OECD principles, including the transparency and explainability, human-centered AI, and building human capacity for AI.[1276] The Chilean Transparency Council's recommendations on algorithmic transparency[1277] published in August 2024 align with OECD guidelines to promote accountability and the ethical use of algorithms in public services by introducing guidelines for proactive transparency, effective monitoring, and use of clear language in the delivery of information about automated and semi-automated decision-making systems. MinCiencia's AI system public registry implements these principles.[1278]

Council of Europe AI Treaty

Chile has not signed the Council of Europe Framework Convention on Artificial Intelligence and Human Rights, Democracy, and the Rule of Law, the first legally binding international AI treaty.[1279]

UNESCO Recommendation on AI Ethics

Chile is a UNESCO member and adopted the UNESCO Recommendation on the Ethics of AI in November 2021.[1280]

[1275] OECD AI Policy Observatory, *AI Policies in Chile* (2026), https://oecd.ai/en/dashboards/national/chile

[1276] Chamber of Deputies, *Legislature 372: Regulation of Artificial Intelligence Systems* [Spanish only] (May 7, 2024), https://www.camara.cl/legislacion/ProyectosDeLey/tramitacion.aspx?prmID=17429&prmBOLETIN=16821-19; for an overview in English, see Guillermo Carey, José Ignacio Mercado, and Ricardo Alonso, *Bill to Regulate Artificial Intelligence Systems Is Introduced to the Chilean Chamber of Deputies*, Carey (May 13, 2024), https://www.carey.cl/en/bill-to-regulate-artificial-intelligence-systems-is-introduced-to-the-chilean-chamber-of-deputies/

[1277] Council for Transparency (CPLT), *Resolution No. 372: Approval Text of the Council for Transparency's Recommendations on Algorithmic Transparency*, Official Journal of the Republic of Chile, Ministry of the Interior and Public Safety (Aug. 30, 2024), https://www.consejotransparencia.cl/wp-content/uploads/instruccion/2024/09/Informe-recomendaciones-congreso-CPLT.pdf

[1278] Diario Oficial, *Updates to the "National Artificial Intelligence Policy" Approved [Aprueba Actualización de la "Política Nacional de Inteligencia Artificial"]* (Jan 28, 2025), https://www.diariooficial.interior.gob.cl/publicaciones/2025/01/28/44060/01/2600578.pdf

[1279] Council of Europe Treaty Office, *Chart of Signatures and Ratifications of Treaty 225* (Mar. 5, 2026), https://www.coe.int/en/web/conventions/full-list?module=signatures-by-treaty&treatynum=225

[1280] UNESCO, *Ethics of Artificial Intelligence: The Recommendation* (2026), https://www.unesco.org/en/artificial-intelligence/recommendation-ethics

Chile's Ministry of Science, in partnership with UNESCO and CAF,[1281] the Latin American development bank, convened the first Regional Forum on AI Ethics in November 2023. The aim of the event was to shape a collective strategy among Latin American and Caribbean (LAC) countries aligned with the UNESCO Recommendation. Chile signed the resulting Santiago Declaration to Promote Ethical Artificial Intelligence.[1282] The Delcaration establishes fundamental principles that should guide public policy on AI, including proportionality, security, fairness, non-discrimination, gender equality, accessibility, sustainability, privacy and data protection. The Declaration also constitutes a milestone toward establishing a Regional Council on AI for the LAC region. Chile will preside over the working group created to this end.[1283]

Chile among the first countries to complete the UNESCO Readiness Assessment Methodology (RAM),[1284] a tool to support the effective implementation of the Recommendation. The RAM helps countries and UNESCO identify and address any institutional and regulatory gaps.[1285] Chile implemented the RAM with the launch of an updated National AI Policy and action plan in May 2024.[1286] Chile also proposed a new bill aimed at regulating the ethical and responsible development of AI based on the recommendations from the RAM.[1287] Chile has also updated the privacy law, as recommended on the RAM report.

Evaluation

Chile has made considerable progress toward developing legislative and policy frameworks in data protection and AI and aligning them with international

---

[1281] Ángel Melguizo and Gabriela Ramos, *Ethical and Responsible Artificial Intelligence: From Words to Actions and Rights*, Somos Iberoamérica (Feb. 1, 2023), https://www.somosiberoamerica.org/pt-br/tribunas/inteligencia-artificial-etica-e-responsavel-das-palavras-aos-fatos-e-direitos/

[1282] Cumbre Ministerial y de Altas Autoridades de América Latina y el Caribe, *Declaracion de Santiago "Para promover una inteligencia artificial ética en América Latina y el Caribe"* (Oct. 2023), https://minciencia.gob.cl/uploads/filer_public/40/2a/402a35a0-1222-4dab-b090-5c81bbf34237/declaracion_de_santiago.pdf

[1283] UNESCO, *UNESCO and Leading Ministry in Santiago de Chile host Milestone Regional LAC Forum on Ethics of AI* (Dec. 5, 2023), https://www.unesco.org/en/articles/unesco-and-leading-ministry-santiago-de-chile-host-milestone-regional-lac-forum-ethics-ai?hub=387

[1284] UNESCO Global AI Ethics and Governance Observatory, *Chile* (Oct. 2024), https://www.unesco.org/ethics-ai/en/chile

[1285] UNESCO Global AI Ethics and Governance Observatory, *Readiness Assessment Methodology* (2026), https://www.unesco.org/ethics-ai/en/ram

[1286] UNESCO, *Chile Launches National AI Policy and Introduces AI Bill Following UNESCO's Recommendations* (May 4, 2024), https://www.unesco.org/en/articles/chile-launches-national-ai-policy-and-introduces-ai-bill-following-unescos-recommendations

[1287] Chamber of Deputies, *Legislature 372: Regulation of Artificial Intelligence Systems [Regula Sistemas de Inteligencia Artificial]* (May 7, 2024), https://www.camara.cl/legislacion/ProyectosDeLey/tramitacion.aspx?prmID=17429&prmBOLETIN=16821-19

standards, including the UNESCO Recommendation on the Ethics of AI and OECD AI Principles. Chile's efforts refining its National Al Policy in accordance with international Al governance standards and best practices reflects the country's recognition of the need for global cooperation to govern AI for the benefit of humanity. Joining the Cpuncil of Europe AI group as an observer and endorsing the AI treaty would further demonstrate Chile's commitment to international cooperation.

Updates to the data protection law establish a regulatory oversight agency with adequate enforcement powers and create enforceable standards in algorithmic transparency.[1288] The AI Bill under deliberation addresses the issue of regulatory oversight and includes provisions for the establishment of a National Commission on Artificial Intelligence,[1289] which will enforce and monitor compliance with the legislation. Chile is employing public consultations to gather public input on AI Policy updates through the Digital Government Secretariat.[1290]

The data protection law entering into force in December 2026 addresses some concerns over the proliferation of intrusive facial recognition technologies in public spaces with protections for biometric data, though the government has continued to implement pilot projects.

[1288] Chamber of Deputies, *Law 19.628 on Personal Data Protection* (Aug. 29, 2024), https://www.carey.cl/wp-content/uploads/2024/08/Consolidado-PDL-datos-personales-aprobado-29.Aug_.2024.pdf; unofficial English translation, https://www.carey.cl/wp-content/uploads/2024/08/Texto-ley-ingles-clean.pdf

[1289] Digital Policy Alert, *Chile: Establishment of National Commission on Artificial Intelligence in Bill to Regulate AI systems, Robotics, and Related Technologies (Bill 15869-19)* (Apr. 26, 2023), https://digitalpolicyalert.org/change/8827-establishment-of-national-commission-on-artificial-intelligence-in-chile-nr-15869-19

[1290] MinCiencia, *Public Consultation: National Policy on Artificial Intelligence: Results Report 2024* (2024), https://www.minciencia.gob.cl/areas/inteligencia-artificial/politica-nacional-de-inteligencia-artificial/

China

In 2025, China progressed on national AI governance by releasing guidance and standards on data and the AI industry and amendments to strengthen AI ethics in the Cybersecurity Law. China also launched a Global AI Governance Action Plan aligned with the UN Global Digital Compact.

National AI Strategy

Since 2013, the Chinese government has published several national-level policies, guidelines, and action plans, which reflect the intention to develop, deploy, and integrate AI in various sectors. In 2015, Prime Minister Li Keqiang launched the "Made in China" (MIC 2025) initiative aimed at turning the country into a production hub for high-tech products within the next few decades.[1291] In the same year, the State Council released guidelines on China's Internet + Action Plan.[1292] The Action Plan sought to integrate the internet into all elements of the economy and society. The document emphasized the importance of cultivating emerging AI industries and investing in research and development.

The Central Committee of the Communist Party of China's 13th 5-year plan is another notable example. The document mentioned AI as one of six critical areas for developing the country's emerging industries and as an essential factor in stimulating economic growth.[1293] The Robot Industry Development Plan[1294] and Artificial Intelligence Innovation Action Plan for Higher Education Institutions[1295] illustrate detailed action plans and guidelines concerning specific sectors.

Most notable of all is the New Generation Artificial Intelligence Development Plan (AIDP)—an ambitious strategy to make China the world leader in AI by 2030 and the most transparent and influential indication of China's AI strategy's driving

---

[1291] State Council [国务院], *Circular of the State Council on the Issuance of Made in China 2025* [国务院关于印发《中国制造2025》的通知] (May 8, 2015), https://www.gov.cn/zhengce/content/2015-05/19/content_9784.htm

[1292] State Council, *China Unveils "Internet Plus Action Plan" to Fuel Growth* [**国务院关于积极推进"互联网+"行动的指导意见**] (Jul. 4, 2015), https://www.gov.cn/zhengce/content/2015-07/04/content_10002.htm

[1293] Central Committee of the CPC, *The 13th Five Year Plan for Economic and Social Development 2016–2020* (2016), https://en.ndrc.gov.cn/policies/202105/P020210527785800103339.pdf

[1294] National Development and Reform Commission, *Robot Industry Development Plan (2016–2020) Released* [机器人产业发展规划（2016-2020年）发布] (2016), https://www.ndrc.gov.cn/xxgk/zcfb/ghwb/201604/t20160427_962181.html

[1295] Ministry of Education, Interpret the "Action Plan for Artificial Intelligence Innovation in Higher Education Institutions" and Introduce the Relevant Work [高等学校人工智能创新行动计划] http://www.moe.gov.cn/jyb_xwfb/xw_fbh/moe_2069/xwfbh_2018n/xwfb_20180608/201806/t20180608_338911.html

forces. China's State Council issued the AIDP in 2017. According to the Plan, AI should be used in a broad range of sectors, including defense and social welfare. The AIDP also indicates the need to develop standards and ethical norms for the use of AI. The actual innovation and transformation are expected to be driven by the private sector and local governments. The Chinese government has handpicked three major tech giants to focus on developing specific sectors of AI: Baidu, Alibaba, and Tencent.[1296] These companies receive preferential contract bidding, more convenient access to finance, and sometimes market share protection.

Following the AIDP, the Ministry of Science and Technology published the New Generation Artificial Intelligence Development Code of Ethics (Code of Ethics).[1297] The New Generation AI Governance Professional Committee, consisting of experts from academics and the industry,[1298] created the Code of Ethics and is responsible for its implementation. The Code of Ethics aims to "thoroughly implement the AIDP, refine the principles of AIDP, enhance society's awareness and behavioural consciousness of AI ethics, actively guide responsible AI research, development, and application activities, as well as promote the healthy development of AI." The Code of Ethics highlights six fundamental ethical requirements: (1) "improving human well-being"; (2) "promoting fairness and justice"; (3) "protecting privacy and security"; (4) "ensuring controllability, and trustworthiness"; (5) "strengthening responsibility"; and (6) "improving ethical literacy," taking issues like privacy, biases, discrimination, and fairness into consideration.[1299]

Regarding local governments, there is a system of incentives for fulfilling national government policy aims. For this reason, local governments often become a testing ground for the central government's policies. Chinese cities and provinces, as well as regional administrations, compete for new AI incentives. Large metropolises, such as Tianjin and Shanghai, have already launched multi-billion-dollar AI city Venture Capital funds and converted entire districts and islands for new AI

1296 Meng Jing and Sarah Dai, *China Recruits Baidu, Alibaba and Tencent to AI 'national team,'* South China Morning Post (Nov. 21, 2017), https://www.scmp.com/tech/china-tech/article/2120913/china-recruits-baidu-alibaba-and-tencent-ai-national-team

1297 Ministry of Science and Technology of the People's Republic of China, *New Generation Artificial Intelligence Development Code of Ethics* ["新一代人工智能伦理规范" 发布] (Sept. 26, 2021), https://perma.cc/RC4V-Q2FX; Center for Security and Emerging Technology, *Ethical Norms for New Generation Artificial Intelligence Released* [English translation] (Oct. 12, 2021), https://cset.georgetown.edu/wp-content/uploads/t0400_AI_ethical_norms_EN.pdf

1298 Intelligent Hardware, *The National Artificial Intelligence Governance Professional Committee Was Established, Composed of Academic and Business Personnel* [国家人工智能治理专业委员会成立 其中有学界与企业界人员组成] (Jun. 9, 2020), https://www.21ic.com/article/775428.html

1299 Ministry of Science and Technology of the People's Republic of China, *New Generation Artificial Intelligence Development Code of Ethics* ["新一代人工智能伦理规范" 发布] (Sept. 26, 2021), https://perma.cc/RC4V-Q2FX; Center for Security and Emerging Technology, *Ethical Norms for New Generation Artificial Intelligence Released* [English translation] (Oct. 12, 2021), https://cset.georgetown.edu/wp-content/uploads/t0400_AI_ethical_norms_EN.pdf

companies. Shenzhen and Shanghai have taken the lead to create policies and standards for the Artificial Intelligence industry in China at the provincial level. The Regulations on the Promotion of Artificial Intelligence Industry in Shenzhen from 2022,[1300] marked the birth of the country's first regulation on the AI industry. Shanghai, in the meantime, published Regulations of Shanghai on Promoting the Development of Artificial Intelligence Industry, coming into effect on October 1, 2022.[1301] The two Regulations aim to promote the innovation and development of the AI industry in their provinces for the benefit of citizens, society, and the economy. Both Regulations establish mechanisms and Committees of AI Ethics to develop AI ethics policy and standards as well as provide guidance on AI ethics.

The establishment of the New Generationa AI Governance Professional Committee at the national level and Committees of AI Ethics at the provincial level embodies the creation of a monitoring regime for the guidance and supervision of AI and AI ethics. The impact of the oversight of AI in policy implementation processes as well as the level of independence of these mechanisms remains unclear.

Cybersecurity Law

China's Cybersecurity Law of 2017 has broad provisions to "ensure cybersecurity; safeguard cyberspace sovereignty and national security, and social and public interests; protect the lawful rights and interests of citizens, legal persons, and other organizations; and promote the healthy development of informatization of the economy and society." The law applies to those constructing, operating, maintaining, and using networks on the mainland and aims to support research and development in artificial intelligence without sacrificing security.[1302]

[1300] Shenzhen Municipal People's Congress Standing Committee, *Regulations on the Promotion of Artificial Intelligence Industry in Shenzhen* [深圳经济特区人工智能产业促进条例] (Sept. 5, 2022), [archived] https://perma.cc/6VKQ-NJZF; Center for Security and Emerging Technology, *Translation: Regulations for the Promotion of the Artificial Intelligence Industry in Shenzhen Special Economic Zone* (Dec. 7, 2022), https://cset.georgetown.edu/wp-content/uploads/t0480_Shenzhen_AI_regs_EN.pdf; ses also, Justice Bureau of Shenzhen Municipality, *527 Regulations of Shenzhen Special Economic Zone on Promoting the Artificial Intelligence Industry* (Mar. 28, 2024), https://sf.sz.gov.cn/fggzywyb/content/post_11216296.html

[1301] Shanghai Municipal People's Congress Standing Committee, *Regulations of Shanghai on Promoting the Development of Artificial Intelligence*, International Services Shanghai (Sept. 22, 2022), https://english.shanghai.gov.cn/en-LocalRules/20240913/4395d1563dd548b39406cbe8833c9ef6.html; [Chinese] https://perma.cc/PQ8J-ARCA

[1302] DIGICHINA, *Translation: Cybersecurity Law of the People's Republic of China (Effective June 1, 2017)* (Jun. 29, 2018), https://digichina.stanford.edu/work/translation-cybersecurity-law-of-the-peoples-republic-of-china-effective-june-1-2017/; Cyberspace Administration of China, 中华人民共和国网络安全法 (Nov. 7, 2016), https://www.cac.gov.cn/2016-11/07/c_1119867116.htm

Revisions in 2025 aim to strengthen the regulatory regime, including risk monitoring, assessment, and safety oversight as well as improving AI ethics rules.[1303] The amendments define liability and specify penalties for individuals as well network operators responsible for "serious" failures such as data breaches or infrastructure failure. Individuals and operators responsible for disseminating or failing to stop the spread of prohibited information can also be found liable and face penalties, including fines.

National AI Standards

China's Cyberspace Administration in partnership with several other technology-focused government offices, including the National Development & Reform Commission and the National Standardization Administration, released official guidelines for a National Comprehensive Standardization System for the AI industry in 2024.[1304] Primary standards cover security and governance as well as concerns related to products, services, industrialization, and technical matters. None of the Standards are substantive, providing a roadmap more than a strict framework.

The National Development and Reform Commission issued guidelines for a National Data Standards System in 2024 to activate data's potential, an adjacent but important area for AI strategy. These guidelines build on the National Artificial Intelligence Industry Comprehensive Standardization System and target areas such as Data aggregation technical standards and 5G data transmission network standards.[1305] The Artificial Intelligence Standardization Technical Committee established under the Ministry of Industry and Information Technology (MIIT) in December 2024 furthers China's strategy to strengthen regulatory oversight as a means to support industrial growth and global AI norms and standards.[1306]

The National Cybersecurity Standardization Technical Committee 260 (TC260) adopted the AI Safety Governance Framework, a voluntary AI risk management framework that "outlines control measures to address different types of AI safety risks through technological and managerial strategies." The document stresses that "comprehensive AI safety and security risk governance mechanisms and

---

[1303] National People's Congress China Daily, *AI Support at Heart of Cybersecurity Revision* (Oct. 30, 2025), http://en.npc.gov.cn.cdurl.cn/2025-10/30/c_1136472.htm

[1304] Cyberspace Administration of China, *Notice of four departments on issuing the Guidelines for the Construction of the National Artificial Intelligence Industry Comprehensive Standardization System (2024 Edition)* [四部门关于印发国家人工智能产业综合标准化体系建设指南（2024版）的通知] (Jul. 3, 2024), https://www.cac.gov.cn/2024-07/03/c_1721686809220407.htm

[1305] Cyberspace Administration of China, *Notice of the National Development and Reform Commission and Other Departments on Issuing the "Guidelines for the Construction of the National Data Standards System* [国家发展改革委等部门关于印发《国家数据标准体系建设指南》的通知] (Oct. 11, 2024), https://www.cac.gov.cn/2024-10/11/c_1730337907916423.htm

[1306] State Council, *China Establishes Technical Committee to Further AI Standardization* (Dec. 31, 2024), https://english.www.gov.cn/news/202412/31/content_WS67736228c6d0868f4e8ee657.html

regulations" should "engage multi-stakeholder participation, including technology R&D institutions, service providers, users, government authorities, industry associations, and social organizations."[1307]

Management of Algorithmic Recommendations

China has started to adopt a comprehensive set of regulations on AI. The Provisions on the Management of Algorithmic Recommendations in Internet Information Services came into force in March 2022.[1308] The Provisions apply to any entity that uses algorithm recommendation technologies to provide Internet information services within Mainland China. The Provisions require companies to register algorithms to an online database to the extent that the algorithm has public opinion properties or social mobilization capabilities. Companies shall also complete an algorithm security self-assessment report. Users shall have the right to opt out of recommendation algorithms. The use of recommendation algorithms for illegal or harmful purposes is prohibited.

Deep Synthesis Regulations

In January 2023, the Cyberspace Administration of China (CAC) introduced regulations prohibiting the production of deepfakes without user consent and requires specific identification that the content had been generated using AI. The regulations were formulated to "carry forward the core Socialist value vision, to safeguard national security and the societal public interest."[1309] Any content that was created using an AI system must be clearly labeled with a watermark, i.e., text or image visually superimposed on the video indicating that the content had been edited. Deep synthesis services cannot use the technology to disseminate fake news. Content that goes against existing laws is prohibited, as is content that endangers national security and interests, damages the national image or disrupts the economy.

[1307] National Cybersecurity Standardization Technical Committee 260, *AI Safety Governance Framework* (Sept. 6, 2024), https://www.tc260.org.cn/upload/2024-09-09/1725849192841090989.pdf; archived at https://perma.cc/JNQ9-AG59

[1308] Cyber Administration of China, the Ministry of Information and Industry Technology, the Ministry of Public Security of the People's Republic of China, and State Administration for Market Regulation, *Provisions on the Management of Algorithmic Recommendations in Internet Information Services* (Dec. 31, 2021), http://www.cac.gov.cn/2022-01/04/c_1642894606364259.htm

[1309] Cyberspace Administration of China, *Regulations on the In-depth Synthesis Management of Internet Information Services* [国家互联网信息办公室等三部门发布"互联网信息服务深度合成管理规定"] (Dec. 22, 2022), http://www.cac.gov.cn/2022-12/11/c_1672221949318230.htm; Stanford University DigiChina, *Translation: Internet Information Service Deep Synthsis Management Provisions (Draft for Comment, Jan. 2022)*, (Feb. 4, 2022), https://digichina.stanford.edu/work/translation-internet-information-service-deep-synthesis-management-provisions-draft-for-comment-jan-2022/

The Cyberspace Administration of China published Measures for the Identification of Synthetic Content Generated by AI in March 2025.[1310] The instrument aims to standardize the labeling of AI-generated content by internet information service providers with requirements for production, display, and distribution. Service providers must embed explicit labels in files containing AI-generated content and include implicit labels with production details as per regulatory standards. If services include downloading, reproducing, or exporting generated synthetic content, service providers "shall ensure that the files contain explicit labels that satisfy these requirements." Additionally, service providers are required to inform users of labeling specifications in service agreements.

Generative AI Measures

The Provisional Administrative Measures of Generative Artificial Intelligence Services (Generative AI Measures) were published by the Cyberspace Administration of China (CAC), together with six other authorities, in July 2023 and took effect that August. The Generative AI Measures apply to "the use of generative AI technology to provide services for generating text, pictures, sounds, videos and other content within the territory of China." The Measures will apply to domestic companies and to overseas generative AI service providers offering AI services in China to the public.[1311]

AI Ethics

The AIDP outlines China's desire to become a world leader in defining ethical norms and standards for AI.[1312] Government bodies and private companies have attempted to define ethical standards. For example, China's Artificial Intelligence Industry Alliance (AIIA) released a draft joint pledge on self-discipline in the AI industry emphasizing AI ethics, safety, standardization, and international engagement.[1313]

[1310] Cyberspace Administration of China, *Notice on Issuing the "Measures for Identifying Artificial Intelligence-Generated and Synthetic Content"* [关于印发"人工智能生成合成内容标识办法"的通知] (Mar. 14, 2025), https://www.cac.gov.cn/2025-03/14/c_1743654684782215.htm

[1311] Cyberspace Administration of China. *Interim Measures for Generative Artificial Intelligence Service Management* (Jul. 13, 2023), http://www.cac.gov.cn/2023-07/13/c_1690898327029107.htm

[1312] *China's New Generation Artificial Intelligence Development Plan* [English translation] (Jul. 20, 2017), https://www.newamerica.org/cybersecurity-initiative/digichina/blog/full-translation-chinas-new-generation-artificial-intelligence-development-plan-2017/; [original Chinese], https://www.gov.cn/zhengce/content/2017-07/20/content_5211996.htm

[1313] *Chinese AI Alliance Drafts Self-Discipline "Joint Pledge"* [English translation] (Jun. 17, 2019), https://www.newamerica.org/cybersecurity-initiative/digichina/blog/translation-chinese-ai-alliance-drafts-self-discipline-joint-pledge/; [original Chinese], https://mp.weixin.qq.com/s/x7HTx4AR6oNBWwWxUpnSuQ

The National New Generation Artificial Intelligence Governance Expert Committee established by China's Ministry of Science and Technology prepared the Governance Principles for Developing Responsible Artificial Intelligence in 2019.[1314] The document outlines eight principles for the governance of AI: harmony and friendliness; fairness and justice; inclusivity and sharing; respect for human rights and privacy; security; shared responsibility; open collaboration; and agility to deal with new and emerging risks. The document emphasizes that AI development should begin by enhancing the common well-being of humanity.

Another important document is a white paper on AI standards[1315] released in 2018 by the Standardization Administration of the People's Republic of China, the national-level body responsible for developing technical standards. Three key principles for setting the ethical requirements of AI technologies are (1) the ultimate goal of AI is to benefit human welfare; (2) transparency and the need to establish accountability as a requirement for both the development and the deployment of AI systems and solutions; (3) protection of intellectual property.

In April 2022, China published "Opinions on Strengthening the Management of Science and Technology Ethics,[1316] which is the first national policy on tech ethics. The document proposes five principles to supervise and review technology ethics and calls for the establishment of the China Science and Technology Ethics Society and ethical supervision in AI research.

China issued the Global AI Governance Action Plan in 2025, calling on all parties to take concrete and effective actions in advancing global AI development and governance based on the objectives and principles of promoting AI for good and in service of humanity, respecting national sovereignty, aligning with development goals, ensuring safety and controllability, upholding fairness and inclusiveness, and fostering open cooperation. The plan contains 13 points, including: Creating a diverse, open, and innovative ecosystem; Addressing energy and environmental issues; Advancing the governance of AI safety; and Jointly implementing the Global Digital Compact, taking the United Nations as the main channel.[1317]

[1314] *Chinese Expert Group Offers 'Governance Principles' for 'Responsible AI'* [English translation] (Jun. 17, 2019), https://www.newamerica.org/cybersecurity-initiative/digichina/blog/translation-chinese-expert-group-offers-governance-principles-responsible-ai/; [original Chinese], https://mp.weixin.qq.com/s/JWRehPFXJJz_mu80hlO2kQ

[1315] Jeffrey Ding and Paul Triolo, *Translation: Excerpts from China's White Paper on Artificial Intelligence Standardization,* New America (Jun. 20, 2018), https://www.newamerica.org/cybersecurity-initiative/digichina/blog/translation-excerpts-chinas-white-paper-artificial-intelligence-standardization/

[1316] General Office of the CPC Central Committee General Office of the State Council, *Opinions on Strengthening the Management of Science and Technology Ethics* [中共中央办公厅 国务院办公厅印发"关于加强科技伦理治理的意见"] (Mar. 20, 2022), http://www.gov.cn/gongbao/content/2022/content_5683838.htm

[1317] Ministry of Foreign Affairs, *Global AI Governance Action Plan* (Jul. 26, 2025), https://www.fmprc.gov.cn/mfa_eng/xw/zyxw/202507/t20250729_11679232.html

Public Participation

The National People's Congress of China operates a permanent online public-comment system through which it regularly publishes draft laws and regulations for feedback.[1318]

Between 2021 and 2025, several AI-related public consultations were conducted by China's main regulatory authorities. These include the CAC's draft Regulations on the Management of Internet Information Service Algorithm Recommendations,[1319] the draft Regulations on the Management of Deep Synthesis of Internet Information Services,[1320] and the draft Measures for the Management of Generative AI Services.[1321] In addition, the Ministry of Industry and Information Technology (MIIT) sought comments on the draft Guidelines for the Construction of a Comprehensive AI Industry Standard System.[1322] Most recently, the MIIT consulted the public on the draft Measures for the Management of Artificial Intelligence Science & Technology Ethics Services (Trial).[1323]

Scholarly research and findings from international organizations suggest that while China provides structured avenues for public input, these mechanisms function

---

[1318] National People's Congress (NPC) Observer, *China NPC Drafts for Public Comment – Searchable List* (Oct. 2025), https://npcobserver.com/china-npc-draft-law/

[1319] Cyberspace Administration of China (CAC), *Notice on Public Solicitation of Comments on the Draft Regulations on the Management of Internet Information Service Algorithm Recommendation Activities* [互联网信息服务算法推荐管理规定（征求意见稿）公开征求意见的通知] (Aug. 27, 2021), https://www.cac.gov.cn/2021-08/27/c_1631652502874117.htm

[1320] Cyberspace Administration of China (CAC), *Notice on Public Solicitation of Comments on the Draft Regulations on the Management of Internet Information Service Deep Synthesis* [互联网信息服务深度合成管理规定（征求意见稿）公开征求意见的通知] (Jan. 28, 2022), https://www.cac.gov.cn/2022-01/28/c_1644970458520968.htm

[1321] Cyberspace Administration of China (CAC), *Notice on Public Solicitation of Comments on the Draft Measures for the Management of Generative Artificial Intelligence Services* [生成式人工智能服务管理办法（征求意见稿）公开征求意见的通知] (Apr. 11, 2023), https://www.cac.gov.cn/2023-04/11/c_1682854275475410.htm?trk=public_post_comment-text

[1322] Ministry of Industry and Information Technology (MIIT), *Notice on Public Solicitation of Comments on the Draft Guidelines for the Construction of a Comprehensive Standardization System for the National Artificial Intelligence Industry* [公开征求对 国家人工智能产业综合标准化体系建设指南（征求意见稿）的意见] (Jan. 17, 2024), https://wap.miit.gov.cn/jgsj/kjs/gzdt/art/2024/art_956f95c93db8432e824b5e68dcc7d2fc.html

[1323] Ministry of Industry and Information Technology (MIIT), *Notice on Public Solicitation of Comments on the Draft Measures for the Management of Artificial Intelligence Science & Technology Ethics Services (Trial)* [人工智能科技伦理管理服务办法（试行）（公开征求意见稿] (Aug. 7, 2025), https://www.miit.gov.cn/jgsj/kjs/jscx/gjsfz/art/2025/art_092a447008f340d3abd55819b8c8e5cf.html

more as formal consultations than as effective tools for meaningful participation in national AI governance.[1324]

At the 2022 World Artificial Intelligence Conference, the Shanghai Artificial Intelligence Laboratory Governance Research Center, Tsinghua University, and Fudan University jointly launched an open platform for artificial intelligence ethics: OpenEGLab.[1325] There are five major sections in the platform including rule sets, governance maps, risk displays, evaluation frameworks, and industry solutions. The rule set is dedicated to building a global governance knowledge base. Currently, about 1,500 marked rule documents have been included, including ethical principles, policy strategies, laws, regulations, and standards.

Data Protection

In recent years, China has introduced major data protection laws, including the Personal Information Protection Law (PIPL) (effective from November 1, 2021) and the Data Security Law (DSL) (effective from September 1, 2021), together with a series of implementation regulations and administrative rules. The Cyberspace Administration of China (CAC) enforces data protection on a national level, though the CAC shares this responsibility with various local-level offices and the public security authority. China is not a member of the Global Privacy Assembly (GPA).[1326]

The PIPL[1327] establishes a new comprehensive regulatory framework for personal information protection in China, requiring consent as the principal basis for data collection and handling, introducing provisions with extraterritorial effect, restricting cross border data transfers, and imposing significant revenue-based fines

---

[1324] Steven J. Balla et al., *Notice the Comment?: Chinese Government Responsiveness to Public Participation in the Policymaking Process*, GW Regulatory Studies Center, George Washington University (Nov. 2022), https://regulatorystudies.columbian.gwu.edu/sites/g/files/zaxdzs4751/files/2022-11/chinese_government_responsiveness_rsc_working_paper_zxie_11-30-2022.pdf; International Service for Human Rights (ISHR), *China: Constraints on Civil Society Participation and Access to the United Nations – 2025 Report* (Apr. 2025), https://ishr.ch/wp-content/uploads/2025/04/ISHR-2025-Report_China-UN-Access-2.pdf

[1325] Fudan University, *Respond to the Challenges of Global AI Governance and Systematically Promote the Implementation of Governance Principles: "Dandelion" Artificial Intelligence Governance Open Platform Debuted at the World Artificial Intelligence Conference* [应对全球AI治理挑战 系统推进治理原则落地："蒲公英"人工智能治理开放平台亮相2022世界人工智能大会] (Sept. 4, 2022), https://cs.fudan.edu.cn/f9/18/c24280a456984/page.htm

[1326] Global Privacy Assembly (GPA), *List of Accredited Members* (2025), https://globalprivacyassembly.com/participation-in-the-assembly/list-of-accredited-members/

[1327] National People's Congress (NPC), *Personal Information Protection Law of the People's Republic of China* (Dec. 29, 2021), http://en.npc.gov.cn.cdurl.cn/2021-12/29/c_694559.htm; China Briefing, *The PRC Personal Information Protection Law (Final): A Full Translation* (Aug. 24, 2021), https://www.china-briefing.com/news/the-prc-personal-information-protection-law-final-a-full-translation/

for non-compliant conduct. The law seeks to regulate how private companies collect and use personal rather than granting data subjects control over their data or providing redress mechanisms than data protection laws such as the EU GDPR.[1328] For example, the PIPL, allows the collection of unlimited amounts of personal data, including images, so long as it is done "for the purpose of safeguarding public security."[1329] That is to say, the law does not limit the government's ability to collect or store biometric data obtained through facial recognition.

The PIPL does not assign responsibilities to government entities that collect personal data or guidance for who will be held responsible for data leaks. This becomes increasingly important with the rise of recent incidents of government leaks of the personal information of its citizens.[1330] Nevertheless, as the big data industry has been rapidly growing in China, the law provides more protection to users against unwanted data collection by private companies.

In March 2023, China's Ministry of Science and Technology established the National Data Administration (NDA). The NDA will be "responsible for the coordination and advancement of building the data factor system; for overall planning of the integrated sharing and development and use of data resources; for overall planning and advancement of Digital China, digital economy; and digital society plans and construction."[1331] The new regulator will have responsibilities that were previously performed by the Cyberspace Administration of China and by the National Development and Reform Commission.[1332]

The National Development and Reform Commission (NDRC) and National Data Bureau launched a high-quality data set initiative in 2025 aimed at supporting

---

[1328] Gabriela Zanfir-Fortuna, *China's Draft Personal Information Protection Law in 13 Key Points,* The StartUp (Nov. 3, 2020), https://medium.com/swlh/chinas-draft-personal-information-protection-law-in-13-key-points-5a9b9cdcf02c; Gil Zhang and Kate Yin, *A Look at China's Draft of Personal Data Protection Law*, IAPP (Oct. 26, 2020), https://iapp.org/news/a/a-look-at-chinas-draft-of-personal-data-protection-law/

[1329] National People's Congress (NPC), *Personal Information Protection Law of the People's Republic of China* (Dec. 29, 2021), http://en.npc.gov.cn.cdurl.cn/2021-12/29/c_694559.htm

[1330] Financial Times, *China, Coronavirus and Surveillance: The Messy Reality of Personal Data* (Apr. 2, 2020), https://www.ft.com/content/760142e6-740e-11ea-95fe-fcd274e920ca; Paul Mozur, *China, Desperate to Stop Coronavirus, Turns Neighbor Against Neighbor*, New York Times (Feb. 3, 2020), https://www.nytimes.com/2020/02/03/business/china-coronavirus-wuhan-surveillance.html; Chris Udemans, *Personal Data Leaks Spread along with Coronavirus Panic,* TechNode, (Feb. 3, 2020), https://technode.com/2020/02/03/wuhan-data-leak-coronavirus/

[1331] State Council & Central Committee of the Communist Party of China, *Party and State Institutions Reform Plan* [党和国家机构改革方案] (Mar. 16, 2023), https://www.gov.cn/zhengce/2023-03/16/content_5747072.htm; Graham Webster, *Translation: Establishing the National Data Administration*, DIGICHINA (Mar. 7, 2023), https://digichina.stanford.edu/work/translation-establishing-the-national-data-administration-march-2023/

[1332] Matt Sheehan, *China's AI Regulations and How They Get Made*, Carnegie Endowment for International Peace (July 10, 2023), https://carnegieendowment.org/2023/07/10/china-s-ai-regulations-and-how-they-get-made-pub-90117

the AI+ action plan. This initiative focuses on curating structured data to enhance AI applications across public administration, finance, healthcare, and industrial automation. China seeks to establish greater reliability in AI-driven decision-making while reinforcing data sovereignty policies by providing AI models government-approved, high-integrity data for training.[1333]

Algorithmic Transparency

Article 4 of the Interim Measures for Generative Artificial Intelligence Services refers to "effective measures to increase transparency in generative AI services and to increase the accuracy and reliability of generated content" based on the service type. The National Information Security Standardization Technical Committee (T260) then released the Basic Requirements for Security of Generative Artificial Intelligence Service[1334] with specific requirements for providing users with information about the service and its functioning. Enforcement of those requirements is currently rolled out as an AI-content oversight campaign by Cyberspace Administration of China (CAC).[1335] The Ministry of Industry and Information Technology's target is that at least 60% of these prospective standards will be implemented in "general key technologies and application development projects," with over 1,000 Chinese companies championing early adoption.[1336]

Complementing these AI-content oversight efforts, the CAC announced new measures under the Clean Internet Platforms: Governance of Typical Algorithm Issues campaign in May 2025. The measures respond to public concerns over algorithm-driven content risks such as information cocoons and polarization by enhancing transparency and accountability in algorithmic recommendation systems

---

[1333] National Development and Reform Commission, *The Kick-Off Meeting for the Construction of High-Quality Data Sets Was Held in Beijing* [高质量数据集建设工作启动会在京召开] (Feb. 21, 2025), https://www.ndrc.gov.cn/fzggw/wld/llh/zyhd/202502/t20250221_1396252.html

[1334] Center for Security and Emerging Technology, *Translation: [...] Basic Safety Requirements for Generative Artificial Intelligence Services* (Apr. 4, 2024), https://cset.georgetown.edu/publication/china-safety-requirements-for-generative-ai-final/; TC260, *Basic Security Requirements for Generative Artificial Intelligence Service* [生成式人工智能服务安全基本要求] (Feb 29, 2024), https://r20.rs6.net/tn.jsp; [permalink: https://perma.cc/GU3Q-GAJ3]

[1335] Global Times, *China's Cyberspace Regulator Launches Campaign to Oversee AI-Generated Content* (Mar. 15, 2024), https://www.ecns.cn/news/sci-tech/2024-03-15/detail-ihcyptyk0579386.shtml

[1336] Ministry of Industry and Information Technology, *Guidelines for the Construction of a National Comprehensive Standardization System for the Artificial Intelligence Industry (Draft Open for Public Comments)* [公开征求对"国家人工智能产业综合标准化体系建设指南"（征求意见稿）的意见] (Jan. 17, 2024), https://www.miit.gov.cn/jgsj/kjs/gzdt/art/2024/art_956f95c93db8432e824b5e68dcc7d2fc.html; Josh Ye, *China Issues Draft Guidelines for Standardising AI Industry*, Reuters (Jan. 18, 2024), https://www.reuters.com/technology/china-issues-draft-guidelines-standardising-ai-industry-2024-01-17/

across major platforms such as Douyin, Xiaohongshu, Weibo, Kuaishou, and WeChat Video Channel.[1337]

The Provisions on the Management of Algorithmic Recommendations in Internet Information Services[1338] highlight the principles of "openness" and "transparency" while establishing legal liabilities for violations of these principles. Article 4 Chapter I emphasizes that "the provision of algorithm recommendation services shall comply with laws and regulations, respect social morality and ethics, abide by business ethics and professional ethics, and follow the principles of fairness, openness and transparency, scientific reasonableness, and good faith." Companies shall be transparent regarding how their recommendation algorithms are trained and deployed, including which datasets the algorithm is trained on. Article 12 states, "Algorithmic recommendation service providers are encouraged to [...] optimize the transparency and understandability of search, ranking, selection, push notification, display, and other such norms, to avoid creating harmful influence on users, and prevent or reduce controversies or disputes." Chapter V establishes clear legal liability for non-compliance to and violation of the Provisions, including "the suspension of information updates and a fine of between 10,000 and 10,000 RMB, administration punishments or sanctions such as ordering the closure of websites, revoking relevant business permits, or revoking business licenses."

Article 24 Chapter II, Section 1 of the PIPL requires information processors using personal information in automatic decision-making to ensure "the transparency of decision-making and the fairness and justice of the results" and "shall not impose unreasonable differential treatment on individuals in terms of transaction price and other transaction conditions." Article 24 also requires alternatives to algorithmic business marketing and information push be provided or "a convenient way for individuals to reject." The article provides individuals adversely affected with the right an explanation and "to reject the decision made by the personal information processor only through automatic decision-making,"[1339]

Social Governance

Social governance is another area in which AI is promoted as a strategic opportunity for China. The Chinese authorities focus on AI as a way of overcoming

---

[1337] Cyberspace Administration of China (CAC), *The Cyberspace Administration of China Continues to Strengthen the Governance of Information Recommendation Algorithms* [中央网信办持续加强信息推荐算法治理] (May 22, 2025), https://www.cac.gov.cn/2025-05/22/c_1749536203490537.htm

[1338] Cyberspace Administration of China, Ministry of Information and Industry Technology, Ministry of Public Security, and State Administration for Market Regulation, *Provisions on the Management of Algorithmic Recommendations in Internet Information Services* [互联网信息服务算法推荐管理规定] (Dec. 31, 2021), http://www.cac.gov.cn/2022-01/04/c_1642894606364259.htm

[1339] National People's Congress (NPC), *Personal Information Protection Law of the People's Republic of China* (Dec. 29, 2021), http://en.npc.gov.cn.cdurl.cn/2021-12/29/c_694559.htm

social problems and improving the welfare of citizens,[1340] specifically on healthcare reform,[1341] environmental protection[1342], the administration of justice,[1343] and the implementation of the Social Credit System or Social Score.[1344] Another concrete example of how China is using AI in social governance can be seen in the sphere of internal security and policing.

China has been at the forefront of the development of smart cities equipped with digital infrastructure cloud computing and computer vision. The National Development and Reform Commission (NDRC), together with the National Data Administration (NDA) and other ministries, issued guidance to deepen smart city development and outlining China's plans to accelerate the integration of AI, big data, and cloud computing into urban governance and public services, with the goal of building livable, resilient, and efficiently managed smart cities nationwide by 2027.[1345] While the government emphasizes the convenience of smart cities, they also raise privacy and human rights concerns given their potential use for surveillance.[1346]

Facial Recognition

As early as the 2008 Beijing Olympics, China began to deploy new technologies for mass surveillance.[1347] China put in place more than two million

---

[1340] Heilmann, *Big Data Reshapes China's Approach to Governance*, Financial Times (2017), https://www.ft.com/content/43170fd2-a46d-11e7-b797-b61809486fe2

[1341] Andy Ho, *AI Can Solve China's Doctor Shortage. Here's how*, World Economic Forum (Sept. 17, 2018), https://www.weforum.org/agenda/2018/09/ai-can-solve-china-s-doctor-shortage-here-s-how/

[1342] Genia Kostka and Chunman Zhang, *Tightening the Grip: Environmental Governance under Xi Jinping*, Environmental Politics, 27(5):769–781 (2018), https://doi.org/10.1080/09644016.2018.1491116; People's Daily Online, *AI-Powered Waste Management Underway in China* (Feb. 26, 2019), http://en.people.cn/n3/2019/0226/c98649-9549956.html

[1343] Susan Finder, *China's Master Plan for Remaking Its Courts*, The Diplomat (Mar. 26, 2015), https://thediplomat.com/2015/03/chinas-master-plan-for-remaking-its-courts/; Anthony H.F. Li, *Centralization of Power in the Pursuit of Law-Based Governance: Legal Reform in China under the Xi Administration*, China Perspectives (2016), https://journals.openedition.org/chinaperspectives/6995

[1344] Severine Arsene, *China's Social Credit System: A Chimera with Real Claws*, Asie Visions (Nov. 2019), https://www.ifri.org/sites/default/files/migrated_files/documents/atoms/files/arsene_china_social_credit_system_2019.pdf

[1345] State Council Information Office, *China Unveils Guideline to Promote Development of Smart Cities* (May 21, 2024), http://english.scio.gov.cn/pressroom/2024-05/21/content_117202227.htm

[1346] Umberto Bacchi, *'I know your favorite drink': Chinese Smart City to Put AI in Charge,* Reuters (Dec. 5, 2020), https://www.reuters.com/article/technology/i-know-your-favourite-drink-chinese-smart-city-to-put-ai-in-charge-idUSL8N2IJ24L/

[1347] EPIC, *Privacy International, Privacy and Human Rights: An International Survey of Privacy Laws and Developments* (Report on People's Republic of China) (2006), http://www.worldlii.org/int/journals/EPICPrivHR/2006/PHR2006-People_s.html

CCTV cameras in Shenzen, making it the most-watched city in the world.[1348] In recent years, the techniques for mass surveillance have expanded rapidly, most notably in Shenzen, also to oversee the Muslim minority group the Uyghurs, and in Hong Kong. Modern systems for mass surveillance rely on AI techniques for such activities as facial recognition, communications analysis, and location tracking. As one industry publication has reported, "In the world of surveillance, no country invests more in its AI-fueled startups and growth-stage businesses than China. And no technology epitomizes this investment more than facial recognition—a technology that courts more controversy than almost any other."[1349] Forbes continues, "But a thriving domestic tech base has done nothing to quell the concerns of citizens. China is held up as a Big Brother example of what should be avoided by campaigners in the West, but that doesn't help people living in China."

However, the Supreme People's Court adopted and released provisions on civil cases relating to the use of facial recognition technology, which came into effect on August 1, 2021.[1350] The Provisions aim to "hear civil cases relating to the use of facial recognition technology in processing personal information, protect the lawful rights and interests of litigants, and promote the healthy development of the digital economy," in accordance with related laws and codes.

In particular, the Supreme People's Court highlights that the Provisions apply to civil cases arising from violations of the provisions of laws, administrative regulations, or the agreement of both parties to process facial information, which refers to biometrics information or facial information generated through facial recognition technologies.[1351] In this regard, the Provisions establish civil liabilities for the infringement of the rights and interests of natural persons in the process of biometrics data and information in the application of facial recognition technologies, which was republished and redistributed by local People's Courts.[1352] Before the enforcement of the Provisions regulating facial recognition technology, the China Academy for Information and Communications Technology (CAICT), a national

---

[1348] Naomi Wolf, *China's All-Seeing Eye*, Rolling Stone (May 15, 2018), https://www.commondreams.org/views/2008/05/15/chinas-all-seeing-eye

[1349] Zak Doffman, *Hong Kong Exposes Both Sides of China's Relentless Facial Recognition Machine*, Forbes (Aug. 26, 2019), https://www.forbes.com/sites/zakdoffman/2019/08/26/hong-kong-exposes-both-sides-of-chinas-relentless-facial-recognition-machine/

[1350] Supreme People's Court, *Provisions on Several Issues Concerning the Application of Law in the Trial of Civil Cases Relating to the Use of Facial Recognition Technology to Process Personal Information* [中华人民共和国最高人民法院] (Jul. 27, 2021), http://gongbao.court.gov.cn/Details/118ff4e615bc74154664ceaef3bf39.html

[1351] Ibid

[1352] Henan Provincial Higher People's Court, *Provisions of the Supreme People's Court on Several Issues Concerning the Application of Law in the Trial of Civil Cases Related to the Use of Facial Recognition Technology to Process Personal Information* [最高人民法院关于审理使用人脸识别技术处理个人信息相关民事案件适用法律若干问题的规定] (Jan. 29, 2022), http://pdszhfy.hncourt.gov.cn/public/detail.php?id=3181

think tank administrated by the Ministry of Industry and Information Technology, initiated the Trustworthy Facial Recognition Protection Plan in response to issues including "privacy leakage, technology abuse, biases and discrimination" in the development and application of facial recognition technologies while encouraging self-discipline of the industry.[1353]

There are many reports on China's use of facial recognition technology against ethnic minorities.[1354] The discriminatory ways in which state organs, companies, and academics have researched, developed, and implemented facial recognition in China would seem not to comply with the OECD AI Principles or the Governance Principles for the New Generation Artificial Intelligence. The deployment of facial recognition has also provoked opposition within China.[1355] This gap between stated ethical principles and on-the-ground applications of AI demonstrates the weakness of unenforceable ethics statements.[1356]

In September 2019, China's information-technology ministry announced that telecom carriers must scan the face of anyone applying for mobile and internet service.[1357] There are over 850 million mobile Internet users in China. Meanwhile, the Hong Kong government invoked emergency powers in October 2019 to ban demonstrators from wearing face masks.[1358]

Protests in Hong Kong over the use of facial surveillance are widespread. Umbrellas once used to deflect pepper spray, are now deployed to shield protester activities from the digital eyes of cameras.[1359] It is notable that the battle over the use of facial surveillance in Hong Kong began with widespread public protests about a national security law that extended police authority over the semi-autonomous region.[1360] According to the AP, "Young Hong Kong residents protesting a proposed

---

[1353] Biometric Update, *Telpo Joins Chinese Facial Recognition Best Practices Group* (Feb. 14, 2022), https://www.biometricupdate.com/202202/telpo-joins-chinese-facial-recognition-best-practices-group

[1354] Joi Ito, *My Talk at the MIT-Harvard Conference on the Uyghur Human Rights Crisis* (May 2, 2019), https://doi.org/10.31859/20190502.0901

[1355] Seungha Lee, *Coming into Focus: China's Facial Recognition Regulations,* Center for Strategic and International Studies (May 4, 2020), https://www.csis.org/blogs/trustee-china-hand/coming-focus-chinas-facial-recognition-regulations

[1356] See the section of this country report, AI and Surveillance.

[1357] Jane Li, *Getting a New Mobile Number in China Will Involve a Facial Recognition Test*, Quartz (Oct. 3, 2019), https://qz.com/1720832/china-introduces-facial-recognition-step-to-get-new-mobile-number/

[1358] Ilara Maria Sala, *Hong Kong Is Turning to a 1922 Law That Was Used to Quell a Seamen's Strike to Ban Face Masks,* Quartz (Oct. 4, 2019), https://qz.com/1721951/anti-mask-law-the-1922-origins-of-hong-kongs-emergency-powers/

[1359] Paul Mozur and Lin Qiqing, *Hong Kong Takes Symbolic Stand Against China's High-Tech Controls*, New York Times (Oct. 3, 2019), https://www.nytimes.com/2019/10/03/technology/hong-kong-china-tech-surveillance.html

[1360] BBC, *Hong Kong Security Law: What Is It and Is It Worrying*? (Jun. 30, 2020), https://www.bbc.com/news/world-asia-china-52765838

extradition law that would allow suspects to be sent to China for the trial are seeking to safeguard their identities from potential retaliation by authorities employing mass data collection and sophisticated facial recognition technology."[1361]

China is also exporting the model of mass surveillance by facial recognition to other parts of the world. A detailed report, published in *The Atlantic* in September 2020, stated that "Xi Jinping is using artificial intelligence to enhance his government's totalitarian control—and he's exporting this technology to regimes around the globe."[1362] According to *The Atlantic,* "Xi's pronouncements on AI have a sinister edge. Artificial intelligence has applications in nearly every human domain, from the instant translation of spoken language to early viral outbreak detection. But Xi also wants to use AI's awesome analytical powers to push China to the cutting edge of surveillance. He wants to build an all-seeing digital system of social control, patrolled by precog algorithms that identify potential dissenters in real-time."

In September 2020, the United States State Department issued voluntary guidelines for American companies "to prevent their products or services […] from being misused by government end-users to commit human rights abuses."[1363] The report came amid growing concern that China is rapidly exporting its own surveillance capabilities to authoritarian regimes around the world, as part of its Belt and Road Initiative (BRI).[1364] But the Washington Post highlighted the ongoing role of US-made technology in the sweeping surveillance of China, and notably the Uighur Muslim minority.[1365] The Washington Post explained that "the aim is to monitor cars, phones, and faces—putting together patterns of behavior for 'predictive policing' that justifies snatching people off the street for imprisonment or so-called reeducation. This complex opened four years ago, and it operates on the power of chips manufactured by U.S. supercomputer companies Intel and Nvidia." The Post editorial

---

1361 Christopher Bodeen, *Hong Kong Protesters Wary of Chinese Surveillance Technology* (Jun. 13, 2019), https://apnews.com/article/028636932a874675a3a5749b7a533969

1362 Ross Anderson, *The Panopticon Is Already Here*, The Atlantic (Sept. 2020), https://www.theatlantic.com/magazine/archive/2020/09/china-ai-surveillance/614197/

1363 U.S. Department of State, Bureau of Democracy, Human Rights, and Labor, *U.S. Department of State Guidance on Implementing the "UN Guiding Principles" for Transactions Linked to Foreign Government End-Users for Products or Services with Surveillance Capabilities* (Sept. 30, 2020), https://2017-2021.state.gov/key-topics-bureau-of-democracy-human-rights-and-labor/due-diligence-guidance

1364 Abhijnan Rej, *US Issues Human Rights Guidelines for Exporters of Surveillance Tech: The Directions to American Businesses Come Amid Growing Concern Around China's Export of Advanced Mass-Surveillance Capabilities to More Than 60 Countries*, The Diplomat (Oct. 2, 2020), https://thediplomat.com/2020/10/us-issues-human-rights-guidelines-for-exporters-of-surveillance-tech/

1365 Washington Post, Editorial, *US-Made Technologies Are Aiding China's Surveillance of Uighurs. How Should Washington Respond?* (Nov. 28, 2020), https://www.washingtonpost.com/opinions/us-made-technologies-are-aiding-chinas-surveillance-of-uighurs-how-should-washington-respond/2020/11/26/0218bbb4-2dc9-11eb-bae0-50bb17126614_story.html

followed a New York Times investigation which found extensive involvement by U.S. firms in the Chinese surveillance industry.[1366]

Facial recognition technology will also be pushed forward with the development of humanoid robots to be deployed across all sectors of social life by 2025.[1367] The draft Shenzhen Accelerates High-level Application Action Plan for Promoting the High-quality Development of Artificial Intelligence (2023–2024) released in May 2023 mentions the deployment of "general-purpose embodied intelligent robots" and "large-scale application of humanoid robots".[1368]

The Cyberspace Administration of China and Ministry of Public Security finalized regulations to standardize facial recognition usage and bolster personal data protection in March 2025. The regulations stipulate that the deployment of facial recognition must be lawful and not compromise national security, public welfare, or individual rights. Users must establish the necessity of the technology's use, advocate for minimal intrusion on personal rights, and mandate strict protection of collected data. Prior to collecting facial data, individuals must be clearly informed about the data collector's identity, the purpose of collection, processing methods, and data retention period. The regulations prohibit organizations and individuals from using misleading or coercive tactics to verify identities and from installing facial recognition equipment in private spaces such as hotel rooms, bathrooms, changing rooms, and restrooms. Additionally, they call for measures like data encryption, security auditing, access control, and intrusion detection to protect collected facial recognition information.[1369]

AI and Public Health

In China, the ultimate ambition of AI is to liberate data for public health purposes. The AIDP outlines the ambition to use AI to "strengthen epidemic

---

[1366] Paul Mazur and Don Clark, *China's Surveillance State Sucks Up Data. US Tech Is Key to Sorting It: Intel and Nvidia Chips Power a Supercomputing Center that Tracks People in a Place where Government Suppresses Minorities, Raising Questions about the Tech Industry's Responsibility* (Nov. 22, 2020), https://www.nytimes.com/2020/11/22/technology/china-intel-nvidia-xinjiang.html

[1367] Ministry of Industry and Information Technology, *Guiding Opinions on the Innovation and Development of Humanoid Robots* [“人形机器人创新发展指导意见” 解读] (Nov. 2, 2023), https://www.miit.gov.cn/zwgk/zcjd/art/2023/art_e3f5686c2f0d49f9968b7ae011d558e1.html

[1368] General Office of the Shenzhen Municipal Committee of the Communist Party of China and the General Office of the Shenzhen Municipal People's Government, *Shenzhen Action Plan to Accelerate the High-Quality Development and High-Level Application of Artificial Intelligence (2023-2024)* [“深圳市加快推动人工智能高质量发展高水平应用行动方案（2023—2024年）”发布] (May 31, 2023), http://sz.people.com.cn/n2/2023/0531/c202846-40438647.html

[1369] Ministry of Ecology and Environment Information Center, *Administrative Measures for the Safe Management of Facial Recognition Technology Applications* [人脸识别技术应用安全管理办法] (Mar. 25, 2025), http://www.chinaeic.net/xxgk/zcfg/202503/t20250325_1104616.html; State Council, *Facial Recognition Concerns Addressed*, China Daily (Mar. 22, 2025), https://english.www.gov.cn/news/202503/22/content_WS67de1eebc6d0868f4e8f10f5.html

intelligence monitoring, prevention and control," and to "achieve breakthroughs in big data analysis, Internet of Things, and other key technologies" for the purpose of strengthening intelligent health management.[1370] The State Council's 2016 official notice on the development and use of big data in the healthcare sector also explicitly states that health and medical big data sets are a national resource and that their development should be seen as a national priority to improve the nation's health.[1371] However, there is a rising concern that relaxed privacy rules and the transfer of personal data between government bodies will promote the collection and aggregation of health data without the need for individual consent.[1372]

Beijing Municipal Health Commission strictly prohibited the use of AI for automatically generating medical prescriptions and regulated various online healthcare activities with the draft set of 41 rules, with the goal that AI "shall not replace the doctors to provide diagnosis and treatment services."[1373]

Use of AI in COVID-19 Response

In June 2020, the State Council released a White Paper, entitled "Fighting COVID-19: China in Action," which provides that China has "fully utilized" artificial intelligence to not only research, analyze, and forecast COVID-19 trends and developments, but also to track infected persons, identify risk groups, and facilitate the resumption of normal business operations."[1374] During the pandemic, China has used AI for surveillance of infected individuals and medical imaging. China also sought to reduce human interaction by using computers and robots for various purposes and has proven to be very effective in reducing exposure, providing necessary services such as assistance for healthcare professionals, improving

---

[1370] State Council, *China's Next Generation Artificial Intelligence Development Plan* [国务院关于印发 新一代人工智能发展规划的通知] (Jul. 8, 2017), https://www.gov.cn/zhengce/content/2017-07/20/content_5211996.htm; DIGICHINA, *Full Translation: China's "New Generation Artificial Intelligence Development Plan" (2017)* (Aug. 1, 2017), https://digichina.stanford.edu/work/full-translation-chinas-new-generation-artificial-intelligence-development-plan-2017/

[1371] Zhang Zhihao, *Scientist: China to Focus on Innovation to Boost Economy, Lives*, China Daily (Sept. 1, 2018), https://www.chinadaily.com.cn/a/201801/09/WS5a543bd5a31008cf16da5fa9.html

[1372] Huw Roberts et al., *The Chinese Approach to Artificial Intelligence: An Analysis of Policy, Ethics, and Regulation*, AI and Society (Jun. 17, 2020), https://link.springer.com/article/10.1007/s00146-020-00992-2

[1373] OECD, *Beijing Drafts Regulations to Limit Use of A.I. in Health Care* (Aug. 25, 2023), https://oecd.ai/en/incidents/39033

[1374] State Council Information Office, *Full Text: Fighting COVID-19: China in Action*, Xinhua News via the State Council (June 7, 2020), https://archive.vn/NYJQg

efficiency in hospitals, and precautionary measures for returning to normal business operations.[1375]

Environmental Impact of AI

Policy initiatives in 2025 began to implement standards established in 2024 to steer industrial development toward more sustainable practices.[1376] The National Development and Reform Commission (NDRC), along with the Ministry of Industry and Information Technology (MIIT) 2024 Action Plan aims to reduce the average power usage effectiveness (PUE) of data centers to below 1.5 by 2025. The plan also seeks to increase the utilization rate of renewable energy in data centers by 10 percent annually.[1377] These measures include enhanced fiscal support for green technologies, streamlined foreign investment in strategic sectors, and a focus on digital transformation—all aimed at boosting energy efficiency and accelerating the transition to renewable energy sources in industries such as high-end manufacturing and data centers.[1378]

At the municipal level, the city of Shenzhen, known as the 'Chinese Silicon Valley' for its vibrant technology ecosystem, has implemented regulations to promote the development of green AI data centers. The city's government is tasked with improving the evaluation system for AI data centers and setting higher energy efficiency standards to build environmentally friendly facilities that support AI development.[1379]

The Ministry of Foreign Affairs 2024 Shanghai Declaration on Global AI Governance affirmed China's commitment to protecting the health of the environment while pursuing its National AI Strategy.[1380]

[1375] Emily Weinstain, *China's Use of AI in Its COVID-19 Response,* Center for Security and Emerging Technology (Aug. 2020), https://cset.georgetown.edu/research/chinas-use-of-ai-in-its-covid-19-response/

[1376] Cyberspace Administration of China, *Notice of Four Departments on Issuing the Guidelines for the Construction of the National Artificial Intelligence Industry Comprehensive Standardization System (2024 Edition)* [四部门关于印发国家人工智能产业综合标准化体系建设指南（2024版）的通知] (Jul. 3, 2024), https://www.cac.gov.cn/2024-07/03/c_1721686809220407.htm

[1377] State Council, *China Sets Green Targets for Data Centers* (Jul. 24, 2025), https://english.www.gov.cn/news/202407/24/content_WS66a0b167c6d0868f4e8e96ba.html

[1378] China Briefing, *The Policies Shaping China's Industry Landscape in 2025*, https://www.china-briefing.com/news/chinas-policy-driven-industrial-initiatives-2025/; see also, National Development and Reform Commission & National Energy Administration, *China Unveils AI–Energy Integration Plan to Spur Green, Low-Carbon Transition* (Sept. 8, 2025), https://english.www.gov.cn/news/202509/08/content_WS68be8c3ec6d0868f4e8f566d.html

[1379] Justice Bureau of Shenzhen, *527 Regulations of Shenzhen Special Economic Zone on Promoting the Artificial Intelligence Industry* (Aug. 30, 2022), https://sf.sz.gov.cn/fggzywyb/content/post_11216296.html

[1380] Chinese Ministry of Foreign Affairs, *Full Text: Shanghai Declaration on Global AI Governance* (Jul. 4, 2024), https://www.mfa.gov.cn/eng/xw/zyxw/202407/t20240704_11448351.html

Lethal Autonomous Weapons

China has pursued AI development in the military since 2017 while calling on countries to "exercise restraint"[1381] in developing AI weapons to ensure they comply with international humanitarian law. The AIDP states that "the development of AI [is] a major strategy to enhance national competitiveness and protect national security" and that China will "promote all kinds of AI technology to become quickly embedded in the field of national defense innovation."[1382] Military and defense leaders echoed these sentiments that AI in the military was inevitable both on the battlefield and in decision-making,[1383] and their commitment to ensuring China's military leads these innovations.[1384] These sentiments are shared by academics from the People's Liberation Army (PLA) who believe that AI will be used to predict battlefield situations and outpace human decision-making.[1385]

In 2018, China's Ministry of National Defense established two major research organizations focused on AI and unmanned systems: the Unmanned Systems Research Center (USRC) and the Artificial Intelligence Research Center (AIRC).[1386] According to some experts, China is pursuing the most aggressive strategy for developing AI for military uses among the major military powers.[1387]

In the official position on Lethal Autonomous Weapons, China maintains that both the common security and the dignity of mankind should be safeguarded through an international treaty. China's contribution to the UN Secretary General report on LAWS continued this emphasis and called for "an agreed understanding on key elements and technical characteristics" of LAWS to "formulate targeted measures and

---

[1381] Permanent Mission of the People's Republic of China to UN, *Statement by Chinese Delegation at the Thematic Debate on Conventional Weapons at the 80th Session of the UNGA First Committee* (Oct. 23, 2025), https://un.china-mission.gov.cn/eng/chinaandun/disarmament_armscontrol/202510/t20251024_11739691.htm

[1382] State Council, *China's Next Generation Artificial Intelligence Development Plan* [国务院关于印发 新一代人工智能发展规划的通知] (Jul. 8, 2017), https://www.gov.cn/zhengce/content/2017-07/20/content_5211996.htm

[1383] Rajeev Ranjan Chaturvedy, *Beijing Xiangshan Forum and the New Global Security Landscape*, East Asia Forum (Dec. 1, 2018), https://www.eastasiaforum.org/2018/12/01/beijing-xiangshan-forum-and-the-new-global-security-landscape/

[1384] Elsa Kania, *AlphaGo and Beyond: The Chinese Military Looks to Future 'Intelligentized' Warfare*, Lawfare (Jun. 5, 2017), https://www.lawfaremedia.org/article/alphago-and-beyond-chinese-military-looks-future-intelligentized-warfare

[1385] Elsa B. Kania, 杀手锏 and 跨越发展: *Trump Cards and Leapfrogging*, Strategy Bridge (Sept. 6, 2017), https://thestrategybridge.org/the-bridge/2017/9/5/-and-trump-cards-and-leapfrogging

[1386] Gregory C. Allen, *Understanding China's AI Strategy: Clues to Chinese Strategic Thinking on Artificial Intelligence and National Security*, Center for a New American Security (Feb. 6, 2019), https://www.cnas.org/publications/reports/understanding-chinas-ai-strategy

[1387] Adrian Pecotic, *Whoever Predicts the Future Will Win the AI Arms Race*, Foreign Policy (Mar. 5, 2019), https://foreignpolicy.com/2019/03/05/whoever-predicts-the-future-correctly-will-win-the-ai-arms-race-russia-china-united-states-artificial-intelligence-defense/

negotiate a legally binding international instrument" while avoiding "blanket bans or restrictions" that could "undermine States' legitimate defence capabilities" and "their right to the peaceful use of these technologies."[1388]

China recently stressed that the CCW is the best place for these discussions.[1389] China has consistently supported agree-upon definitions as the basis for an international treaty and the need to avoid an AI arms race in discussions at the UN,[1390] GGE on LAWS,[1391] and Convention on Certain Conventional Weapons (CCW). China's position that international negotiations for a treaty related to LAWS should happen in the CCW may account for the country's abstention from the widely accepted Resolution on Lethal Autonomous Weapons (L.56) at the 78th UN General Assembly.[1392]

China attended the first summit on Responsible AI in the Military Domain—REAIM 2023 hosted by the Netherlands. The summit provided a platform for all stakeholders (governments, industry, civil society, academia, and think tanks) to forge a common understanding of the opportunities, dilemmas, and vulnerabilities associated with military AI.[1393] China is one of the 57 endorsing countries that underlined the need to put the responsible use of AI higher on the political agenda and to further promote initiatives that make a contribution to this respect. However, China did not endorse the "Blueprint for AI Action" to govern responsible use of artificial intelligence in the military that was supported by 61 countries.[1394]

[1388] UN General Assembly, *Lethal Autonomous Weapons Systems: Report of the Secretary-General* (Jul. 1, 2024), https://documents.un.org/doc/undoc/gen/n24/154/32/pdf/n2415432.pdf

[1389] Permanent Mission of the People's Republic of China to UN, *Statement by Chinese Delegation at the Thematic Debate on Conventional Weapons at the 80th Session of the UNGA First Committee* (Oct. 23, 2025), https://un.china-mission.gov.cn/eng/chinaandun/disarmament_armscontrol/202510/t20251024_11739691.htm

[1390] Chinese Ambassador Shen Jian, Statement by H.E. Ambassador Shen Jian at the Thematic Debate on Conventional Weapons of the 78th Session of UNGA First Committee (Oct. 25, 2023), http://geneva.china-mission.gov.cn/eng/dbtxwx/202310/t20231025_11167472.htm; UN General Assembly, *Lethal Autonomous Weapons Systems: Report of the Secretary-General* (Jul. 1, 2024), https://documents.un.org/doc/undoc/gen/n24/154/32/pdf/n2415432.pdf; United Nations, *Fourteen New Drafts, Including on Implications of Artificial Intelligence in Military Domain, Approved in First Committee by 34 Votes* (Nov. 6, 2024), https://press.un.org/en/2024/gadis3757.doc.htm

[1391] China, *Working Paper of the People's Republic of China on Lethal Autonomous Weapons Systems* (Jul. 1, 2022), https://documents.unoda.org/wp-content/uploads/2022/07/Working-Paper-of-the-Peoples-Republic-of-China-on-Lethal-Autonomous-Weapons-Systems%EF%BC%88English%EF%BC%89.pdf

[1392] UN General Assembly, *Resolution L.56: Lethal Autonomous Weapons*, A/C.1/78/L.56 (Oct. 12, 2023), https://reachingcriticalwill.org/images/documents/Disarmament-fora/1com/1com23/resolutions/L56.pdf

[1393] REAIM 2023 *Endorsing Countries* (Feb. 15–16, 2023), https://www.government.nl/documents/publications/2023/02/16/reaim-2023-endorsing-countries

[1394] Reuters, *Sixty Countries Endorse 'Blueprint' for AI Use in Military; China Opts Out* (Sept. 10, 2024), https://www.reuters.com/technology/artificial-intelligence/south-korea-summit-announces-blueprint-using-ai-military-2024-09-10/

AI Literacy

The Beijing Municipal Education Commission announced in March 2025 that AI education will become mandatory across all levels of compulsory schooling, from elementary to high school, beginning Fall 2025. Schools will offer at least 8 hours of AI curriculum per year, with methods adapted for the grade level. Courses may be integrated with others subjects or stand alone. The plan responds to the rapid expansion of the AI industry in China and is an effort "to nurture future-oriented and innovative talent."[1395]

Human Rights

China has endorsed the Universal Declaration of Human Rights. As a party to the UDHR, China shall recognize "the inherent dignity" of all human beings and secure their fundamental rights to privacy. Privacy rights are guaranteed to Chinese citizens under the Constitution. However, Article 40 of the constitution justifies the invasion of privacy "to meet the needs of State security." Furthermore, the provisions of the Constitution are not directly enforceable since there is neither a constitutional court nor any possibility to assert constitutional rights.[1396] Relatedly, problematic exemptions for the collection and use of data, when it is related to security, health, or the flexibly interpretable "significant public interests"[1397] contribute to weak data protection in China.

Freedom House classifies China as Not Free with a total score of 9/100[1398] and a negative score of -2/40 for political rights. Freedom House Report cites increasing repression as the ruling Chinese Communist Party (CCP) tightens control over all aspects of life and governance, including the state bureaucracy, the media, online speech, religious practice, universities, businesses, and civil society associations. Human Rights Watch concurs with the assessment of government repression, citing mass detentions in Xinjiang, suppression of civil society, and the continued dismantling of freedoms in Hong Kong under the 2020 national security legislation.[1399] According to Decoding China, while the United Nations defines human rights as universal and intrinsic to all human beings, China argues that

---

[1395] State Council, *Beijing to Introduce AI Course across Primary, Secondary Schools*, China Daily (Mar. 12, 2025), https://english.www.gov.cn/news/202503/12/content_WS67d18e9ec6d0868f4e8f0c40.html

[1396] Graham Greenleaf, *Data Privacy Laws in Asia: Context and History*, p. 196 (2014); Anja Geller, *How Comprehensive Is Chinese Data Protection Law from a European Perspective?* GRUR International 69, no. 12, pp. 1191–1203 (Dec. 2020), https://academic.oup.com/grurint/article/69/12/1191/5909207

[1397] Samm Sacks, *New China Data Privacy Standard Looks More Far-Reaching than GDPR*, Center for Strategic and International Studies (2018), https://www.csis.org/analysis/new-china-data-privacy-standard-looks-more-far-reaching-gdpr

[1398] Freedom House, *Freedom in the World 2025: China* (2025), https://freedomhouse.org/country/china/freedom-world/2025

[1399] Human Rights Watch, *China and Tibet* (2025), https://www.hrw.org/asia/china-and-tibet

"sovereignty and non-interference trump the notion of universal human rights" and maintains rights are internal affairs rather than a matter of international oversight.[1400]

Global AI Governance

In October of 2023, China suggested the establishment of a Global AI Governance Initiative.[1401] In its announcement, the Ministry of Foreign Affairs called on countries to strengthen cooperation on the governance of AI. After stressing the "unpredictable risks and complicated challenges" posed by AI, the Ministry highlighted the need to "uphold a people-centered approach in developing AI," to "respect other countries' national sovereignty and strictly abide by their laws when providing them with AI products and services." The Ministry stated, "We oppose using AI technologies for the purposes of manipulating public opinion, spreading disinformation, intervening in other countries' internal affairs, social systems and social order, as well as jeopardizing the sovereignty of other states." The Ministry indirectly called out US restrictions regarding export controls on AI chips and chipmaking tools, stating: "We oppose drawing ideological lines or forming exclusive groups to obstruct other countries from developing AI. We also oppose creating barriers and disrupting the global AI supply chain through technological monopolies and unilateral coercive measures. [...] We should increase the representation and voice of developing countries in global AI governance, and ensure equal rights, equal opportunities, and equal rules for all countries in AI development and governance." [1402]

China hosted the World Artificial Intelligence Conference and High-Level Meetings on Global AI Governance in July 2025, where it released the Global AI Governance Action Plan (2025).[1403] The plan calls for global cooperation to ensure AI's "safety, reliability, controllability, and fairness," while aligning with the UN Pact for the Future and the Global Digital Compact. The plan advances 13 priority actions, including promoting sustainable AI, improving risk assessment and safety governance, strengthening technical standards, empowering the Global South through capacity building, and developing an inclusive multi-stakeholder governance model.

1400 Decoding China, *Human Rights* (2025), https://decodingchina.eu/key-term/human-rights/

1401 Chinese Ministry of Foreign Affairs, *Global AI Governance Initiative* (Oct. 20, 2023), https://www.mfa.gov.cn/eng/zy/gb/202405/t20240531_11367503.html

1402 See most recently, Reuters, *US Updates Export Curbs on AI Chips and Tools to China* (Mar. 30, 2024), https://www.reuters.com/technology/us-commerce-updates-export-curbs-ai-chips-china-2024-03-29/

1403 Ministry of Foreign Affairs of the People's Republic of China, *Global AI Governance Action Plan Released at the 2025 World Artificial Intelligence Conference* (Jul. 26, 2025), https://www.fmprc.gov.cn/mfa_eng/xw/zyxw/202507/t20250729_11679232.html

AI Safety Summit

In November 2023, China participated in the first AI Safety Summit and endorsed the Bletchley Declaration.[1404] Endorsing parties affirmed that for the good of all, AI should be designed, developed, deployed, and used, in a manner that is safe, in such a way as to be human-centric, trustworthy and responsible."

At the 2025 AI Safety Summit, rebranded as the AI Action Summit, President Xi Jinping's special representative, Vice Premier Zhang Guoqing, stated that China is prepared to work with other nations to promote AI development, safeguard security, and share technological achievements.[1405] He noted that AI is driving a new technological revolution and referenced the Global Initiative for AI Governance introduced in October 2023. Zhang called for international collaboration to develop AI for good, enhance innovation, and improve global governance. China was among the China was among the approximately 60 countries and international organizations to sign the outcome document, Statement on Inclusive and Sustainable Artificial Intelligence for People and the Planet. [1406]

OECD / G20 AI Principles

As a member of the G20, China is bound by the G20 AI Guidelines, which were drawn from the OECD Principles. [1407] China also adopted the G2O Ministerial Declaration in September 2024, reaffirming its commitment to ethical AI for inclusive sustainable development, aiming to drive growth, reduce inequality, and mitigate risks.[1408] China's Global Artificial Intelligence Governance Initiative[1409] and the

---

[1404] UK Department for Science, Innovation & Technology, Foreign, Commonwealth & Development Office, *Prime Minister's Office, The Bletchley Declaration by Countries Attending the AI Safety Summit* (Nov. 2023), https://www.gov.uk/government/publications/ai-safety-summit-2023-the-bletchley-declaration/the-bletchley-declaration-by-countries-attending-the-ai-safety-summit-1-2-november-2023

[1405] State Council, *China Willing to Promote AI Development with Other Countries: Chinese Vice Premier* (Feb. 12, 2025), https://english.www.gov.cn/news/202502/12/content_WS67abd3cbc6d0868f4e8ef92e.html

[1406] Élysée Palace (Government of the French Republic), *Statement on Inclusive and Sustainable Artificial Intelligence for People and the Planet* (Feb. 11, 2025), https://www.elysee.fr/en/emmanuel-macron/2025/02/11/statement-on-inclusive-and-sustainable-artificial-intelligence-for-people-and-the-planet

[1407] OECD, *Recommendation of the Council on Artificial Intelligence*, Adherents (May 3, 2024), https://legalinstruments.oecd.org/en/instruments/OECD-LEGAL-0449#adherents

[1408] G7G20 Documents Database, *G20 DEWG Maceio Ministerial Declaration* (Sept. 13, 2024), https://g7g20-documents.org/database/document/2024-g20-brazil-sherpa-track-digital-economy-ministers-ministers-language-g20-dewg-maceio-ministerial-declaration

[1409] Cyberspace Administration of China (CAC), *Global Artificial Intelligence Governance Initiative* [全球人工智能治理倡议] (Oct. 18, 2023), https://www.cac.gov.cn/2023-10/18/c_1699291032884978.htm

Global AI Governance Action Plan[1410] share several core themes with the OECD AI Principles, including safety, fairness, inclusiveness, and international cooperation.

Council of Europe AI Treaty

China neither signed the Council of Europe AI Treaty nor participated in its elaboration.[1411] China will be eligible to accede to the Framework Convention on Artificial Intelligence and Human Rights, Democracy, and the Rule of Law only after this first international treaty enters into force.[1412]

UNESCO Recommendation on AI Ethics

As Member of the UNESCO Ad Hoc Expert Group on AI Ethics, Zeng Yi, Director of the Sino-British Research Centre for AI Ethics and Governance, Institute of Automation, at the Chinese Academy of Sciences as well as Member of the National New Generation AI Governance Professional Committee, actively participated in the drafting and formulation of the UNESCO Recommendation on the Ethics of AI.[1413] According to Zeng Yi, "the Recommendation is the most widely reached consensus worldwide at the governmental level, which will become a benchmark for reference to the development of AI ethics related international standards and international law."

The value of "harmonious coexistence," as Zeng Yi pointed out in the formulation of the Recommendation, contributes to the harmonious development of human society coexisting with AI technologies as well as the empowerment of human development, society, and ecology by AI.[1414] "Harmonious coexistence" is reflected in China's New Generation Artificial Intelligence Governance Principles[1415] (Governance Principles) and New Generation Artificial Intelligence Code of Ethics[1416] (Code of Ethics) formulated by the National New Generation AI

1410 Ministry of Foreign Affairs, *Global AI Governance Action Plan* (Jul. 26, 2025), https://www.mfa.gov.cn/eng/xw/zyxw/202507/t20250729_11679232.html

1411 Council of Europe, *Chart of Signatures and Ratifications of Treaty 225* (Dec. 19, 2025), https://www.coe.int/en/web/conventions/full-list?module=signatures-by-treaty&treatynum=225

1412 Council of Europe, *Framework Convention on Artificial Intelligence* (2025), https://www.coe.int/en/web/artificial-intelligence/the-framework-convention-on-artificial-intelligence

1413 Chinese Academy of Science, *Artificial Intelligence Ethics Is on a New Journey towards a Global Consensus*, [人工智能伦理迈向全球共识新征程], Chinese Journal of Science and Technology (Dec. 2, 2021), https://www.cas.cn/cm/202112/t20211203_4817066.shtml

1414 Ibid

1415 Ministry of Science and Technology, *Developing Responsible AI: Next Generation AI Governance Principles Released* (Jun. 17, 2019), https://www.most.gov.cn/kjbgz/201906/t20190617_147107.html

1416 Ministry of Science and Technology, *New Generation Artificial Intelligence Development Code of Ethics* ["新一代人工智能伦理规范"发布] (Sept. 26, 2021), https://perma.cc/RC4V-Q2FX; Center for Security and Emerging Technology, *Ethical Norms for New Generation Artificial Intelligence*

Governance Professional Committee, which shows similar visions with the Recommendation.

Another shared vision concerns sustainable development. Zhang Yi's proposal for the sustainable development of AI was accepted by the Recommendation in the policy section, which provided suggestions for achieving sustainability through deploying AI in the field of environment. At the national level, both the Governance Principles and the Code of Ethics issued by China consider sustainable development as the overall vision and objective for the New Generation Artificial Intelligence.[1417]

Since the publication of the UNESCO Recommendation on AI Ethics, Chinese mainstream news media, think tanks, research institutions, and organizations actively interpreted, reposted, and communicated the Recommendation to the wider public in China. China has not implemented nor is in the process of implementing the UNESCO's Readiness Assessment Methodology (RAM).[1418]

Evaluation

China is one of the first AI superpowers implementing the ambitious plan of leading the world in AI by 2030. The country has endorsed principles on AI ethics, enforced a new law on data protection, and adopted a comprehensive set of regulations on AI. However, concerns exist that the main objective of China's legislative frenzy is not to protect human rights but to make sure that the government controls and monitors the digital world as much as it does the physical world. The validation of the use of facial recognition for public security purposes in the data protection law and the enforcement mechanisms and sanction regimes to regulate AI are some cases in point. The use of AI technology and facial recognition against minorities and dissidents is also of concern.

China has participated in international efforts to create an ethical AI framework, including representatives contributing to the UNESCO Recommendation on the Ethics of AI and forming the Global AI Governance Initiative. These efforts suggest a willingness to establish some common understandings in international AI governance or, absent meaningful action to implement international frameworks, signal China's international status as an AI superpower.

Released [English translation] (Oct. 12, 2021), https://cset.georgetown.edu/wp-content/uploads/t0400_AI_ethical_norms_EN.pdf

[1417] Chinese Academy of Science, *Artificial Intelligence Ethics is on a New Journey towards a Global Consensus,* Chinese Journal of Science and Technology (Dec. 2, 2021), https://www.cas.cn/cm/202112/t20211203_4817066.shtml

[1418] UNESCO Global AI Ethics and Governance Observatory, *Global Hub* (Oct. 2025), https://www.unesco.org/ethics-ai/en/global-hub

Colombia

In 2025, Colombia approved the National AI Policy to foster ethical and sustainable use and development of AI. The country also became the first country to formally adopt the UNESCO Guidelines for the Use of AI in the Judicial System.

National AI Strategy

"In an increasingly digital world, artificial intelligence is presented as a fundamental tool that can positively shape the future of our nation. However, we recognize that this power must be guided by solid ethical principles and a strategic vision that guarantees the well-being of all Colombians" said the Minister of Science, Technology, and Innovation Yesenia Olaya, during the launch of the AI Roadmap to ensure the ethical and sustainable adoption of AI in Colombia in February 2024.[1419]

The Roadmap centers ethical and sustainable adoption and identifies challenges to that objective. Ethics and Governance, including a dedicated section on algorithmic transparency, is one of five key themes.[1420] The goal is to establish and strengthen the legal and regulatory AI framework to tackle ethical challenges posed by AI. The other strategic centers are Education, research, and innovation; Data and Organizations; Innovative and emerging industries; Privacy, cybersecurity, and defense. The AI Roadmap not only promotes scientific integrity in the development of technological solutions but also seeks to establish effective governance structures that ensure its ethical application of artificial intelligence in solving social, economic, and environmental challenges at the national level.

The Roadmap serves as the basis for the National AI Policy by the National Council of Economic and Social Policy (*Consejo Nacional de Política Económica y Social, CONPES)* approved in 2025.[1421] The National AI Policy aims to "foster capacity in research, development, adoption, and the ethical and sustainable use of AI systems to promote social and economic transformation in Colombia."[1422] The strategy outlines 106 actionable measures across six specific objectives: 1) Strengthen

---

[1419] Ministerio de Ciencia, Tecnología e Innovacion (MinCiencia), *Colombia ya cuenta con una Hoja de Ruta en Inteligencia Artificial* (Feb. 12, 2024), https://minciencias.gov.co/sala_de_prensa/colombia-ya-cuenta-con-una-hoja-ruta-en-inteligencia-artificial

[1420] MinCiencia, *Roadmap for the Development and Application of Artificial Intelligence in Colombia [Hoja de Ruta para el Desarrollo y Aplicación de la Inteligencia Artificial en Colombia]*, p. 16 (Feb. 2024), https://minciencias.gov.co/sites/default/files/upload/noticias/hoja_de_ruta_adopcion_etica_y_sostenible_de_inteligencia_artificial_colombia_0.pdf

[1421] Consejo Nacional de Política Económica y Social (CONPES), *Document 4144: National Artificial Intelligence Policy [Documento 4144: Política Nacional de Inteligencia Artificial]*, approved version (Feb. 14, 2025), https://colaboracion.dnp.gov.co/CDT/Conpes/Económicos/4144.pdf

[1422] Ibid, p. 3

governance and application of ethical principles; 2) Strengthen ethics and governance, data infrastructure, research and innovation, talent development, risk mitigation, and sectoral AI adoption. The policy proposes regulatory sandboxes and ethical frameworks to safely test AI technologies in real-world contexts and identifies work in progress under the executive branch to develop a legal framework for governing AI.[1423] Training and mentorship programs aimed at increasing women's participation in AI policymaking are intended to alleviate the gender gap in technology.

The national policy builds on earlier actions and policies across sectors and strategic work areas to create the conditions to support ethical and sustainable AI.[1424] These initiatives began with the National Policy for Digital Transformation and AI[1425] in 2019, led by the National Planning Department (*Departamento Nacional de Planeación*, DNP), the Ministry of Information and Communications Technologies (*Ministerio de Tecnologías de la Información y las Comunicaciones*, MinTIC), and the Office of the President. The Strategy introduced social and economic conditions that facilitate the development of AI by creating a flexible framework of principles and guidelines rather than a rule-based structure. These consultations helped define goals of the National Development Plan 2022–2026 "Colombia World Power of Life," adopted in May 2023[1426] and the National Digital Strategy 2023–2026, issued in February 2024.[1427]

Colombia's Ethical Framework for AI,[1428] according to President Iván Duque, provides "tools that strengthen the principle of democracy, free competition and equity."[1429] The principles in the Framework—transparency, explainability, privacy,

---

[1423] Ibid, p. 83

[1424] Ibid, pp. 17–34

[1425] Ministerio de Tecnologías de la Información y las Comunicaciones (MinTIC), Departamento Nacional de Planeación y la Presidencia, *National Policy for Digital Transformation and Artificial Intelligence (CONPES No. 3975)* (Nov. 8, 2019), https://colaboracion.dnp.gov.co/CDT/Conpes/Econ%C3%B3micos/3975.pdf

[1426] Luis Eudoro Vammejo Zamudio, *The National Development Plan 2022–2026: Columbia, World Power of Life*, Editorial, Apuntes del Cenes (Nov. 20, 2023), https://doi.org/10.19053/01203053.v42.n76.2023.16467

[1427] Departamento Nacional de Planeación, *The Government of Change Presents the National Digital Strategy 2023–2026 [El Gobierno del Cambio presenta la Estrategia Nacional Digital 2023–2026]* (Feb. 7, 2024), https://www.dnp.gov.co/Prensa_/Noticias/Paginas/gobierno-del-cambio-presenta-estrategia-nacional-digital-2023-2026.aspx

[1428] Ministerio de Tecnologías de la Información y las Comunicaciones (MinTIC), *Colombia Early Adopter in the Region of the UNESCO Recommendations on the Ethics of Artificial Intelligence [Colombia adopta de forma temprana recomendaciones de ética en Inteligencia Artificial de la Unesco para la region]* (Mar. 9, 2022), https://mintic.gov.co/portal/inicio/Sala-de-prensa/Noticias/208109:Colombia-adopta-de-forma-temprana-recomendaciones-de-etica-en-Inteligencia-Artificial-de-la-Unesco-para-la-region

[1429] Diario La Libertad, *With the Ethical Framework for Artificial Intelligence, Colombia Is at the Forefront in Latin America: Duque,* Diario la Libertad (Nov. 25, 2020), https://diariolalibertad.com/sitio/2020/11/25/con-el-marco-etico-de-inteligencia-artificial-colombia-se-pone-a-la-vanguardia-en-america-latina-duque/

human control over decisions, security, responsibility, non-discrimination, and inclusion—also underline the National AI Policy.[1430]

The National AI Policy's first action item details that the MinTIC; Ministry of Science, Technology, and Innovation (*Ministerio de Ciencia, Tecnología e Innovación*, MinCiencias); and National Planning Department to coordinate with the Administrative Department of the Presidency (*Departamento Administrativo de la Presidencia de la República*, DAPRE) "will define and implement the governance model for Artificial Intelligence (AI) and then formalize it through a normative instrument."[1431] A bill for governing AI submitted to the Senate[1432] by the MinTIC and MinCiencia in May 2025 names the MinCiencias the National Authority for Artificial Intelligence, charged with implementing the law. MinCiencias would also be in charge of overseeing and enforcing the standards and mechanisms in coordination with MinTIC and other entities. A draft law by representatives in the House (*Cámara de Representantes*) was submitted in July.[1433]

Ethical Framework for Artificial Intelligence

In Colombia, the aim of the Ethical Framework for AI is to address concerns arising from the implementation of emerging technologies such as AI.[1434] The Ethical Framework for AI was developed around ethical principles that can serve as a criterion for evaluating the different uses and challenges that arise in this respect. The Ethical Framework for AI was developed as a tool that can be applied to different sectors taking into consideration the diversity of interests and opinions around the use of AI. The ten principles provided by the framework to guide the design, development, implementation and evaluation of AI systems include transparency, explainability, privacy, human oversight over AI decisions, security, responsibility, non-

---

[1430] Consejo Nacional de Política Económica y Social, *Document 4144: National Artificial Intelligence Policy [Documento 4144: Política Nacional de Inteligencia Artificial]]*, approved version (Feb. 14, 2025), https://colaboracion.dnp.gov.co/CDT/Conpes/Económicos/4144.pdf
[1431] Ibid, p. 83
[1432] Ministerio de la Ciencia, Tecnología e Innovación y Ministerio de Tecnologías de la Información y las Comunicaciones (MinTIC), *Proyecto de Ley "Por medio del cual se regula la inteligencia artificial en Colombia para garantizar su desarrollo ético y responsable y se dictan otras disposiciones"]*, pp. 70–73 (May 7, 2025), https://minciencias.gov.co/sites/default/files/upload/noticias/pl_ia_finalizado.pdf
[1433] Cámara de Representantes, *Bill for Artificial Intelligence Law for Psychosocial Protection and Digital Equity [Proyecto de Ley: Ley de Inteligencia Artificial para la protección psicosocial y la equidad digital]* (Jul. 20 2025), https://www.camara.gov.co/wp-content/uploads/2025/10/proyectos-ley/documentos/proyecto-33950/p_1_098_2025sc_marco_regulatorio_de_inteligencia_artificial-f6c87c61.pdf
[1434] The Development Bank of Latin America (CAF), *Why Has Colombia Positioned Itself as a Regional Leader on Artificial Intelligence* (Sept. 14, 2021), https://www.caf.com/es/conocimiento/visiones/2021/09/por-que-colombia-se-ha-posicionado-como-lider-regional-en-inteligencia-artificial/

discrimination, inclusion, prevalence of the rights of children and adolescents, and social benefit.[1435]

The framework also proposes an ethical algorithm register in which entities periodically report what their AI project is about, how they are implementing the AI ethics principles, and the ethical risks to the use of AI in their project. The register allows for the monitoring of progress in the implementation of AI principles and reinforces citizen participation by inviting their comments or questions on policies, good practices and projects related to AI.[1436] The DAPRE and the Superintendency of Industry and Commerce (*Superintendencia de Industria y Comercio*, SIC) are responsible for monitoring AI projects. The Superintendency of Industry and Commerce monitors AI projects by carrying out the Data privacy regulatory Sandbox.[1437] The ethical algorithm register of Colombia is based on the models of Amsterdam and Helsinki.[1438]

Regulatory Sandboxes and Beaches

Colombia has adopted a smart regulation approach to AI through regulatory sandboxes and beaches. This controlled environment was set up to experiment and test AI systems in the local context, to identify technical and governance flaws while promoting innovation.[1439]

At first, a Draft Model Concept for the Design of Regulatory Sandboxes and Beaches in AI was published in August 2020 with public comments received from various stakeholders.[1440] The purpose of this policy was to understand technology before trying to regulate it, by balancing precaution with experimentation and learning. The document suggests a process of implementation that includes (1)

[1435] Presidency of the Republic of Colombia, *Ethical Framework for Artificial Intelligence in Colombia* [English draft version] (Aug. 2020), https://cyber.harvard.edu/sites/default/files/2020-12/Colombia_AI_Ethical_Framework.pdf

[1436] Presidencia, *Ethical Framework for Artificial Intelligence in Colombia [Marco ético para la Inteligencia Artificial en Colombia]* (May 2021), https://minciencias.gov.co/sites/default/files/marco-etico-ia-colombia-2021.pdf

[1437] Superintendencia de Industria y Comercio, *Sandbox on Privacy by Design and Default in Artificial Intelligence Projects [Sandbox sobre privacidad desde el diseño y por defecto en proyecto de inteligencia artificial]*, https://www.sic.gov.co/sandbox-microsite

[1438] Center for Technology and Society Studies (CETyS) of the University of San Andrés, *The Colombian Case: Adopting collaborative governance as a path for implementing ethical artificial intelligence* (2021), https://repositorio.udesa.edu.ar/items/c2d4a6b9-0268-4d44-b348-6c84018b8c79

[1439] Global Policy Journal, *The Colombia Case: A New Path for Developing Countries Addressing the Risks of Artificial Intelligence* (May 2021), https://www.globalpolicyjournal.com/sites/default/files/pdf/Mu%C3%B1oz,%20Uribe%20and%20Espa%C3%B1ol%20-%20The%20Colombian%20Case,%20A%20New%20Path%20for%20Developing%20Countries%20Addressing%20the%20Risks%20of%20Artificial%20intelligence_0.pdf

[1440] Armando Guío Español, *Model Concept for the Design of Regulatory Sandboxes & Beaches in AI* (Aug. 2020), https://dapre.presidencia.gov.co/AtencionCiudadana/DocumentosConsulta/consulta-200820-MODELO-CONCEPTUAL-DISENO-REGULATORY-SANDBOXES-BEACHES-IA.pdf

defining a policy leadership to implement public policy; (2) defining emerging technologies and preliminary problems to be addressed; (3) completing a regulatory mapping of the impacted sector; (4) selecting a public entity to perform inspection and surveillance functions; (5) capacity building and training; (6) creating working groups; (7) designing a risk model and defining possible risks; (8) setting out a selection criteria for the risks; (9) designing the sandbox; (10) sharing the project for comment; (11) publishing and implementing; and (12) reporting on the findings and evidence.[1441]

A policy proposing a regulatory sandbox for Privacy and AI was open for public comments by the DAPRE and the Superintendency of Industry and Commerce until November 30 2021Superintendency. The regulatory sandbox here is meant to be preventive, so AI systems related to e-commerce, advertising and marketing protect personal data from the stage of design to execution, using tools like privacy impact assessments and privacy by design. It proposes (1) criteria to ensure compliance with regulation on data processing in AI; (2) proper processing of personal data in all stages of an AI project; (3) creation of AI products that respects individual rights to personal data; (4) advice to companies on the protection of personal data in AI systems; (5) adoption of a preventive approach to protect human rights in AI projects; (6) suggestion of amendments or modifications to Colombian regulations on technological advances.[1442]

The Superintendency of Industry and Commerce released a final document governing privacy by design and by default in AI projects in April 2021 and the government approved the first proposals under the sandbox program in August 2021.[1443] The Superintendency of Industry and Commerce's sandbox for Privacy and AI started operating during 2021, and on January 2022, selected a first project.[1444]

Public Participation

The Development Bank of Latin America (CAF) with the authorship of international expert, Armando Guío Español, has developed AI policy and implementation documents, including the ethical framework for AI, a model concept for the design of regulatory sandboxes and beaches in AI, a data governance model, a task force on the development and implementation of AI, and the outline of an international council for the implementation of AI policy.[1445]

Draft AI policies and legislations of Colombia have been opened for public comment from academia, national, regional and international civil society actors,

---

[1441] Ibid

[1442] Superintendencia de Industria y Comercio, *Sandbox on Privacy by Design and by Default in Artificial Intelligence Projects* (Apr. 2021), https://www.sic.gov.co/sandbox-microsite

[1443] Ibid

[1444] Ibid

[1445] Development Bank of Latin America, *AI Experience: Data and Artificial Intelligence in the Public Sector* (Sept. 15, 2021), https://www.caf.com/en/currently/events/ai-experience/

intergovernmental organizations and the private sector. These consultations have taken on various forms.[1446] Previously, the Ministry of Science, Technology and Innovation (MinCiencias) conducted a public consultation on the Roadmap to ensure the ethical and sustainable adoption of AI in Colombia[1447] and plans on organizing subsequent consultations to flesh it out further.[1448]

The MinCiencias expanded the participatory framework through the initiative Artificial Intelligence, Education and Social Sciences: A Strategic Dialogue in CLACSO.[1449] The initiative aims to foster Colombia's commitment to inclusive AI governance by integrating social-science perspectives into public policy design and ensuring that marginalized voices are represented in AI-related consultations. The program also encourages collaboration between research institutions and public authorities to enhance digital literacy and critical understanding of algorithmic decision-making. This program brought together representatives from academia, policymakers, and civil society to examine the ethical, social, and cultural implications of AI adoption in Latin America.

[1446] See for example, Berkman Klein Center for Internet and Society at Harvard University, *Summary Report of Expert Roundtable on Colombia's Draft AI Ethical Framework* (Jan. 2021), https://cyber.harvard.edu/sites/default/files/2021-01/Colombia_Roundtable_Report.pdf

[1447] Ministerio de Ciencia, Tecnologia e Innovacion (MinCiencias), *Roadmap for the Development and Application of Artificial Intelligence in Colombia [Hoja de ruta para el desarrollo y aplicación de la Inteligencia Artificial en Colombia]* (Feb. 2024), https://inteligenciaartificial.minciencias.gov.co/wp-content/uploads/2024/02/Hoja-de-Ruta-Adopcion-Etica-y-Sostenible-de-Inteligencia-Artificial-Colombia-1.pdf

[1448] Ministerio de Ciencia, Tecnologia e Innovacion (MinCiencias), *Colombia ya cuenta con una Hoja de Ruta en Inteligencia Artificial* (Feb. 12, 2024), https://minciencias.gov.co/sala_de_prensa/colombia-ya-cuenta-con-una-hoja-ruta-en-inteligencia-artificial.

[1449] Ministerio de Ciencia, Tecnologia e Innovacion (MinCiencias), *Artificial Intelligence, Education, and Social Sciences: A Strategic Dialogue in CLACSO [Inteligencia Artificial, Educación y Ciencias Sociales: Un Diálogo Estratégico en CLACSO]* (Jun. 11, 2025), https://inteligenciaartificial.minciencias.gov.co/inteligencia-artificial-educacion-y-ciencias-sociales-un-dialogo-estrategico-en-clacso/

Data Protection

Data Protection in Colombia is governed by Article 15 of the Constitution.[1450] Colombia also regulates financial credit, commercial and services information,[1451] and personal data processing and databases.[1452]

For the implementation and monitoring of regulations regarding privacy, the Personal Data Authority (DPA) was established under the Division of Data Protection of the Superintendency of Industry and Commerce (SIC), according to Article 19 of Law 1581 of 2012.[1453] This authority functions as an oversight body, providing instructions and setting mandates, along with receiving complaints on the handling of data. The National Government introduced a bill in late 2025 to modernize Colombia's data protection framework.[1454] The proposal updates Law 1581 of 2012 to address data governance in artificial intelligence, algorithmic profiling, and cross-border data transfers. The bill expands the powers of the *S*uperintendency of Industry and Commerce to supervise automated decision-making systems and introduces mandatory AI Data Impact Assessments. Law 1581 aligns national law with international standards such as the OECD principles and UNESCO recommendation.

On automated decision-making, Law 1581 establishes that personal data processing must be for a legitimate purpose under the Constitution and other laws, it must be notified to the subject, and the purpose must be specific. As a result, if automated decision-making is the purpose of processing data, then it must be (1) legitimate as per the Constitution and other laws of Colombia, (2) specific in purpose, and (3) data subject should be informed of it.[1455]

Secondary decrees, decisions, and regulations provide a better understanding of data protection, particularly regarding the application of automated mechanisms to

---

1450 Congreso de la República, *Colombia's Constitution of 1991 with Amendments through 2015* (subsequently amended), https://www.constituteproject.org/constitution/Colombia_2015.pdf?lang=en

1451 Congreso de la República, *Law 1266 of 2008 on the processing of financial data, credit records, and commercial information collected in Colombia or abroad [Ley Estatutaria 1266 de 2008 por la cual de dictan las disposiciones generales del hábeas data y se regula el manejo de la información contenida en bases de daots personales, en especial la financiera, crediticia, comercial, de servicios y la proveniente de terceros países y se dictan otras disposiciones]* (Dec. 15, 2025), http://www.secretariasenado.gov.co/senado/basedoc/ley_1266_2008.html

1452 Congreso de la República, *Law 1581 of 2012 on the Protection of Personal Data* (Oct. 17, 2012), https://www.funcionpublica.gov.co/eva/gestornormativo/norma.php?i=49981

1453 Congreso de la República, *Law 1266 of 2008 on the processing of financial data, credit records, and commercial information collected in Colombia or abroad* (Dec. 15, 2025), http://www.secretariasenado.gov.co/senado/basedoc/ley_1266_2008.html

1454 Ministerio de Ciencia, Tecnología e Innovación (MinCiencias), *National Government Submitted a Bill to Update the Current Regulations on Personal Data Protection in Colombia [Gobierno Nacional radico proyecto de ley para la protección de datos personales en Colombia]* (Aug. 27, 2025), https://inteligenciaartificial.minciencias.gov.co/gobierno-nacional-radico-proyecto-de-ley-para-actualizar-la-normatividad-vigente-en-la-proteccion-de-datos-personales-en-colombia/

1455 Dejusticia, *Accountability of Google and Other Businesses in Colombia: Personal Data Protection in the Digital Age* (2019), https://www.dejusticia.org/wp-content/uploads/2019/01/Accountability-of-Google-and-other-Businesses-in-Colombia.pdf

databases. Databases have been defined under Article 3 of Law 1581 as an organized set of personal data which is treated in the same as personal data. Decree 886 that regulates Article 25 of Law 1581 explains this further by stating that, when automation is applied to databases containing personal data, it should be registered in the public directory of databases called the National Register of Databases.[1456] Drawing from the necessity to register information from databases and the protection of the right to habeas data,[1457] the Constitutional Court concluded that the administrator of a database has specific obligations regarding the quality of data being transmitted and allows data subjects to authorize how their information in an automated system is handled.[1458] Additionally, Article 26 of Decree 1377 establishes the principle of proven liability, according to which those responsible for handling personal data have an obligation to prove that they have taken sufficient and effective measures to abide by regulations, even when the data is processed by an automated method.[1459]

In July 2020, through Resolution 38281, the Superintendency of Industry and Commerce concluded that Law 1581 is thematically and technologically neutral.[1460] Thus, the provisions of Law 1581 apply to the processing of any data regardless of techniques or technologies used. The protection of personal data extends to all techniques and tools, including AI in its use for predictive dialing, robocalls and nuisance calls. Nelson Remolina, the Superintendent for the Protection of Personal Data, elaborated on this by stating that, while Colombian law allows for the creation, design, and use of technological innovations to process data, it must be done in a way that respects the legal system by complying with all the rules pertaining to the processing of personal data.[1461]

Colombian data protection law differs in scope from the GDPR, since it only applies to data processing carried out by data processors and data controllers within

---

[1456] Función Pública, *Decree 886 of 2014 that regulates Article 25 of Law 1581 of 2021, related to the National Register of Databases [Por el cual se reglamenta el artículo 25 de la Ley 1581 de 2021, relativo al Registro Nacional de Bases de Datos]* (May 13, 2014), https://www.suin-juriscol.gov.co/viewDocument.asp?id=1184150

[1457] *Habeas data* is a fundamental right and tool to provide legal protection to owners of personal data, particularly when faced with undue or illegal processing of their personal data by databases, or public or private registries.

[1458] Constitutional Court of Colombia, *Sentence C-1011/08: Habeas Data in Statutory Law and the Handling of Information Contained in Personal Databases* (2008), https://www.corteconstitucional.gov.co/relatoria/2008/C-1011-08.htm

[1459] Presidencia, *Decree 1377 of 2013* (2013) http://www.suin-juriscol.gov.co/viewDocument.asp?ruta=Decretos/1276081

[1460] Superintendencia de Industria y Comercio, *Resolution 38281* (Jul. 14, 2020), https://www.sic.gov.co/sites/default/files/files/Proteccion_Datos/Res%2038281%20del%2014VII2020%20Mervicol%20marcadores%20predictivos%20robocalls%20IA.pdf

[1461] Ibero-American Data Protection Network (RIPD), *Colombian Data Protection Authority Concluded that Predictive Dialing, Robocalls and Artificial Intelligence Must Comply with Regulation Regarding the Processing of Personal Data* (Jul. 20, 2020), https://www.redipd.org/en/news/colombian-data-protection-authority-concluded-predictive-dialing-robocalls-and-artificial

the country or to those who have a legal obligation under international law and treaties. Unlike the GDPR, the Colombian privacy law does not set out conditions under which data profiling is allowed. However, Colombian privacy law protects individual privacy through habeas data rights, guaranteed by Article 15 of the Colombian Constitution. Article 15 is intended to protect an individual's right to know, update, and rectify information gathered about them in online files or databases.[1462]

In January 2022, President Duque issued Decree 092, modifying the structure of the Superintendency of Industry and Commerce. The modifications include the creation of the Habeas Data Department, under the Deputy Superintendency for Personal Data Protection. The Decree tasks the Habeas Data Department with ensuring that all entities covered by the data protection regime comply with data protection law. The Habeas Data Department may also resolve any complaint or claim submitted by data subjects seeking to enforce their data rights. The Department may issue orders to enforce its statutory mandates.[1463]

The Superintendency of Industry and Commerce, as a member of the Ibero-American Network for the Protection of Personal Data (*Red Iberoamericano de Protección de Datos*, RIPD), which comprises 16 data protection authorities from 12 countries, also endorsed the General Recommendations for the Processing of Personal Data in Artificial Intelligence[1464] and the accompanying Specific Guidelines for Compliance with the Principles and Rights that Govern the Protection of Personal Data in Artificial Intelligence Projects.[1465] Both have been framed in accordance with the RIPD Standards for Personal Data Protection for Ibero-American States.[1466] The adoption of the Standards signals Colombia's recognition of guiding principles and rights for the protection of personal data that RIPD members can adopted and developed in their national legislation to guarantee a proper treatment of personal data facilitate homogeneous rules in the region. The guiding principles of personal data protection are: legitimation, lawfulness, loyalty, transparency, purpose, proportionality, quality, responsibility, safety and confidentiality. Controllers must also guarantee the exercise of the following rights by data subjects: right of access, right to correction, right to cancellation, right to opposition, right not to be subject to

---

[1462] Privacy International, *State of Privacy Colombia* (Jan. 2019), https://privacyinternational.org/state-privacy/58/state-privacy-colombia

[1463] Government of Colombia, *Decreto 092 de 2022*, https://www.funcionpublica.gov.co/eva/gestornormativo/norma.php?i=176826

[1464] Red Iberoamericana de Protección de Datos (RIPD), *General Recommendations for the Processing of Personal Data in Artificial Intelligence* (Jun. 2019), https://www.redipd.org/documentos/guia-recomendaciones-generales-tratamiento-datos-ia

[1465] Red Iberoamericana de Protección de Datos (RIPD), *Specific Guidelines for Compliance with the Principles and Rights that Govern the Protection of Personal Data in Artificial Intelligence Projects* (Jun. 2019), https://www.redipd.org/documentos/guia-orientaciones-especificas-proteccion-datos-ia

[1466] Red Iberoamericana de Protección de Datos (RIPD), *Standards for Personal Data Protection for Ibero-American States* (2017), https://www.redipd.org/sites/default/files/2022-04/standars-for-personal-data.pdf

automated individual decisions, right to portability of personal data and right to the limitation of treatment of personal data.

The RIPD data protection authorities initiated a coordinated action regarding ChatGPT, developed by OpenAI, in May 2023 over concerns regarding the risk of misinformation and harmful biases in responses.[1467] Colombia's Superintendency of Industry and Commerce also launched an independent investigation in May 2023 to assess whether the application ChatGPT complies with Colombian data protection law.[1468]

The Superintendency of Industry and Commerce sponsored the 2020 GPA Resolution on Accountability in the Development and Use of Artificial Intelligence[1469] and co-sponsored the 2022 GPA Resolution on Principles and Expectations for the Appropriate Use of Personal Information in Facial Recognition Technology.[1470] However, the organization did not co-sponsor the 2020 GPA Declaration on Ethics and Data Protection in AI[1471] or the 2023 GPA Resolution on Generative AI.[1472]

Algorithmic Transparency

Colombian data protection law for partial algorithmic transparency with the requirement that the data subjects know when their data is processed automatically. The right to habeas data requires the administrator of a database to ensure the quality of data being transmitted and allows data subjects to authorize how their information

---

[1467] Red Iberoamericana de Protección de Datos (RIPD), *RIPD Authorities Initiative a Coordinated Action in Relation to the ChatGPT Service [Las autoridades de la Red Iberoamericana de Protección de Datos Personales inician una acción coordinada en relación con el servicio ChatGPT]* (May 8, 2023), https://www.redipd.org/noticias/autoridades-red-iberoamericana-de-proteccion-de-datos-personales-inician-accion-chatgpt

[1468] Superintendencia de Industria y Comercio, *Sic Puts The Spotlight On The Application "Chat-Gpt" To Determine If It Complies With The Regulation Of Personal Data Protection* (Jun. 9, 2023), https://www.sic.gov.co/noticias/sic-puts-spotlight-application-chat-gpt-determine-if-it-complies-regulation-personal-data-protection.

[1469] Global Privacy Assembly, *Resolution on Accountability in the Development and Use of Artificial Intelligence* (Oct. 2020), https://globalprivacyassembly.org/wp-content/uploads/2020/10/FINAL-GPA-Resolution-on-Accountability-in-the-Development-and-Use-of-AI-EN-1.pdf

[1470] Global Privacy Assembly, *Resolution on Accountability in the Development and Use of Artificial Intelligence* (Oct. 2022), https://globalprivacyassembly.org/wp-content/uploads/2022/11/15.1.c.Resolution-on-Principles-and-Expectations-for-the-Appropriate-Use-of-Personal-Information-in-Facial-Recognition-Technolog.pdf

[1471] Global Privacy Assembly, *Declaration on Ethics and Data Protection in Artificial Intelligence* (Oct. 20), http://globalprivacyassembly.org/wp-content/uploads/2018/10/20180922_ICDPPC-40th_AI-Declaration_ADOPTED.pdf

[1472] Global Privacy Assembly, *Resolution on Generative Artificial Intelligence Systems* (Oct. 2023), https://globalprivacyassembly.org/wp-content/uploads/2023/10/5.-Resolution-on-Generative-AI-Systems-101023.pdf

in an automated system is handled.[1473] The 2022 Ethical Framework for AI[1474] provides for both transparency and explainability. The 2024 AI Roadmap also refers to algorithmic transparency.[1475]

The RIPD Specific Guidelines offers some clarity on the expectations for explainability as related to the transparency principle: "The information provided regarding the logic of the AI model must include at least basic aspects of its operation, as well as the weighting and correlation of the data, written in a clear, simple and easily understood language, it will not be necessary to provide a complete explanation of the algorithms used or even to include them. The above always looking not to affect the user experience."[1476]

The Colombian Constitutional Court made a landmark ruling emphasizing the need for human oversight in AI use within the judiciary. The ruling cited UNESCO's AI tools and guidelines, highlighting the importance of transparency, privacy, and human judgment in judicial processes.[1477] The court directed Colombia's Consejo de la Judicatura to develop comprehensive guidelines on AI use, aligning with best practices to ensure ethical application that respects human rights.

Data Infrastructure

Colombia facilitates data access for those designing and developing AI systems, achieved by removing unnecessary and unjustified barriers to access information. To facilitate such data access and to generate social and economic well-being, the Colombian government has developed data infrastructure policies with a shared dynamic and standardized resources across different actors. Thus, data infrastructure is used to strengthen institutional capacity to provide better quality

---

[1473] Constitutional Court of Colombia, *Sentence C-1011/08: Habeas Data in statutory law and the handling of information contained in personal databases* (2008), https://www.corteconstitucional.gov.co/relatoria/2008/C-1011-08.htm

[1474] Ministerio de Tecnologías de la Información y las Comunicaciones (MinTIC), *Colombia adopta de forma temprana recomendaciones de ética en Inteligencia Artificial de la UNESCO para la region* (Mar. 9, 2022), https://mintic.gov.co/portal/inicio/Sala-de-prensa/Noticias/208109:Colombia-adopta-de-forma-temprana-recomendaciones-de-etica-en-Inteligencia-Artificial-de-la-Unesco-para-la-region

[1475] Ministerio de Ciencia, Tecnología e Innovación (MinCiencias) *Roadmap for the Development and Application of Artificial Intelligence in Colombia [Hoja de Ruta para el Desarrollo y Aplicación de la Inteligencia Artificial en Colombia]*, pp. 33–34, https://minciencias.gov.co/sites/default/files/upload/noticias/hoja_de_ruta_adopcion_etica_y_sostenible_de_inteligencia_artificial_colombia_0.pdf

[1476] Red Iberoamericana de Protección de Datos(RED), *Specific Guidelines for Compliance with the Principles and Rights that Govern the Protection of Personal Data in Artificial Intelligence Projects* (Jun. 2019), p. 17, https://www.redipd.org/documentos/guia-orientaciones-especificas-proteccion-datos-ia

[1477] UNESCO, *AI in the Courtroom: Colombian Constitutional Court's Landmark Ruling Cites UNESCO's AI Tools* (Aug. 16, 2024), https://www.unesco.org/en/articles/ai-courtroom-colombian-constitutional-courts-landmark-ruling-cites-unescos-ai-tools

services to citizens, to include citizens and the private sector in data governance, to drive innovation in governance, and to guide decision-making.

The first policy document on data infrastructure is the National Policy on Data Exploitation (CONPES 3920), developed by the National Council on Economic and Social Policy, the National Planning Department, and the Office of the President.[1478] This policy uses data within a legal, ethical, and institutional framework to generate social and economic value; to increase the availability and interoperability of government data; to promote data culture in public entities, academia and the private sector; to promote data ethics and AI; and to provide test environments through data sandboxes, sandboxes on privacy and AI, and conceptual models for regulatory sandboxes and beaches in AI. To achieve this target, CONPES 3920 sets out 45 action steps with indicators, responsible parties, budgets and a timeline.

The second policy document is the National Data Infrastructure Plan (PNID) developed by MinTIC, the National Planning Department, and the Office of the President, with the support of the World Economic Centre for the Fourth Industrial Revolution.[1479] The draft of the PNID was shared for public comment until 17 September 2021. The PNID presents an approach to data as infrastructure, defines the components of data infrastructure, and provides a roadmap with concrete actions to implement data infrastructure in the country. This roadmap identifies 6 elements including governance, data, data leveraging, infrastructure interoperability, data security and privacy, as well as technical and technological input for data management.

For the successful integration of the PNID into the data regulation ecosystem, the government intends to create between 2022 and 2025 PNID guidelines. The aim is to identify priority data and create guidelines to ensure data quality; to develop a data infrastructure governance model; to identify indicators for monitoring; and to draw up a collaborative participation strategy for different actors in the data ecosystem.

CONPES 3920 and the PNID both emphasized the need to develop an institutional framework to accompany the development of data infrastructure. The governance model responds to this need. The governance model outlines five objectives to guide its design: institutional coordination, private sector participation, confidence building, technical modelling, and international impact. Under each of these objectives, responsible parties or entities and specific tasks have been provided.

[1478] National Council on Economic and Social Policy, the National Planning Department, and the Office of the President, *National Policy on Data Exploitation (Big Data) [Política Nacional de Explotación de Datos* (Apr. 17, 2018), https://colaboracion.dnp.gov.co/CDT/Conpes/Económicos/3920.pdf

[1479] Ministry of Information and Communications Technology (MinTIC), National Planning Department and the Office of the President, *National Data Infrastructure Plan (Plan Nacional de Infraestructura de Datos, PNID)* (Sept. 2021), https://mintic.gov.co/portal/715/articles-179710_recurso_2.pdf

The Ministry of Information and Communications Technologies introduced a bill aimed at developing a data infrastructure to enhance data availability, usage, and exchange, improve decision-making, and foster a data-driven culture.[1480] The bill applies to public administration entities and individuals performing public functions, including the legislative, judicial, and control bodies. Key provisions include obligating public entities to implement recommendations from the National Data Committee, ensuring compliance with the Digital Government Policy for those using emerging technologies like AI, and guaranteeing interoperability among data infrastructure across public entities.

Data Scraping

Colombia's Superintendency of Industry and Commerce and international data protection and privacy counterparts released a follow-up joint statement laying out expectations for industry related to data scraping and privacy.[1481] The initial statement in August 2023 established that "[social media companies] and other websites are responsible for protecting personal information from unlawful scraping."[1482] The follow-up, which incorporates feedback from engagement with companies such as Alphabet (YouTube), X Corp (formerly Twitter), and Meta Platforms (Facebook, Instagram, WhatsApp), adds that organizations must "comply with privacy and data protection laws when using personal information, including from their own platforms, to develop AI large language models"; must regularly review and update privacy protection measures to keep pace with technological advances; and ensure that permissible data scraping, such as that for research, "is done lawfully and following strict contractual terms."[1483]

Data scraping generally involves the automated extraction of data from the web. Data protection authorities are seeing increasing incidents involving data

[1480] El Congreso De Colombia, *Bill throgh which Dispositions for the Provision, Exchange, and Use of Data Infrastructure in the State of Colombia (IDEC) and the Interoperability of Information Systems for Public Entities, and Others, are Declared [Proyecto Ley por medio de la cual se dictan disposiciones para el suministro, intercambio y aprovechamiento de la infraestructura de datos del Estado colombiano (IDEC) y la interoperabilidad de los sistemas de información de las entidades públicas y se dictan otras disposiciones]* (Oct. 2025), https://www.camara.gov.co/wp-content/uploads/2025/10/proyectos-ley/documentos/proyecto-33228/pl_447_2024c_infraestructura_de_datos_del_estado_colombiano_0-14f1fca8.pdf

[1481] UK Information Commissioner's Office, *Global Privacy Authorities Issue Follow-Up Joint Statement on Data Scraping after Industry Engagement* (Oct. 28, 2024), https://ico.org.uk/about-the-ico/media-centre/news-and-blogs/2024/10/global-privacy-authorities-issue-follow-up-joint-statement-on-data-scraping-after-industry-engagement/

[1482] Office of the Privacy Commissioner of Canada, *Joint Statement on Data Scraping and the Protection of Privacy* (Aug. 24, 2023), https://www.priv.gc.ca/en/opc-news/speeches-and-statements/2023/js-dc_20230824/

[1483] Office of the Privacy Commissioner of Canada, *Concluding Joint Statement on Data Scraping and the Protection of Privacy* (Oct. 2024), https://www.priv.gc.ca/en/opc-news/speeches-and-statements/2024/js-dc_20241028/

scraping, particularly from social media companies and the operators of other websites that host publicly accessible data. Scraped personal information can be exploited for targeted cyberattacks, identity fraud, monitoring, profiling and surveillance purposes, unauthorized political or intelligence gathering purposes, unwanted direct marketing or spam.

Use of AI in Public Administration

As a member of the Latin American Centre for Development Administration (CLAD),[1484] Colombia approved the Ibero American Charter on Artificial Intelligence in Civil Service. The Charter aims to provide a roadmap and common framework for CLAD member states to adapt their AI policy strategies and laws for the challenges and opportunities involved in the implementation of AI in public administration. The guiding principles are human autonomy; transparency, traceability and explainability; accountability, liability and auditability; security and technical robustness; reliability, accuracy, and reproducibility; confidence, proportionality, and prevention of damage; privacy and protection of personal data; data quality and safety; fairness, inclusiveness, and non-discrimination; human-centering, public value, and social responsibility; and sustainability and environmental protection.

In addition to common principles, the Charter recommends specific tools for monitoring AI impact as well as a risk classification mechanism that guardrails for high-risk use cases and prohibitions against uses that violate human rights. Prohibited uses include facial recognition systems; systems aimed at behavior and cognitive manipulation of specific individuals or vulnerable groups (children or the elderly); and social ranking systems based on certain personality traits, individual behaviors, socio-demographic features, or economic status.[1485]

Use of AI in Courts

The Colombian government introduced the use of technology in the administration of justice through Article 95 of Law 270 in 1996, while mandating protection of confidentiality, privacy, and security of personal data.[1486] As a result, several government entities use AI in judicial aspects of their work. This includes the Constitutional Court (*Corte Constitucional*), the Office of the Attorney General, and the Superintendency of Companies.

Concerning the Constitutional Court of Colombia, where thousands of case documents are received daily, their processing has been expedited using an AI system

---

[1484] Centro Latinoamericano de Administración para el Desarrollo (CLAD), *Ibero American Charter on Artificial Intelligence in Civil Service* (Nov. 2023), https://web-api-backend.clad.org/uploads/ciia-en-03-2024.pdf

[1485] Ibid, p. 21

[1486] Congress of Colombia, *Law 270 of 1996* (Mar. 7, 1996), https://www.funcionpublica.gov.co/eva/gestornormativo/norma.php?i=6548#:~:text=Expide%20la%20Ley%20Estatutaria%20de,las%20jurisdicciones%20y%20altas%20Cortes

called Prometea. Applying machine learning abilities, this system investigates, analyses, identifies and suggests priority cases on health-related aspects within a few minutes.[1487] Additionally, it produces statistical reports, automates documentation, systematizes and synthesizes case law across the country, and improves security by integrating blockchain technology. This system that includes human oversight, improved the efficiency of case processing by 937%.[1488]

The Office of the Attorney General makes use of an AI system called Fiscal Watson, which consolidates criminal cases across different databases and regional offices to analyze similar evidentiary elements such as modus operandi, physical attributes, types of weapon and other aspects.[1489] By accelerating the processing of case information, Fiscal Watson has helped connect and solve similar cases across the country. The Attorney General has also suggested that Fiscal Watson can be used to identify irregularities in government contracts made during the COVID-19 pandemic, since all data and documentation is available online.[1490]

The Superintendency of Companies, an administrative body, employs a robot assistant called Siarelis (System based on AI for the Resolution of Company Litigation) to exercise its discretionary judicial powers in corporate cases related for example to the piercing of the corporate veil or insolvency.[1491] Using Case Based Reasoning (CBR), Siarelis helps officials identify relevant case law for a specific case and also provides users with possible decisions that could be reached in their case. The outcome reached by the system is decided based on the judicial history and precedent relevant to a specific case.

[1487] Juan Camilo Rivandeniera, *Prometea, inteligencia artificial para la revisión de tutelas en la Corte Constitutional [Artificial Intelligence for the Review of Guardianships in the Constitutional Court* (Mar. 22, 2019), https://www.ambitojuridico.com/noticias/informe/constitucional-y-derechos-humanos/prometea-inteligencia-artificial-para-la

[1488] Juan Corvalan, *Prometea: Artificial Intelligence to Transform Justice and Public Organizations* (Apr. 13, 2020), https://ojs.imodev.org/?journal=RIDDN&page=article&op=view&path%5B%5D=342&path%5B%5D=523

[1489] Pablo Medina Uribe and Luisa Fernanda Gomez, *Watson, the intelligent investigator with which the Prosecutor's Office seeks to block crime* (Jul. 25, 2020), https://www.elpais.com.co/judicial/watson-el-investigador-inteligente-con-el-que-la-fiscalia-busca-cerrarle-el-paso-al-crimen.html

[1490] Vanguardia, *Fight against Corruption in Santander Will Be Done with Artificial Intelligence* (Jun. 24, 2020), https://www.vanguardia.com/politica/lucha-contra-la-corrupcion-en-santander-se-hara-con-inteligencia-artificial-XC2532257

[1491] Center for Technology and Society Studies (CETyS) of the University of San Andrés, *Readiness of the judicial sector for Artificial Intelligence in Latin America – Analytical and Exploratory Framework, Republic of Colombia* (2021), https://cetys.lat/wp-content/uploads/2021/09/colombia-ENG.pdf

Law 2213 of 2022[1492] and the General Procedure Code[1493] allow the use of different technologies in the different stages of Civil, Labor, Family, Administrative, Constitutional, Disciplinary and Criminal procedures in Colombia. The Law specifies that such technologies should only be used when they are suitable for the task. In this regard, Professor Gutierrez noted, "If ChatGPT and other LLMs currently available are evidently unreliable, since their outputs tend to include incorrect and false information, then judges would require significant time to check the validity of the AI-generated content, thereby undoing any significant "time savings". As it happens with AI in other areas, under the narrative of supposed "efficiencies," fundamental rights can be put at risk."[1494]

In 2023, ChatGPT was used in Colombian Courts in two different cases. The answers provided by ChatGPT were determinant in both cases. In the first case, the judge asked questions with regard to key legal issues, specific to the Colombian legal system. In the second case, the questions touched upon issues of access to justice and due process pertaining to carrying out a hearing in the metaverse. There was no evidence that the answers received by ChatGPT were corroborated by other sources. Professor Florez also raised the risk of over-reliance on "the outputs of automated systems."[1495]

These cases led the Constitutional Court of Colombia to issue a "landmark"[1496] ruling indicating that AI tools like ChatGPT may be used to assist with judicial tasks but cannot replace human judgment in legal decisions.[1497] The Court emphasized that while AI can help manage workload and draft documents, excessive reliance on these systems could threaten due process and judicial independence. The ruling established key principles including transparency, privacy protection, and the requirement for human oversight.

In addition, the Court required the issue of guidelines for the use of AI in courts within four months after the ruling. To this end, the Superior Council of the Judiciary (*Consejo Superior de la Judicatura*) established a partnership with UNESCO to tackle four areas: development of judicial AI guidelines, capacity building, algorithmic

---

[1492] *Law 2213 of 2022*, https://www.funcionpublica.gov.co/eva/gestornormativo/norma.php?i=187626

[1493] Public Function, *Law 1564 of 2012, on Issuance of the General Procedure Code and Other Provisions* (Jul. 12, 2012), https://www.funcionpublica.gov.co/eva/gestornormativo/norma.php?i=48425

[1494] Juan David Gutierrez, *ChatGPT in Colombian Courts: Why We Need to Have a Conversation about the Digital Literacy of the Judiciary*, Verfassungsblog (Feb. 23, 2023), https://verfassungsblog.de/colombian-chatgpt/#104

[1495] Maria Lorena Florez Rojas, *A judge in Cartagena (Colombia) claims to have use ChatGPT as a support tool to resolve a guardianship for health care neglect*, Foro Administración, Gestion y Política Pública (Feb. 2, 2023) https://forogpp.com/2023/02/03/a-judge-in-cartagena-colombia-claims-to-have-use-chatgpt-as-support-tool-to-resolve-a-guardianship-for-health-care-neglect/.

[1496] UNESCO, *AI in the Courtroom: Colombian Constitutional Court's Landmark Ruling Cites UNESCO's AI Tools* (Aug. 16, 2024), https://www.unesco.org/en/articles/ai-courtroom-colombian-constitutional-courts-landmark-ruling-cites-unescos-ai-tools

[1497] Constitutional Court of Colombia, *Ruling T-323 of 2024* (Aug. 2, 2024), https://forogpp.com/wp-content/uploads/2024/08/t_323_2024_t_9301656-caso-chatgpt-cartagena.pdf

impact assessment, and transfer of knowledge.[1498] The Superior Council adopted the Guidelines for the Responsible Use of Artificial Intelligence in the Colombian Judiciary in 2024.[1499] Colombia is the first country to adopt these Guidelines, developed with UNESCO.[1500] The Guidelines introduce a risk-based classification system for AI tools, requires mandatory human supervision, and sets out obligations of transparency, explainability, and data protection. The Guidelines explicitly state that AI systems may support administrative and analytical tasks but cannot replace human reasoning or decision-making in judicial proceedings. The guidelines also establish procedures for algorithmic audits, staff training, and ethical review before deployment.

Facial Recognition

In Colombia, facial recognition technology is used extensively by the State. The Ministry of Transport is integrating a network of cameras with facial recognition technology throughout Bogota. The system, which is meant to prevent and reduce road accidents, has become operational in December 2021 and provides the location of cameras out of transparency.[1501] A facial recognition system to improve surveillance was also introduced by local authorities in September 2021 at a stadium in Barranquilla. This system combines cameras with access to unlimited databases to record and track individuals.[1502] The system will be used to identify and detain anyone with cases pending before the judicial system or any other relevant authority. The Atanasio Girardot stadium in Medellín also has 170 smart cameras installed for surveillance, since 2016.[1503] Expanding the reach of surveillance in the city, Medellín introduced 40 security robots with facial recognition capability and an integrated AI security system to patrol the city.[1504] The Border Control Agency located at El Dorado

---

[1498] UNESCO, *UNESCO and Colombia: Leaders in the Ethical and Responsible Use of AI in the Judiciary* (Oct. 10, 2024), https://www.unesco.org/en/articles/unesco-and-colombia-leaders-ethical-and-responsible-use-ai-judiciary

[1499] Consejo Superior Judicatura, *Agreement PCSJA24-12243 by which Guidelines Are Adopted for the Respectful, Responsible, Safe and Ethical Us and Exploitation of Artificial Intelligence in the Judicial Branch* (Dec. 16, 2024), https://rm.coe.int/colombia-guidelines-for-the-use-of-artificial-intelligence-in-the-judi/1680b53484

[1500] UNESCO *Justice Meets Innovation: Colombia's Groundbreaking AI Guidelines for Courts* (Feb. 25, 2025), https://www.unesco.org/en/articles/justice-meets-innovation-colombias-groundbreaking-ai-guidelines-courts

[1501] El Tiempo, *Do you agree with life-saving cameras recognizing your face?* (Aug. 2, 2021), https://www.eltiempo.com/bogota/camaras-salvavidas-tendrian-reconocimiento-facial-en-bogota-movilidad-607508

[1502] El Tiempo, *Colombia vs Chile match will have facial recognition system* (Sept. 9, 2021), https://www.eltiempo.com/colombia/barranquilla/el-metropolitano-cuenta-con-sistema-de-reconocimiento-facial-616845

[1503] NEC Corporation, *Integrated Surveillance and Security System for Atanasio Girardot Stadium - Medellín* (2017), https://www.nec.com/en/case/medellin/es/pdf/brochure.pdf

[1504] El Tiempo, *Mayor Daniel Quintero revealed robots to track crime in Medellín* (Aug. 11, 2021), https://www.eltiempo.com/colombia/medellin/daniel-quintero-presento-robot-para-predecir-delitos-

International Airport in Bogota uses the Iris recognition system, with the system expected to reach other airports in the country within the next few years.

Fundación Karisma, a civil society organization dedicated to supporting the responsible use of tech, highlights the pitfalls of these systems. In a report titled "Discreet Cameras," Fundacion Karisma points out that surveillance technology and biometric identification systems in Colombia only take into consideration technical and impact considerations while assessing systems. There is no analysis using necessity, proportionality or the possible effect of the technology on human rights.[1505] However, Colombia's co-sponsorship of the 2022 GPA Resolution on Principles and Expectations for the Appropriate Use of Personal Information in Facial Recognition Technology indicates a move towards official recognition of the importance of parameters such as reasonableness, necessity, proportionality, and the effect of the technology on human rights.[1506] Although the government tries to ensure transparency by sharing the location of video surveillance systems that use facial recognition technology, the right to privacy and other fundamental rights of individuals are still ignored.[1507]

Use of AI during COVID-19 Pandemic

Despite this situation during the emergency of the COVID-19 the government released a mobile app named CoronApp.[1508] At first, CoronApp's main objective was to allow people to stay updated on the progression of the pandemic in Colombia. In the days that followed, however, the narrative around the app changed. Its purpose became more ambitious. It soon became a digital tool to keep the population informed and "save as many lives as possible."

Concerns were raised about the app's data privacy implications, as it collected personal information such as location data and potentially exposes users to security risks. The Colombian government stated that the app is voluntary, and that users' data will be kept confidential, but it is important for individuals to carefully review the app's privacy policy and consider the potential risks before downloading and using it.

---

en-medellin-609912; El Tiempo, *How is facial recognition done in Colombia?* (May 17, 2019), https://www.eltiempo.com/tecnosfera/dispositivos/colombia-que-usos-de-reconocimiento-facial-hay-en-el-pais-362220

[1505] Fundación Karisma, *Discreet Cameras Report* (Feb. 2, 2018), https://web.karisma.org.co/camaras-indiscretas/

[1506] Global Privacy Assembly, *Resolution on Accountability in the Development and Use of Artificial Intelligence* (Oct. 2022), https://globalprivacyassembly.org/wp-content/uploads/2022/11/15.1.c.Resolution-on-Principles-and-Expectations-for-the-Appropriate-Use-of-Personal-Information-in-Facial-Recognition-Technolog.pdf

[1507] Fundación Karisma, *Discreet Cameras Report* (Feb. 2, 2018), https://web.karisma.org.co/camaras-indiscretas/

[1508] Government of Colombia, *CoronApp—Colombia, the Application to Follow the Evolution of the Coronavirus in the Country, Press Release No. 051* (Mar. 7. 2020), https://www.minsalud.gov.co/English/paginas/coronapp---colombia,-the-application-to-follow-the-evolution-of-the-coronavirus-in-the-country.aspx

In response to these concerns, the Colombian human rights organization Dejusticia called for greater transparency and accountability in the use of personal data collected through this app and another implemented in the future.[1509] Dejusticia had also called for the government to implement strict measures to protect the privacy of individuals and ensure that their data is only used for the purposes that the app was created.

Dejusticia has also urged the government to establish clear guidelines for the collection, use, and storage of personal data, and to provide users with the ability to opt out of the app(s) at any time. Additionally, they have called for the government to establish a system of oversight and accountability to ensure that data collected through the app(s) is not misused or mishandled.

CoronApp had the potential to be a useful tool in guaranteeing the health of the population, but it is important that the government takes steps to ensure that the personal data of users is protected and used responsibly. In April 2022, the Constitutional Court reviewed and determined that the plaintiffs' rights to privacy should be respected, even during a national state of emergency. Furthermore, the Court declared that authorities had a duty to "avoid the abusive and arbitrary use of personal data" and the National Data Authority was ordered to erase the plaintiffs' data.[1510]

Environmental Impact of AI

The growing adoption of AI technologies in Colombia has prompted concerns about their environmental impact, leading to the inclusion of sustainability considerations in recent AI policy proposals. Colombia's particular vulnerability to climate phenomena, coupled with its 70% dependence on hydroelectric power, makes environmental sustainability a critical aspect of AI governance.[1511]

The national AI policy roadmap centers the ethical and sustainable use of AI and aims to develop AI in support of the UN Sustainable Development Goals. However, the specific recommendations related to environmental impact center on using AI tools to monitor water resource usage and quality and air quality and to assist with sustainable urban planning.[1512] The National AI Policy (CONPES 4144) notes

---

[1509] Dejustica, *New Report: Misuse of Technologies in Emergency Responses* (Feb. 8, 2023), https://www.dejusticia.org/en/new-report-misuse-of-technologies-in-emergency-responses/

[1510] European Center for Not-for-Profit Law (ECNL), International Network of Civil Liberties Organizations, and Privacy International, *Under Surveillance: (Mis)use of Technologies in Emergency Responses* (Dec. 2022), https://ecnl.org/publications/under-surveillance-misuse-technologies-emergency-responses

[1511] Departamento Nacional de Planeación, *Artificial Intelligence as a Catalyst for Social and Economic Development in Colombia [La inteligencia artificial como motor clave para el desarrollo social y económico de Colombia]* (Aug. 2, .2024), https://colaboracion.dnp.gov.co/CDT/PublishingImages/Planeacion-y-desarrollo/2024/Septiembre/PDF/inteligencia-artificial.pdf

[1512] Miniterio de Ciencia, Tecnología e Innovación (MinCiencias), *Roadmap for the Development and Application of Artificial Intelligence in Colombia [Hoja de Ruta para el Desarrollo y Aplicación de la Inteligencia Artificial en Colombia]*, p. 80 (Feb. 2024),

that Colombia lacks a way to measure the environmental impact of AI systems.[1513] The document charges the MinCiencias with creating hyperscale data centers capable of supporting AI development "fed with renewable energy sources and technologies that optimize water use, minimizing the environmental impact, thereby guaranteeing the competitivity and sustainability of the country."[1514]

Lethal Autonomous Weapons

During the 2015 meeting of the Group of Governmental Experts (GGE) on lethal autonomous weapons (LAWS), Colombia issued a statement calling for multilateral regulation to ensure human control over autonomous weapons at all times, so that no machine makes an autonomous decision.[1515] Colombia has called for a pre-emptive ban on all LAWS[1516] and for an international treaty that will ensure meaningful human control over any use of force.[1517]

Colombia was among the 70 countries that endorsed a joint statement on autonomous weapons systems at the 2022 UN General Assembly meeting. In this joint statement, States urged "the international community to further their understanding and address these risks and challenges by adopting appropriate rules and measures, such as principles, good practices, limitations and constraints. We are committed to upholding and strengthening compliance with International Law, in particular International Humanitarian Law ("IHL"), including through maintaining human responsibility and accountability in the use of force."[1518]

---

https://minciencias.gov.co/sites/default/files/upload/noticias/hoja_de_ruta_adopcion_etica_y_sostenible_de_inteligencia_artificial_colombia_0.pdf

[1513] Consejo Nacional de Política Económica y Social (CONPES), *Document 4144: National Artificial Intelligence Policy [Documento 4144: Política Nacional de Inteligencia Artificial]*, approved version, p. 72 (Feb. 14, 2025), https://colaboracion.dnp.gov.co/CDT/Conpes/Económicos/4144.pdf

[1514] Ibid, p. 88

[1515] Government of Colombia, *Statement at the Convention on Certain Convention Weapons – Informal meeting of experts on Lethal Autonomous Weapons Systems* (Apr. 2015), https://www.reachingcriticalwill.org/images/documents/Disarmament-fora/ccw/2015/meeting-experts-laws/statements/17April_Colombia.pdf

[1516] World Summit of Nobel Peace Laureates, *Final Declaration of the 16th World Summit of Nobel Peace Laureates* (Feb. 4, 2017), http://www.nobelpeacesummit.com/final-declaration-of-the-16th-world-summit-of-nobel-peace-laureates/

[1517] Government of Colombia, *Statement at the Convention on Certain Conventional Weapons – Group of Government al Experts on Lethal Autonomous Weapons Systems* (Apr. 13, 2018), https://documents.un.org/doc/undoc/gen/g18/323/29/pdf/g1832329.pdf

[1518] United Nations (UN) General Assembly, First Committee, *Joint Statement on Lethal Autonomous Weapons Systems First Committee, 77th United Nations General Assembly Thematic Debate – Conventional Weapons* (Oct. 21, 2022), https://estatements.unmeetings.org/estatements/11.0010/20221021/A1jJ8bNfWGlL/KLw9WYcSnnAm_en.pdf

At the 78th UN General Assembly First Committee in 2023, Colombia voted in favor[1519] of resolution L.56[1520] on autonomous weapons systems, along with 163 other states. The Resolution emphasized the "urgent need for the international community to address the challenges and concerns raised by autonomous weapons systems," and mandated the UN Secretary-General to prepare a report reflecting the views of member and observer states on autonomous weapons systems. The report should analyze ways to address the challenges and concerns autonomous weapon systems raise from humanitarian, legal, security, technological and ethical perspectives and reflect on the role of humans in the use of force.

The Secretary-General issued the report in the 79th UN General Assembly and analyzed ways to address the challenges and concerns autonomous weapon systems raise from humanitarian, legal, security, technological and ethical perspectives and reflect on the role of humans in the use of force.[1521] Colombia contributed to statements from the Ibero-American states[1522] and as part of a group High Contracting Parties to the CCW. Both reports reiterated support for a legally binding instrument to prohibit the use of autonomous weapons outside human control and restrict other uses. The latter statement also proposed a draft protocol VI originally submitted to the GGE in July 2022 as a starting point for a legally binding instrument.[1523]

Colombia also endorsed, along with more than 30 other Latin American and Caribbean states, the Belén Communiqué of 2023,[1524] which calls for "urgent negotiation" of a binding international treaty to regulate and prohibit the use of autonomous weapons to address the grave concerns raised by removing human control from the use of force.

AI Literacy

The National AI Policy[1525] includes an axis (5.3.4) dedicated to Capacity Development and Digital Talent with action items to develop the capabilities and

---

[1519] Isabelle Jones, *164 States Vote against the Machine at the UN General Assembly*, Stop Killer Robots (Nov. 1, 2023), https://www.stopkillerrobots.org/news/164-states-vote-against-the-machine/

[1520] General Assembly, *Resolution L56: Lethal Autonomous Weapons* (Oct. 12, 2023), https://reachingcriticalwill.org/images/documents/Disarmament-fora/1com/1com23/resolutions/L56.pdf

[1521] UN General Assembly, 79th session, *Lethal Autonomous Weapons Systems, Report of the Secretary-General* (Jul. 1, 2024), https://documents.un.org/doc/undoc/gen/n24/154/32/pdf/n2415432.pdf

[1522] Ibid, pp. 20–21

[1523] Ibid, pp. 34–36

[1524] Latin American and the Caribbean Conference of Social and Humanitarian Impact of Autonomous Weapons, *Communiqué*, La Ribera, Belén (Feb. 24, 2023), https://www.rree.go.cr/files/includes/files.php?id=2261&tipo=documentos

[1525] Consejo Nacional de Política Económica y Social (CONPES), *Document 4144: National Artificial Intelligence Policy [Documento 4144: Política Nacional de Inteligencia Artificial]*, approved version, pp. 93–96 (Feb. 14, 2025), https://colaboracion.dnp.gov.co/CDT/Conpes/Económicos/4144.pdf

digital talent necessary for the design, use and adoption of AI, in addition to promoting the social appropriation of knowledge of this technology.[1526] The policy charges the Ministry of National Education (*Ministerio de Educación Nacional*) to add a section on AI to the Safe Use of Information Technology guide through 2026 and, over five years, to implement strategies in the curriculum to strengthen students' digital competencies, including with AI. The policy also charges the Ministry of Education, with the help of the institute for education evaluation, to recommend a way to evaluate digital competencies through the national standardize tests by the end of 2026. More generally, the MinCiencias, MinTIC, and National Planning Department will increase general levels of social understanding of AI.

Human Rights

Colombia is a signatory to many international human rights treaties and conventions. Colombia has ranked as "free" in the Freedom in the World Report since 2023,[1527] with a score in 2025 of 70/100.[1528] Colombia ranks as "partly free" on Internet Freedom with a score of 64/100, in part because of violence against digital journalists and cyberattacks on media and government entities.[1529]

According to Article 93 of Colombia's Constitution, rights and duties in the national system are interpreted according to international treaties and conventions that have been ratified by its Congress.[1530] Thus, Colombia has made powerful commitments backed by strong action that encourages legal certainty, with an entire implementation, regulatory and monitoring ecosystem for AI. This is strengthened by expert contributions and public participation at the national, regional and international level.

OECD / G20 AI Principles

Colombia has endorsed the OECD AI Principles[1531] and is part of the Global Partnership on Artificial Intelligence (GPAI).[1532] Colombia's Ethical Framework for

---

[1526] OECD AI Policy Observatory, *National Policy on Artificial Intelligence (CONPES 4144)* (Dec. 24, 2025), https://oecd.ai/en/dashboards/policy-initiatives/national-policy-on-artificial-intelligence-conpes-4144-3012

[1527] Freedom House, *Freedom in the World 2023: Colombia* (2023), https://freedomhouse.org/country/colombia/freedom-world/2023

[1528] Freedom House, *Freedom in the World 2025: Colombia* (2025), https://freedomhouse.org/country/colombia

[1529] Freedom House, *Freedom on the Net 2025: Colombia* (2025), https://freedomhouse.org/country/colombia/freedom-net/2025

[1530] Congreso de la República, *Colombia's Constitution of 1991 with Amendments through 2015* (subsequent amendments) (1991), https://www.constituteproject.org/constitution/Colombia_2015.pdf?lang=en

[1531] OECD Legal Instruments, *Recommendation of the Council on Artificial Intelligence*, Adherents (May 3, 2024), https://legalinstruments.oecd.org/en/instruments/OECD-LEGAL-0449#adherents

[1532] OECD AI Policy Observatory, *About the Global Partnership on Artificial Intelligence (GPAI)* (2026), https://oecd.ai/en/about/about-gpai

AI introduced as a guideline for trustworthy AI, provides standards for the ethical use and governance of AI and is aligned with the OECD AI principles.[1533] Additionally, Colombia has developed policy intelligence tools and a follow up plan to monitor the implementation of the OECD AI Principles, while identifying good practices to determine if OECD's recommendations to Colombia have been implemented.

Council of Europe AI Treaty

Colombia is not a member of the Council of Europe and has not signed the Council of Europe Framework Convention on Artificial Intelligence and Human Rights, Democracy, and the Rule of Law.[1534]

UNESCO Recommendations on AI Ethics

Colombia endorsed the UNESCO Recommendation on the Ethics of AI in November 2021. Even before the UNESCO recommendation was completed, the Presidency cited the example of OECD and UNESCO principles to acknowledge "the importance of ethical considerations for the design, development, and implementation of artificial intelligence in Colombia and more precisely the need to adopt an Ethical Framework as a non-binding guide for the implementation of artificial intelligence in the country."[1535]

Colombia is part of the Latin America and Caribbean Regional Council of national and local governments established by CAF, the development bank of Latin America, and UNESCO to implement the UNESCO Recommendation in Latin America and the Caribbean.[1536] At the inaugural meeting in 2023, the Council focused on the collaborative exchange and formulation of proposals, encompassing both political and technical perspectives, aimed at guiding the ethical development of Artificial Intelligence in the region.[1537] Colombia signed the resulting 2023 Santiago

[1533] Presidential Advisory for Economic Affairs and Digital Transformation (DAPRE), *Ethical Framework for Artificial Intelligence* (Aug. 2020), https://cyber.harvard.edu/sites/default/files/2020-12/Colombia_AI_Ethical_Framework.pdf

[1534] Council of Europe, *Chart of Signatures and Ratifications of Treaty 225* (Nov. 14, 2024), https://www.coe.int/en/web/conventions/full-list?module=signatures-by-treaty&treatynum=225

[1535] Gobierno de Colombia, *Ethical Framework for Artificial Intelligence in Colombia [Marco ético para la inteligencia artificial en Colombia]*, version 1, p. 9 [Spanish version] (May 2021), https://minciencias.gov.co/sites/default/files/marco-etico-ia-colombia-2021.pdf

[1536] Ángel Melguizo and Gabriela Ramos, *Ethical and Responsible Artificial Intelligence: From Words to Actions and Rights*, Somos Iberoamérica (Feb. 1, 2023), https://www.somosiberoamerica.org/pt-br/tribunas/inteligencia-artificial-etica-e-responsavel-das-palavras-aos-fatos-e-direitos/

[1537] UNESCO, *Chile Will Host the First Latin American and Caribbean Ministerial and High Level Summit on the Ethics of Artificial Intelligence* (Sept. 25, 2023), https://www.unesco.org/en/articles/chile-will-host-first-latin-american-and-caribbean-ministerial-and-high-level-summit-ethics

Declaration to Promote Ethical Artificial Intelligence.[1538] The Santiago Declaration reflects UNESCO's Recommendation on the Ethics of AI and establishes fundamental principles that should guide public policy on AI. These include proportionality, security, fairness, non-discrimination, gender equality, accessibility, sustainability, privacy and data protection.

UNESCO published the report on Colombia's Readiness Assessment Methodology (RAM) exercise in 2025.[1539] The country does not yet have a profile on the UNESCO Global Hub, however.[1540]

Evaluation

Colombia has anchored the National AI Policy in the Ethical Framework for Artificial Intelligence and international frameworks and other national models. Colombia is part of the Regional Council for the implementation of the UNESCO Recommendation on AI Ethics and has endorsed the Ibero-American Charter on AI in Civil Service. Colombia also adopted the UNESCO Guidelines on use of AI in courts, the first country to do so. The Constitution of Colombia established a right to data protection, and the country has a comprehensive data protection law and an active data protection agency. Draft bills for ethical and sustainable development and application of AI are being discussed in Congress. The Colombian data protection authority is the first in the region to have opened an investigation on OpenAI ChatGPT. However, Colombia's laws have not yet codified the right to algorithmic transparency in law. There are also growing concerns regarding the use of facial recognition systems.

[1538] Cumbre Ministerial y de Altas Autoridades de América Latina y el Caribe, *Declaración de Santiago "Para promover una inteligencia artificial ética en América Latina y el Caribe"* (Oct. 2023), https://minciencia.gob.cl/uploads/filer_public/40/2a/402a35a0-1222-4dab-b090-5c81bbf34237/declaracion_de_santiago.pdf.

[1539] UNESCO, *Colombia: Readiness Assessment Methodology [Colombia: Metodología de evaluación del estadio de preparación]* UNESCO Digital Library (2025), https://unesdoc.unesco.org/ark:/48223/pf0000396015?posInSet=1&queryId=257f4eea-9428-49d3-ac27-0e9499d4228d

[1540] UNESCO Global AI Ethics and Governance Observatory, *Global Hub* (Oct. 2025), https://www.unesco.org/ethics-ai/en/global-hub

Costa Rica

In 2025, Costa Rica released the Action Plan for the National AI Strategy centered on ethical, safe, and responsible AI. The Legislative Assembly furthered discussions on three bills covering responsible AI use, development, and implementation.

National AI Strategy

Costa Rica adopted its National Artificial Intelligence Strategy (*Estrategia Nacional de Inteligencia Artificial*, ENIA) developed by the Ministry of Science, Innovation, Technology, and Telecommunications (*Ministerio de Ciencia, Innovación, Tecnología y Telecomunicaciones*, MICITT) in October 2024.[1541] This initiative, drafted with UNESCO's support,[1542] aims to align national efforts toward a long-term vision for AI. ENIA's primary goals include fostering the ethical, safe, and responsible use, adoption, and development of AI technologies to maximize benefits for citizens while minimizing potential harm. The strategy outlines specific action areas to harness AI's advantages and to address associated ethical, social, and technical risks and challenges.[1543]

The MICITT later adopted an Action Plan to implement ENIA, introducing AI ethical and legal frameworks, guidance on technical norms and good practice on data protection.[1544] The Action Plan focuses on seven strategic axes of the ENIA 2024–2027 to ensure the comprehensive development of AI in Costa Rica, including Ethical, Safe, and Responsible AI; Widespread, Responsible Use for Economic Development; Promotion of R&D; Smart Government; Talent Training and Capacity-Building; Digital Infrastructure for Public Services; and International Leadership.

The country has also developed a general human-centered science, technology and innovation plan and a specific digital strategy to address issues related to AI

[1541] Ministerio de Ciencia, Innovación, Tecnología y Telecomunicaciones, *National Artificial Intelligence Strategy 2024–2027* [Simplified English version] (Feb. 3, 2025), https://www.micitt.go.cr/sites/default/files/2025-02/National Artificial Intelligence Strategy of Costa Rica.pdf; for the full strategy in Spanish, see https://www.micitt.go.cr/sites/default/files/2024-10/Estrategia Nacional de Inteligencia Artificial de Costa Rica ESP.pdf

[1542] Mariana Álvarez, *Costa Rica Will Be the First Country in Central America to Have an Artificial Intelligence Strategy*, UNESCO (Feb. 28, 2023), https://www.unesco.org/en/articles/costa-rica-will-be-first-country-central-america-have-artificial-intelligence-strategy

[1543] Ministerio de Ciencia, Innovación, Tecnología y Telecomunicaciones, *National Artificial Intelligence Strategy 2024–2027* [Simplified, English version] (Feb. 3, 2025), https://www.micitt.go.cr/sites/default/files/2025-02/National Artificial Intelligence Strategy of Costa Rica.pdf

[1544] Ministerio de Ciencia, Innovación, Tecnología y Telecomunicaciones, *National Artificial Intelligence Strategy Action Plan, Costa Rica 2024–2027 [Plan de Acción Estrategia Nacional de Inteligencia Artificial Costa Rica 2024–2027]* (Aug. 11, 2025), https://www.micitt.go.cr/sites/default/files/2025-10/Plan de Acción ENIA - Versión 11 agosto 2025 Versión Publicación.pdf

technology. The National Plan on Science, Technology and Innovation 2022–2027 (provides a general overview of the country's goals regarding the use of information and communication technologies for a sustainable, equitable and creative future. Regarding AI, the plan focuses on pursuing developing technical capabilities and competencies across STEM/STEAM curricula in schools and technical colleges, technical certification in AI, the development of understandable and accessible AI systems for all, and ensuring ethical and responsible Ai adoption through the support of the fAIr Latin America and Caribbean (LAC) initiative designed by the Inter-American Development Bank (IDB).[1545]

Adoption of the AI Strategy in 2024 marked progress on a key action item from the Digital Transformation Strategy 2023–2027.[1546] The Digital Transformation Strategy frames AI use as a way to drive the country's economic competitiveness and productivity through innovation and value creation through digital citizenship and good governance.

AI Legislation

Costa Rica's legislature is currently working on at least four bills related to artificial intelligence.[1547] The Draft Bill for the Responsible Promotion of Artificial Intelligence proposed in September 2023[1548] is the furthest along. The National Assembly's Human Rights Commission approved the bill in early 2025 and sent to plenary. Deputies proposed alternative text for consideration in October 2025.[1549] The Bill on Artificial Intelligence Regulation seeks to regulate the development,

---

[1545] Ministerio de Ciencia, Innovación, Tecnología y Telecomunicaciones, *National Plan for Science, Technology, and Innovation 2022–2027 [Plan Nacional de Ciencia, Tecnología e Innovación 2022–2027]*, https://www.micitt.go.cr/sites/default/files/planes_estrategias/Plan_Nacional_Ciencia_Tecnologia_Innovacion_2022-2027.pdf

[1546] Ministerio de Ciencia, Tecnología y Telecomunicaciones, *Digital Transformation Strategy 2023–22027 [Estrategia de Transformación Digital, 2023–2027]* (Aug. 30, 2023), https://www.micitt.go.cr/sites/default/files/GobernanzaDigital/ETD 2023-2027 V FINAL 30-08-2023_v2.pdf

[1547] Erick Murillo, *Study: Costa Rica Leads Central America in AI Governance and Public Policies [Estudio: Costa Rica lidera gobernanza digital y políticas pública de IA en Centroamérica]*, CR Hoy (Feb. 20, 2026), https://crhoy.com/tecnologia/estudio-costa-rica-lidera-gobernanza-digital-y-politicas-publicas-de-ia-en-centroamerica/

[1548] Asamblea Legislativa de La Republica de Costa Rica, *Expediente 23919, Law for the Responsible Promotion of Artificial Intelligence in Costa Rica [Ley Para La Promoción Responsable de la Inteligencia Artificial en Costa Rica]* (Apr. 1, 2025), https://www.asamblea.go.cr/Centro_de_informacion/Consultas_SIL/SitePages/ConsultaProyectos.aspx

[1549] Asamblea Legislativa, *Expediente 23.919, Moción de Texto Sustitutivo vía Artículo 137* (Oct. 1, 2025), https://www.asamblea.go.cr/glcp/Consultas_mociones/MOCIONES DE FONDO VÍA ARTÍCULO 137/23.919/23.919 Primer día 12-2-2026.pdf

implementation, and use of AI in Costa Rica.[1550] The Technology and Education Committee approved the bill in October 2024 but it has not been considered in the plenary. The subcommittee issued a report[1551] on and approved a Law for the Implementation of AI Systems, aimed at integrating AI in the public sector with attention to capacity-building. A law to regulate the use of AI in the electoral process has been studied by a special committee on the electoral process and is awaiting hearing in the plenary session.[1552]

Public Participation

Citizen participation in Costa Rica has a vast and rich history that has given rise to a significant number of democratic innovations. The Constituent Assembly of 1949 was reintroduced into the constitutional text with instruments for citizen participation. With later reforms to the Constitution in 2002 and 2003, public participation was enhanced through the inclusion of the Referendum and the Popular Initiative. In 2003, Article 9 of the Constitution of Costa Rica was further amended, supplementing the various instruments of citizen participation that had already been incorporated into the Constitution. Costa Rica's democratic institutions provide a robust legal framework to support public participation.

The National Commission for Open Government is using its online platform to enable public consultations on relevant national policies. The MICITT publishes all its public consultation opportunities on its website[1553] as well as on social media through its Facebook page.[1554] However, the MICITT does not include comments from the consultation on the site and the current site lists consultations back to only 2022.[1555]

[1550] Asamblea Legislativa, *Ex. 23771, Bill on the Regulation of Artificial Intelligence in Costa Rica [Ley de Regulación de la Inteligencia Artificial en Costa Rica]* (Sept. 3, 2024), https://www.asamblea.go.cr/Centro_de_informacion/Consultas_SIL/SitePages/ConsultaProyectos.aspx

[1551] Asamblea Legislativa, Comisión Permanente Especial de Ciencia, Tecnología y Educación, *Ex. 24484, Law for the Implementation of Artificial Intelligence (AI) Systems [Ley para la Implementación de sistemas de Inteligencia Artificial (IA)* (Oct. 20, 2025), https://www.asamblea.go.cr/glcp/mociones_total/24484 INFORME DE SUBCOMISIÓN.pdf

[1552] Asamblea Legislativa, *Ex. 24875, Law to Regulate the Use of Artificial Intelligence in Electoral Processes [Ley para regular el uso de Inteligencia Artificial en los procesos electorales]* (Mar. 30, 2025), https://www.asamblea.go.cr/Centro_de_informacion/Consultas_SIL/SitePages/ConsultaProyectos.aspx

[1553] Ministerio de Ciencia, Innovación, Tecnología y Telecomunicaciones, *Consultas Públicas* (Dec. 2025), https://www.micitt.go.cr/transparencia/consultas-publicas

[1554] Ministerio de Ciencia, Innovación, Tecnología y Telecomunicaciones de Costa Rica, *Facebook* https://www.facebook.com/micitcr

[1555] Ministerio de Ciencia, Innovación, Tecnología y Telecomunicaciones, *Consultas Públicas* (Dec. 2025), https://micitt.go.cr/transparencia/consultas-publicas

Costa Rica has made important progress in aligning its regulations and practices regarding public participation with OECD standards.[1556] Costa Rica also takes part in the CoST initiative to implement the Infrastructure Transparency Index (ITI) for improving transparency and accountability across public infrastructures. These commitments consolidate the transformation of Costa Rica towards an open government.[1557]

The Ministry of Science, Innovation, Technology and Telecommunications publishes consultation opportunities on a dedicated website.[1558] However, consultations held before 2022 are no longer available. The ministry held consultations on the national Cybersecurity Strategy in 2022.[1559]

Costa Rica launched a public national consultation on the National Artificial Intelligence Strategy, inviting citizens to participate in shaping the country's approach to artificial intelligence.[1560] The consultation, which was open March to April 11, 2024, sought input to promote the ethical, safe, and responsible adoption and development of AI in the nation. Participants were encouraged to review the National Artificial Intelligence Strategy document and submit feedback using a public consultation form via email.

The Legislative Assembly's Science and Technology Committee and the Chamber of Information and Communication Technologies (CAMTIC) held a working group on three bills related to AI regulation.[1561] The session was limited to experts. The Minister of Science and Technology expressed fears that any attempts to regulate artificial intelligence would become obsolete even before publication. A technology attorney called requirements in the laws for prior authorization and registries "an outdated form of regulation" and noted that even the EU AI Act, "which is the most advanced in the world in AI, does not have these controls."

[1556] OECD, *Public Governance in Costa Rica* (2021), https://web-archive.oecd.org/pdfViewer?path=/2021-07-29/595534-costa-rica-public-governance-evaluation-accession-review.pdf; Ministerio de Comercio Exterior, *Costa Rica busca fortalecer la participación ciudadana con políticas de impacto regulatorio*, Comunicados Desarrollo Social Economía (Dec. 8, 2017), https://www.comex.go.cr/sala-de-prensa/comunicados/2017/diciembre/costa-rica-busca-fortalecer-la-participación-ciudadana-con-políticas-de-impacto-regulatorio/

[1557] CoST Costa Rica, *CoST Infrastructure Transparency Initiative*, https://infrastructuretransparency.org/where/cost-costa-rica/

[1558] Ministerio de Ciencia, Innovación, Tecnología y Telecomunicaciones, *Consultas Públicas* (Dec. 2025), https://www.micitt.go.cr/transparencia/consultas-publicas

[1559] Ministerio de Ciencia, Innovación, Tecnología y Telecomunicaciones, *Consulta Pública no vinculante Estrategia Nacional de Ciberseguridad—Costa Rica* (2022), https://www.micitt.go.cr/node/1046

[1560] Ministerio de Ciencia, Innovación, Tecnología y Telecomunicaciones, *Estrategia Nacional de Inteligencia Artificial 2024-2027, Consulta Pública No Vinculante* (Mar. 22, 2024), https://www.micitt.go.cr/node/1574

[1561] Mario Bermúdez Vives, *MICITT Opposes Projects to Regulate Artificial Intelligence [MICITT se opone a proyectos para legislar sobre Inteligencia Artificial]*, Seminario Universidad (Jan. 29, 2025), https://semanariouniversidad.com/pais/micitt-se-opone-a-proyectos-para-legislar-sobre-inteligencia-artificial/

Data Protection

Costa Rica has two comprehensive laws that protect personal data. Law No. 8968 protects the personal data handling of individuals. Law No. 7975, the Undisclosed Information Law, specifies that the unauthorized disclosure of confidential and/or personal information is a crime.[1562] These laws, together with their by-laws, were enacted to regulate activities of companies that administer databases containing personal information.[1563] The laws require data subjects' consent before the processing or use of their data. Consent must cover in particular the purpose for data collection and how the data will be processed. The identity of all recipients and parties with access to personal data must be disclosed. Transfer of personal data to third countries is also conditional upon data subjects' consent.

The Agency for the Protection of Inhabitants' Data (*Agencia de Protección de Datos de los Habitantes*, Prodhab) is a fully independent body[1564] to "guarantee to any person, regardless of their nationality, residence, or home, the respect of their right to control information about their private life or activity and other rights of their personhood like the defense of their freedom and equality with respect to automated or manual treatment of the data related to their person or property."

Additional laws include Costa Rica's Executive Decree No. 37554-JP for data breach and the General Telecommunications Law No. 8642 (Article 42) that protects the privacy of communications and personal information. The General Telecommunications Law is supplemented by administrative regulation Nº 35205-MINAET that guarantees the secrecy of communications, the right to privacy, and the protection of personal data of subscribers and users.[1565]

After proposed amendments to align the data protection law with the GDPR stalled,[1566] the legislative assembly is discussing a new proposed law that would repeal Law 8968 and replace it.[1567] The proposed new legislation would establish a set of principles and rights for personal data protection to ensure proper handling of residents' personal data and enhance the level of protection for individuals regarding data processing. These revisions would align Costa Rica with international standards

---

[1562] Asamblea Legislativa, *No. 7975, Law of Undisclosed Information [Ley de Información No Divulgada]*, Gaceta Oficial No. 12 (Jan. 18, 2000), https://www.wipo.int/wipolex/es/text/208114

[1563] Asamblea Legislativa, *No. 8968, Protección de la Persona frente al tratamiento de sus datos personales* (Jul. 7, 2011), https://micitt.go.cr/sites/default/files/marco_juridico_legal/08. Ley n.° 8968 Ley de Protección de la Persona frente al tratamiento de sus datos personales..pdf; to latest version, https://micitt.go.cr/node/370

[1564] Agencia de Protección de Datos de los Habitantes, Prodhab, *Acerca de PRODHAB*, https://prodhab.go.cr/acercade/index.aspx

[1565] Sistema Costarricense de Información Jurídica, *Ley General de Telecomunicaciones*, https://www.pgrweb.go.cr/scij/Busqueda/Normativa/Normas/nrm_texto_completo.aspx?param1=NRTC&nValor1=1&nValor2=63431&nValor3=91176&strTipM=TC; https://micitt.go.cr/node/368

[1566] Giró and Martínez, *Costa Rica: Towards a Comprehensive Reform on Data Privacy* (Feb. 2021), https://www.giromartinez.com/news/costa-rica-comprehensive-reform-on-data-privacy/

[1567] Asamblea Legislative, *Ex. 23097, Pesonal Data Protection Law [Ley de Protección de Datos Personales]* (Oct. 17, 2024), https://delfino.cr/asamblea/proyecto/23097

on data privacy in a society increasingly shaped by information and communication technologies. The proposed law also aims to promote the development of mechanisms for international cooperation among control authorities in Ibero-American states, non-regional control authorities, and international entities specializing in data protection. The first reading of the law was approved and a letter from the Costa Rica Institute of Technology in June 2025 indicated the law was still under consideration.[1568]

The Ministry of Science, Innovation, Technology, and Communications included data protection as a main objective under the Ethical, Safe, and Responsible AI and Smart Government axes under the AI Action Plan.[1569]

As a member of the Ibero-American Network for the Protection of Personal Data (RIPD) which comprises 16 data protection authorities from 12 countries, the Prodhab endorsed the General Recommendations for the Processing of Personal Data in Artificial Intelligence[1570] and the accompanying Specific Guidelines for Compliance with the Principles and Rights that Govern the Protection of Personal Data in Artificial Intelligence Projects.[1571] The RIPD authorities initiated a coordinated action regarding ChatGPT, developed by OpenAI, in May 2023 on the basis that it may entail risks for the rights and freedoms of users in relation to the processing of their personal data and because it may "generate responses with cultural, racial or gender biases, as well as false ones."[1572]

Despite being an accredited member of the Global Privacy Assembly (GPA) since 2012,[1573] the Prodhab has not endorsed the 2018 GPA Resolution on Ethics and

---

[1568] Institutional Council of the Costa Rica Institute of Technology, *Pronouncement on the Draft Law "Ley de Protección de Datos Personales,"* SCI-467-2025, Session No. 3411, Article 10 (Jun. 11, 2025), https://www.tec.ac.cr/sites/default/files/media/doc/s3411-art-10-pronunciamiento-proyecto-ley-23097-sci-467-2025.pdf

[1569] Ministerio de Ciencia, Innovación, Tecnología y Telecomunicaciones, *National Artificial Intelligence Strategy Action Plan, Costa Rica 2024–2027 [Plan de Acción Estrategia Nacional de Inteligencia Artificial Costa Rica 2024–2027]* (Aug. 11, 2025), https://www.micitt.go.cr/sites/default/files/2025-10/Plan de Acción ENIA - Versión 11 agosto 2025 Versión Publicación.pdf

[1570] Ibero-American Network for the Protection of Personal Data (RIPD), *General Recommendations for the Processing of Personal Data in Artificial Intelligence* (Jun. 2019), https://www.redipd.org/en/document/guide-general-recommendations-processing-personal-data-ai-en.pdf

[1571] Ibero-American Network for the Protection of Personal Data (RIPD), *Specific Guidelines for Compliance with the Principles and Rights that Govern the Protection of Personal Data in Artificial Intelligence Projects* (Jun. 2019), https://www.redipd.org/en/document/guide-specific-guidelines-ai-projects-en.pdf

[1572] Ibero-American Network for the Protection of Personal Data (RIPD), *Las autoridades de la Red Iberoamericana de Protección de Datos Personales inician una acción coordinada en relación con el servicio ChatGPT* (May 8, 2023), https://www.redipd.org/noticias/autoridades-red-iberoamericana-de-proteccion-de-datos-personales-inician-accion-chatgpt

[1573] Global Privacy Assembly, *List of Accredited Members* (2026), https://globalprivacyassembly.com/participation-in-the-assembly/list-of-accredited-members/

Data Protection in AI,[1574] 2020 GPA Resolution on Accountability in th Development and Use of AI,[1575] 2022 GPA Resolution on the Appropriate use of personal Information in Facial Recognition Technology,[1576] or 2023 GPA Resolution on Generative AI Systems.[1577]

Algorithmic Transparency

The current data protection laws of Costa Rica acknowledge the right to equality and data control whether processing is automatic or manual.[1578] Bill 23097 under consideration by the Legislative Assembly (*Asamblea Legislativa*) does not refer to algorithms or AI but contains a general transparency principle, a right to data access, and the right to not be object to automated decisions with significant impact or without human intervention. [1579] Costa Rica's AI strategy emphasizes transparency in AI systems as necessary for building trust. The Strategy would require developers to ensure that AI processes are understandable and that automated decisions are explainable, especially in sensitive sectors like public administration.[1580] Regular audits and clear information on how AI algorithms work are recommended to avoid biases and maintain accountability, particularly for AI models impacting the public. Transparency is also one of the key principles of the 2022 National Code of Digital Technologies.[1581]

[1574] International Conference on Data, *Declaration on Ethics and Data Protection in Artificial Intelligence* (Oct. 2018), https://globalprivacyassembly.org/wp-content/uploads/2018/10/20180922_ICDPPC-40th_AI-Declaration_ADOPTED.pdf

[1575] Global Privacy Assembly, *Adopted Resolution on Accountability in the Development and Use of Artificial Intelligence* (Oct. 2020), https://globalprivacyassembly.org/wp-content/uploads/2020/10/FINAL-GPA-Resolution-on-Accountability-in-the-Development-and-Use-of-AI-EN-1.pdf

[1576] Global Privacy Assembly, *2022 GPA Resolution on Principles and Expectations for the Appropriate Use of Personal Information in Facial Recognition Technology* (Oct. 2022), https://globalprivacyassembly.com/wp-content/uploads/2022/11/15.1.c.Resolution-on-Principles-and-Expectations-for-the-Appropriate-Use-of-Personal-Information-in-Facial-Recognition-Technolog.pdf

[1577] Global Privacy Assembly, *Resolution on Generative Artificial Intelligence Systems* (Oct. 2023), https://globalprivacyassembly.com/wp-content/uploads/2023/10/5.-Resolution-on-Generative-AI-Systems-101023.pdf

[1578] Agencia de Protección de Datos de los Habitantes (PRODHAB), *Acerca de PRODHAB*, https://prodhab.go.cr/acercade/index.aspx

[1579] Asamblea Legislative, *Ex. 23097, Pesonal Data Protection Law [Ley de Protección de Datos Personales]* (Oct. 17, 2024), https://delfino.cr/asamblea/proyecto/23097

[1580] Ministerio de Ciencia, Innovación, Tecnología y Telecomunicaciones, *National AI Strategy (ENIA)*, pp. 38–77 (Oct. 27, 2024), https://www.micitt.go.cr/sites/default/files/2024-10/Estrategia Nacional de Inteligencia Artificial de Costa Rica ESP.pdf

[1581] Ministerio de Ciencia, Innovación, Tecnología y Telecomunicaciones, *National Code of Digital Technologies [Código Nacional de Tecnologías Digitales] (Current Versión)* (2024), https://www.micitt.go.cr/sites/default/files/GobernanzaDigital/CNTD version 2024 15-1-2024 EMA 25-1-2024GRMC-firmado.pdf

The RIPD Specific Guidelines provide clarity on the definition and requirements of "transparency," noting: "The information provided regarding the logic of the AI model must include at least basic aspects of its operation, as well as the weighting and correlation of the data, written in a clear, simple and easily understood language, it will not be necessary to provide a complete explanation of the algorithms used or even to include them. The above always looking not to affect the user experience."[1582]

Use of AI in Public Administration

The ENIA highlights AI's potential to streamline public administration by automating tasks, improving public service delivery, and enhancing decision-making efficiency.[1583] For instance, AI is used to manage health services, such as the EDUS platform by Costa Rica's Social Security Fund, which leverages AI for chronic disease management and prevention. The document underlines the need for ethical standards and effective human oversight in these applications to align with national values and ensure responsible deployment. The Smart Government axis (Axis 4) of the Action Plan for the ENIA[1584] establishes specific actions to promote the use of AI in the public sector. Milestones include publishing a digital catalog of AI projects in the public sector, developing the capacity and skills of public officials in the use and management of AI, and implementing systems that utilize AI to enhance the quality of public services and decision-making.

As a member of the Latin American Centre for Development Administration (CLAD), Costa Rica approved the Ibero American Charter on Artificial Intelligence in Civil Service in November 2023.[1585] The Charter aims to provide a roadmap and common framework for CLAD member states to learn about the challenges and opportunities involved in the implementation of AI and adapt their AI policy strategies and laws accordingly. The guiding principles are human autonomy; transparency, traceability and explainability; accountability, liability and auditability; security and technical robustness; reliability, accuracy, and reproducibility; confidence,

---

[1582] Ibero-American Network for the Protection of Personal Data (RIPD), *Specific Guidelines for Compliance with the Principles and Rights that Govern the Protection of Personal Data in Artificial Intelligence Projects*, p. 17 (Jun. 2019), https://www.redipd.org/en/document/guide-specific-guidelines-ai-projects-en.pdf

[1583] Ministerio de Ciencia, Innovación, Tecnología y Telecomunicaciones, *National AI Strategy Estrategia Nacional de Inteligencia Artificial (ENIA)*, p. 39 (Oct. 27, 2024), https://www.micitt.go.cr/sites/default/files/2024-10/Estrategia Nacional de Inteligencia Artificial de Costa Rica ESP.pdf

[1584] Ministerio de Ciencia, Innovación, Tecnología y Telecomunicaciones, *Action Plan of the National Artificial Intelligence Strategy [Plan de Acción de la Estrategía Nacional de Inteligencia Artificial]* (Aug. 14, 2025), https://www.micitt.go.cr/sites/default/files/2025-10/Plan de Acción ENIA - Versión 11 agosto 2025 Versión Publicación.pdf

[1585] Latin American Centre for Development Administration (CLAD), *Ibero American Charter on Artificial Intelligence in Civil Service* (Nov. 2023), https://web-api-backend.clad.org/uploads/ciia-en-03-2024.pdf

proportionality, and prevention of damage; privacy and protection of personal data; data quality and safety; fairness, inclusiveness, and non-discrimination; human-centering, public value, and social responsibility; and sustainability and environmental protection. In addition to common principles, the Charter recommends specific tools for monitoring AI impact as well as a risk classification mechanism that guardrails for high-risk use cases and prohibitions against uses that violate human rights.[1586]

Use of AI in Courts

While Costa Rica's judiciary has not yet fully integrated AI, the ENIA envisions a future where AI could assist with case management, evidence analysis, and even support decision-making. The strategy stresses that any use of AI in the legal system must be transparent, accountable, and subject to human oversight to uphold justice and prevent discrimination or errors in judgment.[1587]

A new AI system called Prometea developed by the Innovation and AI Laboratory of the School of Law of the University of Buenos Aires and the Public Prosecutor's Office of the Autonomous City of Buenos Aires, has been implemented in the judicial system of the Inter-American Court of Human Rights in Costa Rica.[1588] Regardless of whether judicial cases are simple (e.g. minor infractions, traffic accidents, and taxi license disputes) or complex (e.g. murder trials), controversy lies in the inevitable biases of the AI-enabled system and the lack of algorithmic transparency behind the decision-making process. Transparency of the AI system's algorithm, therefore, must be assured to protect due process and the rule of law.

AI and Hiring

The use of data analytics and AI techniques for hiring, firing, and promoting employees in Costa Rica has also been very controversial due to AI systems' bias and lack of transparency. Companies recruiting in Costa Rica must adapt their recruiting processes to accommodate Costa Rican law, especially with regard to discrimination. However, controversy persists because Costa Rican law does not directly regulate the impact or effects of AI on recruitment.[1589]

[1586] Ibid, p. 21

[1587] Ministerio de Ciencia, Innovación, Tecnología y Telecomunicaciones, *National AI Strategy [Estrategia Nacional de Inteligencia Artificial] (ENIA)*, p. 78 (Oct. 27, 2024), https://www.micitt.go.cr/sites/default/files/2024-10/Estrategia Nacional de Inteligencia Artificial de Costa Rica ESP.pdf

[1588] F. Giandana and D. Morar, *Victor Frankenstein's Responsibility? Determining AI Legal Liability in Latin America,* Global Information Society Watch (2019), https://giswatch.org/ar/node/6178

[1589] Alvaro Aguilar, Aguilar Castillo Love, *Costa Rica - Artificial Intelligence Bias and Data Transparency in the Legal Workforce: The Use of Data Analytics for Hiring, Firing, and Promotion in Costa Rica* (Jun. 3, 2019), https://www.americanbar.org/groups/labor_law/publications/ilelc_newsletters/issue-june-2019/artificial-intelligence-bias-and-data-transparency/

Costa Rica has piloted a skills and training platform that uses AI to "provide skill development suggestions to participants, based on their abilities, interests and experiences." [1590]

The ENIA addresses the role of AI in the hiring process, acknowledging its potential to make recruitment more efficient and objective.[1591] However, it also highlights the importance of preventing algorithmic biases that could affect vulnerable groups. Costa Rica's AI strategy encourages transparency in AI-driven hiring tools so applicants understand how decisions are made. The ENIA also supports providing opt-out capability to ensure ethical AI practices in employment.

According to a September 2024 ManpowerGroup survey, Costa Rica ranks among the top countries with strong hiring intentions, with 36% of employers planning to increase their workforce. This positive outlook is partly driven by advancements in technology and the integration of AI in various sectors.[1592]

Environmental Impact of AI

Costa Rica's ENIA centers sustainability and well-being as guiding principles,[1593] stipulating that AI systems must continuously evaluate the impacts of these technologies on society, animal life, and the environment. Given the environmental impact of AI represents one of the key risks associated with AI in Costa Rica, the country highlights the urgency of developing AI sustainably, prioritizing energy efficiency, and adopting renewable energy sources. Axis 1 of the AI Strategy Action Plan calls for technical recommendations to incorporate sustainability and energy efficiency criteria into AI development.[1594] Axis 4, Smart Government, calls for training, awareness, and resource exchange to promote the "use, development, and sustainable management of AI."[1595]

[1590] OECD Public Governance Reviews, *The Strategic and Responsible Use of Artificial Intelligence in the Public Sector of Latin America and the Caribbean*, p. 51 (2022), https://oecd-opsi.org/wp-content/uploads/2022/03/lac-ai.pdf

[1591] Ministerio de Ciencia, Innovación, Tecnología y Telecomunicaciones, *Estrategia Nacional de Inteligencia Artificial (ENIA)*, p. 43 (Oct. 27, 2024), https://www.micitt.go.cr/sites/default/files/2024-10/Estrategia Nacional de Inteligencia Artificial de Costa Rica ESP.pdf

[1592] Reuters, *Global Hiring Intentions Hold Steady in Q4, ManpowerGroup Survey Shows* (Sept. 10, 2024), https://www.reuters.com/markets/global-hiring-intentions-hold-steady-q4-manpowergroup-survey-shows-2024-09-10/

[1593] Ministerio de Ciencia, Innovación, Tecnología y Telecomunicaciones, *The National Artificial Intelligence Strategy (ENIA) of Costa Rica 2024-2027 [Simplified English Version]* (Oct. 24, 2024), https://www.micitt.go.cr/sites/default/files/2025-02/National Artificial Intelligence Strategy of Costa Rica.pdf

[1594] Ministerio de Ciencia, Innovación, Tecnología y Telecomunicaciones, *National Artificial Intelligence Strategy Action Plan, Costa Rica 2024–2027 [Plan de Acción Estrategia Nacional de Inteligencia Artificial Costa Rica 2024–2027]*, p. 6 (Aug. 11, 2025), https://www.micitt.go.cr/sites/default/files/2025-10/Plan de Acción ENIA - Versión 11 agosto 2025 Versión Publicación.pdf

[1595] Ibid, p. 22

Lethal Autonomous Weapons

Costa Rica is one of the 126 member states signatory to the Convention on Certain Conventional Weapons (CCW) and participated in CCW meetings. During the UN General Assembly in October 2013, Costa Rica proposed that critical functions of weapons systems be subject to meaningful human control highlighting the risks and harms identified with the use of armed drones and robotic weapons. In 2016 Costa Rica called for a preemptive ban on lethal autonomous weapons systems.[1596]

Costa Rica has continued to advocate for the ban on lethal autonomous weapons. In 2020 the Ministry of Foreign Affairs, together with the Foundation for Peace and Democracy (FUNPADEM), called for the prohibition of the use of Lethal Autonomous Weapons (LAWs).[1597] Still in 2020, Costa Rica advocated for the development of international legally binding agreements providing for prohibitions and regulations on autonomous weapons systems at the Group of Governmental Experts (GGE) on emerging technologies in the area of LAWS.[1598]

Costa Rica was among the 70 countries that endorsed a joint statement on autonomous weapons systems at the 2022 UN General Assembly meeting. In this joint statement, States urged "the international community to further their understanding and address these risks and challenges by adopting appropriate rules and measures, such as principles, good practices, limitations and constraints. We are committed to upholding and strengthening compliance with International Law, in particular International Humanitarian Law (IHL), including through maintaining human responsibility and accountability in the use of force."[1599]

In February 2023, the Costa Rican Ministry of Foreign Relations, together with the Foundation for Peace and Democracy, organized the Latin American and Caribbean Conference on the social and humanitarian impact of autonomous weapons. The Conference resulted in the endorsement by Costa Rica and more than

---

[1596] Human Right Watch, *Stopping Killer Robots: Country Positions on Banning Fully Autonomous Weapons and Retaining Human Control* (2020), https://www.hrw.org/report/2020/08/10/stopping-killer-robots/country-positions-banning-fully-autonomous-weapons-and - :~:text=Costa Rica,-At the UN&text=%5B74%5D It called for a,killer robots in 2016-2019

[1597] *Fundación Arias y Cancillería de Costa Rica firman Convenio de Cooperación sobre el TCA y POA*, https://att-assistance.org/es/node/6601

[1598] UN Office of Disarmament Affairs, *Intervention from Costa Rica*, CCW GGE LAWS (Sept. 21, 2020), https://documents.unoda.org/wp-content/uploads/2020/09/Intervention-of-Costa-Rica-GGE-Laws-21.09.2020-ENG-PDF.pdf

[1599] United Nations (UN) General Assembly, First Committee, *Joint Statement on Lethal Autonomous Weapons Systems First Committee, 77th United Nations General Assembly Thematic Debate – Conventional Weapons* (Oct. 21, 2022), https://estatements.unmeetings.org/estatements/11.0010/20221021/A1jJ8bNfWGlL/KLw9WYcSnnAm_en.pdf

30 other Latin American and Caribbean states, of the Belén Communiqué.[1600] The Belén Communiqué calls for "urgent negotiation" of a binding international treaty to regulate and prohibit the use of autonomous weapons to address the grave concerns raised by removing human control from the use of force.

At the 78th UN General Assembly First Committee in 2023, Costa Rica co-authored draft resolution L.56[1601] on autonomous weapons systems. The Resolution emphasized the "urgent need for the international community to address the challenges and concerns raised by autonomous weapons systems," and mandated the UN Secretary-General to prepare a report reflecting the views of member and observer states on autonomous weapons systems. Costa Rica's submission to the report reiterated its support for a legally binding instrument.[1602]

Costa Rica voted in favor of Draft Resolution L.77 on lethal autonomous weapons systems at the 79th UN General Assembly First Committee in October 2024. The draft drew attention to the "negative consequences and impact of autonomous weapon systems on global security and regional and international stability" and marked the urgent necessity of human interventions to ensure accountability, responsibility, and compliance toward international law. L.77 also noted the Secretary-General's proposal on enforcing legally binding prohibitions and regulations for autonomous weapons systems.[1603] Speaking at the assembly, Costa Rica reiterated the potential harms caused by lethal autonomous weapons and called for legally binding instruments to prevent the deployment of such weapons.[1604]

AI Literacy

The ENIA[1605] identifies AI literacy and training as essential to ensure that all members of society can fully benefit from the technology, and tasks the Ministry of

---

[1600] *Communiqué of the Latin American and the Caribbean Conference of Social and Humanitarian Impact of Autonomous Weapons* (Feb. 24, 2023), https://www.rree.go.cr/files/includes/files.php?id=2261&tipo=documentos

[1601] General Assembly, *Lethal Autonomous Weapons*, Resolution L56 (Oct. 12, 2023), https://reachingcriticalwill.org/images/documents/Disarmament-fora/1com/1com23/resolutions/L56.pdf

[1602] UN Office for Disarmament Affairs, *Costa Rica: Posición Nacional sobre los sistemas de armas autónomas* (May 2024), https://docs-library.unoda.org/General_Assembly_First_Committee_-Seventy-Ninth_session_(2024)/78-241-Costa_Rica-SP.pdf

[1603] United Nations General Assembly First Committee, *General and Complete Disarmament: Lethal Autonomous Weapons Systems* (Oct. 18, 2024), https://documents.un.org/doc/undoc/ltd/n24/305/45/pdf/n2430545.pdf

[1604] Misión Permanente de Costa Rica, *United Nations General Assembly First Committee Conventional Weapons* (Oct 23, 2024), https://reachingcriticalwill.org/images/documents/Disarmament-fora/1com/1com24/statements/23Oct_Costa_Rica.pdf

[1605] Ministerio de Ciencia, Innovación, Tecnología y Telecomunicaciones, *The National Artificial Intelligence Strategy (ENIA) of Costa Rica 2024-2027 [Simplified English Version]*, pp. 22–23 (Oct. 24, 2024), https://www.micitt.go.cr/sites/default/files/2025-02/National Artificial Intelligence Strategy of Costa Rica.pdf

Science, Innovation, Technology, and Telecommunications with coordinating public AI literacy efforts and training for public officials. The ENIA identifies improvements to communications infrastructure and adaptation of training to local and regional languages and dialects as core dependencies of implementing AI training nationwide.

The AI Strategy Action Plan incorporates objectives for training and talent development into Axis 5, [1606] with the expected result of formulating and executing an AI literacy strategy through Intelligent Community Centers (*Centro Comunitario Inteligentes, CECI*). This strategy includes awareness campaigns, accessible educational resources, and training activities for the community. This axis also targets the implementation of workforce reskilling and upskilling programs, directed to engineers and personnel from various areas who can benefit from AI technologies in their roles. Other efforts are directed to end-users, through guidance from trained personnel on the application of AI technologies to improve their tasks.

Human Rights

Freedom House ranked Costa Rica "Free" with an overall score of 91/100 in the 2026 Freedom in the World report. [1607] Costa Rica also ranked as "Free" with a score of 86/100 for freedom on the internet.[1608] However, the report cited "worsening online intimidation," especially against journalists critical of the president. "Politicalization in the allocation of state advertising" to media outlets also threatens Costa Ricans' online freedom. Human rights in Costa Rica predominantly stem from the UDHR, the country's Constitution, and the Inter-American Human Rights System.

OECD / G20 AI Principles

Costa Rica endorsed the OECD AI Principles for designing safe, fair, trustworthy and robust AI systems[1609] before becoming a member of OECD on May 25, 2021.[1610]

Several policies proposed in Costa Rica's national digital transformation strategy, in which AI is included, align with the OECD AI principles. These include

---

[1606] Ministerio de Ciencia, Innovación, Tecnología y Telecomunicaciones, *Action Plan of the National Artificial Intelligence Strategy [Plan de Acción de la Estrategía Nacional de Inteligencia Artificial]* (Aug. 14, 2025), https://www.micitt.go.cr/sites/default/files/2025-10/Plan de Acción ENIA - Versión 11 agosto 2025 Versión Publicación.pdf

[1607] Freedom House, *Freedom in the World 2025: Costa Rica* (2026), https://freedomhouse.org/country/costa-rica/freedom-world/2026

[1608] Freedom House, *Freedom on the Net 2024: Costa Rica* (2025), https://freedomhouse.org/country/costa-rica/freedom-net/2025

[1609] OECD Legal Instruments, *Recommendation of the Council on Artificial Intelligence, Adherents* (May 3, 2024), https://legalinstruments.oecd.org/en/instruments/OECD-LEGAL-0449#adherents

[1610] Banco Interamericano de Desarrollo, *La importancia de establecer un marco orientador de política pública para el uso responsable y ético de la inteligencia artificial y su aplicación en Costa Rica* (Nov. 2021), https://publications.iadb.org/es/la-importancia-de-establecer-un-marco-orientador-de-politica-publica-para-el-uso-responsable-y

a focus on inclusive growth; human-centered values and fairness; transparency; robustness, security, and safety; and accountability.[1611] The National AI Strategy[1612] and related Action Plan[1613] further these principles and Costa Rica's commitment to ethical, reliable human-centered AI for sustainable development.

Council of Europe AI Treaty

Costa Rica contributed as a non-member[1614] to the negotiations of the Council of Europe Framework Convention on AI, Human Rights, Democracy and the Rule of Law. Despite this participation, Costa Rica has not signed this first legally binding international AI treaty.[1615]

UNESCO Recommendation on AI Ethics

Costa Rica is one of the 193 countries that adopted the UNESCO Recommendation on the Ethics of AI during the 41st General Conference in November 2021. The instrument is the first of its kind that considers ethics as a basis for normative evaluation and guidance of AI technologies with reference to human dignity and wellbeing, and the prevention of harm.[1616]

CAF, the development bank of Latin America, and UNESCO pledged to create a Regional Council composed of national and local governments in the region to support implementation of the UNESCO Recommendation. Costa Rica is a member of the Regional Council,[1617] which held the inaugural meeting in October 2023 in

---

1611 Ministerio de Ciencia, Innovación, Tecnología y Telecomunicaciones, *Estrategia de Transformación Digital, Costa Rica 2023-2027* (Aug. 2023), https://www.micitt.go.cr/sites/default/files/GobernanzaDigital/ETD%202023-2027%20V%20FINAL%2030-08-2023_v2.pdf

1612 Ministerio de Ciencia, Innovación, Tecnología y Telecomunicaciones, *The National Artificial Intelligence Strategy (ENIA) of Costa Rica 2024-2027 [Simplified English Version]*, pp. 22–23 (Oct. 24, 2024), https://www.micitt.go.cr/sites/default/files/2025-02/National Artificial Intelligence Strategy of Costa Rica.pdf

1613 Ministerio de Ciencia, Innovación, Tecnología y Telecomunicaciones, *Action Plan of the National Artificial Intelligence Strategy [Plan de Acción de la Estrategía Nacional de Inteligencia Artificial]* (Aug. 14, 2025), https://www.micitt.go.cr/sites/default/files/2025-10/Plan de Acción ENIA - Versión 11 agosto 2025 Versión Publicación.pdf

1614 Council of Europe, *Framework Convention on Artificial Intelligence* (2026), https://www.coe.int/en/web/artificial-intelligence/the-framework-convention-on-artificial-intelligence

1615 Council of Europe Treaty Office, *Chart of Signatures and Ratifications of Traty 225* (Mar. 19, 2025), https://www.coe.int/en/web/Conventions/full-list/?module=signatures-by-treaty&treatynum=225

1616 UNESCO, *Ethics of Artificial Intelligence: The Recommendation*, https://www.unesco.org/en/artificial-intelligence/recommendation-ethics

1617 Ángel Melguizo and Gabriela. Ramos, *Inteligência Artificial ética e responsável: das palavras aos fatos e direitos*, Somos Ibero-America (Feb. 1, 2023), https://www.somosiberoamerica.org/pt-br/tribunas/inteligencia-artificial-etica-e-responsavel-das-palavras-aos-fatos-e-direitos/

Santiago de Chile.[1618] Costa Rica signed the resulting Santiago Declaration to Promote Ethical Artificial Intelligence.[1619] The Declaration aligns with the UNESCO Recommendation and establishes fundamental principles that should guide public policy on AI: proportionality, security, fairness, non-discrimination, gender equality, accessibility, sustainability, privacy and data protection.[1620]

Costa Rica also participated in the Second Ministerial and High Authorities Summit on the Ethics of Artificial Intelligence in Latin America and the Caribbean held in Montevideo, Uruguay in 2024.[1621] The summit consolidated the work proposed in the Ministerial Summit in Santiago, seeking to advance collaborative and sustainable AI-regulations efforts for Latin America and the Caribbean in the Montevideo Declaration.[1622]

The Ministry of Science, Innovation, Technology and Telecommunications in Costa Rica developed the National AI Strategy released in 2024 through collaboration with UNESCO to align the strategy with the UNESCO Recommendation.[1623]

Costa Rica is currently in process of completing the UNESCO Readiness Assessment Methodology (RAM),[1624] a tool to support the effective implementation of the Recommendation. The RAM helps countries and UNESCO identify and address any institutional and regulatory gaps.[1625]

[1618] UNESCO, *Chile will host the First Latin American and Caribbean Ministerial and High Level Summit on the Ethics of Artificial Intelligence* (Sept. 25, 2023), https://www.unesco.org/en/articles/chile-will-host-first-latin-american-and-caribbean-ministerial-and-high-level-summit-ethics

[1619] Cumbre Ministerial y de Altas Autoridades de América Latina y el Caribe, *Declaracion de Santiago "Para promover una inteligencia artificial ética en América Latina y el Caribe"* (Oct. 2023), https://minciencia.gob.cl/uploads/filer_public/40/2a/402a35a0-1222-4dab-b090-5c81bbf34237/declaracion_de_santiago.pdf

[1620] UNESCO, *UNESCO and Leading Ministry in Santiago de Chile Host Milestone Regional LAC Forum on Ethics of AI* (Dec. 5, 2023), https://www.unesco.org/en/articles/unesco-and-leading-ministry-santiago-de-chile-host-milestone-regional-lac-forum-ethics-ai?hub=387.

[1621] Development Bank of Latin America and the Caribbean, *Second Ministerial Summit on Ethics of Artificial Intelligence in Latin America and the Caribbean* (Oct. 2024), https://www.caf.com/en/currently/events/2024/10/online-second-ministerial-summit-on-the-ethics-of-artificial-intelligence-in-latin-america-and-the-caribbean/

[1622] Cumbre sobre la Ética de la Inteligencia Artificial de América Latina y el Caribe, *Declaration of Montevideo "For the construction of a regional approach on the governance of Artificial Intelligence and its impacts on our society"* (Oct. 4, 2024), https://www.gub.uy/agencia-gobierno-electronico-sociedad-informacion-conocimiento/sites/agencia-gobierno-electronico-sociedad-informacion-conocimiento/files/documentos/noticias/EN - Montevideo Declaration approved.pdf

[1623] Mariana Alvarez, *Costa Rica Will Be the First Country in Central America to Have an Artificial Intelligence Strategy* (Mar. 1, 2023), https://www.unesco.org/en/articles/costa-rica-will-be-first-country-central-america-have-artificial-intelligence-strategy

[1624] UNESCO Global AI Ethics and Governance Observatory, *Global Hub* (Oct. 2025), https://www.unesco.org/ethics-ai/en/global-hub

[1625] UNESCO Global AI Ethics and Governance Observatory, *Readiness Assessment Methodology* https://www.unesco.org/ethics-ai/en/ram

Evaluation

Costa Rica adopted a national AI strategy aligned with the UNESCO Recommendation through collaboration with UNESCO and then created an Action Plan for implementation. Costa Rica has specifically endorsed worldwide initiatives focused on the ethical and responsible use of AI, including the OECD AI principles and the UNESCO Recommendation on the Ethics of AI, and has also participated in regional efforts to implement the frameworks. Costa Rica has an independent data protection agency charged with enforcing the comprehensive data protection law regime. The extensive use of AI both in public and private sectors calls, however, for protection of the right to algorithmic transparency.

Czechia

In 2025, Czechia approved a roadmap for the AI Strategy and submitted legislation to implement the EU AI Act. Czechia also launched a national AI skills initiative to build basic AI knowledge and skills in the public administration and across society.

National AI Strategy

The Czech Ministry of Industry and Trade (MIT) announced[1626] the approval of an updated National AI Strategy of the Czech Republic 2030[1627] (NAIS 2030) by government resolution on July 24, 2024. This updated strategy represents a fundamental shift from Czechia's original 2019 National Artificial Intelligence Strategy's general aspirational guidelines approach to a more structured and accountable framework. The updated NAIS moves from voluntary guidelines to specific requirements with designated oversight and enforcement mechanisms, reflecting Czechia's commitment to following through on EU regulatory standards and to ensuring the cohesive implementation of AI ethics and funding mechanisms for AI in the country. This includes a mandatory implementation framework with regular reporting requirements, specific timelines for implementation, clear accountability measures, and concrete funding commitments.[1628]

The NAIS 2030 specifically refers, multiple times, across multiple contexts, to: 1) Fairness (*spravedlnost*); 2) Accountability (*odpovědnost*); 3) Transparency (*transparentnost*); 4) Rule of Law (*právnístát*); 5) Legal Force (*právnísíla*); and 6) Enforcement (*vymáhání*).[1629] The 2019 NAIS established a standard of "maintaining a high level of protection of fundamental and other rights, in line with the European approach of human-centric AI."[1630]

The updated NAIS 2030 includes a restructuring around seven key areas: 1) AI in Research, Development and Innovation, 2) Education and Expertise in AI, 3) AI Skills and Labor Market Impacts; 4) Ethical and Legal Aspects of AI, 5) Security Aspects of AI, 6) AI in Industry and Business, and 7) AI in Public Administration and

---

[1626] Ministry of Industry and Trade, *Czechia as a Technological Leader: Government Approved the National Strategy for Artificial Intelligence of the Czech Republic 2030* (Jul. 24, 2024), https://mpo.gov.cz/en/guidepost/for-the-media/press-releases/czechia-as-a-technological-leader--government-approved-the-national-strategy-for-artificial-intelligence-of-the-czech-republic-2030--282278

[1627] Ministry of Industry and Trade, *National AI Strategy of the Czech Republic 2030 [Národní strategii umělé inteligence čr 2030]* (Jul. 24, 2024), https://mpo.gov.cz/assets/cz/podnikani/digitalni-ekonomika/umela-inteligence/2024/8/Narodni-strategie-umele-intelience-CR-2030.pdf

[1628] Ibid, pp. 47–48

[1629] Ibid, pp. 5–61

[1630] Ministry of Industry and Trade of the Czech Republic, *National Artificial Intelligence Strategy of the Czech Republic*, pp. 3, 36 (2019), https://www.mpo.cz/assets/en/guidepost/for-the-media/press-releases/2019/5/NAIS_eng_web.pdf

Public Services.[1631] International Cooperation is integrated across all areas, and Ethics and Legal Aspects of AI is now a standalone are of focus, after previously being combined as two among five topics in the 2019 NAIS.[1632]

The Deputy Prime Minister and Minister of Industry and Trade are directly responsible for the NAIS implementation, coordinating through the newly established AI Committee. The AI committee is a subcommittee of the Digital Czech Republic Steering Committee. The Committee primarily includes responsible ministries who are competent to coordinate each specific key area of the Strategy and will be in charge of the NAIS operational management. Working groups are affected to each of the NAIS seven key objectives. Once a year, the Steering Committee and the Czech Government will receive a progress report on the NAIS implementation. The 2025 Action Plan attached to NAIS 2030 earmarks significant public funding to priority projects across R&D, skills, public administration use, and SME adoption, strengthening the credibility of delivery commitments.[1633]

The 2019 NAIS provided context for the updated version.[1634] The NAIS follows up on the commitment the Czech Republic undertook by signing, together with other EU Member States the 2018 EU Declaration of Cooperation on Artificial Intelligence.[1635] The NAIS is part of the Innovation Strategy of the Czech Republic 2019–2030[1636] and the Digital Czech Republic strategy.[1637] The 2019 NAIS referenced work by the UN, OECD, and Council of Europe as guides for addressing the legal, societal, and ethical aspects of AI. Introducing the 2019 NAIS, the prime minister stated, "We are going to focus on protecting every person and consumer, their rights and privacy, especially the weakest ones. We are going to prevent discrimination, manipulation and misuse of AI, we are going to set the rules for decision-making of algorithms about people in everyday life."[1638]

The National AI Strategy of the Czech Republic 2030 continues the path first established with the 2019 NAIS to embody, and now put into more specific action,

---

[1631] Ibid, pp. 32–61

[1632] Ibid, p. 5

[1633] Ministry of Industry and Trade, *The Government of the Czech Republic Approved the 2025 Action Plan of the National AI Strategy 2030 [Vláda ČR schválila Akční plán Národní strategie umělé inteligence ČR 2030 pro rok 2025]* (Apr. 2, 2025), https://mpo.gov.cz/cz/podnikani/digitalni-ekonomika/vlada-cr-schvalila-akcni-plan-narodni-strategie-umele-inteligence-cr-2030-pro-rok-2025--287000/

[1634] Ministry of Industry and Trade, *National Artificial Intelligence Strategy of the Czech Republic* (2019), https://www.mpo.cz/assets/en/guidepost/for-the-media/press-releases/2019/5/NAIS_eng_web.pdf

[1635] European Commission, *EU Member States Sign Up to Cooperate on Artificial Intelligence* (Apr. 10. 2018), https://digital-strategy.ec.europa.eu/en/news/eu-member-states-sign-cooperate-artificial-intelligence

[1636] European Commission AI Watch, *Czech Republic AI Strategy Report* (Sept. 1, 2021), https://ai-watch.ec.europa.eu/countries/czech-republic/czech-republic-ai-strategy-report_en

[1637] Ministry of Industry and Trade, *Digital Czech Republic* (Feb. 12, 2019), https://www.mpo.cz/en/business/digital-society/digital-czech-republic--243601/

[1638] Ibid, p. 3

the Government's "commitment to becoming one of Europe's innovative leaders and a country of the technological future within twelve years."[1639] This translates into a position cautioning the EU against overregulation in view of the adoption of an EU AI Act, following the release of the European Commission's White Paper on AI.[1640] In November 2019, the Czech Republic prepared a position non-paper for the EU on the "Regulatory Framework for AI in the EU."[1641] In the non-paper, the Czech Republic recommended to refrain from initial overregulation of AI, to promote self-regulation and soft-law based on best practices, and to define horizontal red lines as a means of ensuring the protection of fundamental rights as well as legal certainty. The Czech Republic also emphasized that "securing the safety of citizens is the very precondition for the true implementation of fundamental human rights and freedoms."

The Czech Republic is also one of 14 EU Member States that urged the Commission in October 2020 to push for as little regulation as possible in the AI field to find a balance between setting up rules and ensuring fast AI development. In their non-paper, Czechia and other EU Member States set out two visions for the EU's development of AI: (1) promoting innovation, while managing risks through a clear framework and (2) establishing trustworthy AI as a competitive advantage.[1642]

EU Digital Services Act

As an EU member state, Czechia shall apply the EU Digital Services Act (DSA).[1643] The DSA regulates online intermediaries and platforms to prevent illegal and harmful activities online and the spread of disinformation.

Czechia designated the Czech Telecommunication Office (ČTÚ) as the Digital Services Coordinator (DSC)[1644] charged with monitoring platform compliance, transparency, and online content governance. Although trusted-flagger systems and penalty mechanisms are still being finalized, the coordination framework is fully

---

[1639] Ibid, p. 3

[1640] Union of Industry and Transport of the Czech Republic, *Joint Position on AI White Paper* (Jun. 16, 2020), https://www.spcr.cz/images/Joint_Position_on_AI_White_Paper_CEEcountries.pdf

[1641] AI Observatory and Forum, *Regulatory Framework for Artificial Intelligence in the European Union: Non-paper of the Czech Republic* (Feb. 15, 2020), http://observatory.ilaw.cas.cz/index.php/2020/02/15/czech-republics-non-paper-on-ai-regulatory-framework-in-the-eu/

[1642] Denmark Ministry of Industry, Business and Financial Affairs, *Innovative and Trustworthy AI: Two Sides of the Same Coin*, Position Paper on behalf of Denmark, Belgium, the Czech Republic, Finland, France Estonia, Ireland, Latvia, Luxembourg, the Netherlands, Poland, Portugal, Spain and Sweden (Oct. 25, 2023), https://www.em.dk/media/15214/non-paper-innovative-and-trustworthy-ai-two-side-of-the-same-coin.pdf

[1643] EUR-Lex, *Regulation (EU) 2022/2065 of the European Parliament and of the Council of 19 October 2022 on a single market for digital services and amending Directive 2000/31/EC (Digital Services Act) (Text with EEA relevance)*, L 277, 1–102 (Oct. 21, 2022), https://eur-lex.europa.eu/legal-content/EN/TXT/?uri=celex%3A32022R2065

[1644] Český Telekomunikační Úřad (ČTÚ), *The Role of the Digital Services Coordinator* (2025), https://ctu.gov.cz/en/role-digital-services-coordinator

functional and demonstrates growing regulatory maturity. However, in 2025 the European Commission referred Czechia to the Court of Justice of the EU for not equipping the DSC with all required enforcement powers and penalties, indicating institutional capacity gaps to be addressed.[1645]

EU AI Act

As an EU member State, Czechia is bound by the EU AI Act.[1646] The EU AI Act is a risk-based market regulation that aims to promote a human-centric approach to AI and make the EU a global leader in the development of secure, trustworthy, and ethical AI.

Czechia designated the Personal Data Protection Office (*Úřad pro ochranu osobních údajů*) and Public Defender of Rights (Veřejný ochránce práv) as national competent authorities to ensure the protection of fundamental rights.[1647] The AI Committee under the Ministry of Industry and Trade, responsible for implementing the NAIS,[1648] is also planning implementation of the AI Act.[1649] The Ministry of Industry and Trade submitted a bill to Parliament[1650] in September 2025 that designates the Telecommunications Office (ČTÚ) as the single point of contact.[1651]

[1645] European Commission, *Commission Decides to Refer Czechia, Spain, Cyprus, Poland and Portugal to the Court of Justice of the European Union Due to Lack of Effective Implementation of the Digital Services Act* (May 7, 2025), https://digital-strategy.ec.europa.eu/en/news/commission-decides-refer-czechia-spain-cyprus-poland-and-portugal-court-justice-european-union-due

[1646] European Parliament, Artificial Intelligence Act, European Parliament legislative resolution of 13 March 2024 on the proposal for a regulation of the European Parliament and of the Council on laying down harmonised rules on Artificial Intelligence (Artificial Intelligence Act) and amending certain Union Legislative Acts (COM(2021)0206 – C9-0146/2021 – 2021/0106(COD)), P9_TA(2024)0138, https://www.europarl.europa.eu/RegData/seance_pleniere/textes_adoptes/definitif/2024/03-13/0138/P9_TA(2024)0138_EN.pdf

[1647] European Commission, *Fundamental Rights Protection Authorities with Special Powers under the AI Act* (Jul. 25, 2025), https://digital-strategy.ec.europa.eu/en/policies/fundamental-rights-protection-authorities-ai-act

[1648] Ministry of Industry and Trade, *Czechia as a Technological Leader: Government Approved the National Strategy for Artificial Intelligence of the Czech Republic 2030* (Jul. 24, 2024), https://mpo.gov.cz/en/guidepost/for-the-media/press-releases/czechia-as-a-technological-leader--government-approved-the-national-strategy-for-artificial-intelligence-of-the-czech-republic-2030--282278

[1649] Ministry of Industry and Trade, *Artificial Intelligence Committee Discussed Further Development of AI in the Czech Republic* (May 14, 2025), https://mpo.gov.cz/en/guidepost/for-the-media/press-releases/artificial-intelligence-committee-discussed-further-development-of-ai-in-the-czech-republic--287662/

[1650] ODok Portal, *Draft Act on Artificial Intelligence and on Amendments to Act No. 87/2023 Coll., on Market Surveillance of Products and on Amendments to Certain Related Acts, as Amended [Návrh zákona o umělé inteligenci a o změně zákona č. 87/2023 Sb., o dozoru nad trhem s výrobky a o změně některých souvisejících zákonů, ve znění pozdějších předpisů]* (Oct. 24, 2025), https://odok.gov.cz/portal/veklep/material/KORNDLSJSEUC/

[1651] Ministry of Industry and Trade, *The Ministry of Industry and Trade Has Prepared a Draft Law on Artificial Intelligence* (Sept. 26, 2025), https://mpo.gov.cz/en/guidepost/for-the-media/press-

The National Bank and Office for Personal Data Protection also have oversight duties while the Office for Technical Standardization will be the notifying authority and oversee conformity assessments.

Public Participation

Public participation to formulate the 2024 National AI Strategy was formalized and inclusive. The public participation phase for preparing the update took place from June 28–August 20, 2023, and included 517 respondents from the general public, public administration representatives, private sector stakeholders, academic institutions, research organizations, and non-profit organizations.[1652] "To ensure broad-based support and relevance, the MPO consulted with public and private representatives. […] We were inspired by the approaches of countries like Canada, the USA, the United Kingdom, and the Netherlands. We organized an online roundtable where experts from these countries presented their approaches to AI. The roundtable served as a platform for exchanging information and best practices," said Petr Očko, Chief Director for Digitalization and Innovation at the MPO.[1653]

This approach demonstrates more active commitment to public participation, including the use of Expert Working Groups, organized by the Aspen Institute in cooperation with prg.ai, under the auspices of the Ministry of Industry and Trade. However, it is unclear exactly how the public was involved in the consultation, which appears to possibly have focused mostly on experts, industry, and academics with some inclusion of NGOs and social partners. Despite the lack of clear insights into this process, it is notable that there appears to be a dedicated effort by key Czech AI leaders to embrace the need for, and execution of, robust public participation, perhaps in response to the lack of civil-society engagement, such as by human rights organizations, in the 2019 strategy.[1654]

In February 2020, Czechia launched the AI Observatory and Forum, an expert platform on legal aspects of AI to create a favorable social and legal environment for research, development and use of responsible AI.[1655] The platform is tasked with identifying legislative obstacles for research, development and use of AI and offering recommendations on their removal, developing legal and ethical recommendations for

---

releases/the-ministry-of-industry-and-trade-has-prepared-a-draft-law-on-artificial-intelligence----289865/

[1652] Ministry of Industry and Trade, *National AI Strategy of the Czech Republic 2030* [Czech only], pp. 11–12 (Jul. 24, 2024), https://www.mpo.gov.cz/assets/cz/podnikani/digitalni-ekonomika/umela-inteligence/2024/8/Narodni-strategie-umele-intelience-CR-2030.pdf

[1653] CZ Daily, *Czech Republic Unveils New AI Strategy for 2030* (Jul. 25, 2024), https://czechdaily.cz/czech-republic-unveils-new-ai-strategy-for-2030

[1654] International Center for Not-for-Profit Law, *Being AIware: Incorporating Civil Society into National Strategies on Artificial Intelligence: Czech Republic* (Dec. 2020), https://ecnl.org/sites/default/files/2021-02/ECNL ICNL NAIS Czech Republic Dec 2020.pdf

[1655] AI Observatory and Forum, *Czech Republic's Expert Platform and Forum for Monitoring Legal and Ethical Rules for Artificial Intelligence*, http://observatory.ilaw.cas.cz/

practice, providing space for public debate and engaging Czechia in international discussions on AI regulation and data economy. Audits in the public and private sector will also detect the existence of legal barriers. The core of the expert platform and forum consists of a team of independent experts that continuously monitors trends in research and development of AI, its social impact, the development of legal and ethical rules in Czechia, in other countries and at international level. The platform cooperates closely with the AI Committee in charge of coordinating the implementation of the NAIS.

Despite functioning for several years, however, the forum has not published reports of findings in the public domain.[1656] The latest news uploaded to the platform website are from 2020. Given the inactivity of the AI Observatory and Forum since 2020, the standing mechanism for sustained public co-governance remains limited; while the 2023 consultation engaged experts and stakeholders, broader citizen participation could be strengthened in the next update cycle.[1657]

Data Protection

Since the Czech Republic is an EU Member State, the General Data Protection Regulation (GDPR)[1658] is directly applicable in Czechia and to Czechs. The aim of the GDPR is to "strengthen individuals' fundamental rights in the digital age and facilitate business by clarifying rules for companies and public bodies in the digital single market. A single law will also do away with the current fragmentation in different national systems and unnecessary administrative burdens."[1659] The GDPR has been applicable since May 25, 2018. The EU Data Protection Law Enforcement Directive (LED)[1660] protects citizens' fundamental right to data protection in handling

---

[1656] OECD AI Policy Observatory, *AI Strategies and Policies in Czechia* (2024), https://oecd.ai/en/dashboards/countries/CzechRepublic

[1657] Institute of State and Law, Czech Academy of Sciences, *AI Observatory and Forum—Czech Republic's Expert Platform for Monitoring Legal and Ethical Rules for Artificial Intelligence* (Jul. 9, 2025), http://observatory.ilaw.cas.cz/

[1658] European Parliament and Council of the European Union, *Regulation (EU) 2016/679 of the European Parliament and of the Council of 27 April 2016 on the protection of natural persons with regard to the processing of personal data and on the free movement of such data (General Data Protection Regulation)*, L 119, 1–88 (2016), https://eur-lex.europa.eu/legal-content/EN/TXT/?uri=CELEX%3A32016R0679

[1659] European Commission, *Legal Framework of EU Data Protection*, https://commission.europa.eu/law/law-topic/data-protection/legal-framework-eu-data-protection_en

[1660] European Parliament and Council of the European Union. *Directive (EU) 2016/680 of the European Parliament and of the Council of 27 April 2016 on the protection of natural persons with regard to the processing of personal data by competent authorities for the purposes of the prevention, investigation, detection or prosecution of criminal offences or the execution of criminal penalties, and on the free movement of such data, and repealing Council Framework Decision 2008/977/JHA*. Official Journal of the European Union, L 119, 89–131 (2016), https://eur-lex.europa.eu/eli/dir/2016/680/oj/eng

by criminal law enforcement authorities.[1661] The LED provides for the prohibition of any decision based solely on automated processing, unless it is provided by law, and of profiling that results in discrimination.[1662]

The 2019 Personal Data Processing Act (ZZOÚ) supplements the GDPR and implements the LED in Czech law.[1663] The Office of Personal Data Protection (*Úřad pro ochranu osobních údajů*) serves as the Czech Data Protection Authority under the ZZOÚ.[1664]

The Czech Republic is also a member of the Council of Europe and has signed but has not yet ratified[1665] the Council of Europe's Convention 108+ for the protection of individuals with regard to the processing of personal data.[1666]

Czechia's Office of Personal Data Protection joined the Global Privacy Assembly (GPA) as an accredited member in 2002.[1667] The Czech DPA did not endorse or co-sponsor the 2018 Declaration on Ethics and Data Protection in Artificial Intelligence,[1668] 2020 Resolution on Accountability in the Development and Use of Artificial Intelligence,[1669] 2022 Resolution on Principles and Expectations for the Appropriate Use of Personal Information in Facial Recognition Technology,[1670] or 2023 Resolution on Generative Artificial Intelligence Systems.[1671]

[1661] European Commission, *Legal Framework of EU Data Protection*, https://commission.europa.eu/law/law-topic/data-protection/legal-framework-eu-data-protection_en

[1662] Article 11 (1) and (2) of the LED, https://eur-lex.europa.eu/legal-content/EN/TXT/?uri=celex%3A02016L0680-20160504

[1663] *Czech Personal Data Protection Act* (Mar. 12, 2019), https://uoou.gov.cz/media/act-no-110-2019-coll.pdf

[1664] Office for Personal Data Protection, *About the Czech DPA* (2026), https://uoou.gov.cz/en/about-the-czech-dpa

[1665] Council of Europe, *Chart of Signatures and Ratifications of Treaty 223* (Mar. 9, 2025), https://www.coe.int/en/web/conventions/full-list?module=signatures-by-treaty&treatynum=223

[1666] Council of Europe, *Modernised Convention for the Protection of Individuals with Regard to the Processing of Personal Data* (May 18, 2018), https://www.coe.int/en/web/data-protection/convention108-and-protocol

[1667] Global Privacy Assembly, *List of Accredited Members* (2026), https://globalprivacyassembly.com/participation-in-the-assembly/list-of-accredited-members/

[1668] International Conference of Data Protection & Privacy Commissioners, *Declaration on Ethics and Data Protection in Artificial Intelligence* (Oct. 23, 2018), https://globalprivacyassembly.com/wp-content/uploads/2018/10/20180922_ICDPPC-40th_AI-Declaration_ADOPTED.pdf

[1669] Global Privacy Assembly, *Adopted Resolution on Accountability in the Development and Use of Artificial Intelligence* (Oct. 2020), https://globalprivacyassembly.com/wp-content/uploads/2020/11/GPA-Resolution-on-Accountability-in-the-Development-and-Use-of-AI-EN.pdf

[1670] Global Privacy Assembly, *Resolution on Principles and Expectations for the Appropriate Use of Personal Information in Facial Recognition Technology* (Oct. 2022), https://globalprivacyassembly.com/wp-content/uploads/2022/11/15.1.c.Resolution-on-Principles-and-Expectations-for-the-Appropriate-Use-of-Personal-Information-in-Facial-Recognition-Technolog.pdf

[1671] Global Privacy Assembly, *Resolution on Generative Artificial Intelligence Systems* (Oct. 2023), https://globalprivacyassembly.com/wp-content/uploads/2023/10/5.-Resolution-on-Generative-AI-Systems-101023.pdf

Algorithmic Transparency

The Czech Republic has ratified Convention 108, which includes provisions for algorithmic transparency, but has not yet ratified the Protocol that amends it (Convention 108+) and expands algorithmic transparency requirements. [1672] However, the Czech Republic is subject to the GDPR as an EU member state. Czechs have a general right to obtain access to information about automated decision-making and to the factors and logic of an algorithm.[1673] The EU AI Act, effective from August 2, 2024,[1674] also introduces transparency obligations for AI systems in the Czech Republic. The 2020 Recommendation of the Council of Europe Committee of Ministers on human rights impacts of algorithm systems[1675] specifically emphasizes requirements on transparency, accountability and effective remedies.

AI Oversight

The Czech DPA has been proactive in the deployment of GDPR protections, having fined companies for data leaks and conducted bilateral meetings with the Austrian oversight body[1676] on the risks of AI in social media.[1677] Until the oversight authorities for the EU AI Act are designated by law, the DPA remains the primary oversight body for algorithmic-processing.

Environmental Impact of AI

The Czech Republic NAIS does not specifically address policies aimed at mitigating the environmental impact of AI technologies.[1678] Still, the Czech Republic complies the EU Net Zero policies and the European Green Deal and has obligations under the ESG "Do no harm" principle, which prescribes an overall careful approach to the environment before the deployment of new technologies. The Digital Czech

---

[1672] Council of Europe, *Chart of Signatures and Ratifications of Treaty 223* (Mar. 9, 2025), https://www.coe.int/en/web/conventions/full-list?module=signatures-by-treaty&treatynum=223

[1673] See Recital 63 and Article 22 of the GDPR.

[1674] European Commission, *AI Act*, https://digital-strategy.ec.europa.eu/en/policies/regulatory-framework-ai; European Commission, *Third Draft General Purpose AI Code of Practice* (Mar. 2025), https://digital-strategy.ec.europa.eu/en/library/third-draft-general-purpose-ai-code-practice-published-written-independent-experts

[1675] Committee of Ministers, *Recommendation CM/Rec(2020)1 of the Committee of Ministers to member States on the human rights impacts of algorithmic systems* (Apr. 8, 2020), https://search.coe.int/cm/pages/result_details.aspx?objectid=09000016809e1154

[1676] Office for Personal Data Protection, *Czech DPA Imposed Fine of 351 Million CZK for GDPR Infringement* (2025), https://uoou.gov.cz/en/news/business-communication/czech-dpa-imposed-fine-of-351-million-czk-for-gdpr-infringement

[1677] Office for Personal Data Protection, *Czech and Austrian Personal Data Protection Face Similar Challenges* (Mar. 2025), https://uoou.gov.cz/ochrana-osobnich-udaju-v-ceske-republice-a-rakousku-celi-obdobnym-vyzvam-umela-inteligence-a-socialni-site-jsou-spojovany-s-radou-rizik-pro-obcany

[1678] Ministry of Industry and Trade, *National AI Strategy of the Czech Republic 2030 [Národní strategii umělé inteligence čr 2030]* (Jul. 24, 2024), https://www.mpo.gov.cz/assets/cz/podnikani/digitalni-ekonomika/umela-inteligence/2024/8/Narodni-strategie-umele-intelience-CR-2030.pdf

Republic[1679] strategy highlights the importance of a public and political debate to find new and innovative ways of adapting to the emerging climate change through digital and AI-enhanced technologies and tools.

Lethal Autonomous Weapons

According to the Czech Republic, "it is indispensable for the CCW High Contracting Parties to have sufficient guidance on how to ensure that any new weapon, means or methods of warfare are in compliance with the International Humanitarian Law, which is our main objective" and "welcome[s] the work on Lethal Autonomous Weapons Systems."[1680]

The Czech Republic is one of the 70 countries that endorsed a joint statement on autonomous weapons systems at the 2022 United Nations General Assembly. The joint statement urged "the international community to further their understanding and address these risks and challenges by adopting appropriate rules and measures, such as principles, good practices, limitations and constraints. We are committed to upholding and strengthening compliance with International Law, in particular International Humanitarian Law, including through maintaining human responsibility and accountability in the use of force."[1681] Czechia voted in favor[1682] of resolution L.56[1683] on autonomous weapons systems, emphasizing the "urgent need for the international community to address the challenges and concerns raised by autonomous weapons systems." The Czech Republic continued this support for discussions on LAWS in the UN General Assembly with the goal of an international agreement with support for Resolution 79/62[1684] on lethal autonomous weapons systems in the 2024 UN General Assembly.

Czechia participated in the REAIM 2023 international summit on the responsible application of artificial intelligence in the military domain hosted by the Netherlands. At the end of the Summit, Czechia endorsed a joint call for action stressing "the paramount importance of the responsible use of AI in the military

---

[1679] Ministry of Industry and Trade of the Czech Republic, *Digital Czech Republic* (2024), https://mpo.gov.cz/en/business/digital-society/digital-czech-republic--243601/

[1680] Hani Stolina, *Statement by Mr. Hani Stolina at the Thematic Discussion on Conventional Weapons of the First Committee of the 74th Session of the General Assembly of the United Nations* (Oct. 23, 2019), https://perma.cc/NC9W-YDM2

[1681] UN General Assembly, *Joint Statement on Lethal Autonomous Weapons Systems First Committee, 77th United Nations General Assembly Thematic Debate—Conventional Weapons* (Oct. 21, 2022), https://estatements.unmeetings.org/estatements/11.0010/20221021/A1jJ8bNfWGlL/KLw9WYcSnnAm_en.pdf

[1682] Stop Killer Robots, *164 States Vote against the Machine at the UN General Assembly*, https://www.stopkillerrobots.org/news/164-states-vote-against-the-machine/

[1683] General Assembly, *Resolution L56: Lethal Autonomous Weapons* (Oct. 12, 2023), https://reachingcriticalwill.org/images/documents/Disarmament-fora/1com/1com23/resolutions/L56.pdf

[1684] UN General Assembly, *General and Complete Disarmament, Report of the first committee*, Draft resolution A/C.1/79/L.77, p. 103 (Nov. 15, 2024), https://docs.un.org/en/A/79/408

domain, employed in full accordance with international legal obligations and in a way that does not undermine international security, stability and accountability."[1685]

AI Literacy

AI literacy is embedded in NAIS 2030 under Education and Expertise in AI and AI Skills and Labour Market Impacts.[1686] The AI National Skilling Plan[1687] launched in June 2025 with Microsoft aims to train more than 350,000 people in basic AI skills over the next year. Target groups include public administration, teachers, students, women, job-seekers, and non-profit organizations. Charles University is a partner in the program and participated in an AI Skills for Public Administration. The AI for Kids program is training teachers to support students' literacy on what AI is, how to use it "meaningfully and sensibly," and to understand the impact of AI on individuals and society. A brochure on tips for using AI in education and beyond was one of the first program outputs. The initiative explicitly aligns to commitments in the NAIS 2030.

Human Rights

Czechia adopted the United Nations Declaration on Human Rights. As a member of the European Union and of the Council of Europe, Czechia is committed to upholding the EU Charter of Fundamental Rights and the European Convention on Human Rights. Freedom House ranks the Czech Republic "Free" with a score of 95/100, recognizing the country's positive standing on political and civil rights.[1688]

In a 2020 Recommendation to member States on the human rights impacts of algorithmic systems, the Council of Europe Committee of Ministers recalled that their commitment to the Convention for the Protection of Human Rights and Fundamental Freedoms required member state to "ensure that any design, development and ongoing deployment of algorithmic systems occur in compliance with human rights and fundamental freedoms, which are universal, indivisible, inter-dependent and

---

[1685] Government of Netherlands, *Call to Action on Responsible Use of AI in the Military Domain* (Feb. 16, 2023), https://www.government.nl/latest/news/2023/02/16/reaim-2023-call-to-action

[1686] Ministry of Industry and Trade, *National AI Strategy of the Czech Republic 2030 [Národní strategii umělé inteligence čr 2030]* (Jul. 24, 2024), https://www.mpo.gov.cz/assets/cz/podnikani/digitalni-ekonomika/umela-inteligence/2024/8/Narodni-strategie-umele-intelience-CR-2030.pdf

[1687] Microsoft, *Microsoft Launches AI National Skilling Plan in the Czech Republic* (Jun. 27, 2025), https://news.microsoft.com/cs-cz/2025/06/26/microsoft-spousti-ai-national-skilling-plan-v-ceske-republice-v-prvni-fazi-projektu-investuje-pres-10-milionu-korun-do-vzdelavani-vice-nez-350-000-lidi/

[1688] Freedom House, *Freedom in the World 2025: Czechia* (2025) https://freedomhouse.org/country/czechia/freedom-world/2025

interrelated, with a view to amplifying positive effects and preventing or minimising possible adverse effects."[1689]

OECD / G20 AI Principles

The Czech Republic has been a member of OECD since 1995 and has endorsed the OECD AI Principles.[1690] The original NAIS emphasizes the importance of cooperation with the OECD for the implementation of the Strategy measures. Exchange of information in expert groups on AI at the OECD and coordination of preparations for negotiations within the OECD are amongst those measures.[1691] The OECD AI Policy Observatory acknowledges that the NAIS addresses "inclusive growth, sustainable development and well-being; robustness, security and safety; accountability; fostering a digital ecosystem for AI; providing an enabling policy environment for AI; building human capacity and preparing for labour market transition, international co-operation for trustworthy AI."[1692]

The Czech Republic joined the Global Partnership on AI, a multi-stakeholder initiative which aims to foster international cooperation on AI research and applied activities, in 2022.[1693] The GPAI is "built around a shared commitment to the OECD Recommendation on Artificial Intelligence."

Council of Europe AI Treaty

The Czech Republic contributed as a Council of Europe and EU Member State in the negotiations of the Council of Europe Framework Convention on AI, Human Rights, Democracy and the Rule of Law.[1694] The Czech Republic is party to the Convention as an EU member state but has not signed independently.[1695]

[1689] Committee of Ministers, *Recommendation CM/Rec(2020)1 of the Committee of Ministers to member States on the human rights impacts of algorithmic systems* (Apr. 8, 2020), https://search.coe.int/cm/pages/result_details.aspx?objectid=09000016809e1154

[1690] OECD AI Policy Observatory, *OECD AI Principles Overview, Countries Adhering to the AI Principles*, https://oecd.ai/en/ai-principles

[1691] Ministry of Industry and Trade of the Czech Republic, *National Artificial Intelligence Strategy of the Czech Republic*, pp. 36, 38–39 (2019), https://www.mpo.cz/assets/en/guidepost/for-the-media/press-releases/2019/5/NAIS_eng_web.pdf

[1692] OECD AI Policy Observatory, *AI Policies in Czechia: National AI Strategy of the Czech Republic* (Jul. 10, 2025), https://oecd.ai/en/dashboards/policy-initiatives/national-ai-strategy-of-the-czech-republic-2030-6563

[1693] Ministry of Foreign Affairs of the Czech Republic. *Czech Republic Joins Global Partnership for Artificial Intelligence* (Jan. 20, 2022), https://mzv.gov.cz/jnp/en/issues_and_press/archive/events_and_issues/x2022/czech_republic_joins_global_partnership$219729.html

[1694] Council of Europe, *Framework Convention on Artificial Intelligence* (2026), https://www.coe.int/en/web/artificial-intelligence/the-framework-convention-on-artificial-intelligence

[1695] Council of Europe Treaty Office, *Chart of Signatures and Ratifications of Treaty 225* (Mar. 9, 2025), https://www.coe.int/en/web/conventions/full-list?module=signatures-by-treaty&treatynum=225

UNESCO Recommendation on AI Ethics

A UNESCO member state since 1993,[1696] Czechia is among the 193 countries that endorsed the UNESCO Recommendation on AI, the first ever global agreement on the ethics of AI.[1697] The Czech Republic supported the implementation of the UNESCO Recommendation by hosting the first ever Global Forum on the Ethics of AI under the Czech Presidency of the Council of the European Union and under the patronage of UNESCO in December 2022. The forum placed a spotlight on "ensuring inclusion in the AI world," and took stock of the implementation of the recommendation.[1698]

Czechia has not initiated the Readiness Assessment Methodology (RAM),[1699] a tool developed by UNESCO to assist in the implementation of the Recommendation.

Evaluation

The Czech Republic's ambition to become a European leader in the digital field by 2030[1700] has shaped its position on AI policy. The release in 2024 of an updated National Artificial Intelligence Strategy 2030 builds on that ambition by shifting to a focus on implementation.[1701] The updated National AI Strategy of the Czech Republic 2030, effective as of July 2024, shifted transitioned from the broad aspirational approach in the 2019 NAIS to a structured, more enforceable framework.

Although the country adheres to a human centric approach to AI through its commitment to implementing the OECD AI Principles and the UNESCO Recommendation on the Ethics of AI, Czechia has consistently stood against perceived overregulation, notably through the EU AI Act, and strives with its national AI strategy to remove what it considers as legal impediments to AI development. The Czech Republic has not yet ratified the modernized Convention 108+ for the protection of individuals with regard to the processing of personal data. Czechia has legislation pending to establish a national supervisory mechanism for AI aligned with the AI Act.

[1696] UNESCO, *Member States*, https://www.unesco.org/en/countries#

[1697] UNESCO, *UNESCO Member States Adopt the First Ever Global Agreement on the Ethics of Artificial Intelligence* (Nov. 2021), https://www.unesco.org/en/articles/unesco-member-states-adopt-first-ever-global-agreement-ethics-artificial-intelligence

[1698] UNESCO, *First Global Forum on Ethics of AI Held in Prague, One Year after the Adoption of UNESCO's Recommendation* (2022), https://www.unesco.org/en/articles/first-global-forum-ethics-ai-held-prague-one-year-after-adoption-unescos-recommendation

[1699] UNESCO Global AI Ethics and Governance Observatory, *Global Hub* (Oct. 2025), https://www.coe.int/en/web/artificial-intelligence/the-framework-convention-on-artificial-intelligence

[1700] Ministry of Industry and Trade of the Czech Republic, *National Artificial Intelligence Strategy of the Czech Republic*, p. 3 (2019), https://www.mpo.cz/assets/en/guidepost/for-the-media/press-releases/2019/5/NAIS_eng_web.pdf

[1701] Ministry of Industry and Trade, *National AI Strategy of the Czech Republic 2030 [Národní strategii umělé inteligence čr 2030]* (Jul. 24, 2024), https://www.mpo.gov.cz/assets/cz/podnikani/digitalni-ekonomika/umela-inteligence/2024/8/Narodni-strategie-umele-intelience-CR-2030.pdf

Denmark

In 2025, Denmark passed legislation to implement the EU AI Act by designating national supervisory authorities. The Ministry of Culture proposed the Copyright Act as a means to combat Deepfakes and proposed measures to tighten control over training AI systems on copyrighted material.

National AI Strategy

The Danish government unveiled the national AI strategy in March 2019.[1702] The Danish strategy on AI development outlines the issues that must be tackled and defines specific policy efforts and key initiatives. The national AI strategy intends to establish Denmark as a leader in responsible AI development. There are four objectives to accomplish this goal 1) Establish a consistent ethical and human-centered foundation for artificial intelligence; 2) Prioritize and promote research in artificial intelligence; 3) Encourage the growth of Danish firms through the development and use of artificial intelligence; and 4) Ascertain that the public sector utilizes AI to provide world-class services to citizens and society.

The strategy covers both the public and the private sector and aims to create a framework to improve the level of trust in AI. Six principles for AI guide this objective: self-determination, human dignity equality and justice, and addresses also responsibility and explainability.[1703] The national strategy establishes healthcare, energy and utilities, agriculture, and transport as priority areas to lift the work of AI within Denmark.

The Agency for Digital Government (*Digitaliserings-styrelsen*) published three AI Guides in 2024 directed to companies, citizens, and authorities testing and working with Generative AI.[1704] The Agency also launched an AI Sandbox in cooperation with the Data Protection Agency to facilitate compliance with GDPR and, over time the EU AI Act.

The 2024 Strategic Approach to Artificial Intelligence[1705] by the Ministry of Digital Affairs (*Regeringen*) aims to provide a "more robust framework for the responsible development and use of AI in Denmark." The new strategy document clarifies the Danish vision for AI, centered on three guiding principles, the first of

[1702] Danish Government, *National Strategy for Artificial Intelligence* (2019), https://en.digst.dk/media/19337/305755_gb_version_final-a.pdf

[1703] Ibid

[1704] Agency for Digital Government, *The Agency for Digital Government Publishes AI Guides* (May 3, 2024), https://en.digst.dk/news/news-archive/2024/maj/the-agency-for-digital-government-publishes-ai-guides/

[1705] Ministry of Digital Affairs, *Strategic Approach to Artificial Intelligence: A More Robust Foundation for the Responsible Development and Use of AI in Denmark* (Dec. 2024), https://www.english.digmin.dk/Media/638719220318136690/Stategic%20Approach%20to%20Artificial%20Intelligence.pdf

which is "Developing and using AI must be centred on citizens' fundamental rights and in accordance with Danish values." The other principles center on supporting businesses "to develop, apply and sell solutions and business models based on the responsible use of AI" and leading the world in using AI in the public sector to benefit citizens.[1706]

Led by the Ministry of Culture (*Kulturminister*),[1707] political representatives are also working to protect citizens' rights to their images and bodies with an agreement to create legislation that will make it illegal to share digital imitations of personal characteristics such as deepfakes.[1708] The bill will also make it illegal to share imitations of performances, providing protection for performing artists, while leaving room for paradies and satire.

Nordic-Baltic and Nordic Cooperation on AI

As for the regional landscape, the Danish Minister for digitalization signed the declaration on "AI in the Nordic-Baltic region" establishing a collaborative framework on "developing ethical and transparent guidelines, standards, principles and values to guide when and how AI applications should be used" and "on the objective that infrastructure, hardware, software and data, all of which are central to the use of AI, are based on standards, enabling interoperability, privacy, security, trust, good usability, and portability."[1709]

The ministerial declaration Digital North 2.0[1710] builds on the common priorities of the Nordic-Baltic countries, and follows the previous ministerial declaration, Digital North 2017-2020. "In order to promote work with digitalisation, co-ordinate efforts, and follow up on the goals of the declaration, a council of ministers for digitalisation (MR-DIGITAL) was established in 2017. The aim is to promote development in three areas: (1) Increase mobility and integration in the Nordic and Baltic region by building a common area for cross-border digital services; (2) Promote green economic growth and development in the Nordic-Baltic region through data-driven innovation and a fair data economy for efficient sharing and re-

---

[1706] Ibid, p. 9

[1707] Kulturministeriet, *Deepfakes, Production Discounts and Less Bureaucracy Are on This Year's Legislative Agenda* (Oct. 7, 2025), https://kum.dk/aktuelt/nyheder/deepfakes-produktionsrabat-og-mindre-bureaukrati-er-paa-aarets-lovprogram

[1708] Kulturministeriet, *Broad Agreement on Deepfakes Gives Everyone the Right to Their Own Body and Voice* (Jun. 26, 2025), https://kum.dk/aktuelt/nyheder/bred-aftale-om-deepfakes-giver-alle-ret-til-egen-krop-og-egen-stemme; *Værn mod deling af digitale efterligninger af personalige kendetegn* [Danish only], https://kum.dk/fileadmin/_kum/1_Nyheder_og_presse/2025/Aftale.pdf

[1709] Nordic and Baltic Ministers of Digitalization, *AI in the Nordic-Baltic Region* (May 14, 2018), https://www.norden.org/en/declaration/ai-nordic-baltic-region

[1710] Nordic and Baltic Ministers of Digitalization, *Ministerial Declaration Digital North 2.0* (Sept. 29, 2020), https://www.norden.org/en/declaration/ministerial-declaration-digital-north-20

use of data; and (3) Promote Nordic-Baltic leadership in the EU/EEA and globally in a sustainable and inclusive digital transformation of our societies."[1711]

The Nordic and Baltic ministers for digitalization released another joint statement in November 2021, announcing a focus on digital inclusion, striving to implement measures to make digital services more accessible to all Danish inhabitants and ensuring that those who do not possess the necessary level of skills get the opportunity to acquire them. [1712]

In September 2022, the Nordic and Baltic ministers for digitalization issued a common statement on the importance of cooperation on digital security in the Nordic-Baltic region following the COVID-19 pandemic and the war in Ukraine. The ministers stressed that this "rapid transformation has challenged everyone to adapt to new, digital ways of doing business, learning and accessing public authorities." The ministers declared that they "have committed to ensuring that our region maintains its position as a leader in digitalisation, and that everyone in the region benefit from digitalisation regardless of age, wealth, education or level of digital skills. [...] Robust and secure digital services, safeguarding users' privacy and ensuring that personal data are stored and processed in a trustworthy way, are crucial to the citizens' sustained trust in digital services."[1713]

As part of its action plan for Vision 2030 (2021–2024), the Nordic Council of Ministers identified innovation, digital integration, the safe use of artificial intelligence, data development and open data, education and digitalization as key objectives.[1714] The Nordic Council of Ministers also emphasizes the involvement of civil society in efforts relating to our vision for 2030 thanks to "a Nordic civil society network and public consultations."[1715]

Nordic Project 2024 will continue toward this vision to support collaboration among Nordic and Baltic countries in developing and using AI in a responsible way. The project aims to support Nordic businesses' development and application of

---

[1711] Nordic Co-operation, *Nordic-Baltic Co-operation on Digitalisation*, https://www.norden.org/en/information/nordic-baltic-co-operation-digitalisation

[1712] Nordic and Baltic Ministers of Digitalization, *Common statement on the importance of promoting digital inclusion as a central part of the digital transformation in the Nordic-Baltic region* (Nov. 26, 2021), https://www.norden.org/en/declaration/common-statement-importance-promoting-digital-inclusion-central-part-digital

[1713] Nordic and Baltic Ministers of Digitalization, *Common statement on the importance of cooperation on digital security in the Nordic-Baltic region* (Sept. 6, 2022), https://www.norden.org/en/declaration/common-statement-importance-cooperation-digital-security-nordic-baltic-region

[1714] Nordic Council of Ministers, *The Nordic Region—Towards Being the Most Sustainable and Integrated Region in the World, Action Plan for 2021–2024* (Dec. 14, 2020), https://www.norden.org/en/publication/nordic-region-towards-being-most-sustainable-and-integrated-region-world

[1715] Nordic Council of Ministers, *Guidelines for Involving Civil Society in Work Relating to Our Vision 2030* (Feb. 12, 2021), https://www.norden.org/en/publication/guidelines-involving-civil-society-work-relating-our-vision-2030

responsible and ethical AI to achieve critical mass by pooling and coordinating resources through a collaboration platform, center, or hub.[1716] The Nordic-Baltic Co-operation Programme for Digitalisation 2025 to 2030 builds on this initiative with a subgoal (3.3) dedicated to "trustworthy, secure, responsible, and sustainable use of technology, including Artificial Intelligence."[1717]

EU Digital Services Act

As an EU member state, Denmark applies the EU Digital Services Act (DSA)[1718] regulating online intermediaries and platforms. The main objective is to prevent illegal and harmful activities online and the spread of disinformation by requiring online intermediaries and platforms to implement ways to prevent and remove posts containing illegal goods, services, or content while giving users the means to report or flag this type of content.

The Agency for Digital Governance (*Digitaliserings-styrelsen*) serves as Denmark's Digital Services Coordinator.[1719]

EU AI Act

As an EU member State, Denmark is bound by the EU AI Act.[1720] The EU AI Act is a risk-based market regulation which supports the objective of promoting a human-centric approach to AI and making the EU a global leader in the development of secure, trustworthy and ethical AI.

Denmark designated the Agency for Digital Government the national competent authority charged with coordinating with supervisory authorities and the European AI Office.[1721] The Agency is part of the Ministry of Digital Affairs and

---

[1716] Nordic Innovation, *New Nordic AI-Preparation Project* (Aug. 30, 2024), https://www.nordicinnovation.org/programs/new-nordics-ai-preparation-project

[1717] Nordic Council, *Nordic-Baltic Co-operation Programme for Digitalisation 2025 to 2030* (Oct. 2024), https://pub.norden.org/politiknord2024-754/

[1718] *Regulation (EU) 2022/2065 of the European Parliament and of the Council of 19 October 2022 on a Single Market for Digital Services and amending Directive 2000/31/EC (Digital Services Act)*, (Oct. 21, 2022), https://eur-lex.europa.eu/legal-content/EN/TXT/?uri=celex%3A32022R2065

[1719] Digitaliserings-styrelsen, *Digital Services—DSA [Digitale tjenester—DSA]*, https://digst.dk/tilsyn/digitale-tjenester-dsa/

[1720] European Parliament, *Artificial Intelligence Act, European Parliament legislative resolution of 13 March 2024 on the proposal for a regulation of the European Parliament and of the Council on laying down harmonised rules on Artificial Intelligence (Artificial Intelligence Act) and amending certain Union Legislative Acts (COM(2021)0206 – C9-0146/2021 – 2021/0106(COD)), P9_TA(2024)0138*, https://www.europarl.europa.eu/RegData/seance_pleniere/textes_adoptes/definitif/2024/03-13/0138/P9_TA(2024)0138_EN.pdf

[1721] Ministry of Digital Affairs, *The Agency for Digital Government Is Appointed as the Supervisory Authority for Artificial Intelligence [Digitaliseringsstyrelsen udpeges som tilsynsynsmyn-dighed for kunstig intelligens]* (Apr. 10. 2024),

implements digital and AI policies in the public sector for citizens and businesses. For example, the agency implemented the regulatory sandboxes and created AI guides for citizens, businesses, and authorities.[1722]

The Act on Supplementary Provisions to the AI Regulation implements the EU AI Act in Denmark.[1723] The Act designates the Agency for Digital Government as the central contact point and supervisory authority for most prohibited practices and the Danish Data Protection Agency and the Danish Court Administration as competent authorities for specific areas. In line with this role,[1724] the Agency for Digital Government published a six-part guidance package on prohibited AI uses.[1725]

Public Participation

Denmark provides programs that enable non-governmental actors (e.g., the academic community, business, civil society, and regional and local governments) to express their perspectives or provide expert advice that informs policy-making processes. These policy initiatives enable stakeholders or experts to engage in public discussions to share information and foster collaboration. Public awareness campaigns and civic engagement activities include informing and consulting with members of the public. The government also has a consultation portal (*Høring Portalen*) for receiving comments, as in the 2025 consultation on amendments to the copyright law.[1726]

The Ministry of Foreign Affairs Tech for Democracy Initiative launched in 2021 brought together government members, organizations, industries and civil society to ensure that technology works for humans and not the other way round.[1727] At the launch event, the Copenhagen Pledge on Tech for Democracy was signed by

---

https://www.digmin.dk/digitalisering/nyheder/nyhedsarkiv/2024/apr/digitaliseringsstyrelsen-udpeges-som-tilsynsmyndighed-for-kunstig-intelligens

[1722] Agency for Digital Government, *The Agency for Digital Government Publishes AI Guides* (May 3, 2024), https://en.digst.dk/news/news-archive/2024/maj/the-agency-for-digital-government-publishes-ai-guides/

[1723] Restinformation [Legal Information], *Law no. 467 of 14/05/2025, Act on Supplementary Provisions to the Regulation on Artificial Intelligence [Lov om supplerende bestemmelser til forordningen om kunstig intelligens]* (May 14, 2025), https://www.retsinformation.dk/eli/lta/2025/467

[1724] Agency for Digital Government, *New Guidance on Prohibited AI Uses* (Oct. 20, 2025), https://digst.dk/nyheder/nyhedsarkiv/2025/oktober/vejledning-om-forbudt-ai/

[1725] Agency for Digital Government, *Prohibited AI Practices [Forbudte former for AI-praksis]*, https://digst.dk/tilsyn/ai-forordningen/reglerne-i-ai-forordningen/forbudte-former-for-ai-praksis/

[1726] Hørling Portalen, *Consultation on the Proposal for Act amending the Copyright Act (Introduction of performance protection and protection against digitally generated imitations, etc.)[Forslag til lov om ændring af lov om ophavsret (Indførelse af en præstationsbeskyttelse og beskyttelse mod digitalt genererede efterligninger mv)]* (Jul. 7, 2025), https://hoeringsportalen.dk/Hearing/Details/70269

[1727] Ministry of Foreign Affairs, *The Tech for Democracy Initiative*, https://techamb.um.dk/impact/tech-for-democracy; see also, Tech for Democracy, https://techfordemocracy.dk/

over 200 stakeholders.[1728] Throughout the ensuing Year of Action, ten Action Coalitions convened to deliberate on issues such as transparency, content moderation, and online harassment.[1729] In March 2023, Denmark, alongside Norway, supported the EC in launching the Digital Democracy Initiative (DDI) as a Team Europe Initiative, with an implementation period that stretches to 2026.[1730]

The Danish AI strategy for its part is based on proposals from a Digital Growth Panel[1731] and the Danish Government's Disruption Committee.[1732]

Data Protection

Denmark, like other European countries, has enacted laws to supplement the EU General Data Protection Regulation (GDPR).[1733] In Denmark, the GDPR and its Danish supplementary act, the Data Protection Act (DPA)[1734] are the primary regulations governing the processing of personal data. The DPA provides for certain exceptions to the GDPR, most notably regarding the processing of personal data in the employment sector and the processing of national registration numbers. The independent Danish Data Protection Agency, the Datatilsynet, enforces data protection in Denmark.

In 2002, the Danish Act on Personal Data Processing came into force, implementing Directive 95/46 EC. However, despite the fact that the Danish data protection regulation is more than two decades old, until the GDPR was implemented in 2016, little attention was paid to data protection in Denmark.

The Danish and other European supervisory authorities have released several recommendations and decisions interpreting the GDPR and relevant national legislation, allowing Danish businesses to conduct significantly more targeted and resource-efficient compliance operations. Denmark has lagged behind the majority of

---

[1728] Digital Democracy Initiative, *Tech for Democracy* (2024) https://digitaldemocracyinitiative.net/underside-tech-for-democracy

[1729] Danish Institute for International Studies, *Tech for Democracy: Learnings from the Year of Action* (2023), https://pure.diis.dk/ws/portalfiles/portal/20707338/Tech_for_Democracy_2023_Report.pdf

[1730] Digital Democracy Initiative, *About the DDI* (2024), https://digitaldemocracyinitiative.net/underside-about

[1731] European Commission, Directorate-General for Communications Networks, Content and Technology, *Digital Growth Strategy 2025* (2021), https://digital-skills-jobs.europa.eu/en/actions/national-initiatives/national-strategies/denmark-digital-growth-strategy-2025

[1732] The Danish Government, Ministry for Economic Affairs and the Interior, *Denmark's National Reform Programme* (2019), https://commission.europa.eu/content/2019-european-semester-national-reform-programmes-and-stabilityconvergence-programmes_en

[1733] *Regulation (EU) 2016/679 of the European Parliament and of the Council of 27 April 2016 on the protection of natural persons with regard to the processing of personal data and on the free movement of such data, and repealing Directive 95/46/EC (General Data Protection Regulation)* (Apr. 27, 2016), http://data.europa.eu/eli/reg/2016/679/oj

[1734] Data Protection Act, *Act No. 502 on supplementary provisions to the regulation on the protection of natural persons with regard to the processing of personal data and on the free movement of such data* (May 23, 2018), https://www.datatilsynet.dk/media/7753/danish-data-protection-act.pdf

other EU Member States when it comes to data protection knowledge and compliance. The levels of fines (100.000 DKK and 50.000 DKK) were very low compared to fines in other countries (with fines up to millions). A case in August 2024 marked a shift.[1735] The Danish Data Protection Authority imposed a DKr 200,000 ($29,560) fine on a local authority for insufficient security measures. Furthermore, on February 13, 2025, the European Court of Justice ruled that fines for GDPR violations should be calculated based on the worldwide turnover of the entire corporate group, not just the violating entity.[1736] This decision, stemming from a case involving Danish company ILVA, could lead to significantly higher fines in Denmark and across the EU. According to some privacy law professionals, there is a significant risk for Denmark to be considered a "safe haven" in relation to fines, which could cause companies that do not want to comply with the GDPR to locate in Denmark.[1737]

Regarding the activities of law enforcement authorities, Denmark transposed the EU Data Protection Law Enforcement Directive (LED)[1738] through the Danish Law Enforcement Act.[1739] The EDPB has produced guidelines on the use of facial recognition technologies in the area of law enforcement, stressing that law enforcement must "satisfy the requirements of necessity and proportionality."[1740] One of the first publicly known applications of facial recognition technology in Denmark under the framework of the LED and national data protection rules was approved in January 2025. The Danish Data Protection Agency authorized the use of automated facial recognition by Football Club Copenhagen to enhance public safety and prevent

---

[1735] Martina Lindberg, *Danish Municipality Vejen Fined for Lack of Data Encryption* (Aug. 16, 2024), https://www.grip.globalrelay.com/danish-municipality-vejen-fined-for-lack-of-data-encryption/

[1736] InfoCuria Case-law, *JUDGMENT OF THE COURT (Fifth Chamber)*, (Feb. 14th, 2025), https://curia.europa.eu/juris/document/document.jsf?text=&docid=295319&pageIndex=0&doclang=EN&mode=lst&dir=&occ=first&part=1&cid=21586718

[1737] Claas Thöle, *Four Years of GDPR: The Danish Approach to Data Protection, or Absence Thereof?* (Sept. 15, 2022), https://inplp.com/latest-news/article/four-years-of-gdpr-the-danish-approach-to-data-protection-or-absence-thereof/

[1738] *Directive (EU) 2016/680 of the European Parliament and of the Council of 27 April 2016 on the protection of natural persons with regard to the processing of personal data by competent authorities for the purposes of the prevention, investigation, detection or prosecution of criminal offences or the execution of criminal penalties, and on the free movement of such data, and repealing Council Framework Decision 2008/977/JHA*, https://eur-lex.europa.eu/legal-content/EN/TXT/?uri=celex%3A02016L0680-20160504

[1739] *Danish Law Enforcement Act*, https://www.datatilsynet.dk/Media/637998758521368022/The%20Danish%20Law%20Enforcement%20Act.pdf

[1740] European Data Protection Board, *Guidelines 05/2022 on the Use of Facial Recognition Technology in the Area of Law Enforcement,* Final Version (May 17, 2023), https://www.edpb.europa.eu/our-work-tools/our-documents/guidelines/guidelines-052022-use-facial-recognition-technology-area_en

entry by banned individuals.[1741] The system may be deployed at Parken Stadium and during away games at other venues. While the LED sets strict compliance requirements for all uses of facial recognition, this decision is an early example of a high-risk, security-driven deployment deemed to meet the thresholds of necessity and proportionality. It may serve as a precedent for similar applications in other public spaces.

The European Data Protection Board (EDPB) launched a Coordinated Enforcement Framework action focusing on the right to erasure under the GDPR in March 2025. The Danish Data Protection Agency is participating in this initiative to evaluate how data controllers manage and respond to erasure requests, ensuring compliance with GDPR standards.[1742]

The EU Charter of Fundamental Rights more generally provides that EU citizens have the right to protection of their personal data. Article 8 of the Charter states that: "Everyone has the right to the protection of personal data concerning him or her. Such data must be processed fairly for specified purposes and on the basis of the consent of the person concerned or some other legitimate basis laid down by law."

Despite being a member of the Global Privacy Assembly (GPA) since 2002, the Datatilsynet has not endorsed the 2018 GPA Declaration on Ethics and Data Protection in Artificial Intelligence,[1743] the 2020 GPA Resolution on AI Accountability,[1744] the 2022 GPA Resolution on Facial Recognition Technology,[1745] or the 2023 GPA Resolution on Generative AI.[1746]

Algorithmic Transparency

Denmark is party to the Council of Europe Convention 108 on automatic processing information, but has not signed the updated Convention 108+, which

---

[1741] Masha Burak, *FC Copenhagen Will Use Facial Recognition to Stop Violent Football Fans*, Biometric Update (Jan. 17, 2025), https://www.biometricupdate.com/202501/fc-copenhagen-will-use-facial-recognition-to-stop-violent-football-fans

[1742] Datatilsynet, *EDPB Launches Coordinated Action on the Right to Erasure* (Mar. 3, 2025), https://www.datatilsynet.dk/internationalt/internationalt-nyt/2025/mar/edpb-igangsaetter-koordineret-indsats-om-retten-til-sletning

[1743] Global Privacy Assembly, *Declaration on Ethics and Data Protection in Artificial Intelligence* (Oct. 23, 2018), https://globalprivacyassembly.org/wp-content/uploads/2018/10/20180922_ICDPPC-40th_AI-Declaration_ADOPTED.pdf

[1744] Global Privacy Assembly, *Resolution on Accountability in the Development and Use of Artificial Intelligence* (Oct. 2020), https://globalprivacyassembly.org/wp-content/uploads/2020/10/FINAL-GPA-Resolution-on-Accountability-in-the-Development-and-Use-of-AI-EN-1.pdf

[1745] Global Privacy Assembly, *Resolution on Principles and Expectations for the Appropriate Use of Personal Information in Facial Recognition Technology* (Oct. 2022), https://globalprivacyassembly.org/wp-content/uploads/2022/11/15.1.c.Resolution-on-Principles-and-Expectations-for-the-Appropriate-Use-of-Personal-Information-in-Facial-Recognition-Technolog.pdf

[1746] Global Privacy Assembly, *Resolution on Generative Artificial Intelligence Systems* (Oct. 2023), https://globalprivacyassembly.org/wp-content/uploads/2023/10/5.-Resolution-on-Generative-AI-Systems-101023.pdf.

explicitly provides for algorithmic transparency.[1747] Convention 108+ treaty will enter into force after Denmark and all other parties to the original convention ratify. Denmark is subject to the GDPR, which provides a general right to obtain access to information about automated decision-making and to the factors and logic of an algorithm.[1748]

Data Ethics Council

The Danish government established the Data Ethics Council to "support the development in an ethically sound manner that takes into account citizens' fundamental rights, legal certainty or fundamental societal values."[1749] The Council advises the public and private sectors on data-related ethical issues and provide a venue for debate and public awareness. The Council publishes responses to consultations on various data-related issues, such as the Copyright Act,[1750] which protects against digitally generated production imitations, and applications of AI such as in health. The Council also issued recommendations on ensuring the responsible use of generative AI in the public sector in 2025.[1751]

Danish Labeling Program for Digital Accountability

Another key initiative is the "D-seal." The Danish government established an independent labelling scheme in collaboration with a consortium of the Confederation of Danish Industry, the Danish Chamber of Commerce, SMEdenmark, and the Danish Consumer Council. It is supported by the Danish Business Authority and financed by the Danish Industry Foundation. "All companies, regardless of their size, must meet the D-seal's data ethics requirements if they develop software and if they use or develop algorithms and AI." [1752] "The purpose of the D-seal is to promote data security, data protection and data ethics at Danish companies, so that customers and

---

[1747] Council of Europe Treaty Office, *Chart of Signatures and Ratifications of Treaty 223* (Feb. 6, 2026), https://www.coe.int/en/web/conventions/full-list;?module=signatures-by-treaty&treatynum=223

[1748] See Recital 63 and Article 22 of the GDPR

[1749] Dataetisk Råd, *About the Danish Data Ethics Council: Background and Purpose [Om Dataestisk Råd: Baggrund og formål]* (2026), https://www.dataetiskraad.dk/om-dataetisk-raad/baggrund-og-formaal

[1750] Dataetisk Råd, *The Danish Data Ethics Council's Response to the Consultation Regarding the Copyright Act [Dataetisk Råds høringssvar vedr. ophavsretsloven]* (Sept. 12, 2025), https://www.dataetiskraad.dk/hoeringssvar/ophavsretsloven

[1751] Dataetisk Råd, *The Danish Data Ethics Council focuses on Generative AI in the Public Sector [Dataetisk Råd sætter fokus på generativ AI i det offentlige]* (Feb. 5, 2025), https://www.dataetiskraad.dk/nyheder/dataetisk-raad-saetter-fokus-paa-generativ-ai-i-det-offentlige

[1752] Birgitte Kofod Olsen, *Danish Labeling Program for Digital Accountability* (Sept. 28, 2021), https://dataethics.eu/danish-labeling-program-for-digital-accountability/; Frederick Weiergang Larsen, *Denmark: An Independent Council and a Labelling Scheme to Promote the Ethical Use of Data*, OECD The AI Wonk (Jun. 15, 2020), https://oecd.ai/wonk/an-independent-council-and-seal-of-approval-among-denmarks-measures-to-promote-the-ethical-use-of-data

consumers can feel safe when using the companies' products and services."[1753] The idea is for the seal to create a market incentive for actors to be more data ethical. The D-seal was launched in September 2021.

The D-seal is based on 8 criteria: A management system for data accountability; Awareness and safe behavior; Technical IT security; Requirements for suppliers' IT security and digital accountability; Transparency and control of data; Privacy & security by design & default; Reliable algorithms & AI; Data ethics.[1754]

In 2020, ahead of the negotiations regarding the draft EU AI Act, Denmark, together with 13 other EU Member States, issued a non-paper responding to the European Commission's White Paper on AI.[1755] Through this non paper, Denmark called for a "flexible" framework with voluntary self-labelling schemes. Denmark argued for a risk-based approach towards AI, highlighting that trustworthy and human-centric AI goes hand in hand with innovation, economic growth and competitiveness. The non-paper concludes with a call to the EU for a voluntary European labelling scheme that would make it visible for potential users, for example, citizens, businesses as well as public administrations, which applications are based on secure, responsible and ethical AI and data.[1756]

AI in Public Administration

In October 2023, the Danish Data Protection Authority (*Datatilsynet*) published guidance on the development and use of AI by public authorities.[1757] The report contained the results of the mapping of the use of AI across the public sector longer term. The Danish Data Protection Authority stated that it will look at more guidance on how organizations can handle the risks that may be associated with the use of AI, such as bias and lack of transparency.

Denmark introduced an AI compliance framework on the use of AI Assistants in the private and public sectors in 2024.[1758] A consortium that included Netcompany, the Agency for Digital Governance, Danish Business Authority and other entities

---

[1753] Ibid

[1754] D-seal, *Criteria* (2024), https://d-seal.eu/en/get-d-sealed/criteria

[1755] Denmark, Belgium, the Czech Republic, Finland, France, Estonia, Ireland, Latvia, Luxembourg, the Netherlands, Poland, Portugal, Spain and Sweden, *Non-paper: Innovative and Trustworthy AI: Two Sides of the Same Coin*, Netherlands and You (Aug. 10, 2020), https://www.netherlandsandyou.nl/web/pr-eu-brussels/documents

[1756] Melissa Heikkla, *6 Key Battles ahead for Europe's AI Law* (Apr. 21, 2021), https://www.politico.eu/article/6-key-battles-europes-ai-law-artificial-intelligence-act/,

[1757] Datatilsynet, *New Guidance on Public Authorities' Use of AI and Mapping of AI across the Public Sector [Ny vejledning om offentlige myndigheders brug af AI og kortlægning af AI på tværs af den offentlige sektor]* (Oct. 5, 2023), https://www.datatilsynet.dk/presse-og-nyheder/nyhedsarkiv/2023/okt/ny-vejledning-om-offentlige-myndigheders-brug-af-ai-og-kortlaegning-af-ai-paa-tvaers-af-den-offentlige-sektor

[1758] Dansk Industri, *Whitepaper: Responsible Use of AI Assistants in the Public and Private Sectors* (2024), https://www.danskindustri.dk/globalassets/brancher/di-digital/2024/dokumenter/di-digitaliseringspolitik-2024-ai-hvidbog-lang-version_uk.pdf?v=251028

published the whitepaper with the framework, with support from major financial entities and Microsoft. The framework provides structured methodologies for risk and bias management, secure data handling, and staff training. The guidance is designed to help organizations align with the EU AI Act and GDPR, ensuring transparency and efficiency in AI adoption across sectors."

Amnesty International's Algorithmic Accountability Lab (AAL) released an investigative report on Denmark's welfare agency (*Udbetaling Danmark*, UDK) scrutinizing the use of algorithmic decision-making in welfare distribution.[1759] The report highlights concerns over potential discriminatory practices, where automated systems disproportionately target individuals seeking benefits. These findings align with broader European concerns regarding the risks of algorithmic bias in public sector decision-making.

Autonomous Vehicles

Amendments to Denmark's Road Traffic Act in 2017 imposed restrictions on the testing of self-driving motor vehicles.[1760] Effective 1 July 2025, Denmark's Executive Order on Trials and Testing of Self-Driving Units issued by the Denmark Road Authority eases many of those restrictions.[1761] The Order allows municipalities to establish designated test zones, where trials may proceed with a permit from the Danish Road Traffic Authority. The order introduces updated technical requirements, clearer sanction rules, and more flexible conditions for approving tests, addressing earlier concerns that regulations were overly restrictive.

Mass Surveillance

The Danish Data Ethics Council raised concerns regarding the Danish Parliament's draft law on the Danish Security and Intelligence Service (PET),[1762] which was introduced without a data ethics impact assessment. The Council deems the impact assessment essential for legislation involving such extensive intrusions into citizens' privacy. The proposed law seeks to expand PET's surveillance powers by allowing large-scale, automated analyses of all Danish citizens' digital communications and online activity, ostensibly to strengthen national security and

---

[1759] Amnesty International, *Coded Injustice* (Nov. 14, 2024), https://cdn.amnesty.at/media/eeadvmyt/amnesty-report_coded-injustice_surveillance-and-discrimination-in-denmarks-automated-welfare-state_november-2024.pdf

[1760] Folketinget, *Act Amending the Road Traffic Act [Lov om ændring af færdselsloven]* (May 13, 2017), https://www.ft.dk/ripdf/samling/20161/lovforslag/l120/20161_l120_som_vedtaget.pdf

[1761] Retsinformation, *Executive Order on Trials and Testing of Self-Propelled Units [Bekendtgørelse om forsøg med og test af selvkørende enheder]* (Jun. 24, 2025), https://www.retsinformation.dk/eli/lta/2025/925

[1762] Dataetisk Råd, *Data Ethics Impact Assessment of the Bill on the Danish Police Intelligence Service (PET) [Dataetisk konsekvensvurdering af lovforslag om Politiets Efterretningstjeneste (PET)]* (Jul. 11, 2025), https://dataetiskraad.dk/alle-hoeringssvar/dataetisk-konsekvensvurdering-af-lovforslag-om-politiets- efterretningstjeneste-pet

counter-terrorism capabilities in response to Denmark's elevated threat level. The Councils assessment uncovered serious concerns. The Council cautioned that the draft law grants PET sweeping access to citizens' digital footprints without sufficient evidence of the measure's effectiveness or adequate safeguards to protect individual rights and freedoms.

Environmental Impact of AI

The Danish National Strategy for Artificial Intelligence emphasizes responsible and ethical AI use across sectors, including the green transition. This strategy is a part of Denmark's broader commitment to sustainability. It includes measures such as promoting the use of renewable energy in AI-related infrastructures and supporting the development of efficient data processing methods to reduce carbon footprints.[1763]

This commitment gained urgency as projections indicated that data centers, critical infrastructure for AI development, could account for one-fifth of the total Danish energy budget by 2030. To mitigate this impact and enhance sustainability, the government announced an agreement in 2025 to remove the existing pricing cap on excess heat generated by businesses.[1764] Removing the price cap, which requires a legislative amendment, provides an incentive for businesses, particularly data centers, to invest in heat recovery projects that would promote market-driven utilization and integration of waste heat into existing district heating systems.

Lethal Autonomous Weapons

At a 2015 informal meeting of the Convention on Certain Convention Weapons (CCW) Group of Global Experts (GGE) on Lethal Autonomous Weapons Systems, Denmark affirmed that weapons must remain under "meaningful human control.[1765] In 2020, during the General Debate of the First Committee of the 75th UN General Assembly, the Permanent Representative of Denmark declared, "Denmark supports the work of the GGE on Lethal Autonomous Weapons Systems (LAWS), in particular the 11 guiding principles. In our work on these principles we should in particular aim to develop an understanding of the type and degree of human machine interaction."[1766]

[1763] Agency for Digital Government, *The Danish National Strategy for Artificial Intelligence*, p. 65 (Mar. 2019), https://en.digst.dk/strategy/the-danish-national-strategy-for-artificial-intelligence/

[1764] Invest in Denmark, *Denmark Removes Price Cap on Surplus Heat, Unlocking Investment and Sustainability Potential* (Mar. 18, 2025), https://investindk.com/insights/denmark-removes-price-cap-on-surplus-heat-unlocking-investment-og-sustainability-potential

[1765] Government of Denmark, *Statement to the Convention on Conventional Weapons Informal Meeting of Experts on Lethal Autonomous Weapons Systems* (Apr. 13, 2015), http://www.reachingcriticalwill.org/images/documents/Disarmament-fora/ccw/2015/meeting-experts-laws/statements/13April_Denmark.pdf

[1766] Statement by H.E. Mr. Martin Bille Hermann, *Permanent Representative of Denmark General Debate First Committee of the 75th UN General Assembly* (Oct. 19, 2020),

Denmark has consistently supported measures to discuss the implications of LAWS in the UN General Assembly,[1767] and to collaborate on a framework to protect against a loss of control.[1768] Denmark supports a two-tier approach whereby applications of LAWS that cannot comply with international law or international humanitarian law (IHL) would be explicitly prohibited and other uses would be regulated to ensure compliance. The country laid out this position in a working paper submitted with Bulgaria, France, Germany, Italy, Luxembourg, and Norway to the GGE on LAWS in 2024.[1769] UN Secretary-General cited the working paper multiple times in the Report on LAWS prepared under direction from the General Assembly resolution L.56.[1770] During the GGE session in August 2024, the Danish delegation emphasized the importance of adhering to the principles of distinction and proportionality under International Humanitarian Law (IHL), reinforcing the need for meaningful human control over autonomous military technologies.[1771] Denmark was one of 39 High Contracting Parties that supported a joint statement declaring that the GGE's rolling text was a sufficient basis for launching negotiations on a legally binding instrument on lethal autonomous weapons systems.[1772]

In February 2023, Denmark participated in the REAIM international summit on the responsible application of artificial intelligence in the military domain hosted by the Netherlands. At the end of the Summit, Denmark endorsed a joint call for action

---

https://reachingcriticalwill.org/images/documents/Disarmament-fora/1com/1com20/statements/19Oct_Denmark.pdf

[1767] United Nations (UN) General Assembly, First Committee, *Joint Statement on Lethal Autonomous Weapons Systems First Committee, 77th United Nations General Assembly Thematic Debate – Conventional Weapons* (Oct. 21, 2022), https://estatements.unmeetings.org/estatements/11.0010/20221021/A1jJ8bNfWGlL/KLw9WYcSnnAm_en.pdf

[1768] General Assembly, *Resolution L56: Lethal Autonomous Weapons* (Oct. 12, 2023), https://reachingcriticalwill.org/images/documents/Disarmament-fora/1com/1com23/resolutions/L56.pdf

[1769] GGE on Emerging Technologies in the Area of LAWS, *Working Paper Submitted by Bulgaria, Denmark, France, Germany, Italy, Luxembourg, and Norway,* CCW/GGE.1/2024/WP.3 (Mar. 4, 2024), https://docs-library.unoda.org/Convention_on_Certain_Conventional_Weapons_-Group_of_Governmental_Experts_on_Lethal_Autonomous_Weapons_Systems_(2024)/CCW-GGE.1-2024-WP.3.pdf

[1770] UN Digital Library, *Lethal Autonomous Weapons Systems: Report of the Secretary-General* (2024), https://digitallibrary.un.org/record/4059475?v=pdf

[1771] UN Office for Disarmament Affairs, *Group of Governmental Experts on Emerging Technologies in the Area of Lethal Autonomous Weapons Systems: Intervention by Denmark* (Aug. 26, 2024), https://docs-library.unoda.org/Convention_on_Certain_Conventional_Weapons_-Group_of_Governmental_Experts_on_Lethal_Autonomous_Weapons_Systems_%282024%29/GGE_LAWS_session_August_2024._Danish_statement._General_Statement.pdf

[1772] UN Office of Disarmament Affairs, *Joint Statement to the September 2025 Session of the CCW GGE LAWS* (Sept. 2025), https://docs-library.unoda.org/Convention_on_Certain_Conventional_Weapons_-Group_of_Governmental_Experts_on_Lethal_Autonomous_Weapons_Systems_(2025)/Joint_statement_to_the_September_2025_session_of_the_CCW_GGE_LAWS_-_As_delivered.pdf

on the responsible development, deployment and use of artificial intelligence in the military domain.[1773] Denmark also endorsed the resulting Political Declaration on Responsible Military Use of AI and Autonomy issued in November 2023.[1774]

AI Literacy

Denmark employs a comprehensive, human centered approach to AI literacy, viewing it as crucial for public trust and effective governance. This effort is primarily driven by the AI Competence Pact (*AI Kompetence Pagten*), a major public-private partnership established to upskill one million people in AI to future-proof the public sector workforce.[1775] The focus is on providing foundational awareness for citizens and building functional competence within institutions. The Agency for Digital Governance's specialized AI Guides inform citizens on the responsible use of generative AI,[1776] marking further progress in this regard. The National Agency for Education and Quality has also published recommendations on the use of generative AI in upper secondary education "to support a practice where generative AI is used in a balanced way to support learning, and where pupils are taught to critically reflect on their use of the technology."[1777] These guidelines mirror those published in 2024 for upper secondary education.[1778]

Human Rights

Denmark receives high scores for Political Rights and Civil Liberties in the Freedom House Freedom in the World 2025 Report, earning a score of 97/100 to rate as "Free." The report highlights the Danish government's protections of free expression and association and the judiciary's independence but notes a lack of full rights and protections for immigrants and other newcomers. The Danish Parliament established the Danish Centre for Human Rights in 1987, which was renamed the Danish Institute for Human Rights in 2002.[1779] Denmark is a signatory to major human

---

[1773] Responsible AI in the Military domain Summit, *REAIM Call to Action* (Feb. 16, 2023), https://www.government.nl/documents/publications/2023/02/16/reaim-2023-call-to-action

[1774] US Department of State, *Political Declaration on Responsible Military Use of Artificial Intelligence and Autonomy*, Endorsing States (Nov. 27, 2024), https://www.state.gov/political-declaration-on-responsible-military-use-of-artificial-intelligence-and-autonomy/

[1775] Digital Hub Denmark, *AI Decoding: The AI Competence Pact for a Digitally Proficient Workforce* (2025), https://www.digitalhubdenmark.dk/post/decoding-1

[1776] Agency for Digital Government, *The Agency for Digital Government Publishes AI Guides* (May 3, 2024), https://en.digst.dk/news/news-archive/2024/maj/the-agency-for-digital-government-publishes-ai-guides/

[1777] Eurydice, *Recommendations for the Use of Generative Artificial Intelligence in Primary and Lower Secondary Schools*, Denmark: National Reforms in General School Education (2025), https://eurydice.eacea.ec.europa.eu/eurypedia/denmark/national-reforms-general-school-education

[1778] Eurydice, *Recommendations for the Use of Generative Artificial Intelligence in Upper Secondary Education* (2024), https://eurydice.eacea.ec.europa.eu/eurypedia/denmark/national-reforms-general-school-education

[1779] Danish Institute for Human Rights, *Human Rights in Danish Law*, https://www.humanrights.dk/

rights treaties, including the Universal Declaration of Human Rights which has also ratified several European human rights instruments, including the European Convention on Human Rights (ECHR).[1780]

In a 2020 Recommendation to member States on the human rights impacts of algorithmic systems, the Council of Europe Committee of Ministers reminded member states that their commitment "to ensuring the rights and freedoms enshrined in the Convention for the Protection of Human Rights and Fundamental Freedoms to everyone within their jurisdiction" requires them to "ensure that any design, development and ongoing deployment of algorithmic systems occur in compliance with human rights and fundamental freedoms, which are universal, indivisible, interdependent and interrelated, with a view to amplifying positive effects and preventing or minimising possible adverse effects."[1781]

OECD / G20 AI Principles

Denmark is a member of the OECD and has endorsed the OECD AI Principles.[1782] A 2021 report on the state of implementation of the OECD AI Principles noted Denmark's progress toward implementation of the OECD AI principles.[1783] The 2024 Strategic Approach to Artificial Intelligence reinforces Denmark's commitment to the OECD AI Principles[1784] with the guiding principles centered on developing AI to aligned with citizens' fundamental rights and Danish values, supporting Danish companies to be competitive through responsible AI innovation, and modeling responsible public-sector use of AI.

Denmark is also a member of the Global Partnership on AI, a multi-stakeholder initiative that aims to foster international cooperation on AI research and applied activities and is "built around a shared commitment to the OECD Recommendation on Artificial Intelligence."[1785]

[1780] European Court of Human Rights, *The European Convention on Human Rights (ECHR)* (2021), https://www.echr.coe.int/documents/d/echr/convention_ENG

[1781] *Recommendation CM/Rec(2020)1 of the Committee of Ministers to Member States on the Human Rights Impacts of Algorithmic Systems* (Apr. 8, 2020), https://search.coe.int/cm/pages/result_details.aspx?objectid=09000016809e1154

[1782] OECD Legal Instruments, *Recommendation of the Council on Artificial Intelligence, Adherents* (May 3, 2024), https://legalinstruments.oecd.org/en/instruments/OECD-LEGAL-0449#adherents

[1783] OECD, *State of Implementation of the OECD AI Principles: Insights from National AI Policies* (Jun. 18, 2021), https://www.oecd.org/digital/state-of-implementation-of-the-oecd-ai-principles-1cd40c44-en.htm

[1784] OECD AI Policy Observatory, *Overarching National AI Strategy* (2024), https://oecd.ai/en/dashboards/policy-initiatives/overarching-national-ai-strategy-4375

[1785] OECD AI Policy Observatory, *About the Global Partnership on Artificial Intelligence (GPAI)* (2026), https://oecd.ai/en/about/about-gpai

Council of Europe AI Treaty

Denmark participated in negotiations for the Framework Convention on Artificial Intelligence and Human Rights, Democracy, and the Rule of Law as a Council of Europe member. Denmark became party to this first legally binding international AI treaty with the European Commission signature in September 2024 but has not acted independently on the treaty.[1786]

UNESCO Recommendation on AI Ethics

Denmark has endorsed the UNESCO Recommendations on AI, the first ever global agreement on the ethics of AI.[1787] Denmark's AI Competence Pact[1788] aligns with the Recommendation's policy action on Education and Research while the Data Ethics Council[1789] provides advice and reviews to encourage ethical use of data and the products resulting from this use, as exemplified by the Council's review of the Danish Parliament's draft law on the Danish Security and Intelligence Service (PET).[1790] Denmark has not initiated the UNESCO Readiness Assessment Methodology (RAM),[1791] a tool for implementing the Recommendation.

Evaluation

Denmark's National AI strategy, released in 2019, sets out an ambitious agenda emphasizing responsible AI development. Denmark has acted on this agenda, establishing an independent Data Ethics Council, endorsing the OECD AI Principles, and promoting opportunities for public participation in the development of AI policy. Denmark has also introduced certification seals to promote trustworthy AI. Denmark has established an independent national competent authority in accordance with the EU AI Act and even offered guides on safe and responsible use of generative AI for citizens, companies, and authorities.

[1786] Council of Europe Treaty Office, *Chart of Signatures and Ratifications of Treaty 225* (Feb. 6, 2026), https://www.coe.int/en/web/conventions/full-list?module=signatures-by-treaty&treatynum=225

1787 UNESCO, *UNESCO Member States Adopt the First Ever Global Agreement on the Ethics of Artificial Intelligence* (Nov. 2021), https://www.unesco.org/en/articles/unesco-member-states-adopt-first-ever-global-agreement-ethics-artificial-intelligence.

[1788] Digital Hub Denmark, *AI Decoding: The AI Competence Pact for a Digitally Proficient Workforce* (2025), https://www.digitalhubdenmark.dk/post/decoding-1

[1789] Dataetisk Råd, *About the Danish Data Ethics Council: Background and Purpose [Om Dataestisk Råd: Baggrund og formål]* (2026), https://www.dataetiskraad.dk/om-dataetisk-raad/baggrund-og-formaal

[1790] Dataetisk Råd, *Data Ethics Impact Assessment of the Bill on the Danish Police Intelligence Service (PET) [Dataetisk konsekvensvurdering af lovforslag om Politiets Efterretningstjeneste (PET)]* (Jul. 11, 2025), https://dataetiskraad.dk/alle-hoeringssvar/dataetisk-konsekvensvurdering-af-lovforslag-om-politiets- efterretningstjeneste-pet

[1791] UNESCO Global AI Ethics and Policy Observatory, *Global Hub* (Oct. 2025), https://www.unesco.org/ethics-ai/en/global-hub

As Denmark has endorsed the UNESCO Recommendation on the Ethics of AI and the country's action on the EU AI Act, Council of Europe AI Treaty, and Nordic-Baltic collaboration is a step toward implementation. Ratifying the Convention 108+ would be a strong signal that Denmark prioritizes the fundamental rights of its citizens.

Dominican Republic

In 2025, Dominican legislators introduced the first AI-related bills to Congress. National agencies also progressed on implementing the 2024 Open Government Action Plan to promote public participation in policymaking.

National AI Strategy

The president of the Dominican Republic introduced the National AI Strategy[1792] with these words in October 2023: "We find ourselves at a transformative moment for the future of our nation. AI stands as the emblematic technology of our century, with an impact that will resonate in all facets of our society. A vast territory of new opportunities is opening before us, a horizon of transformation which calls us to immediate action."

The Smart Government Program targeting proactive and ethical digital transformation in public administration is one of three key initiatives in the Strategy.[1793] The other pillars are creating a hub to promote human talent and innovation and building infrastructure to make the Dominican Republic a regional hub for the storage and processing of data. The Strategy takes a regional perspective for all three initiatives, promoting regional collaboration on AI to make the country a catalyst in a regional AI ecosystem. These initiatives align with the Strategy's objectives to insert the country in the Fourth Industrial Revolution, strengthen technological autonomy, and lead the region in AI. "Guaranteeing that the adoption of artificial intelligence in our land is aligned with fundamental ethical principles and reinforces the protection of human rights"[1794] is part of the vision underlying the Strategy.

The National AI Strategy aligns with the Digital Agenda 2030[1795] developed by the Digital Transformation Cabinet (*Gabinete de Transformación Digital de la República Dominicana*), which sits under the executive.[1796] The Digital Agenda 2030 is divided into five main axes: 1) governance and regulatory framework; 2) connectivity and access 3) education and digital skills; 4) digital government; and 5) digital economy. Each of these axes responds to specific objectives through performance measurement indicators. "By 2030, the Dominican Republic expects to

---

[1792] Gobierno de la República Dominicana, *National Artificial Intelligence Strategy of the Dominican Republic [Estrategia Nacional de Inteligencia Artificial de la República Dominicana]*, p. 6 (Oct. 2023), https://agendadigital.gob.do/wp-content/uploads/2023/10/Final_ENIA-Estrategia-Nacional-de-Inteligencia-Artificial-de-la-Republica-Dominicana-.pdf

[1793] Ibid, p. 14

[1794] Ibid, p. 13

[1795] Gabinete de Transformación Digital de la República Dominicana, *Agenda Digital 2030, Versión 1* (2021), https://agendadigital.gob.do/wp-content/uploads/2022/01/Agenda-Digital-2030.pdf

[1796] Luis Abinader, *Decree 571-21* (Aug. 26, 2021), https://presidencia.gob.do/sites/default/files/decree/2021-08/Decreto%20527-21%20Agenda%20Digital%202030.pdf

have reduced the digital divide and ensured access to, and use of, digital technologies in a secure and sustainable environment."[1797] The Digital Agenda has been developed in line with the broader development strategy, more particularly with regard to addressing the digital divide through education.

The Dominican Republic has also been part of the Caribbean Artificial Intelligence Initiative led by the UNESCO Cluster Office for the Caribbean and the Broadcasting Commission of Jamaica (BCJ). The Initiative "aims to develop a sub-regional strategy on the ethical, inclusive and humane use of AI in the Caribbean Small Island Developing States."[1798] The Caribbean AI Policy Roadmap was released in June 2021 and updated in 2024.[1799]

AI Legislation

The Dominican Republic has not yet passed AI-specific legislation, though a number of bills were presented in 2025. One bill to regulate AI systems and applications includes a proposal to establish the National Council for Artificial Intelligence (CONIA) with AI oversight responsibility over the entire AI life cycle.[1800] The bill, which was presented in the Chamber of Deputies (*Cámara de Diputados*), prohibits the use of AI systems for military, defense, or national security purposes.[1801] Other bills propose guidelines for public policies on the development, use, regulation, and implementation of AI in the country[1802] and address the use of AI-generated

[1797] OECD, *Multidimensional Review of the Dominican Republic: Towards Greater Well-Being for All* (Dec. 13, 2022), https://www.oecd-ilibrary.org/development/multi-dimensional-review-of-the-dominican-republic_560c12bf-en

[1798] Caribbean Artificial Intelligence Initiative, *Caribbean Artificial Intelligence Roadmap* (2021), https://ai4caribbean.com/wp-content/uploads/2021/07/Caribbean-Artificial-Intelligence-Policy-Roadmap.pdf

[1799] UNESCO Office in Kingston, *Caribbean Artificial Intelligence Policy Roadmap* (2024), https://unesdoc.unesco.org/ark:/48223/pf0000391996

[1800] Cámara de Diputados Sistema de Información Legislativa, *Bill that Regulates Artificial Intelligence Systems and Their Applications in the Dominican Republic [Proyecto de ley que regula los Sistemas de Inteligencia Artificial y sus aplicaciones en la Republica Dominicana]* (Feb. 19, 2025), https://s-sil.camaradediputados.gob.do:8095/ReportesGenerales/VerDocumento?documentoId=225083

[1801] Dolores Vicioso, *First Artificial Intelligence Bill Presented in Congress*, dr1 (Feb. 20, 2025), https://dr1.com/news/2025/02/20/first-artificial-intelligence-bill-presented-in-congress/

[1802] Senado, *Proyecto de Ley que Establece los Lineamientos para las Políticas Públicas Orientadas al Desarrollo, Uso, Regulación e Implementación de Inteligencia Artificial en la República Dominicana,* Sistema de Información Legislativa (Jan. 23, 2026), http://www.senado.gov.do/mlx/DOCS/1C/2/11/66CE/CE65.pdf

content to commit fraud. [1803] The oversight bill ended the 2025 session in committee and was resubmitted at the start of the 2026 session.[1804]

Public Participation

To engage with stakeholders in development of digital service and policy design, the Dominican Republic used focus groups, public consultations and social media.[1805] The Dominican Republic deepened and systematized the commitment to participatory governance in digital and AI-related policymaking through the Open Government Plan of Action, 2024–2028,[1806] and the National Strategy for Civic Space.[1807]

The Open Government Partnership (OGP) supported the initiatives, which were led by the General Directorate of Ethics and Government Integrity (*Dirección General de Ética e Integridad Gubernamental*, DIGEIG). The Action Plan provides actions steps to be implemented across the government to improve transparency and citizens participation in holding offices accountable.[1808] Participants in creating the plan reported "adequate outreach" and feedback on how suggestions were incorporated, or why they were not.[1809] The DIGEIG also developed the National Strategy for Civic Space through a cocreation process.[1810] The Strategy provides 21 reforms intended to

---

[1803] Cámara de Diputados Sistema de Información Legislativa, *Proyecto de Ley que modifica los articulos 15, 16, 21, 22, 24, 25, 26 y 28 de la Ley 53-07, de Crimenes y Delitos de Alta Tecnología* (Feb. 19, 2025), https://s-sil.camaradediputados.gob.do:8095/ReportesGenerales/VerDocumento?documentoId=225128

[1804] Senado Sistem de Información Legislativa, *Iniciativa 01321-2026-PLO-SE, Proyecto de Ley que esablece los lineamientos para las política públicas orientadas al desarrollo, uso, regulación e implementación de inteligencia artificial en la República Dominicana* (Jan. 23, 2026), http://www.senado.gov.do/wfilemaster/Ficha.aspx?IdExpediente=39012&numeropagina=1&ContExpedientes=2213&Coleccion=53

[1805] OECD, *The Strategic and Responsible Use of Artificial Intelligence in the Public Sector of Latin America and the* Caribbean, p. 155 (2022), https://oecd-opsi.org/wp-content/uploads/2022/03/lac-ai.pdf

[1806] Open Government Partnership, *VI Action Plan of the Dominican Republic under the Open Government Partnership. 2024–2028 [VI Plan de Acción de la República Dominicana ante la Alianza para el Gobierno Abierto 2024–2028* (Feb. 2026), https://www.opengovpartnership.org/wp-content/uploads/2025/01/Dominican-Republic_Action-Plan_2024-2028_Revision-1-Feb-2026_ES.pdf

[1807] Open Government Partnership, *National Strategy for Civic Space* (2025), https://www.opengovpartnership.org/the-open-gov-challenge/dominican-republic-national-strategy-for-civic-space/

[1808] Open Government Partnership, *Dominican Republic: Action Plan Review 2024–2028* (Sept. 30, 2025), https://www.opengovpartnership.org/documents/dominican-republic-action-plan-review-2024-2028/

[1809] Ibid

[1810] Ética e Integridad Gubernamental, *DIGEIG y AGA Cocrean Estrategia de Espacio Cívico con la Participación de la Sociedad en el Segundo Foro Multiactor de RD* (2025), https://digeig.gob.do/noticias/digeig-y-aga-cocrean-estrategia-nacional-de-espacio-civico-con-la-participacion-de-la-sociedad-en-el-segundo-foro-multiactor-de-rd/

"expand the channels of dialog between the State and citizens, protect the rights of expression, association, and participation; and foster sustainable conditions for social collaboration in public governance."[1811]

These initiatives respond to the Agenda Digital 2030 emphasis on public participation as a cross-cutting policy while noting the dearth of digital participation initiatives as an area for improvement.[1812] More recently, the Government Office of Technology, Information and Communication, built a site on Notion to support public consultations on the AI strategy.[1813] The Senate has also hosted meeting with affected private sector representatives.[1814]

The judiciary invited civil society and stakeholders to provide input on how the judicial branch will manage, publish and safeguard data through consultation[1815] on the draft Open Data Policy of the Judiciary.[1816]

Data Protection

The Dominican Constitution enshrines the right to the protection of personal data in public and private records under Section 44(2) and provides that data controllers and processors are to comply with data security, professional secrecy, data quality and data loyalty.[1817]

Law No 172-13 on the Comprehensive Protection of Personal Data of consumers or users[1818] provides the right to access and right to rectification of personal

---

[1811] Ética e Integridad Gubernamental, *RD consolida su liderazgo regional en Gobierno Abierto y Espacio Cívico* (Oct. 9, 2025), https://digeig.gob.do/noticias/rd-consolida-su-liderazgo-regional-en-gobierno-abierto-y-espacio-civico/

[1812] Gabinete de Transformación Digital, *Agenda Digital 2030,* pp. 26, 31 (Feb. 2022) https://agendadigital.gob.do/wp-content/uploads/2022/02/Agenda-Digital-2030-v2.pdf

[1813] Gabnete de Innovación, *Consulta Pública para la Estrategia Nacional de Inteligencia Artificial de la República Dominicana (ENIA) - Foro Comunitario Ciudadano* (Jul. 2023), https://consultapublica.notion.site/Consulta-P-blica-para-la-Estrategia-Nacional-de-Inteligencia-Artificial-de-la-Rep-blica-Dominicana--e58e40a872fc4d79a58766fdcf4d14f3

[1814] Senado, *Comisión de senadores escucha opinión de representantes de Microsoft y GBM-IBM sobre proyecto regula Inteligencia Artificial* (Nov. 27, 2024), https://www.senadord.gob.do/comision-de-senadores-escucha-opinion-de-representantes-de-microsoft-y-gbm-ibm-sobre-proyecto-regula-inteligencia-artificial/

[1815] Poder Judicial, *Consulta Pública: Política de Datos Abiertos del Poder Judicial* (Aug. 11, 2025), https://poderjudicial.gob.do/consulta-publica-politica-de-datos-abiertos-del-poder-judicial/

[1816] Poder Judicial, *Política de Datos Abiertos del Poder Judicial* (Aug. 2025), https://poderjudicial.gob.do/wp-content/uploads/2025/08/Politica-de-Datos-Abiertos-del-Poder-Judicial.pdf

[1817] Constitute, *Dominican Republic's Constitution of 2015* (Apr. 27, 2022), https://www.constituteproject.org/constitution/Dominican_Republic_2015.pdf

[1818] Presidencia, *Law No. 172-13 which Aims at the Comprehensive Protection of Personal Data Recorded in Files, Public Records, Data Banks or Other Technical Means of Data Processing for Reporting Purposes, whether Public or Private [que tiene por objeto la protección integral de los datos personales asentados en archivos, registros públicos, bancos de datos u otros medios técnicos de tratamiento de datos destinados a dar informes, sean estos públicos o privados]*, Arts. 10, 14

data.[1819] Various decisions have upheld these rights in digital formats and with automatic processing.[1820] Still, no institution has official oversight to guarantee these protections for citizens' rights, leaving, in the words of one attorney, "the constitutional recognition solid on paper but weak in practice." The General Directorate of Ethics and Government Integrity is an observer in the Ibero-American Network for Data Protection (*Red Iberoamericana de Protección de Datos*, RIPD).[1821]

The General Law for the Protection of Consumer or User Rights No. 358-05 declared that the National Institute for the Protection of Consumer Rights, "Pro Consumidor," monitors data protection compliance in relation to consumers. Pro Consumidor does not have enforcement powers. Instead, the institution relies on mediation and conciliation competences.[1822]

The Dominican Republic started to draft a data protection law in line with international standards such as the Council of Europe's Convention 108+ in 2019 and received support from the Council of Europe to this end.[1823] However, no such law yet exists.

The 2023 AI Strategy sets revision of Law No 172-13 to include safeguards for the protection of personal data and human rights in the implementation of AI system by 2025.[1824] However, an amended law has not passed the congress. The country developed a Digital Government guide that includes a provision on the documentation and explainability of digital government initiatives, software, and services.[1825]

(Dec. 15, 2013), https://presidencia.gob.do/sites/default/files/statics/transparencia/marco-legal/leyes/Ley-172-13.pdf

[1819] Emely García Aybar, *La legislación dominicana sobre Protección de Datos Personales: Principios, Consentimiento y el proceso de Habeas Data*, Red Iberoamericana de Protección de Datos (RIPD), https://www.redipd.org/documentos/la-legislacion-dominicana-sobre-proteccion-de-datos-personales-principios-consentimiento

[1820] Felicia Santana, *Opinión: República Dominicana y la era digital: ¿Quién protege nuestros datos?* acento (Feb. 21, 2026), https://acento.com.do/opinion/republica-dominicana-y-la-era-digital-quien-protege-nuestros-datos-9626904.html

[1821] Red Iberoamericana de Protección de Datos, *Composición*, https://www.redipd.org/la-red/composicion

[1822] WIPO, *The Dominican Republic Issued Law No 358-05 on the Protection of the Rights of the Consumer or User* (2005), https://www.wipo.int/wipolex/en/legislation/details/11048

[1823] Council of Europe, *GLACY+: The Dominican Republic Works on New Data Protection Law* (Dec. 16–18, 2019), https://www.coe.int/en/web/cybercrime/glacyplusactivities/-/asset_publisher/DD9qKA5QlKhC/content/glacy-the-dominican-republic-works-on-new-data-protection-law

[1824] Gobierno de la República Dominicana, *National Artificial Intelligence Strategy of the Dominican Republic [Estrategia Nacional de Inteligencia Artificial de la República Dominicana]*, pp. 46–57 (Oct. 2023), https://agendadigital.gob.do/wp-content/uploads/2023/10/Final_ENIA-Estrategia-Nacional-de-Inteligencia-Artificial-de-la-Republica-Dominicana-.pdf

[1825] OECD, *The Strategic and Responsible Use of Artificial Intelligence in the Public Sector of Latin America and the Caribbean* (Mar. 22, 2022), https://www.oecd-ilibrary.org/governance/the-strategic-and-responsible-use-of-artificial-intelligence-in-the-public-sector-of-latin-america-and-the-caribbean_1f334543-en

Algorithmic Transparency

The Dominican Republic's data protection regime indirectly addresses automated processing and the right to correct data may provide some protection in the case of automatic decisions. The National AI Strategy includes transparency in the use of AI as one of the guarantees to be established by a proposed code of ethics.[1826] Both the strategy and Digital Government guide propose requirements for government transparency around data use. However, specific guidelines for algorithmic transparency and explainability are not provided.[1827]

Facial Recognition

The Dominican General Director of Immigration, Enrique García, reported in 2021 that facial recognition technology will be used to enforce security in airports and border entry to combat drug trafficking and international crime.[1828] The Dominican Airport Association explained in September 2022 that there are intelligent security cameras in Dominican airports.[1829]

The Director of the General Directorate of Passports (DGP) said in March 2022 that the country will transition from using mechanical passports to electronic passports. The electronic passport would integrate a chip containing all the carrier information and designed to facilitate facial recognition.[1830] The passport agency issued a tender for a passport containing a chip that carries biometric data, such as fingerprints and facial recognition, in May 2024.[1831]

[1826] Gobierno de la República Dominicana, *National Artificial Intelligence Strategy of the Dominican Republic [Estrategia Nacional de Inteligencia Artificial de la República Dominicana]*, p. 39 (Oct. 2023), https://agendadigital.gob.do/wp-content/uploads/2023/10/Final_ENIA-Estrategia-Nacional-de-Inteligencia-Artificial-de-la-Republica-Dominicana-.pdf

[1827] OECD, *The Strategic and Responsible Use of Artificial Intelligence in the Public Sector of Latin America and the Caribbean* (Mar. 22, 2022), https://www.oecd-ilibrary.org/governance/the-strategic-and-responsible-use-of-artificial-intelligence-in-the-public-sector-of-latin-america-and-the-caribbean_1f334543-en

[1828] Arecoa, *DR Airports Will Strengthen Security with Facial Recognition Technology* (Apr. 20, 2021), https://www.arecoa.com/aeropuertos/2021/04/20/aeropuertos-rd-reforzaran-seguridad-tecnologia-reconocimiento-facial/

[1829] Dominican Republic Today, *Airport Association Assures Security Cameras Make Criminal Actions "Extremely Difficult"* (Sept. 16, 2022), https://dominicantoday.com/dr/economy/2022/09/16/airport-association-assures-security-cameras-make-criminal-actions-extremely-difficult/

[1830] Dominican Today, *General Directorate of Passports with a View to Changing to Electronic Passport* (Mar. 10, 2022), https://dominicantoday.com/dr/local/2022/03/10/address-of-passports-with-a-view-to-changing-to-electronic-passport/

[1831] Dirección General de Pasaportes, *Convocatoria para Presentación de Manifestación de Interés* (May 2024), https://www.pasaportes.gob.do/transparencia/phocadownload/Consultas/2024/MANIFESTACIN%20DE%20INTERS%20PASAPORTE-E.pdf

The Director of Migration reported the launch of a pilot plan for biometric facial recognition at formal [land] border crossings in August 2025.[1832] The director is coordinating the installation of technology with the Presidency to test the project ahead of the entry into force of a "resolution that will establish the deadline for mandatory biometric facial recognition registration for foreigners participating in binational markets."

Beyond migration, the police launched an "automated facial recognition system" in the Scientific Policy area in 2023.[1833] The system allows users to find matches in a "database of millions of images collected from various social networks and websites" from a photo of video of "unknown people."

Environmental Impact of AI

The Dominican Republic's Digital Economy Agenda includes an objective to use technology to promote "economic and social well-being as a way to promote sustainable development."[1834] The country has also committed to using mitigating the environmental impact of AI by endorsing the Montevideo Declaration, a regional document centered on implementing the UNESCO Recommendation, that includes Environment and Sustainability as areas of focus for 2024–2025.[1835]

Lethal Autonomous Weapons

The National AI Strategy makes clear that the goal of the Dominican Republic is to pursue the prohibition of the use of AI as a weapon of war.[1836]

[1832] Migración, *Biometric Immigration Controls and the DGM Headquarters Are Advancing in Pedernales* (Aug. 8, 2025), https://migracion.gob.do/en/biometric-immigration-controls-and-the-dgm-headquarters-are-advancing-in-pedernales/

[1833] Policía Nacional, *Mayor General Alberto Then inaugura laboratorio automatizado de reconocimiento facial en la Policía Científica y entrega tres nuevas unidades tipo camionieta* (Nov. 4, 2023), https://www.policianacional.gob.do/mayor-general-alberto-then-inaugura-laboratorio-automatizado-de-reconocimiento-facial-en-la-policia-cientifica-y-entrega-tres-nuevas-unidades-tipo-camioneta/

[1834] Luis Abinader, *Decree 571-21* (Aug. 26, 2021), https://presidencia.gob.do/sites/default/files/decree/2021-08/Decreto%20527-21%20Agenda%20Digital%202030.pdf

[1835] Second Ministerial and High Authorities Summit on the Ethics of Artificial Intelligence in latin America and the Caribbean, *Declaration of Montevideo* (Oct. 4, 2024), https://www.gub.uy/agencia-gobierno-electronico-sociedad-informacion-conocimiento/sites/agencia-gobierno-electronico-sociedad-informacion-conocimiento/files/documentos/noticias/ENG_Montevideo%20Declaration%20approved.pdf

[1836] Gobierno de la República Dominicana, *National Artificial Intelligence Strategy of the Dominican Republic [Estrategia Nacional de Inteligencia Artificial de la República Dominicana]*, p. 40 (Oct. 2023), https://agendadigital.gob.do/wp-content/uploads/2023/10/Final_ENIA-Estrategia-Nacional-de-Inteligencia-Artificial-de-la-Republica-Dominicana-.pdf

The Dominican Republic endorsed 2019 Declaration on Lethal Autonomous Weapons[1837] prepared by France and Germany and opened for endorsement during the Alliance for Multilateralism event in 2019. The 11 principles in the declaration respond to challenges cited by the UN Convention on Certain Conventional Weapons (CCW) relating to the development of LAWS. The principles affirm that "international humanitarian law applies to these systems; a human must always be responsible for the decision to use these systems; [and] States must examine the legality of these new weapons that they are developing or requiring at the design stage."

The Dominican Republic is one of the 70 countries that endorsed a joint statement on autonomous weapons systems at the 2022 United Nations General Assembly urging "the international community to further their understanding and address these risks and challenges by adopting appropriate rules and measures, such as principles, good practices, limitations and constraints. We are committed to upholding and strengthening compliance with International Law, in particular International Humanitarian Law, including through maintaining human responsibility and accountability in the use of force."[1838] The country reiterated this stance with support by co-sponsoring resolution L.56 on autonomous weapons systems[1839] in the First Committee in 2023 and voting in in favor of the UN General Assembly Draft Resolution on Lethal Autonomous Weapons Systems in 2024.[1840] In 2025, Dominican Vice Minister Armando Manzueta warned of AI-related threats to peace and security. He cited lethal autonomous weapons without human oversight, disinformation, and technological asymmetry while reaffirming the call for banning autonomous weapons.[1841]

[1837] France Diplomacy, *11 Principles on Lethal Autonomous Weapons Systems (LAWS),* https://www.diplomatie.gouv.fr/en/french-foreign-policy/united-nations/multilateralism-a-principle-of-action-for-france/alliance-for-multilateralism/article/11-principles-on-lethal-autonomous-weapons-systems-laws#sommaire_2

[1838] UN General Assembly, *First Committee, Joint Statement on Lethal Autonomous Weapons Systems First Committee, 77th United Nations General Assembly Thematic Debate – Conventional Weapons* (Oct. 21, 2022), https://estatements.unmeetings.org/estatements/11.0010/20221021/A1jJ8bNfWGlL/KLw9WYcSnnAm_en.pdf

[1839] UN General Assembly, *Resolution L56: Lethal Autonomous Weapons* (Oct. 12, 2023), https://reachingcriticalwill.org/images/documents/Disarmament-fora/1com/1com23/resolutions/L56.pdf

[1840] UN Meetings Coverage and Press Releases, *In Nearly 50 Separate Recorded Votes, First Committee Approves 15 Drafts on Conventional Weapons, Divergent Approaches to Outer Space Security* (Nov. 5, 2024), https://press.un.org/en/2024/gadis3756.doc.htm

[1841] Ministerio de Administración Pública, *Minister Armando Manzueta Highlights at the UN the Country's Progress in Digital Infrastructure [Viceministro Armando Manzueta destaca en la ONU los avances del país en infraestructuras digitales]* (Oct. 2, 2025), https://map.gob.do/2025/10/02/viceministro-armando-manzueta-destaca-en-la-onu-los-avances-del-pais-en-infraestructuras-digitales/

The Dominican Republic endorsed the Belén Communiqué,[1842] which calls for "urgent negotiation" of a binding international treaty to regulate and prohibit the use of autonomous weapons to address the grave concerns raised by removing human control from the use of force in February 2023.

At the Responsible AI in the Military Domain Summit (REAIM 2023) in the Netherlands, the Dominican Republic joined nearly sixty states in calling for action on the responsible development, deployment and use of AI in the military domain.[1843] The Dominican Republic also endorsed the resulting Political Declaration on Responsible Military Use of AI and Autonomy issued in November 2023.[1844]

AI Literacy

Education and human talent as core components of both the National AI Strategy[1845] and Agenda Digital 2030.[1846] Both documents define objectives to build this talent through pre-university education modules and teacher training and development as well as programs directed to labor-force upskilling and training programs. The AI Strategy proposes collaboration among education and tech experts, parents, and teachers in a proposed HUB of human talent and innovation, #YoSoyFuturoRD. The Agenda Digital 2030 aims to ensure access to technological resources. The objective centers on "Developing the necessary competencies in the Dominican population, ensuring an inclusive focus that will permit the efficient use and adoption of digital technologies and forming the human talent required for sustainable economic and social development."[1847] The Minister of Education proposed digital competencies to be included in the 2026–2027 school year but no

[1842] Latin American and the Caribbean Conference of Social and Humanitarian Impact of Autonomous Weapons, *Communiqué* (Feb. 24, 2023), https://www.rree.go.cr/files/includes/files.php?id=2261&tipo=documentos

[1843] Government of Netherlands, *Call to Action on Responsible Use of AI in the Military Domain,* (Feb. 16, 2023), https://www.government.nl/documents/publications/2023/02/16/reaim-2023-call-to-action

[1844] US Department of State, *Political Declaration on Responsible Military Use of Artificial Intelligence and Autonomy*, Endorsing States (Feb. 12, 2024), https://www.state.gov/political-declaration-on-responsible-military-use-of-artificial-intelligence-and-autonomy/

[1845] Gobierno de la República Dominicana, *National Artificial Intelligence Strategy of the Dominican Republic [Estrategia Nacional de Inteligencia Artificial de la República Dominicana]*, pp. 46–57 (Oct. 2023), https://agendadigital.gob.do/wp-content/uploads/2023/10/Final_ENIA-Estrategia-Nacional-de-Inteligencia-Artificial-de-la-Republica-Dominicana-.pdf

[1846] Gabinete de Transformación Digital, *Agenda Digital 2030*, pp. 46–52 (2021), https://agendadigital.gob.do/wp-content/uploads/2022/01/Agenda-Digital-2030.pdf

[1847] Ibid, p. 48

details have been provided.[1848] Other government agencies have implemented programs for digital literacy on an ad hoc basis.[1849]

Human Rights

In 2025, Freedom House ranked the Dominican Republic as "Partly Free," with a score of 68/100 for political rights and civil liberties.[1850] According to the Freedom House 2024 report, "The Dominican Republic holds regular elections that are relatively free, though recent years have been characterized by controversies around implementing a new electoral framework." "Pervasive corruption undermines state institutions and the use of excessive force by police is a problem. Discrimination against Dominicans of Haitian descent and Haitian migrants, as well as against LGBT+ people, remain serious problems."

Despite these issues, the Human Rights Council Working Group Universal Periodic Review in June 2024 highlighted the country's progress in human rights, noting legislative advances in areas such as gender equality, child protection, and anti-discrimination measures. Key achievements include prohibiting child marriage, expanding educational access, and enhancing social protections for vulnerable groups. The report also discusses ongoing reforms to improve public security, reduce poverty, and protect marginalized communities, particularly through addressing the rights of women, children, and the LGBTQ+ community. Recommendations emphasize the need for further action on issues like statelessness, migration, and gender-based violence, urging the Dominican Republic to continue aligning its policies with international human rights standards.[1851]

[1848] Ministerio de Educación, *Minister Luid Miguel De Camps Presents His Proposal for the Inclusion of Digital Competencies to the CNE [Ministro Luis Miguel De Camps presenta en CNE su apuesta por la inclusión de las competencias digitales]* (Dec. 10, 2025), https://www.ministeriodeeducacion.gob.do/comunicaciones/noticias/ministro-luis-miguel-de-camps-presenta-en-cne-su-apuesta-por-la-inclusion-de-las-competencias-digitales

[1849] Presidencia, *Indotel Launches Digital Literacy Program for 100,000 People [Indotel lanza programa de alfabetización digital para 100,00 personas]* (Oct. 1, 2025), https://presidencia.gob.do/noticias/indotel-lanza-programa-de-alfabetizacion-digital-para-100000-personas

[1850] Freedom House, *Freedom in the World 2025: Dominican Republic* (2025), https://freedomhouse.org/country/dominican-republic/freedom-world/2025

[1851] UN Human Rights Council, *Report of the Working Group on the Universal Periodic Review: Dominican Republic* (Jun. 25, 2024), https://documents.un.org/doc/undoc/gen/g24/090/58/pdf/g2409058.pdf

OECD / G20 AI Principles

The Dominican Republic has not endorsed the OECD AI principles[1852] and is not a member of the Global Partnership on Artificial Intelligence,[1853] which requires a commitment to the OECD AI Principles. The Dominican Republic's participation in international AI governance discussions, especially with peers in Latin America and the Caribbean through institutions such as the Inter-American Development Bank and the fAIr LAC initiative, aligns with OECD AI Principles around international collaboration and sustainable, human-centered AI.[1854] The country's National AI Strategy and Agenda Digital 2030 also implement recommended policy actions including education and workforce training.

Council of Europe AI Treaty

The Dominican Republic has not signed the first legally binding global treaty on AI, the Council of Europe Framework Convention on Artificial Intelligence and Human Rights, Democracy, and the Rule of Law.[1855]

UNESCO Recommendation on AI Ethics

The Dominican Republic is a member state of UNESCO[1856] and endorsed the UNESCO Recommendation on AI Ethics.[1857]

The Dominican Republic participates in the Regional Council incorporating all governments of the region[1858] to implement the UNESCO Recommendation. Following the first summit centered on AI ethics, the Dominican Republic signed the 2023 Santiago Declaration to Promote Ethical Artificial Intelligence.[1859] The

---

[1852] OECD AI Policy Observatory, *OECD AI Principles Overview, Countries Adhering to the AI Principles* (2026), https://oecd.ai/en/ai-principles

[1853] OECD AI Policy Observatory, *About the Global Partnership on Artificial Intelligence (GPAI)* (2026), https://oecd.ai/en/about/about-gpai

[1854] OECD, *State of Implementation of the OECD AI Principles*, p. 76 (Jun. 2021), https://www.oecd-ilibrary.org/docserver/1cd40c44-en.pdf?expires=1679145707&id=id&accname=guest&checksum=4A7E8011553F4B626AD9EF4C04ACCDE1

[1855] Council of Europe Treaty Office, *Chart of Signatures and Ratifications of Treaty 225* (Mar. 10, 2026), https://www.coe.int/en/web/conventions/full-list?module=signatures-by-treaty&treatynum=225

[1856] UNESCO, *List of the Member States and the Associate Members*, https://pax.unesco.org/countries/ListeMS.html

[1857] United Nations, *193 Countries Adopt First-Ever Global Agreement on the Ethics of Artificial Intelligence* (Nov. 25, 2021), https://news.un.org/en/story/2021/11/1106612

[1858] UNESCO, *CAF and UNESCO Will Create a Council to Review Ethical Criteria for Artificial Intelligence in Latin America and Caribbean* (Jun. 23, 2022), https://www.unesco.org/en/articles/caf-and-unesco-will-create-council-review-ethical-criteria-artificial-intelligence-latin-america-and.

[1859] Cumbre Ministerial y de Altas Autoridades de América Latina y el Caribe, *Declaracion de Santiago "Para promover una inteligencia artificial ética en América Latina y el Caribe"* (Oct.

Declaration establishes fundamental principles to guide public policy on AI, including proportionality, security, fairness, non-discrimination, gender equality, accessibility, sustainability, privacy and data protection.[1860] The National AI Strategy commits the country to drafting a code of AI ethics in line with the UNESCO Recommendation.[1861]

Following the second meeting of the Regional Council in 2024, the Dominican Republic signed the Montevideo Declaration,[1862] which emphasizes a commitment to human rights, freedoms, and democracy and outlines five key priorities for countries to pursue in the coming year: Governance and Regulation, Talent and Future Work, Protection of Vulnerable Groups, Environment, and Sustainability. Regional governments and the AI Working Group, with support from UNESCO and CAF, will lead the implementation, with semi-annual reviews to ensure transparency and stakeholder engagement.

The Dominican Republic completed the UNESCO Readiness Assessment Methodology (RAM), a tool to support the effective implementation of the Recommendation in 2024.[1863] Key recommendations focus on advancing the National Artificial Intelligence Strategy (ENIA) with strong institutional and public support. Priorities include updating regulatory frameworks for data protection, cybersecurity, and potential AI-specific regulations, along with strengthening public sector capacities, promoting diversity in AI, and reducing urban-rural technology access disparities.[1864]

Evaluation

The Dominican Republic's endorsement of the UNESCO Recommendation on the Ethics of AI and related regional frameworks such as the Santiago and Montevideo Declarations reflect the country's commitment to ethical AI. The

---

2023), https://minciencia.gob.cl/uploads/filer_public/40/2a/402a35a0-1222-4dab-b090-5c81bbf34237/declaracion_de_santiago.pdf

[1860] UNESCO, *UNESCO and Leading Ministry in Santiago de Chile Host Milestone Regional LAC Forum on Ethics of AI* (Dec. 5, 2023), https://www.unesco.org/en/articles/unesco-and-leading-ministry-santiago-de-chile-host-milestone-regional-lac-forum-ethics-ai?hub=387

[1861] Gobierno de la República Dominicana, *National Artificial Intelligence Strategy of the Dominican Republic [Estrategia Nacional de Inteligencia Artificial de la República Dominicana]*, p. 39 (Oct. 2023), https://agendadigital.gob.do/wp-content/uploads/2023/10/Final_ENIA-Estrategia-Nacional-de-Inteligencia-Artificial-de-la-Republica-Dominicana-.pdf

[1862] Second Ministerial and High Authorities Summit on the Ethics of Artificial Intelligence in Latin America and the Caribbean, *Montevideo Declaration* (Oct. 4, 2024), https://www.gub.uy/agencia-gobierno-electronico-sociedad-informacion-conocimiento/sites/agencia-gobierno-electronico-sociedad-informacion-conocimiento/files/documentos/noticias/ENG_Montevideo%20Declaration%20approved.pdf

[1863] UNESCO Global AI Ethics and Governance Observatory, *Readiness Assessment Methodology* (2026), https://www.unesco.org/ethics-ai/en/ram

[1864] UNESCO, *Dominican Republic: Artificial Intelligence Readiness Assessment Report* (2024), https://unesdoc.unesco.org/ark:/48223/pf0000391573?posInSet=1&queryId=N-EXPLORE-87169348-9aac-4bd8-a172-616945b057f1

adoption of the National AI Strategy and completion of the RAM provide a strong foundation for ethical governance. Success will depend on effective implementation, particularly in ensuring algorithmic transparency and revising data protection laws to provide a strong legal framework to protect citizens' rights.

The country's stance on banning autonomous weapons is commendable though concerns about AI surveillance persist, especially for those whose lack of formal documents threatens their legal status. The introduction of TAINA and other programs designed to modernize public services is promising as are the plans and strategies for improving public participation in policymaking and fostering Civic Spaces.

Ecuador

Ecuador progressed on a national strategy to promote the responsible and ethical development and use of AI in Ecuador aligned to the 2025–2030 public policy for digital transformation. The country also deepened international collaboration by joining the Council of Europe Committee on AI (CAI) and gained recognition from UNESCO for being the first in the region to integrate an AI code of ethics in the public sector after the Superintendency of Economic Competition introduced the code and Guide for AI Use.

National AI Strategy

Ecuador's government institutions and digital innovation agencies initiated AI governance efforts in 2021 with an assessment of the AI landscape.[1865] The Digital Transformation Agenda for Ecuador 2022–2025 followed and defined the responsible use of emerging technologies such as AI among national priorities.[1866] The Agenda created the Artificial Intelligence Committee to coordinate, promote, and evaluate strategies for the development and ethical adoption of AI in Ecuador.[1867] The Committee supports the implementation of the UNESCO Recommendation on the Ethics of Artificial Intelligence and advises on AI initiatives.[1868] These efforts culminated in the Public Policy for the Digital Transformation of Ecuador 2025–2030 to strengthen AI strategy development and ethical oversight.[1869]

The Ministry of Telecommunications and the Information Society (*Ministerio de Telecomunicaciones y de la Sociedad de la Información*, MINTEL) continued these efforts in August 2025 with the launch of the Proposed Strategy for the Promotion of the Development and Ethical and Responsible Use of Artificial

---

[1865] Ministerio de Telecomunicaciones y de la Sociedad de la Información, *Assessment of Artificial Intelligence in Ecuador [Diagnóstico sobre la Inteligencia Artificial en el Ecuador]* (Dec. 2021), https://observatorioecuadordigital.mintel.gob.ec/wp-content/uploads/2022/11/Proyecto-diagnostico-inteligencia-artificial-IA-en-Ecuador-Documento-final-JC-JO-MS-002.pdf

[1866] Ministerio de Telecomunicaciones y de la Sociedad de la Información, *Digital Transformation Agenda of Ecuador 2022–2025 [Agenda de Transformación Digital del Ecuador]* (Aug. 2022), p. 14, https://www.arcotel.gob.ec/wp-content/uploads/2022/08/Agenda-transformacion-digital-2022-2025.pdf

[1867] Ministerio de Telecomunicaciones y de la Sociedad de la Información, *Agreement No. [Acuerdo No.] MINTEL-MINTEL-2023-0019* (Nov. 16, 2023), https://www.telecomunicaciones.gob.ec/wp-content/uploads/2023/12/mintel-mintel-2023-0019-1-1.pdf

[1868] Ecuador Digital Observatory, *Artificial Intelligence Committee (Comité de Inteligencia Artificial)*, https://observatorioecuadordigital.mintel.gob.ec/acerca-comite-ia/

[1869] Ministerio de Telecomunicaciones y de la Sociedad de la Información, *Public Policy for the Digital Transformation of Ecuador 2025-2030 [Política Pública para la Transformación Digital del Ecuador 2025-2030]* (Mar. 2025), https://www.gobiernoelectronico.gob.ec/wp-content/uploads/2025/03/INSTRUMENTO-Politica-Publica-para-la-Transformacion-Digital-Ecuador-2025-2030-MINTEL-signed_f.pdf

Intelligence in Ecuador.[1870] The Proposed Strategy seeks to "promote the development, adoption, and ethical and responsible use of artificial intelligence" that "guarantees the transparent, equitable, and human-centric implementation" through "the strengthening of human and technological capacities; the promotion of research, development, and innovation." This proposed strategy responds to recommendations in the Artificial Intelligence Landscape Assessment in Ecuador conducted by the MINTEL and the UN Development Programme (UNDP)[1871] that the country develop a national AI strategy; strengthen digital infrastructure, data governance, and talent; promote public–private collaboration; and establish ethical, transparent, and inclusive oversight mechanisms.[1872] The Strategy has not yet been implemented.

Building on these efforts, the Superintendence of Economic Competition of Ecuador (*Superintendencia de Competencia Económica*, SCE) introduced the Code of Ethics for the Use of Artificial Intelligence[1873] and the Guide for the Use of AI Tools in 2025.[1874] The Code and the Guide establish a set of core principles fully aligned with the UNESCO Recommendation on the Ethics of Artificial Intelligence. This initiative marked a significant milestone in the region, as UNESCO commended Ecuador's government as the first public institution to implement an ethical AI framework.[1875]

AI Policies

Lawmakers introduced the Organic Law for the Regulation and Promotion of Artificial Intelligence in the National Assembly (*Asamblea Nacional*) in June

---

[1870] Ministerio de Telecomunicaciones y de la Sociedad de la Información, *Proposal for a Strategy for the Promotion of the Promotion of the Development and Ethical and Responsible Use of Artificial Intelligence in Ecuador [Propuesta para una Estrategia para el fomento del desarrollo y uso ético y responsible de la inteligencia artificial en el Ecuador]* (Aug. 1, 2025), https://aportecivico.gobiernoelectronico.gob.ec/system/documents/attachments/000/000/161/original/4419b3369dba6951694e022997bc4db2272e47a5.pdf

[1871] UNDP Ecuador, *Artificial Intelligence (AI) Landscape Assessment [AILA: Evaluación del Panorama de la Inteligencia Artificial en Ecuador]* (May 2025), https://www.undp.org/es/ecuador/publicaciones/evaluacion-del-panorama-de-inteligencia-artificial-ia

[1872] Ibid, pp. 10-11

[1873] Superintendencia de Competencia Económica, *Ethical Guide [Code] for the Use of Artificial Intelligence (AI) [Decálogo Ético para el Uso de la Inteligencia Artificial (IA)]* (Jul. 16, 2025), https://www.sce.gob.ec/sitio/wp-content/uploads/2025/07/DECALOGO-IA-signed.pdf

[1874] Superintendencia de Competencia Económica, *Guide for the Use of Artificial Intelligence (AI) Tools in the Superintendency of Economic Competition (SCE) [Guía de uso de herramientas de inteligencia artificial 'IA' en la Superintendencia de Competencia Económica—SCE]* (Mar. 2025), https://www.sce.gob.ec/sitio/wp-content/uploads/2025/03/Guía-de-uso-de-herramientas-de-inteligencia-artificial-"IA".pdf

[1875] UNESCO, *Ecuador Adopts First Artificial Intelligence (AI) Code of Ethics in a Public Institution* (Oct. 2025), https://www.unesco.org/en/articles/ecuador-adopts-first-artificial-intelligence-ai-code-ethics-public-institution

2024.[1876] The proposed law is the country's first comprehensive framework for AI regulation that aims to uphold fundamental rights alongside responsible innovation. The draft bill adopts a risk-based approach, including explicit prohibitions on AI systems that violate human dignity, fundamental rights, the rule of law, and democratic values.[1877] The draft bill proposes creating a National AI Regulatory Authority to enforce its provisions and oversee a National Registry of High-Risk AI Systems.[1878] Sixteen guiding principles, including human-centric design, privacy protection, transparency and explainability, and environmental sustainability, shape the bill's overarching policies.

A separate Bill for the Promotion and Development of Artificial Intelligence introduced in July 2024 shared many of the same objectives while placing greater emphasis on AI education, digital literacy in schools, and incentives for AI research and investment.[1879] The Legislative Administration Council authorized the Standing Specialized Committee on Education, Culture, Science, Technology, Innovation, and Ancestral Knowledge to merge them into a single Organic Bill on Artificial Intelligence.[1880]

Additionally, the Organic Bill on Digital Utilization and Artificial Intelligence for Children and Adolescents introduced in September 2024 seeks to safeguard children and adolescents in their interaction with AI.[1881] Beyond regulating the use of AI by and for minors, the bill aims to protect minors in their use of the internet, social media, and digital platforms. Mirroring the earlier proposals, the draft bill recommends establishing a National Authority for the Supervision of Artificial Intelligence (*Autoridad Nacional de Supervisión de Inteligencia Artificial*), creating a

---

[1876] Asamblea Nacional, *Organic Law for the Regulation and Promotion of Artificial Intelligence [Proyecto de Ley Orgánica de Regulación y Promoción de la Inteligencia Artificial en Ecuador]* (Jun. 20, 2024), https://www.asambleanacional.gob.ec/sites/default/files/private/asambleanacional/filesasambleanacionalnameuid-19130/2192.%20Proyecto%20de%20Ley%20Org%C3%A1nica%20de%20Regulaci%C3%B3n%20y%20Promoci%C3%B3n%20de%20la%20Inteligencia%20Artificial%20en%20Ecuador%20-pnu%C3%B1ez/pp%20-%20proyecto%20de%20ley%20450889-nu%C3%B1ez.pdf

[1877] Ibid, Title II, Chapter I, Article 9

[1878] Ibid, Title II, Chapter II–III

[1879] Asamblea Nacional, *Bill for the Promotion and Development of Artificial Intelligence [Proyecto de Ley para el Fomento y Desarrollo de la Inteligencia Artificial]* (Jul. 2024), https://observatorioecuadordigital.mintel.gob.ec/wp-content/uploads/2024/08/PP_-_Proyecto-de-ley-453516-subia_c.pdf; CeCo Ecuador, *Ecuador: Three Different AI Regulation Initiatives [Ecuador: Tres iniciativas diferentes de regulación de la IA]* (Jan. 22, 2025), https://centrocompetencia.com/ecuador-tres-iniciativas-diferentes-de-regulacion-de-la-ia/

[1880] Asamblea Nacional, *Resolution CAL-RVVR-2023-2025-0216* (Feb. 13, 2025), https://www.asambleanacional.gob.ec/es/system/files/cal-rvvr-2023-2025-0216.pdf

[1881] Asamblea Nacional, *Organic Law on Digital Use and Artificial Intelligence for Children and Adolescents [Proyecto de Ley Orgánica de Aprovechamiento Digital e Inteligencia Artificial para Niñas, Niños y Adolescentes]* (Sept. 2024), https://observatorioecuadordigital.mintel.gob.ec/wp-content/uploads/2024/09/PP-PRO_1-aprovechamiento-IA-Pierina_Correa_c.pdf

registry of high-risk AI systems, and banning AI systems that manipulate users, process sensitive data, infer emotions without justification, or perform real-time facial recognition outside of public safety necessity.

Public Participation

Ecuador has facilitated several opportunities for public participation with various stakeholders in the development of its AI governance initiatives.

The Artificial Intelligence Committee, for example, promotes official and meaningful engagement across both the public and private sectors.[1882] Committee membership included the Minister of Telecommunications and the Information Society as chair along with representatives from the Telecommunications Regulation and Control Agency, public institutions, private entities, various government levels, and civil society.

In developing the Strategy for the Promotion of the Ethical and Responsible Development and Use of Artificial Intelligence, the Ministry of Telecommunications and the Information Society facilitated a workshop with representatives from public institutions, industry, academia, and civil society to discuss key strategic priorities.[1883] The Ministry completed a public consultation for the proposed Strategy, inviting all members of the public to provide comments.[1884]

Following the recommendation to consolidate the proposed AI bills into a single legislative proposal, the Permanent Specialized Commission on Education, Culture, Science, Technology, Innovation, and Ancestral Knowledge held a National Assembly meeting to discuss the key ethical and technical principles for regulating AI.[1885] Various technical and ethical experts from UNESCO, the European Union

---

[1882] Ministerio de Telecomunicaciones y de la Sociedad de la Información, *Qualification and Selection of Applicants to Form the Artificial Intelligence Committee [Calificación y Selección de postulantes para conformar el Comité de Inteligencia Artificial]* (Jul. 2024), https://observatorioecuadordigital.mintel.gob.ec/wp-content/uploads/2024/07/Acta_de_rectificacio%CC%81n-Comite%CC%81-IA-29-julio-2024-S.pdf

[1883] Ministerio de Telecomunicaciones y de la Sociedad de la Información, *Ecuador Advances in the Participatory Development of Guidelines for the Ethical and Responsible Use and Development of Artificial Intelligence [Ecuador avanza en la construcción participativa de lineamientos para el uso y desarrollo ético y responsable de la inteligencia artificial*] (Aug. 6, 2025), https://www.telecomunicaciones.gob.ec/ecuador-avanza-en-la-construccion-participativa-de-lineamientos-para-el-uso-y-desarrollo-etico-y-responsable-de-la-inteligencia-artificial/

[1884] Diálogo 2.0, *Public Consultation on the Proposed Strategy for the Promotion of the Development and Ethical and Responsible Use of Artificial Intelligence in Ecuador [Consulta pública de la Propuesta de Estrategia para el fomento del desarrollo y uso ético y responsable de la inteligencia artificial en el Ecuador]* (Aug. 2025), https://aportecivico.gobiernoelectronico.gob.ec/legislation/processes/78/draft_versions/99

[1885] Asamblea Nacional, *Resolution CAL-RVVR-2023-2025-0216* (Feb. 13, 2025), https://www.asambleanacional.gob.ec/es/system/files/cal-rvvr-2023-2025-0216.pdf; Marco Rivera, *Education Committee Debates Artificial Intelligence and Ethics Law in Ecuador [Comisión de Educación debate la ley de Inteligencia Artificial y ética en Ecuador]*, Expreso (Sept. 4, 2025),

Intellectual Property Office, and the Latin American Faculty of Social Sciences were invited to share their expertise.

Data Protection

The Constitution of Ecuador recognizes the right to personal data protection under Article 66(19), stating that individuals have the right to the protection of personal data and access to information concerning themselves.[1886] The protection of personal data and data subject rights was further expanded and institutionalized with the Organic Law on the Protection of Personal Data enacted in 2021.[1887] The law sets out data security obligations, introduces data subject rights similar to the EU GDPR, defines lawful processing conditions, and establishes clear rules for obtaining valid consent from data subjects. The personal data protection law also created the Superintendent of Personal Data Protection (*Superintendente de Protección de Datos Personales,* SPDP) as the data protection authority.[1888] The Superintendent and the office of the Superintendency have administrative, technical, and financial autonomy and the authority to investigate and sanction data protection violations. The Presidency issued the Regulation to the Personal Data Protection Organic Law in 2023 to clarify procedures for exercising data rights, reporting security breaches, designating Data Protection Officers, and regulating cross-border data transfers.[1889]

The Superintendency of Personal Data Protection is not a member of the Global Privacy Assembly (GPA) and thus ineligible to sponsor GPA resolutions.[1890] As a member of the Iberoamerican Network for Data Protection (*Red Iberoamericana de Protección de Datos,* RedIPD),[1891] however, Ecuador is party to the Iberoamerican

---

https://www.expreso.ec/actualidad/comision-de-educacion-debate-la-ley-de-inteligencia-artificial-y-etica-en-ecuador-255858.html

[1886] Lexis, *Constitution of the Republic of Ecuador [Constitución de la República del Ecuador]*, Art.66(19) (2008),)https://www.oas.org/juridico/pdfs/mesicic4_ecu_const.pdf

[1887] Asamblea Nacional, *Organic Law on the Protection of Personal Data [Ley Orgánica de Protección de Datos Personales]* (May 26, 2021), https://www.asambleanacional.gob.ec/sites/default/files/private/asambleanacional/filesasambleanacionalnameuid-29/Leyes%202013-2017/920-lmoreno/ro-459-5to-sup-26-05-2021.pdf

[1888] Asamblea Nacional, *Organic Law on the Protection of Personal Data [Ley Orgánica de Protección de Datos Personales]*, Art. 77, pp. 61–62 (May 26, 2021), https://www.asambleanacional.gob.ec/sites/default/files/private/asambleanacional/filesasambleanacionalnameuid-29/Leyes%202013-2017/920-lmoreno/ro-459-5to-sup-26-05-2021.pdf

[1889] Presidencia de Ecuador, *Regulation of the Organic Law on Protection of Personal Data [Reglamento de la Ley Orgánica de Protección de Datos Personales]*, Norma 904, Registro Oficial Suplemento No. 435 (Nov. 13, 2023), https://www.gob.ec/sites/default/files/regulations/2025-01/02%20Reglamento%20General%20a%20la%20Ley%20Orgánica%20de%20Protección%20de%20Datos%20Personales_0.pdf

[1890] Global Privacy Assembly (GPA), *List of Accredited Members* (2025), https://globalprivacyassembly.com/participation-in-the-assembly/list-of-accredited-members/

[1891] Red Iberoamericana de Protección de Datos (RedIDP), *Composition [Composición]*, https://www.redipd.org/la-red/composicion

Standards, which call for transparency in the use of personal data, including in decision-making.[1892] The Iberoamerican Network also provides guidance for ensuring the responsible use of data in developing and deploying AI systems and protecting rights from the impact of automated decisions.[1893]

Algorithmic Transparency

The Artificial Intelligence Landscape Assessment conducted by the Ministry of Telecommunications and UNDP reported that algorithmic transparency is "virtually nonexistent," as there are no public records of algorithms in use, nor legal support ensuring the right to explanations for automated decisions.[1894] The Personal Data Protection Law does not require entities to explain how their algorithms function.[1895]

AI and Crime Mitigation

The Inter-American Development Bank (IDB) supports governments in Chile, Ecuador, and Peru in deploying advanced AI programs to investigate complex organized-crime networks.[1896] The intergovernmental institution, which offers both financial and technical support to address development challenges—cites the cost of crime in the region (3.4 percent of GDP annually) and its impact on public safety as key drivers for adopting AI to address this challenge. Recognizing concerns about privacy and transparency related to the use of AI, IDB leverages its fAIrLAC+ platform—a partnership between the public and private sectors, civil society and academic institutions—to ensure that individual rights are upheld.[1897]

Additionally, EL PACCTO 2.0 is a cooperation between the European Union and Latin American and Caribbean countries (EU-LAC) promoting the use of AI to fight against transnational organized crime with a focus on crimes involving AI.[1898] The joint initiative advances national, regional, and international standards while

---

[1892] Red Iberoamericana de Protección de Datos (RedIDP), *Standards for Personal Data Protection for Ibero-American States* (Jun. 20, 2017), https://www.redipd.org/documentos/estandares-iberoamericanos-2017

[1893] For example, see Red Iberoamericana de Protección de Datos (RedIDP), *Specific Guidelines for Compliance with the Principles and Rights that Govern the Protection of Personal Data in Artificial Intelligence Projects* (Jun. 21, 2019), https://www.redipd.org/documentos/guia-orientaciones-especificas-proteccion-datos-ia

[1894] UNDP Ecuador, *Artificial Intelligence (AI) Landscape Assessment [AILA: Evaluación del Panorama de la Inteligencia Artificial en Ecuador]*, p. 9 (May 2025), https://www.undp.org/es/ecuador/publicaciones/evaluacion-del-panorama-de-inteligencia-artificial-ia

[1895] Ibid, p. 58

[1896] Ilan Goldfajn, *Cracking Crime with AI* (Jan. 21, 2025), https://www.iadb.org/en/blog/cracking-crime-ai

[1897] Ibid; *fAIr LAC*, https://fairlac.iadb.org/en

[1898] EU-LAC, *Artificial Intelligence and Organised Crime* (Nov. 2024), https://www.fiap.gob.es/wp-content/uploads/2024/11/ELPACCTO2-IAyCrimen-EN.pdf

ensuring coherence with legal frameworks, including the Council of Europe Framework Convention on AI and the European Union AI Act.[1899]

Privacy and Surveillance

The National Assembly expanded state surveillance and data interception powers with the approval of the Organic Intelligence Law in 2025.[1900] The law raised concerns and criticisms in the National Assembly and by civil society because of the potential impact on fundamental rights. Civil society groups and digital rights advocates warned that the law may undermine privacy guarantees and conflict with existing data protection principles, especially in the context of AI-driven monitoring systems.[1901]

AI in Education

To address learning gaps caused by the COVID-19 pandemic among Ecuadorian higher education students, the Ministry of Higher Education, Science, Technology and Innovation, with support from World Bank, is providing AI-assisted academic support in math.[1902] The platform reached over 14,000 learners across 400 course modules and delivered adaptive, personalized instruction at scale. Results were significant: student mastery increased from 25 percent to nearly 69 percent within sixteen weeks, which researchers estimated as equivalent to a full year of schooling. Despite these achievements, limited internet connectivity, unequal access to devices, and resistance from some educators and learners remain key barriers to large-scale adoption.

[1899] Fundación para la Internacionalización de las Administraciones Públicas, *EU Promotes Cooperation Model with Latin America and the Caribbean to Combat the Use of AI in Organised Crime* (Feb. 19, 2025), https://www.fiap.gob.es/en/noticias/eu-promotes-cooperation-model-with-latin-america-and-the-caribbean-to-combat-the-use-of-ai-in-organised-crime/

[1900] Registro Oficial, *Organic Intelligence Law [Ley Orgánica de Inteligencia]* (Jun. 10, 2025), https://strapi.lexis.com.ec/uploads/Registro_Oficial_Ano_1_Cuarto_Suplemento_No_57_10_de_junio_de_2025_6565dee8a1.pdf

[1901] María Villacreses Herrera, *Debate on the Draft Organic Law on Intelligence [Debate del Proyecto Ley Orgánica de Inteligencia]* (Jun. 10, 2025), https://www.asambleanacional.gob.ec/es/blogs/maria-villacreses-herrera/107720-sesion-007-asamblea-nacional; Carolina Mella, *Noboa Opens the Door to Mass Surveillance in Ecuador with a New Intelligence Law [Noboa abre la puerta al espionaje masivo en Ecuador con una nueva ley de inteligencia]*, El País (Oct. 2025), https://elpais.com/america/2025-06-23/noboa-abre-la-puerta-al-espionaje-masivo-en-ecuador-con-una-nueva-ley-de-inteligencia.html

[1902] World Bank Group, *In Ecuador, Artificial Intelligence Makes Learning Math Easier* (Oct. 2025), https://www.worldbank.org/en/news/feature/2022/02/10/en-ecuador-aprender-matematicas-es-mas-facil-con-inteligencia-artificial-nivelacion-remediacion-academica

Environmental Impact of AI

Ecuador is the world's first country to grant constitutional rights to nature.[1903] The Constitution of 2008 states that Nature or *Pachamama* "has the right to exist, persist, maintain and regenerate its vital cycles, structure, functions and its processes in evolution."[1904] Ecuador reinforced these rights in 2017 by enacting the Organic Environmental Code, which established a unified regulatory framework for environmental governance.[1905] The Code is based on environmental law principles and establishes rules for impact assessments, permitting, sanctions, and monitoring to safeguard biodiversity as a public interest.

The proposed Organic Law for the Regulation and Promotion of Artificial Intelligence includes provisions for ensuring that AI development upholds environmental sustainability.[1906] The proposal calls for clear environmental standards to reduce carbon footprints, minimize negative ecological impacts, and promote the development and adoption of clean, sustainable technologies.

Lethal Autonomous Weapons

Ecuador has consistently positioned itself as a strong advocate for the regulation of lethal autonomous weapons systems (LAWS), emphasizing the need to maintain human oversight in decisions about the use of force.[1907] As a state party to the Convention on Certain Conventional Weapons (CCW), Ecuador focuses on humanitarian concerns to prevent misuse and protect human rights.[1908] At the 79th UN General Assembly First Committee, Ecuador co-sponsored and voted in favor of

---

[1903] Karen Charman, *Ecuador First to Grant Nature Constitutional Rights*, Capitalism Nature Socialism, 19(4), pp. 131–133 (Nov. 2008), https://doi.org/10.1080/10455750802575828

[1904] Ibid; *Constitution of the Republic of Ecuador* (2008), https://www.oas.org/juridico/pdfs/mesicic4_ecu_const.pdf

[1905] Ministro del Ambiente, Agua y Transición Ecológica [Ministry of Environment, Water, and Ecological Transition], *Organic Environmental Code [Código Orgánico del Ambiente]* (Apr. 12, 2017), https://sustanciasyresiduos.ambiente.gob.ec/producto/codigo-organico-del-ambiente/; Eco Jurisprudence Monitor, *Ecuador Organic Code of the Environment* (Oct. 2025), https://ecojurisprudence.org/initiatives/ecuadors-organic-code-of-the-environment/

[1906] Asamblea Nacional, *Organic Law for the Regulation and Promotion of Artificial Intelligence [Proyecto de Ley Orgánica de Regulación y Promoción de la Inteligencia Artificial en Ecuador]*, Chapter II, Art. 4(1) (Oct. 2025), https://www.asambleanacional.gob.ec/sites/default/files/private/asambleanacional/filesasambleanacionalnameuid-19130/2192.%20Proyecto%20de%20Ley%20Org%C3%A1nica%20de%20Regulaci%C3%B3n%20y%20Promoci%C3%B3n%20de%20la%20Inteligencia%20Artificial%20en%20Ecuador%20-pnu%C3%B1ez/pp%20-%20proyecto%20de%20ley%20450889-nu%C3%B1ez.pdf

[1907] United Nations, *Statement by Ecuador, 2023 CCW GGE on LAWS* (May 15, 2023), https://conf.unog.ch/digitalrecordings/index.html?guid=public/61.0500/093DBBBC-C0F6-4E56-8D11-86ABA380DE10_15h03&position=1585&channel=ENGLISH; Automated Decision Research, Ecuador , https://automatedresearch.org/news/state_position/ecuador/

[1908] Ibid

Draft Resolution L.77, which underscores the risks autonomous weapons pose to global security and stresses the crucial role of humans in their deployment.[1909] The resolution also calls for urgent negotiations toward a legally binding framework and scheduled informal consultations in 2025 to ensure a comprehensive discussion of potential regulatory measures. Ecuador's support for international oversight extends to previous resolutions.[1910]

Ecuador's statements repeatedly emphasize that human responsibility, accountability, and compliance with international law must guide any adoption or deployment of these technologies. At the 2023 CCW Group of Governmental Experts (GGE) on LAWS, for example, Ecuador clarified that negotiations for a binding international instrument are a top priority.[1911] The country views regulation as essential to mitigate the risks these systems pose under international humanitarian law and human rights law. Earlier, during the Sixth Review Conference of the CCW in 2021, Ecuador highlighted the need to move from debate to concrete rules, stressing that binding regulations are necessary to address ethical concerns and the potential dehumanization of warfare.[1912]

Ecuador has called for robust legal frameworks, reinforced human oversight, and collaborative international action to ensure autonomous weapons are governed responsibly across multiple UN and CCW forums, reflecting the country's commitment to ethical, lawful, and stable security practices.[1913]

AI Literacy

Ecuador, in collaboration with the United Arab Emirates, launched a program in August 2025 offering scholarships for 10,000 young people aged 18 to 29 to train in AI.[1914] The fully online training covers AI fundamentals, prompt engineering, chatbots, AI for productivity, and creative applications. This initiative was developed together with the Secretariat of Higher Education, Science, Technology, and

---

[1909] United Nations General Assembly, *Lethal Autonomous Weapons Systems: Draft Resolution A/C.1/79/L.77*, First Committee, 79th Session (Oct. 18, 2024), https://documents.un.org/doc/undoc/ltd/n24/305/45/pdf/n2430545.pdf

[1910] United Nations General Assembly, *Resolution 78/241: Lethal Autonomous Weapons Systems* (A/RES/78/241) (Dec. 22, 2023), https://docs.un.org/en/A/RES/78/241

[1911] Ibid

[1912] United Nations, *Statement by Ecuador, Sixth Review Conference of the CCW* (Dec. 13, 2021), http://149.202.215.129:8080/s2t/UNOG/RCHCP6-13-12-2021-PM_mp3_en.html; link to recording and transcript of the full meeting

[1913] UN General Assembly, *Statement by Ecuador, First Committee* (Oct. 12, 2020), https://front.un-arm.org/wp-content/uploads/2020/10/ecuador-es.pdf

[1914] Ministerio de Telecomunicaciones y de la Sociedad de la Información, *Ecuador Launches 10,000 Artificial Intelligence Scholarships, Thanks to Cooperation with the United Arab Emirates [Ecuador lanza 10 mil becas de Inteligencia Artificial, gracias a la cooperación con Emiratos Árabes Unidos]* (Aug. 2025), https://www.telecomunicaciones.gob.ec/ecuador-lanza-10-mil-becas-de-inteligencia-artificial-gracias-a-la-cooperacion-con-emiratos-arabes-unidos/

Innovation; the Ministry of Telecommunications and the Information Society; and the Ministry of Education.

Human Rights

Ecuador is classified as Partly Free by Freedom House, with an overall score of 65/100.[1915] While the country maintains an electoral system with competitive presidential and legislative elections, political violence and corruption are increasingly affecting the functioning of its institutions and the safety of citizens. In 2023–2024, Daniel Noboa won a runoff election amid high levels of political violence, including the assassination of several candidates. Electoral management bodies generally function independently, though accusations of politicization and irregularities persist.

OECD / G20 AI Principles

Ecuador's efforts in developing AI governance have been broadly consistent with internationally promoted, human-centered AI principles, even though Ecuador is not a member of OECD, the G20, or the Global Partnership on AI (GPAI).[1916] Ecuador has not committed to the OECD AI Principles.[1917] The government's AI initiatives and the national Digital Observatory (*Observatorio Ecuador Digital)* reference responsible public-sector AI, signaling explicit alignment with trustworthiness, human rights, transparency, and accountability in national guidance and capacity-building efforts.[1918]

Council of Europe AI Treaty

Ecuador is neither a member of the Council of Europe nor a signatory to the Council of Europe Framework Convention on Artificial Intelligence.[1919] The Council of Europe's Committee of Ministers granted Ecuador observer status on the Committee on AI (CAI) in 2025. Although Ecuador will be eligible to assent to the

---

[1915] Freedom House, *Freedom in the World 2025: Ecuador* (2025), https://freedomhouse.org/country/ecuador/freedom-world/2025

[1916] OECD AI Policy Observatory, *About the Global Partnership on Artificial Intelligence (GPAI)* (2025), https://oecd.ai/en/about/about-gpai

[1917] OECD Legal Instruments, *Recommendation of the Council on Artificial Intelligence*, Adherents (May 3, 2024), https://legalinstruments.oecd.org/en/instruments/OECD-LEGAL-0449#adherents

[1918] Ministerio de Telecomunicaciones y de la Sociedad de la Información, *Observatorio Ecuador Digital*, https://observatorioecuadordigital.mintel.gob.ec/el-observatorio/

[1919] Council of Europe Treaty Office, *Chart of Signatures and Ratifications of Treaty 225* (Nov. 12, 2025), https://www.coe.int/en/web/Conventions/full-list/?module=signatures-by-treaty&treatynum=225

Framework Convention only after ratification, observer status allows the country to participate in relevant CAI discussions.[1920]

UNESCO Recommendation on AI Ethics

UNESCO has been active in facilitating ethical AI dialogue in Ecuador, with efforts explicitly framed around the Recommendation on the Ethics of Artificial Intelligence.[1921] Experts from UNESCO also contributed to the discussion around the effort to draft a single legislative AI proposal that consolidates AI bills.[1922] UNESCO and the Ministry of Telecommunications and the Information Society (MINTEL) held the first Extraordinary Session of MINTEL's AI Committee to advance Ecuador's progress on the Readiness Assessment Methodology (RAM),[1923] which is still in process.[1924] Experts discussed ethical principles for the responsible and transparent adoption of AI, ensuring alignment with national priorities and human rights.

Evaluation

Ecuador is in the process of drafting a comprehensive national regulatory framework to govern the development and deployment of AI. Early drafts and policy initiatives suggest alignment with the EU AI Act and GDPR and with leading global standards such as the OECD AI Principles and UNESCO Recommendations on the Ethics of AI. The relevant ministries have taken steps to engage diverse stakeholders in AI policy development, and national AI initiatives are publicly available through the country's online Digital Observatory. However, gaps remain in ensuring algorithmic transparency and contestability of automated decisions.

Organized crime continues to exert a significant impact on Ecuador, prompting its government to explore AI for surveillance and crime prevention. Yet, the documented harms of AI in law enforcement—such as wrongful incarceration and biased policing resulting from discriminatory and spurious outputs—underscores the

---

[1920] Council of Europe Artificial Intelligence, *Ecuador Granted Observer Status in the CAI*, News (Sept. 24, 2025), https://www.coe.int/en/web/artificial-intelligence/newsroom/-/asset_publisher/csARLoSVrbAH/content/ecuador-granted-observer-status-in-the-cai

[1921] UNESCO, *Implementation of RAM in Ecuador* (Oct. 9, 2025), https://www.unesco.org/es/articles/implementacion-de-la-ram-en-ecuador

[1922] Asamblea Nacional, *Resolution CAL-RVVR-2023-2025-0216* (Feb. 13, 2025), https://www.asambleanacional.gob.ec/es/system/files/cal-rvvr-2023-2025-0216.pdf; Marco Rivera, *Education Committee Debates Artificial Intelligence and Ethics Law in Ecuador (Comisión de Educación debate la ley de Inteligencia Artificial y ética en Ecuador)*, Expreso (Sept. 4, 2025), https://www.expreso.ec/actualidad/comision-de-educacion-debate-la-ley-de-inteligencia-artificial-y-etica-en-ecuador-255858.html

[1923] UNESCO, *Implementation of RAM in Ecuador* (Oct. 9, 2025), https://www.unesco.org/es/articles/implementacion-de-la-ram-en-ecuador

[1924] UNESCO Global AI Ethics and Governance Observatory, *Global Hub* (Oct. 2025), https://www.unesco.org/ethics-ai/en/global-hub

need for strong safeguards.[1925] These risks are further compounded by Ecuador's increasing political instability and governance challenges. Continuous monitoring is warranted to assess whether adequate safeguard and oversight mechanisms will be implemented to uphold fundamental rights and ethical AI principles.

[1925] Merve Hickok and Evanna Hu, *Don't Let Governments Buy AI Systems That Ignore Human Rights*, Issues in Science and Technology (Apr. 11, 2024), https://issues.org/government-procurement-ai-systems-human-rights-hickok-hu/

Egypt

In 2025, Egypt issued Executive Regulations to enact the 2020 Personal Data Protection Law and issued the Second Edition of the National AI Strategy. The country also completed the UNESCO Readiness Assessment Methodology (RAM) and published the report.

National AI Strategy

The Egyptian Cabinet approved the formation of a National Council for Artificial Intelligence (NCAI) tasked with "outlining, implementing and governing the AI strategy in close coordination with the concerned experts and entities"[1926] in November 2019. The Technical Committee of the National Council for Artificial Intelligence composed of representatives from all relevant government entities and independent experts in the field released the first National AI Strategy in 2021.[1927] On this occasion, the President of the Arab Republic of Egypt, Abdel Fattah Al-Sisi, stated, "We strongly believe that as emerging technologies create opportunities, they also pose challenges that we should be prepared for. Thus, we aim, through the National AI Strategy, to open the door to dialogue with stakeholders and promote international cooperation to exchange views on the best practices for developing and using AI to build the common good."[1928]

The Second Edition of the National AI Strategy for 2025–2030[1929] emphasizes that the main objective of the NCAI is to prepare, formalize, approve, and govern the implementation of Egypt's National AI Strategy. The six strategic objectives are to:

- Ensure ethical and responsible AI use by establishing a comprehensive AI regulatory system, activating the ethical framework, and put a nucleus for a clear regulatory body, actively contributing to global efforts and playing an active role in AI different international fora.
- Enhance quality of life and sectoral efficiency through AI applications.
- Ensure data accessibility and sharing by developing frameworks for national data governance and strengthening life cycle management of domestic data.
- Build a robust scalable AI infrastructure and cloud services, innovate business models, and create a good digital foundation for the development of the AI industry with the support of infrastructure development.

[1926] Ministry of Communications and Information Technology, *Artificial Intelligence, National Council for Artificial Intelligence* (2025), https://mcit.gov.eg/en/Artificial_Intelligence

[1927] Ministry of Communications and Information Technology, *Egypt National Artificial Intelligence Strategy*, paragraph 2 (Jul. 2021), https://mcit.gov.eg/en/Publication/Publication_Summary/9283

[1928] Ibid, paragraph 10

[1929] National Council for Artificial Intelligence, *Egypt National Artificial Intelligence Strategy, Second Edition (2025–2030)* (Jan. 2025), https://ai.gov.eg/SynchedFiles/en/Resources/AIstrategy English 16-1-2025-1.pdf

- Create a healthy AI ecosystem by supporting local startups, small and medium enterprises, and innovation efforts, and strengthening the investment of venture capital institutions in Egypt.
- Strengthen the quantity and quality of local AI talents and experts.[1930]

Egypt's newly announced AI Strategy encourages the responsible use of AI tools and the protection of data, yet recent events signal risks to enforcement. The Surveillance Law of 2025, which allows AI analysis of personal communications, appears incompatible with the principles of fairness, transparency, and privacy protection outlined in the National AI Strategy. Experts note that the law could undercut Egypt's pledge to the ethical and responsible use of AI.[1931]

The National Strategy created Key Performance Indicators (KPIs) to assess the status of ethical AI in Egypt. These include establishing a dedicated track within NCAI for AI Ethics, publishing Guidelines for Responsible and Ethical Development of AI, creating a set of rules and regulations for responsible AI use, and developing Ethics in AI/Technology courses as part of university computing degrees.[1932] The adoption of the Egyptian Charter for Responsible AI in February 2023, which offers "actionable insights and policies for decision makers in government, academia, industry, and civil society,"[1933] marks progress on these KPIs. Egypt's efforts were also witnessed in establishing AI Faculties at Egyptian universities and teaching AI ethics in computer science faculties.[1934]

The first National Strategy was implemented in a phased approach. The first phase (2020–2022) focused domestically on responding to market needs with AI talent and proving the value of AI through pilot projects within government. Internationally, the first phase centered active participation in international organizations on topics such as AI Ethics, AI for SDGs, and the impact of AI on labor markets and education."[1935] In the second phase (2023–2026), the emphasis has been on expanding AI into additional sectors. The government intends to establish a "paperless, collaborative, and smart" government.[1936] Egypt signed a Memorandum of Understanding (MoU) with Thales, a French technology solutions company to integrate AI solutions to governmental services and to build AI capacity.[1937] Priority

---

[1930] Ibid

[1931] Egyptian Front for Human Rights, *Privacy Under Attack: Egypt's Expanding AI Surveillance and Data Monitoring Policies* (Feb. 2025), https://egyptianfront.org/2025/02/privacy-under-attack-egypt-must-reform-its-draft-criminal-procedure-code

[1932] Ibid

[1933] National Council for Artificial Intelligence, *Egyptian Charter for Responsible AI*, v. 1.0 (2023), https://aicm.ai.gov.eg/en/Resources/EgyptianCharterForResponsibleAIEnglish-v1.0.pdf

[1934] Ministry of Communications and Information Technology, *Egypt National Artificial Intelligence Strategy* (Jul. 2021), https://mcit.gov.eg/en/Publication/Publication_Summary/9283

[1935] Ibid, point 2

[1936] Ibid

[1937] Ministry of Communications and Information Technology, *ICT Minister Witnesses Signing MoU between MCIT, Thales to Develop Apps, Build Capacity in AI* (Mar. 16, 2021), https://mcit.gov.eg/en/Media_Center/Press_Room/Press_Releases/63234

initiatives included raising public awareness about AI, supervising the domestic data lifecycle and fostering AI expertise. Egypt's progress in this phase aligns with the OECD AI Principles.[1938]

Egypt is actively working to bring the perspective of developing countries to international discussions, thereby helping to narrow the AI knowledge and development gap between developed and developing countries.[1939] In 2019, Egypt participated in the drafting of the UNESCO Recommendation on the Ethics of AI, serving as the Ad Hoc Expert Group's vice-chair.[1940]

Regional AI Governance Collaboration

Egypt is also positioning itself as a regional leader in the AI and digital policy world. In 2019, Egypt hosted the third session of the African Union (AU) Specialized Technical Committee on Communication and Information Technologies. This meeting was crowned by the adoption of the Sharm El Sheikh Declaration, which recognized and reaffirmed the necessity for a coherent African Digital Transformation Strategy "to guide a common, coordinated response to reap the benefits of the Fourth Industrial Revolution."[1941] Egypt hosted the first-ever DSC MENA 24 AI conference. This conference aimed to gain actionable insights into the international AI landscape through in-depth sessions on generative AI, natural language processing, computer vision, data engineering, cloud data, and AI applications in healthcare.[1942]

Egypt also helped create the AU African Working Group on AI, a group tasked with drafting a continent-wide AI strategy.[1943] This strategy aspires to create a common stance on AI issues, areas of priority, and the role of AI in vital sectors as well as ensure "the governance of AI and the protection and availability of data and developing AI regulations." The strategy also aims to position the African voice at the center of international fora and to "bridge the digital divide between developed

---

[1938] OECD Library, *OECD Artificial Intelligence Review of Egypt* (May 14, 2024), https://www.oecd-ilibrary.org/docserver/2a282726-en.pdf

[1939] Ministry of Communications and Information Technology, *Egypt National Artificial Intelligence Strategy*, point 9 (Jul. 2021), https://mcit.gov.eg/en/Publication/Publication_Summary/9283

[1940] STIP Compass, *Participation in UNESCO Initiatives for Ethical Standards* (Aug. 11, 2025), https://stip.oecd.org/stip/interactive-dashboards/policy-initiatives/2025 data policyInitiatives 26897

[1941] African Union Specialized Technical Committee on Communication and Information Technologies (STC-CICT), *Sharm el Sheikh Declaration* (Oct. 26, 2019), https://au.int/sites/default/files/decisions/37590-2019_sharm_el_sheikh_declaration_-_stc-cict-3_oct_2019_ver2410-10pm-1rev-2.pdf

[1942] Ahramonline, *Egypt to Host First-Ever DSC MENA 24 AI Conference in April* (Apr. 3, 2024), https://english.ahram.org.eg/NewsContent/3/1239/520431/Business/Tech/Egypt-to-host-firstever-DSC-MENA--AI-conference-in.aspx

[1943] François Candelon, Hind El Bedraoui, Hamid Maher, *Developing an Artificial Intelligence Strategy for Africa* (Feb. 9, 2021), https://oecd-development-matters.org/2021/02/09/developing-an-artificial-intelligence-for-africa-strategy/

[States] and African countries."[1944] Such active engagement culminated in Egypt being elected Chair of the African AI Working Group.[1945] In February 2021, Egypt hosted the first meeting of the African AI Working Group, which led to the issuance of the Common Africa Position Paper on the Priority Areas of Africa towards AI, followed by the second meeting, which was also hosted by Egypt in December 2022.[1946] Overall, the meetings of African Working Group on AI aimed to craft an African AI strategy in line with African Agenda 2063.[1947]

Egypt also chairs the Arab League's AI Working Group[1948] and has remained a regional and global player in AI. In 2025, the Minister of Communication and Information Technology, Amr Tallat delivered a keynote address during the opening of the Arab Dialogue Circle on Artificial Intelligence in the Arab World: Innovative Technologies and Ethical Challenges (ADCAI2025).[1949] The Minister also participated in the Ministerial Roundtable of the Global Partnership on Artificial Intelligence (GPAI) members and interested countries held at the headquarters of the French Ministry for Europe and Foreign Affairs, on the sidelines of the AI Action Summit in Paris France, where he highlighted Egypt's active engagement in international forums addressing AI-related issues.[1950]

Public Participation

The Ministry for Communications and Information Technology launched a website under the National Council for Artificial Intelligence[1951] that allows the public to easily access the National Strategy, I news, details about AI events, projects, and capacity-building programs, and information about AI partnerships with governments, international organizations, private sector companies, and academia. The AI Platform also includes a page where researchers can submit academic articles.

[1944] Ministry of Communications and Information Technology, *Egypt Hosts Second Meeting of African AI Working Group* (Dec. 14, 2022), https://mcit.gov.eg/en/Media_Center/Latest_News/News/66696

[1945] Ministry of Communications and Information Technology, *Egypt Chairs AU Working Group on AI* (Feb. 25, 2021), https://mcit.gov.eg/en/Media_Center/Latest_News/News/58203

[1946] Ministry of Communication and Information Technology, *Egypt Hosts Second Meeting of African AI Working Group* (Dec. 14, 2022), https://mcit.gov.eg/en/Media_Center/Latest_News/News/66696

[1947] Ibid

[1948] Ministry of Communications and Information Technology, *Egypt Elected Chair of Arab AI Working Group* (Feb. 16, 2021), https://mped.gov.eg/adminpanel/sharedFiles/OECD_Artificial_Intelligence_Review_of_Egypt_c9a.pdf

[1949] Ministry of Communications and Information Technology, *ICT Minister Delivers Keynote Speech at ADCAI2025* (Feb. 2, 2025), https://mcit.gov.eg/en/Media_Center/Press_Room/Press_Releases/68202

[1950] Ministry of Communications and Information Technology, *ICT Minister Participates in GPAI Ministerial Roundtable* (Feb. 10, 2025), https://mcit.gov.eg/en/Media_Center/Press_Room/Press_Releases/68219

[1951] Ministry of Communications and Information Technology, *Egypt Artificial Intelligence Platform*, https://ai.gov.eg

Egypt has promoted several public events to build internal AI capacity, such as training for citizens on how to build AI apps, hackathons and discuss the future of AI in Egypt. However, spaces dedicated to feedback and active contribution are more reserved to strategic partnerships with private companies and experts.

In line with the second National Artificial Intelligence Strategy (2025–2030), Egypt announced that Cairo will host the AI Everything Middle East & Africa (MEA) 2026 Forum. The event organized by GITEX Global in collaboration with the Ministry of Communications and Information Technology (MCIT) and the Information Technology Industry Development Agency (ITIDA), will convene representatives from over 60 countries, including government officials, private-sector leaders, investors, and academic institutions. The theme spans digital governance, financial technology, healthcare innovation, cybersecurity, and sustainable cloud infrastructure, thereby connecting Egypt's domestic development agenda to regional and global AI ecosystems. In addition to a ministerial-level summit, the program includes start-up exhibitions, investor matchmaking sessions, and youth-oriented capacity-building tracks designed to promote skill development and entrepreneurship in the field of artificial intelligence.

Data Protection

The Egyptian Constitution protects citizens' rights to privacy under Article 57.[1952] Egypt passed Law No. 151 on the Personal Data Protection Law (PDPL) in July 2020.[1953]The Minister of Communications and Information Technology publicized a draft Executive Regulation of the PDPL in June 2022. [1954] The Executive Regulation[1955] implementing the PDPL and operationalizing the national data protection authority were released in November 2025 and will be implemented through 2026.

The PDPL was drafted following the example of the European General Data Protection Regulation (GDPR). Article 2 of the PDPL provides for data subject rights such as the right to erasure, the right to be informed, the right to access, and the right to rectification and to object.[1956] The PDPL also enshrines principles applicable to the collection, storage and processing of personal data. These principles are (1) Data

---

[1952] Constitute Project, *The Egyptian Constitution*, https://www.constituteproject.org/constitution/Egypt_2014.pdf

[1953] Personal Data Protection Center, *Regulations, PDPL No. 151 of 2020* (Nov. 10, 2025), https://pdpc.gov.eg/laws

[1954] Ministry of Communications and Innovation Technology, *Personal Data Protection Law Executive Regulations to Be Issued Late 2022: NTRA VP* (Jun. 21, 2022), https://mcit.gov.eg/Upcont/MediaCenter/MCIT%20in%20Press622202200Personal_Data_Protection_Law_Executive_Regulations_to_Be_Issued_Late_2022_NTRA.pdf

[1955] Personal Data Protection Center, *Regulations, Executive Regulations No. 816 of 2025* (Nov. 10, 2025), https://pdpc.gov.eg/laws

[1956] DLA Piper, *Data Protection Laws of the World: Collection and Processing in Egypt* (Jan. 19, 2024), https://www.dlapiperdataprotection.com/?t=collection-and-processing&c=EG#insight

minimization; (2) Accuracy and security; (3) Lawfulness, and (4) Storage limitation. The PDPL foresees financial sanctions in the event of violations regarding the protection of personal data.

The PDPL created the Personal Data Protection Center (PDPC) as the "competent regulatory authority responsible for overseeing, enforcing, and developing Egypt's data protection framework."[1957] The PDPC will set and apply decisions, regulations, and measures in relation to data protection and foresee an adequate mechanism for law enforcement. The Center is meant to be an independent authority that operates under the Ministry of Communications and Information Technology.[1958]

In January 2025, Egypt amended the Criminal Procedure Code to expand surveillance of personal communications.[1959] The law allows monitoring of phone calls, emails, and social media. Civil society groups raised concerns about the lack of transparency and accountability in AI-powered surveillance systems. Experts noted that the law may present conflicts with Egypt's commitments to ethical AI governance, including transparency (Metric Q9) and fairness (Metric Q8). However, Egypt adopted its first National Open Data Policy in June 2025 to drive digital transformation.[1960]

At a conference on the challenges to the right to privacy given the rapid development of artificial intelligence in July 2022, the Minister of Social Solidarity Nevine El-Qabbaj stated that "artificial intelligence has breached all limits, even our mental privacy."[1961]

The Personal Data Protection Center is not a member of the Global Privacy Assembly (GPA).[1962] Egypt has not endorsed the 2018 GPA Declaration on Ethics and Data Protection in Artificial Intelligence,[1963] the 2020 GPA Resolution on AI

[1957] Personal Data Protection Center, *FAQs*, https://pdpc.gov.eg/faq

[1958] DLA Piper Data Protection, *Data Protection Laws of the World: National Data Protection Authority in Egypt* (Jan. 19, 2024), https://www.dlapiperdataprotection.com/index.html?t=authority&c=EG

[1959] The New Arab, *Egypt Passes Communication Surveillance Law Amid Privacy Concerns* (Jan. 27, 2025), https://www.newarab.com/news/egypt-passes-communication-surveillance-law-amid-privacy-concern

[1960] National Council for Artificial Intelligence, *Open Data Policy, Arab Republic of Egypt* (Jun. 2025), https://ai.gov.eg/SynchedFiles/en/Resources/Open Data Policy.pdf

[1961] Nada Nader, *Egypt Kicks Off Int'l Human Rights Conference on Challenges Facing Right to Privacy* (Jul 21, 2022), https://english.ahram.org.eg/News/471889.aspx

[1962] Global Privacy Assembly, *List of Accredited Members* (2026), https://globalprivacyassembly.com/participation-in-the-assembly/list-of-accredited-members/

[1963] Global Privacy Assembly, *Declaration on Ethics and Data Protection in Artificial Intelligence* (Oct. 23, 2018), https://globalprivacyassembly.org/wp-content/uploads/2018/10/20180922_ICDPPC-40th_AI-Declaration_ADOPTED.pdf

Accountability,[1964] the 2022 GPA Resolution on Facial Recognition Technology,[1965] or the 2023 GPA Resolution on Generative AI.[1966]

Algorithmic Transparency

The PDPL does not refer to algorithmic transparency. However, the principles of "transparency and explainability," including the right of the user to know "when he or she is interacting with an AI system and not a human being" are enshrined in the Egyptian Charter for Responsible AI. [1967] The Charter also emphasizes that: "Developers of AI systems should always strive to provide transparent and explainable AI solutions. The degree of explainability required will vary according to the application domain and project requirements, but project sponsors must be clear on the potential tradeoff between the accuracy/quality and explainability of any given model. When in doubt, developers should opt for simpler models with higher degrees of explainability, without compromising the minimum desired quality and accuracy."

Biometric Recognition

Egypt is increasingly adopting biometric technologies for security and surveillance. A deal between the Arab Organization for Industrialization and Idemia, a leading biometric company, hired the latter to produce biometric devices including facial recognition systems in Egypt.[1968] The Egyptian government agreed with Idemia in early 2020 to build a digital ID system for Egypt Post "backed by fingerprint biometrics and citizen IDs."[1969] In 2021, Fingo, another organization specializing in

---

[1964] Global Privacy Assembly, *Resolution on Accountability in the Development and Use of Artificial Intelligence* (Oct. 2020), https://globalprivacyassembly.org/wp-content/uploads/2020/10/FINAL-GPA-Resolution-on-Accountability-in-the-Development-and-Use-of-AI-EN-1.pdf

[1965] Global Privacy Assembly, *Resolution on Principles and Expectations for the Appropriate Use of Personal Information in Facial Recognition Technology* (Oct. 2022), https://globalprivacyassembly.org/wp-content/uploads/2022/11/15.1.c.Resolution-on-Principles-and-Expectations-for-the-Appropriate-Use-of-Personal-Information-in-Facial-Recognition-Technolog.pdf

[1966] Global Privacy Assembly, *Resolution on Generative Artificial Intelligence Systems* (Oct. 2023), https://globalprivacyassembly.org/wp-content/uploads/2023/10/5.-Resolution-on-Generative-AI-Systems-101023.pdf

[1967] National Council for AI, *Egyptian Charter for Responsible AI*, v. 1.0 (2023), https://aicm.ai.gov.eg/en/Resources/EgyptianCharterForResponsibleAIEnglish-v1.0.pdf

[1968] Ayang MacDonald, *Idemia Renews Mauritania Contract, Signs Deal with AOI for Biometric Device Production in Egypt,* Biometric Update (Nov. 9, 2020), https://www.biometricupdate.com/202011/idemia-renews-mauritania-contract-signs-deal-with-aoi-for-biometric-device-production-in-egypt

[1969] Chris Burt, *Idemia to Build Biometrics-Backed Digital Identity Service in Egypt, Supply TSA Trials, Koins Kantara,* Biometric Update.COM (Mar. 12, 2020), https://www.biometricupdate.com/202003/idemia-to-build-biometrics-backed-digital-identity-service-in-egypt-supply-tsa-trials-joins-kantara

biometrics, announced a partnership with Egypt to develop a vein-based recognition system for the country's national ID program.[1970]

On the sidelines of the COP27 hosted in Sharm El-Sheikh, the Egyptian government signed an MoU with the US-based company, Honeywell International Inc, to "run a pilot model in Sharm El-Sheikh by transforming the new building of the South Sinai Governorate General Assembly and Sharm El-Shaikh Hospital into sustainable smart buildings relying on modern technologies, especially artificial intelligence and data analytics."[1971] This agreement came as the second deal after Egypt contracted Honeywell in 2019 to provide city-wide public safety, security and surveillance system for the new administrative capital, which, according to a press release by the company, will "integrate advanced Internet of Things (IoT) software and [...] also connect video feeds from more than 6,000 IP cameras over a futureproof wireless network, and run sophisticated video analytics to monitor crowds and traffic congestion, detect incidents of theft, observe suspicious people or objects, and trigger automated alarms in emergency situations."[1972]

These deals were concluded in a legal vacuum even though biometric systems could adversely impact several human rights, including the right to freedom of expression, the right to assembly, and the right to privacy.[1973]

A January 2025 surveillance law allows direct monitoring of an individual citizen's personal communications via mobile communications, voice, email, and social networking, as well as the use of biometric tracking technologies. The law has been criticized for its limited accountability and transparency, as well as the potential for automated decision-making without citizen oversight. Egypt's approach to biometric tracking under this law may conflict with its commitment to Algorithmic Transparency (Metric Q9) and Fairness & Accountability (Metric Q8) under the AIDV Index.[1974]

The law includes provisions that could expand the use of biometric recognition tools in public areas, increasing risks of AI misuse. Experts warn that, if unchecked, such policies could restrict civil rights and reduce transparency, making it harder to

---

[1970] Fingo, *Egypt to Unlock Futuristic ID Verification with Finger-Vein Recognition Tech* (Feb. 18, 2021), https://www.fingo.to/media/egypt-to-unlock-futuristic-id-verification-with-finger-vein-recognition-tech/

[1971] North Africa Post, *Egypt: US Honeywell Company to Transform Government Institutions into Smart Buildings* (Nov. 12, 2022), https://northafricapost.com/62581-egypt-us-honeywell-company-to-transform-government-institutions-into-smart-buildings.html

[1972] ZAWYA, *Honeywell to Deploy World-Class Public Safety and Security Infrastructure for Egypt's New Smart City*, https://www.zawya.com/en/press-release/honeywell-to-deploy-world-class-public-safety-and-security-infrastructure-for-egypts-new-smart-city-f8th3lyy

[1973] Institute of Development Studies, University of Sussex, *Surveillance Law in Africa: A Review of Six Countries, Version 2, Egypt Country Report* (Mar. 24, 2026), https://opendocs.ids.ac.uk/articles/report/Surveillance_Law_in_Africa_a_Review_of_Six_Countries/26435920?file=48184978

[1974] ARTICLE 19, *Egypt: Stop Attacks on Privacy and Reform Draft Criminal Procedure Code*, (Feb. 2025), https://www.article19.org/resources/egypt-stop-attacks-on-privacy-and-reform-draft-criminal-procedure-code/

manage AI's ethical implications and posing serious risks to civil liberties if not properly regulated.[1975]

EdTech

In May 2022, Human Rights Watch published a global investigative report on the education technology (EdTech) endorsed by 49 governments, including Egypt, for children's education during the pandemic. Based on technical and policy analysis of 163 EdTech products, Human Rights Watch found that governments' endorsements of the majority of these online learning platforms put at risk or directly violated children's rights.[1976]

This is the case, for example, of the EdTech product "Edmodo" used in Egypt, which, according to Human Rights Watch, can collect Android Advertising IDs that enable advertisers to track children over time and across different apps installed on their devices for advertising purposes. Edmodo also allows shadow profiling by accessing contacts' details and photos, if saved on the phone. Additionally, an Education and skills push elevated to a national priority under Presidential direction could consider AI use as compulsory in pre-university education.[1977]

Environmental Impact of AI

Egypt's 2021 National AI Strategy highlights the use of AI technology to mitigate the environmental cost of agriculture in the AI for Development (AI4D) pillar.[1978] However, the strategy does not address the environmental cost or resources required to develop and deploy AI systems. The Egyptian Charter for Responsible AI requires the environmental impact of Government AI projects to be weighed in "a thorough impact assessment to ensure maximum benefit from the technology, while respecting the guidelines of responsible and ethical AI development."[1979]

The Second Edition of the National Strategy promotes sustainable AI development by encouraging green computing practices and energy-efficient data

---

[1975] The New Arab, *Egypt Passes Communication Surveillance Law Amid Privacy Concern* (Jan. 27, 2025), https://www.newarab.com/news/egypt-passes-communication-surveillance-law-amid-privacy-concern

[1976] Human Rights Watch, *How Dare They Peep into My Private Life* (May 25, 2022), https://www.hrw.org/report/2022/05/25/how-dare-they-peep-my-private-life/childrens-rights-violations-governments

[1977] Sahar Albazar, Member of Parliament, *Governing AI with Inclusion: An Egyptian Model for the Global South* (Sept. 1, 2025), https://oecd.ai/en/wonk/governing-ai-with-inclusion-an-egyptian-model-for-the-global-south

[1978] Ministry of Communications and Information Technology, *Egypt National Artificial Intelligence Strategy*, pp. 29–30 (Jul. 2021), https://mcit.gov.eg/en/Publication/Publication_Summary/9283

[1979] National Council for AI, *Egyptian Charter for Responsible AI*, v. 1.0, p. 5 (2023), https://aicm.ai.gov.eg/en/Resources/EgyptianCharterForResponsibleAIEnglish-v1.0.pdf

centers. Egypt also aims to use AI to support environmental monitoring and conservation in sectors like energy, agriculture, and water management.[1980]

The African Union Continental Artificial Intelligence Strategy to which Egypt contributed identifies emissions and resource demands among the risks of using and developing AI systems.[1981] The Strategy identifies environmental protection as a priority area to apply AI technology but also calls for countries to center sustainability through innovations such as green data centers in developing technology.[1982]

Lethal Autonomous Weapons

While Egypt has signed the Convention on Conventional Weapons (CCW), it has not yet ratified the Convention.[1983] However, Egypt has been actively participating in CCW meetings on killer robots since 2014.[1984] Bassem Yehia Hassan Kassem Hassan, a representative of Egypt, speaking on behalf of the Arab Group, stated that the presence of weapons of mass destruction and their modernization are a grave threat to international security and development, and that international community must develop norms and rules to encourage responsible behavior and increase cooperation to reach concrete progress in dealing with threats posed by lethal autonomous weapons and the use of artificial intelligence in armaments.[1985]

On numerous occasions, Egypt has warned that these weapon systems may have "possible ramifications on the value of human lives [and] the calculation of the cost of war,"[1986] and thus, there must be "specific prohibitions on acquisition, research and development, testing, deployment, transfer, and use [of these systems]."[1987] Egypt has called for a moratorium on lethal autonomous weapons systems until a ban is

---

[1980] National Council for Artificial Intelligence, *Egypt National Artificial Intelligence Strategy, Second Edition (2025–2030)* (Jan. 2025), https://ai.gov.eg/SynchedFiles/en/Resources/AIstrategy English 16-1-2025-1.pdf

[1981] African Union, *Continental Artificial Intelligence Strategy*, p. 25 (Aug. 9, 2024), https://au.int/en/documents/20240809/continental-artificial-intelligence-strategy

[1982] Ibid, pp. 37, 45

[1983] UN Office of Disarmament Affairs, *High Contracting Parties and Signatories CCW* (Mar. 18, 2025), https://disarmament.unoda.org/en/our-work/conventional-arms/convention-certain-conventional-weapons/high-contracting-parties-and-signatories-ccw

[1984] Brian Stauffer, Human Rights Watch, *Stopping Killer Robots Country Positions on Banning Fully Autonomous Weapons and Retaining Human Control* (Aug. 10, 2020), https://www.hrw.org/report/2020/08/10/stopping-killer-robots/country-positions-banning-fully-autonomous-weapons-and#_ftn95

[1985] United Nations, *First Committee Weighs Potential Risks of New Technologies as Members Exchange Views on How to Control Lethal Autonomous Weapons, Cyberattacks* (Oct. 26, 2018), https://press.un.org/en/2018/gadis3611.doc.htm

[1986] Government of Egypt, *Statement to the UN Human Rights Council* (May 30, 2013), http://stopkillerrobots.org/wp-content/uploads/2013/05/HRC_Egypt_10_30May2013.pdf

[1987] Brian Stauffer, Human Rights Watch, *Stopping Killer Robots Country Positions on Banning Fully Autonomous Weapons and Retaining Human Control* (Aug. 10, 2020), https://www.hrw.org/report/2020/08/10/stopping-killer-robots/country-positions-banning-fully-autonomous-weapons-and#_ftn95

achieved, supporting "a legally binding instrument against the development and manufacture of such weapon systems [...] as well as the regulation of existing systems that fall within [the CCW] mandate."[1988]

At the 78th UN General Assembly First Committee in 2023, Egypt voted in favor[1989] of resolution L.56[1990] on autonomous weapons systems, along with 163 other states. The Resolution emphasized the "urgent need for the international community to address the challenges and concerns raised by autonomous weapons systems," and mandated the UN Secretary-General to prepare a report reflecting the views of member and observer states on autonomous weapons systems. Egypt also voted in favor of Resolution L.77 on lethal autonomous weapons systems in 2024.[1991] The resolution, supported by 161 states, raises concerns about the "negative consequences and impact of autonomous weapon systems on global security and regional and international stability" and stresses the "the importance of the role of humans in the use of force to ensure responsibility and accountability and for States to comply with international law."[1992]

AI Literacy

Egypt has adopted a structured approach to AI literacy under its broader Digital Egypt agenda and the Second National Artificial Intelligence Strategy (2025–2030). The strategy identifies "human capital development and AI literacy" as one of six pillars, aiming for an informed, capable, and ethically aware workforce to support national AI deployment.[1993] The Ministry of Communications and Information Technology (MCIT), through its Applied Innovation Centre (AIC) and the Information Technology Institute (ITI), has launched several national programs targeting students, civil servants, and professionals.

To address the growing importance of AI in the public sector, the program trains public officials in understanding algorithmic systems, data governance, and ethical risk assessment. Complementing this initiative, the AI Summer School Initiative provides early-career researchers with exposure to machine learning and

---

[1988] Automated Decision Research, *State Position on Autonomous Weapons Systems: Egypt*, https://automatedresearch.org/news/state_position/egypt/

[1989] Stop Killer Robots, *164 States Vote against the Machine at the UN General Assembly* (Nov. 1, 2023), https://www.stopkillerrobots.org/news/164-states-vote-against-the-machine/

[1990] General Assembly, *Resolution L56: Lethal Autonomous Weapons* (Oct. 12, 2023), https://reachingcriticalwill.org/images/documents/Disarmament-fora/1com/1com23/resolutions/L56.pdf

[1991] Automated Decision Research, *Egypt State Positions* (2024), https://automatedresearch.org/news/state_position/egypt/

[1992] Stop Killer Robots, *161 States Vote against the Machine at the UN General Assembly* (2024), https://www.stopkillerrobots.org/news/161-states-vote-against-the-machine-at-the-un-general-assembly/

[1993] National Council for Artificial Intelligence, *Egypt National Artificial Intelligence Strategy, Second Edition (2025–2030)* (Jan. 2025), https://ai.gov.eg/SynchedFiles/en/Resources/AIstrategy English 16-1-2025-1.pdf

data analytics.[1994] At the institutional level, Egypt's National Council for Artificial Intelligence (NCAI) coordinates AI literacy programs across ministries, ensuring alignment with the Responsible AI Charter (2023) and the UNESCO Recommendation on the Ethics of Artificial Intelligence (2021). This coordination ensures that literacy initiatives incorporate awareness of AI's ethical, societal, and regulatory dimensions.

To expand digital inclusion, the Digital Egypt Cubs Initiative trains students aged 12–17 in foundational AI concepts, programming, and ethics. The Digital Egypt Pioneers Initiative extends this framework by providing professional certification and employment pathways for youth in AI-related fields. Together, these initiatives integrate AI literacy across Egypt's digital-skills ecosystem.[1995]

Human Rights

Egypt has endorsed the Universal Declaration of Human Rights. However, the Freedom in the World report ranks Egypt as "Not Free" with a score of 18/100 for 2025.[1996] The report cites the authoritarian rule of President Abdel Fattah al-Sisi and the lack of meaningful political opposition given the prosecution and imprisonment of dissenters and general restriction of civil liberties such as freedoms of the press, assembly, and expression. Criminalization of dissent, harassment, and surveillance for internet users as well as restrictions to access led to the "Not Free" ranking and 28/100 score on the Freedom on the Net report.[1997]

In March 2021, 31 UN member states penned a joint declaration, supported by numerous NGOs, strongly condemning human rights abuses in Egypt. The declaration highlighted constraints on citizens' freedom of expression, as well as their ability to voice political opposition and to peacefully assemble.[1998]

Egypt is criticized by international human rights organizations and civil society for state surveillance on citizens' communications and censorship on content online.[1999] Enactment of the Anti-Cyber Crime Law aimed to ensure online safety, security and fraud; however, in practice the regulation can be used for surveillance

---

[1994] GovTech Global Alliance, *Applied Innovation Centre—AI for Government Programme* (2024) https://govtechglobal.org/appliedinnovationcenter

[1995] MCIT, *Digital Egypt Cubs and Pioneers Initiatives Overview* (2024), https://deci.mcit.gov.eg

[1996] Freedom House, *Freedom in the World 2025: Egypt* (2025), https://freedomhouse.org/country/egypt/freedom-world/2025

[1997] Freedom House, *Freedom on the Net 2025: Egypt* (2025), https://freedomhouse.org/country/egypt/freedom-net/2025

[1998] Human Rights Watch, *Condemnation of Egypt's Abuses at UN Rights Body: Overdue Action Is a Step Forward* (Mar. 12, 2021), https://www.hrw.org/news/2021/03/12/condemnation-egypts-abuses-un-rights-body

[1999] Wafa Ben Hassine, *Egyptian Parliament Approves Cybercrime Law Legalizing Blocking of Websites and Full Surveillance of Egyptians* (Jun. 20, 2018), https://www.accessnow.org/egyptian-parliament-approves-cybercrime-law-legalizing-blocking-of-websites-and-full-surveillance-of-egyptians/

due to vague definitions and language.[2000] Moreover, Egypt's collaboration with private software and technology solution companies such as Idemia and Thales has been criticized for the lack of transparency and accountability.[2001]

In late 2021, Egypt launched the National Human Rights Strategy.[2002] On this occasion, President Abdel-Fattah El-Sisi declared 2022 as the year of civil society.[2003] The Human Rights Strategy is based on three axes: a) Constitutional Guarantees for Enhancing Human Rights Respect and Protection, b) Egypt's International and Regional Human Rights Obligations, and 3) Sustainable Development Strategy: Egypt Vision 2030. While the Strategy tackled all human rights, particularly political rights, it did not address the positive/adverse impact of technology on these rights or the human rights implications of artificial intelligence. The Strategy referred only to the use of technology to enhance human rights under the "Right to Litigation and Strengthening Guarantees for a Fair Trial."[2004]

In February 2023, Egypt released the Egyptian Charter on Responsible AI. The Charter builds on the OECD AI Principles and UNESCO Recommendation on the Ethics of AI.[2005] The Charter recognizes the significant risks AI might pose, such as bias, data drift, lack of transparency, lack of legal responsibility, and lack of fairness and equality. Therefore, it introduces guidelines and best practices for assessing AI systems trustworthiness with the aim to protect human rights and ensure the responsible, transparent, and fair use of the technology. For example, on the human-centeredness principle, the Charter stresses that "[t]he primary goal of using AI in Government is the well-being of citizens, including combating poverty, hunger, inequality, illiteracy, and corruption; achieving prosperity and inclusion."[2006]

[2000] Killian Balz, Hussam Mujally, *Egypt: The New Egyptian Anti-Cybercrime Law Regulates Legal Responsibility for Web Pages and Their Content* (2018), https://www.mondaq.com/security/820028/the-new-egyptian-anti-cybercrime-law-regulates-legal-responsibility-for-web-pages-and-their-content; Jillian C. York, *Egypt's Draconian New Cybercrime Bill Will Only Increase Censorship* (Jul 12, 2018), https://www.eff.org/deeplinks/2018/07/draconian-new-cybercrime-bills-vietnam-and-egypt-will-only-increase-censorship

[2001] Marceau Sivieude, ed., *Egypt: A Repression Made in France*, International Federation for Human Rights (FIDH), Cairo Institute for Human Rights Studies (CIHRS), Human Rights League (LDH), Armaments Observatory (OBSARM) (Jun. 2, 2018), https://www.fidh.org/en/issues/litigation/egypt-a-repression-made-in-france

[2002] Supreme Standing Committee for Human Rights, *National Human Rights Strategy (2021–2026)*, https://sschr.gov.eg/media/gapb5bq4/national-human-rights-strategy.pdf

[2003] Egypt's State Information Service, *Sisi Declares 2022 as Year of Civil Society* (Jan.18, 2022), https://sis.gov.eg/en/media-center/news/sisi-declares-2022-as-year-of-civil-society/

[2004] *National Human Rights Strategy*, p. 25

[2005] National Council for AI, *Egyptian Charter for Responsible AI*, v. 1.0 (2023), https://aicm.ai.gov.eg/en/Resources/EgyptianCharterForResponsibleAIEnglish-v1.0.pdf

[2006] Ibid

OECD / G20 AI Principles

Although Egypt is not an OECD member, the country contributed to the drafting of the OECD Recommendations on Responsible Use of AI[2007] and was the first Arab and African country to endorse and translate the OECD AI Principles into Arabic.[2008] Egypt also has collaborated with the Global Partnership on AI (GPAI) as an observer and took part in the Ministerial Roundtable of the GPAI[2009] at the AI Action Summit in Paris in 2025, where the ICT Minister presented the main principles of the Second Edition of the National AI Strategy 2025–2030.

As reported by the OECD in the 2021 State of Implementation of the OECD Principles report, Egypt has set up a governing body (the National Council for AI) to oversee the implementation of its AI strategy.[2010] This is a concrete first step toward fulfilling the OECD recommendation of ensuring "a policy environment that will open the way to deployment of trustworthy AI systems."

Egypt has also taken steps toward fulfilling three of the four other OECD recommendations. The creation of both the AI Platform and the new Egyptian Center of Excellence,[2011] a government group that will work with private or academic partners to deliver AI projects on behalf of beneficiaries, help to "foster accessible AI ecosystems with digital infrastructure and technologies and mechanisms to share data and knowledge." Empowering "people with the skills for AI and support workers for a fair transition" will be accomplished through enrollments in the newly created "Faculties of AI" at eight public and private Egyptian universities.[2012] Egypt's cooperation "across borders and sectors to progress on responsible stewardship of trustworthy AI" is evidenced by its participation in and leadership of international AI committees.

The Charter on Responsible AI purports to demonstrate how the country will interpret and implement the OECD AI Principles of human-centeredness, transparency and explainability, fairness, accountability, security and safety.[2013]

[2007] Ministry of Communications and Information Technology, *MCIT Participates Actively in Drafting OECD Recommendations on Responsible Use of AI* (May. 15, 2021), https://mcit.gov.eg/en/Media_Center/Latest_News/News/63363

[2008] OECD Legal Instruments, *Recommendation of the Council on Artificial Intelligence, Adherents* (May. 3, 2024), https://legalinstruments.oecd.org/en/instruments/OECD-LEGAL-0449#adherents

[2009] Ministry of Communications and Information Technology, *ICT Minister Participates in GPAI Ministerial Roundtable* (Feb. 10, 2025), https://mcit.gov.eg/en/Media_Center/Latest_News/News/68219

[2010] OECD, *State of implementation of the OECD AI Principles: Insights from National AI Policies*, p. 10, OECD Digital Economy Papers, No. 311 (Jun. 18, 2021), https://doi.org/10.1787/1cd40c44-en

[2011] National Council for AI, *Egyptian AI Center of Excellence (AIEG)*, https://ai.gov.eg/strategy/center-of-excellence

[2012] Sally Radwan and Samar Sobeih, *Egypt's AI Strategy Is More about Development than AI*, OECD.ai Policy Observatory (May 26, 2021), https://oecd.ai/en/wonk/egypt-ai-strategy

[2013] National Council for AI, *Egyptian Charter for Responsible AI, v. 1.0* (2023), https://aicm.ai.gov.eg/en/Resources/EgyptianCharterForResponsibleAIEnglish-v1.0.pdf

Council of Europe AI Treaty

Egypt did not participate in the negotiations for the Council of Europe Framework Convention on Artificial Intelligence and Human Rights, Democracy, and the Rule of Law, which opened for signature by member and non-member countries in September 2024.[2014] Egypt has not signed the Council of Europe AI Treaty.[2015]

UNESCO Recommendation on AI Ethics

Egypt endorsed the UNESCO Recommendation on AI Ethics[2016] and took an active part in its drafting.[2017] Egypt embedded most of the principles adopted in the UNESCO Recommendations on AI Ethics in its Charter for Responsible AI, albeit no concrete implementation has taken place so far. The Charter draws on the guidelines developed by UNESCO and other organizations by translating these recommendations into steps "to help ensure the responsible development, deployment, management, and use of AI systems in the country."[2018] Although the Charter recognizes the principle of "human-centered AI," it does not acknowledge that "Respect, protection and promotion of human rights and fundamental freedoms" lie at the heart of Ethical AI.

Egypt completed a comprehensive AI Readiness Assessment[2019] in partnership with UNESCO, using UNESCO's Readiness Assessment Methodology (RAM).[2020] The initiative, jointly conducted by the UNESCO Regional Office for Egypt and Sudan and the Ministry of Communications and Information Technology (MCIT), identified priority areas for development, including the need to enhance computing infrastructure, promote ethical innovation, and strengthen regulatory oversight and public accountability mechanisms.[2021] The report also acknowledges Egypt's leadership in regional AI governance, notably through its chairmanship of both the African Union and Arab League AI working groups. The assessment also emphasized

---

[2014] Council of Europe, *Framework Convention on Artificial Intelligence* (2026), https://www.coe.int/en/web/artificial-intelligence/the-framework-convention-on-artificial-intelligence

[2015] Council of Europe Treaty Office, *Chart of Signatures and Ratifications of Treaty 225* (Mar. 29, 2026), https://www.coe.int/en/web/conventions/full-list?module=signatures-by-treaty&treatynum=225

[2016] United Nations, *193 Countries Adopt First-Ever Global Agreement on the Ethics of Artificial Intelligence,* UN News (Nov. 25, 2021), https://news.un.org/en/story/2021/11/1106612

[2017] National Council for Artificial Intelligence, *Partnerships*, https://ai.gov.eg/Partnerships

[2018] National Council for Artificial Intelligence, *Egyptian Charter for Responsible AI*, v. 1.0 (2023), https://aicm.ai.gov.eg/en/Resources/EgyptianCharterForResponsibleAIEnglish-v1.0.pdf

[2019] UNESCO, *Egypt Charts Path towards Ethical and Inclusive AI with UNESCO Support* (Oct. 16, 2025), https://www.unesco.org/en/articles/egypt-charts-path-towards-ethical-and-inclusive-ai-unesco-support

[2020] UNESCO Global AI Ethics and Governance Observatory, *Egypt* (2026), https://www.unesco.org/ethics-ai/en/egypt

[2021] UNESCO, *Egypt: Artificial Intelligence Assessment Report* (2025), https://unesdoc.unesco.org/ark:/48223/pf0000395173

that the transition from assessment to implementation will determine the country's long-term success in ensuring ethical and inclusive AI.

Evaluation

Egypt has been a leading voice in the African and Arab world to foster the regulation of AI. The country chaired African and Arab working groups on the topic and actively participated in the drafting of the UNESCO Recommendation on the Ethics of AI. Nationally, key milestones include the adoption of the Egyptian Charter on Responsible AI and the launch of the Second Edition of the National AI Strategy for 2025–2030. The second edition of the Strategy reflects Egypt's commitment to developing AI Policy to ensure that Egypt implements responsible, ethical AI in a sustainable manner, in line with international standards, in pursuit of the mission to lead the region in AI for development.

Despite developing AI policies in line with international standards, authoritarian government practices and human rights violations undermine implementation of these principles. Surveillance and a lack of supporting institutions continue to undermine democratic principles and rights around public participation, data protection, and civil liberties.

Estonia

In 2025, Estonia designated authorities for the protection of fundamental rights under the EU AI Act and launched the AI Leap Initiative to incorporate AI literacy into secondary- and vocational-school curriculum.

National AI Strategy

The Estonian Cabinet adopted the first National AI Strategy in July 2019.[2022] The Government Chief Information Officer Office, based in the Ministry of Economic Affairs and Communications, was tasked with steering the AI Strategy. The first National AI Strategy built on a May 2019 report of Estonia's AI Taskforce.[2023] The actions detailed in the first AI Strategy were designed to advance the adoption of AI solutions in both the private and public sectors, to increase AI capacities and research and development, and to develop the legal environment to facilitate AI. The first AI Strategy committed to the establishment of a steering group, comprised of government representatives and other stakeholders, to monitor the implementation of the AI Strategy. In addition, the e-Estonia Council was tasked with considering the strategy's implementation annually. The first AI Strategy was conceived as a short-term strategy, intended to apply up until 2021. By adopting a short-term strategy, Estonia intends to gain insight and develop a long-term strategy in response to the experience gained. Estonia aims to monitor the development of the short-term action plan and keep the European Union informed of developments.

In spite of Estonia's national digital adviser initially proposing the adoption of a law granting legal personality to AI, Estonia's AI Taskforce concluded that no substantial legal changes were currently required to address the issues presented by AI.[2024] The Taskforce Report maintained, "Both now and in the foreseeable future, kratts are and will be human tools, meaning that they perform tasks determined by humans and express the intention of humans directly or indirectly." Accordingly, the AI Taskforce Report clarified that the "actions" of AI are attributable to the relevant state body or private party that uses the AI solution.[2025] Minor changes recommended include the removal of obsolete laws and providing additional clarity to facilitate the use of AI. Estonia's Chief Information Officer stated that Estonia wants to "build on

---

[2022] Government, *Estonia's National AI Strategy 2019–2021* (Jul. 2019), https://kratid.ee/en/_files/ugd/980182_8d0df96fd41145739dff2595e0ab3e8d.pdf

[2023] Government Office and Ministry of Economic Affairs and Communications, *Report of Estonia's AI Taskforce* (May 2019) https://f98cc689-5814-47ec-86b3-db505a7c3978.filesusr.com/ugd/7df26f_486454c9f32340b28206e140350159cf.pdf

[2024] Ibid; See also, Astghik Grigoryan, *Estonia: Government Issues Artificial Intelligence Report* (Jul. 31, 2019), https://www.loc.gov/law/foreign-news/article/estonia-government-issues-artificial-intelligence-report/

[2025] Government of the Republic of Estonia, *Estonia's National AI Strategy* 2019–*2021* (Jul. 2019), https://kratid.ee/en/_files/ugd/980182_8d0df96fd41145739dff2595e0ab3e8d.pdf

the EU framework, not to start creating and arguing" for a separate Estonian framework.[2026]

Neither the first AI Strategy nor the AI Taskforce Report provided significant detail on questions related to the ethics of artificial intelligence. Reference was, however, made to guidance provided by the European Commission for the development and implementation of trustworthy artificial intelligence.[2027] The Taskforce Report acknowledged that "trustworthy artificial intelligence must be guided by the principles of human rights, positive rights, and values, thus ensuring the ethics dimension and objective."[2028] The Report recognized the relevance of the EU Charter of Fundamental Rights and referred to the following rights as central according to the Commission guidance on AI: the right to human dignity, the right to freedom, respect of the principles of democracy and the state, based on the rule of law, right to equality, non-discrimination, and acknowledgement of minorities, and civil rights.

To ensure that the development and use of AI is ethical, the Taskforce Report emphasized the importance of ensuring that AI is human-centric; that rights, ethics principles, and values are fundamental; and that AI may bring unintended consequences. The first AI Strategy referenced the EU guidelines that identify the importance of the following values: human agency, technical reliability, privacy and data management, transparency, non-discrimination, social and environmental well-being, and responsibility.

The Estonian Cabinet adopted the Artificial Intelligence Strategy (2022–2023) as a continuation of Estonia's first national AI strategy (2019–2021).[2029] This AI strategy was carried out in line with the objectives laid out by the European Commission in its Coordinated Plan on Artificial Intelligence. Estonia's AI Task

---

[2026] Ronald Liive, *Estonian State IT Manager Siim Sikkut: If There Were 1% in the State Budget for Science, We Could Talk More About Kratind*, DigiGeenius (May 5, 2019), in [US] Library of Congress, *Estonia: Government Issues Artificial Intelligence Report* (Jul. 31, 2019), https://www.loc.gov/item/global-legal-monitor/2019-07-31/estonia-government-issues-artificial-intelligence-report/; in 2018, Estonia signed up to a European Union Declaration of Cooperation on Artificial Intelligence https://digital-strategy.ec.europa.eu/en/news/eu-member-states-sign-cooperate-artificial-intelligence

[2027] Government Office and Ministry of Economic Affairs and Communications, *Report of Estonia's AI Taskforce* (May 2019), https://f98cc689-5814-47ec-86b3-db505a7c3978.filesusr.com/ugd/7df26f_486454c9f32340b28206e140350159cf.pdf; European Commission, *Ethics Guidelines for Trustworthy AI* (Apr. 8, 2019), https://ec.europa.eu/digital-single-market/en/news/ethics-guidelines-trustworthy-ai

[2028] Government Office and Ministry of Economic Affairs and Communications, *Report of Estonia's AI Taskforce* (May 2019), https://f98cc689-5814-47ec-86b3-db505a7c3978.filesusr.com/ugd/7df26f_486454c9f32340b28206e140350159cf.pdf

[2029] Government of the Republic of Estonia, *Estonia's National Artificial Intelligence Strategy, 2019–2021* (Jul. 2019), https://f98cc689-5814-47ec-86b3-db505a7c3978.filesusr.com/ugd/7df26f_27a618cb80a648c38be427194affa2f3.pdf

Force is working to update the national AI strategy and explained the 203 edition will continue until completion.[2030]

Nordic-Baltic Cooperation on AI

As for the international landscape, the Estonian minister responsible for digital development signed the 2018 declaration on "AI in the Nordic-Baltic region" establishing a collaborative framework on "developing ethical and transparent guidelines, standards, principles and values to guide when and how AI applications should be used" and "on the objective that infrastructure, hardware, software and data, all of which are central to the use of AI, are based on standards, enabling interoperability, privacy, security, trust, good usability, and portability."[2031]

The ministerial declaration Digital North 2.0[2032] builds on the common priorities of the Nordic-Baltic countries, and follows the previous ministerial declaration, Digital North 2017–2020. The countries created a Council of Ministers for Digitalization (MR-DIGITAL) "to promote work with digitalisation, co-ordinate efforts, and follow up on the goals of the declaration."[2033]

In November 2021, the Nordic and Baltic ministers for digitalization released another joint statement announcing a focus on digital inclusion, striving to implement measures to make digital services more accessible to all Estonian inhabitants and ensuring that those who do not possess the necessary level of skills get the opportunity to acquire them. [2034]

The Nordic and Baltic ministers for digitalization issued a common statement in 2022 on the importance of cooperation on digital security in the Nordic-Baltic region following the COVID-19 pandemic and the war in Ukraine. In their common statement, the ministers stressed that this "rapid transformation has challenged everyone to adapt to new, digital ways of doing business, learning and accessing public authorities." The ministers declared that they "have committed to ensuring that our region maintains its position as a leader in digitalisation, and that everyone in the region benefit from digitalisation regardless of age, wealth, education or level of digital skills. One important factor that helps ensure a strong level of digitalisation in the region is the trust citizens put in digital services from the public sector – be it at

---

[2030] Ministry of Justice and Digital Affairs, *AI, Vision and Strategies* (Oct. 2025), https:/kratid.ee/en/kratt-visioon/

[2031] Nordic and Baltic Ministers of Digitalization, *AI in the Nordic-Baltic Region* (May 14, 2018), https://www.norden.org/en/declaration/ai-nordic-baltic-region

[2032] Nordic and Baltic Ministers of Digitalization, *Ministerial Declaration Digital North 2.0* (Sept. 29, 2020), https://www.norden.org/en/declaration/ministerial-declaration-digital-north-20

[2033] Nordic Co-operation, *Nordic-Baltic Co-operation on Digitalisation*, https://www.norden.org/en/information/nordic-baltic-co-operation-digitalisation

[2034] Nordic and Baltic Ministers of Digitalization, *Common Statement on the Importance of Promoting Digital Inclusion as a Central Part of the Digital Transformation in the Nordic-Baltic Region* (Nov. 26, 2021), https://www.norden.org/en/declaration/common-statement-importance-promoting-digital-inclusion-central-part-digital

regional, national or local level. In order to keep up this high level of trust, we need to continue our efforts to make our digital public services human centric and accessible. […] Robust and secure digital services, safeguarding users' privacy and ensuring that personal data are stored and processed in a trustworthy way, are crucial to the citizens' sustained trust in digital services."[2035]

The Secretary of the Nordic Council of Ministers published a report in May 2024 presenting the fundamental characteristics of the digitalization of their national public administrations from a legal perspective. The report discusses the impact of proposed EU regulations on AI, such as the AI Act, and their implications for national legislations within the Nordic-Baltic countries. It underscores the need for harmonized legal frameworks that support AI innovation while safeguarding fundamental rights and public trust.[2036] The Nordic Ministers of Digitalisation adopted a Declaration to promote participation of persons with disabilities through inclusive, fair and accessible artificial intelligence in 2025.[2037]

EU Digital Services Act

As an EU member state, Estonia shall apply the EU Digital Services Act (DSA).[2038] The DSA regulates online intermediaries and platforms. The main objective of the DSA is to prevent illegal and harmful activities online and the spread of disinformation. The DSA also bans targeted advertising based on a person's sexual orientation, religion, ethnicity, or political beliefs. The DSA also bans targeted advertising to minors based on profiling.

Signatories to the DSA also commit to take action in several domains, such as demonetizing the dissemination of disinformation; ensuring the transparency of political advertising; empowering users; enhancing the cooperation with fact-checkers; and providing researchers with better access to data, as part of the 2022 DSA mitigation measure Strengthened Code of Practice on Disinformation. [2039]

[2035] Nordic and Baltic Ministers of Digitalization, *Common Statement on the Importance of Cooperation on Digital Security in the Nordic-Baltic Region* (Sept. 6, 2022), https://www.norden.org/en/declaration/common-statement-importance-cooperation-digital-security-nordic-baltic-region

[2036] Nordic Cooperation, *Public Digitalisation in a Legal Perspective* (May 3, 2024) https://pub.norden.org/temanord2024-503/index.html

[2037] Nordic Cooperation, *Promoting Participation of Persons with Disabilities through Inclusive, Fair and Accessible Artificial Intelligence* (May 28, 2025), https://.norden.org/en/declaration/promoting-participation-persons-disabilities-through-inclusive-fair-and-accessible

[2038] EUR-Lex, *Regulation (EU) 2022/2065 of the European Parliament and of the Council of 19 October 2022 on a Single Market for Digital Services and amending Directive 2000/31/EC (Digital Services Act)*, (Oct. 21, 2022), https://eur-lex.europa.eu/legal-content/EN/TXT/?uri=celex%3A32022R2065

[2039] European Commission, *The 2022 Code of Practice on Disinformation*, https://digital-strategy.ec.europa.eu/en/policies/code-practice-disinformation

As part of its obligations under the DSA, Estonia was required to designate a Digital Services Coordinator (DSC) by February 17, 2024.[2040] Although Estonia designated the Consumer Protection and Technical Regulatory Authority (*Tarbijakaitse ja Tehnilise Järelevalve Amet*) as the DSC after the deadline, the country is now in compliance with the act.[2041]

EU AI Act

As an EU member State, Estonia is bound by the EU AI Act[2042] and obligated to designate authorities to oversee the implementation and enforcement of the act. The EU AI Act is a risk-based market regulation that supports the objective of promoting a human-centric approach to AI and making the EU a global leader in the development of secure, trustworthy and ethical AI.

Estonia has designated the Data Protection Inspectorate, the Commission for Gender Equality and Equal Treatment, and the Consumer Protection and Technical Regulatory Authority to protect and supervise fundamental rights in line with Article 77 of the EU AI Act.[2043] While the institutions are not surveillance authorities, they have powers to access documents that providers and deployers are obligated to keep and to be informed by the market surveillance authority of serious incidents related to High Risk AI systems.

Public Participation

In 2018, the Estonian government brought together an expert group to participate in a cross-sectional coordination project on AI.[2044] The three tasks of this expert group were to develop the so-called Estonian artificial intelligence action plan; prepare draft legislation to ensure clarity in the Estonian judicial area and organize the necessary supervision; notify the public about the implementation of kratts and introduce possible options.

[2040] European Commission, *Digital Services Coordinators* (Jan. 26, 2026), https://digital-strategy.ec.europa.eu/en/policies/dsa-dscs#1720699867912-1

[2041] Riigi Teataja, *Act on the Amendments to the Information Society Services Act, the Copyright Act and the Taxation Act [Infoühiskonna teenuse seaduse, autoriõiguse seaduse ja maksukorralduse seaduse muutmise seadus]* (Jun. 4, 2024), https://www.riigiteataja.ee/akt/121062024001

[2042] European Parliament, *Artificial Intelligence Act, European Parliament legislative resolution of 13 March 2024 on the proposal for a regulation of the European Parliament and of the Council on laying down harmonised rules on Artificial Intelligence (Artificial Intelligence Act) and amending certain Union Legislative Acts (COM(2021)0206 – C9-0146/2021 – 2021/0106(COD))*, P9_TA(2024)0138, https://www.europarl.europa.eu/RegData/seance_pleniere/textes_adoptes/definitif/2024/03-13/0138/P9_TA(2024)0138_EN.pdf

[2043] Ministry of Justice and Digital Affairs, *Artificial Intelligence Regulation [Tehisintellekti määrus]* (2024), https://www.kratid.ee/tehisintellektimaarus

[2044] Government Office and Ministry of Economic Affairs and Communications, *Report of Estonia's AI Taskforce*, p. 42 (May 2019), https://f98cc689-5814-47ec-86b3-db505a7c3978.filesusr.com/ugd/7df26f_486454c9f32340b28206e140350159cf.pdf

Participants in the group included representatives from state authorities, the private sector, universities, and sectoral experts. The report collected data through interviews, including with company representatives involved in the development of AI and ICT representatives from universities. Working groups (in the fields of law, education, and the public sector) were also created.[2045] The e-estonia website states, "In these debates, technical and legal expertise goes a long way. But the discussion must also involve the public. Honest, meaningful debate requires that dreamy utopias be balanced with open discussions about AI's controversial attributes and threats. Only this can create user-friendly legislation that's equipped to reduce legal nightmares in the long-term."[2046]

Documents relating to the AI Strategy are accessible on the internet. The website Kratid provides links to the National Artificial Intelligence Strategy, the Report of Estonia's AI Taskforce, the Vision Paper on #KrattAI: The Next Stage of Digital Public Services in #eEstonia, and the #KrattAI Roadmap for 2020.[2047]

The Kratid website was updated with the new AI Strategy (2022–2023) and the new #KrattAI Roadmap for 2021–2022 in 2022. A separate website, the Electronic Information System (EIS),[2048] enables anyone to follow ongoing legislatives procedures, search for documents in the information system, take part in public consultations and comment on a document under inter-agency coordination. The EIS is a working environment for inter-agency coordination, submission of documents to the government and the parliament, and public consultation. In addition to national documents, draft European Union legislation and other documents related to the European Union decision-making process are available in the EIS.

The Ministry of Economic Affairs and Communications published the new national AI strategy (2022–2023) for public consultation for two weeks in 2021. Although the deadline was tight, the public had the opportunity to supplement and comment on the Action Plan.

Bürokratt,[2049] an AI-powered virtual assistant designed by the Estonian government with civil society consultations, saw a significant expansion of its capabilities in January 2024. The government designed the assistant to simplify interactions with government services and promoted public awareness campaigns to increase its adoption among civil society. The government integrated public information in Bürokratt to allow the system to answer questions and find public

---

[2045] Ibid, Annex, p. 42

[2046] e-estonia, *AI and the Kratt Momentum* (Oct. 2018) https://e-estonia.com/ai-and-the-kratt-momentum/

[2047] Kratid, *Artificial Intelligence, Vision and Plans [Visioon ja kavad]* (2024), https://www.kratid.ee/kratt-visioon

[2048] Government Office, *The Draft Information System (EIS, Electronic Information System) [Eelnõude infosüsteem]* (2026), https://eelnoud.valitsus.ee/main#Aok2CJTq

[2049] European Commission, *#Bürokratt Programme and National Virtual Assistant Platform and Ecosystem* (2024), https://commission.europa.eu/projects/burokratt-programme-and-national-virtual-assistant-platform-and-ecosystem_en

information. The State Information Systems Authority (*Riigi Infosüsteemi Amet*) has asked citizens to help test the model by asking questions related to public needs such as social issues, pensions, employment, or citizens and provide feedback on the accuracy of the answers.[2050]

Data Protection

Since Estonia is an EU Member State, the General Data Protection Regulation (GDPR)[2051] is directly applicable in Estonia and to Estonians. The aim of the GDPR is to "strengthen individuals' fundamental rights in the digital age and facilitate business by clarifying rules for companies and public bodies in the digital single market." Estonia's Personal Data Protection Act (PDPA)[2052] and the Personal Data Protection Implementation Act (Implementation Act) were adopted in December 2018 and February 2019 respectively to align Estonian law with the GDPR. They entered into force in January 2019 and March 2019 respectively.

The PDPA also transposed the EU Data Protection Law Enforcement Directive (LED),[2053] which "protects citizens' fundamental right to data protection whenever personal data is used by criminal law enforcement authorities for law enforcement purposes. It will in particular ensure that the personal data of victims, witnesses, and suspects of crime are duly protected and will facilitate cross-border cooperation in the fight against crime and terrorism."[2054] The LED provides for the prohibition of any decision based solely on automated processing, unless it is provided by law, and of profiling that results in discrimination.[2055] The European Data Protection Board (EDPB), of which Estonia is a member, stresses that facial recognition tools should only be used in strict compliance with the Law Enforcement Directive (LED).[2056]

[2050] Riigi Infosüsteemi Amet, *Bureaucrat [Bürokratt]* (Oct. 2025), https://buerokratt.ee/

[2051] *Regulation (EU) 2016/679 of the European Parliament and of the Council of 27 April 2016 on the protection of natural persons with regard to the processing of personal data and on the free movement of such data*, https://eur-lex.europa.eu/eli/reg/2016/679/oj/eng

[2052] Riigi Teatja, *Personal Data Protection Act* (Dec. 12, 2018), https://www.riigiteataja.ee/en/eli/523012019001/consolide

[2053] *Directive (EU) 2016/680 of the European Parliament and of the Council of 27 April 2016 on the protection of natural persons with regard to the processing of personal data by competent authorities for the purposes of the prevention, investigation, detection or prosecution of criminal offences or the execution of criminal penalties, and on the free movement of such data, and repealing Council Framework Decision 2008/977/JHA*, https://eur-lex.europa.eu/legal-content/EN/TXT/?uri=celex%3A02016L0680-20160504

[2054] European Commission, *Legal Framework of EU Data Protection*, https://commission.europa.eu/law/law-topic/data-protection/legal-framework-eu-data-protection_en

[2055] Article 11 (1) and (2) of the LED, https://eur-lex.europa.eu/legal-content/EN/TXT/?uri=celex%3A02016L0680-20160504

[2056] European Data Protection Board, *Guidelines 05/2022 on the Use of Facial Recognition Technology in the Area of Law Enforcement*, version 2 (Apr. 26, 2023), https://www.edpb.europa.eu/system/files/2023-05/edpb_guidelines_202304_frtlawenforcement_v2_en.pdf

The EU Charter of Fundamental Rights more generally provides that EU citizens have the right to protection of their personal data. Article 8 of the Charter states that: "Everyone has the right to the protection of personal data concerning him or her. Such data must be processed fairly for specified purposes and on the basis of the consent of the person concerned or some other legitimate basis laid down by law."

Estonia is also a member of the Council of Europe and ratified the Council of Europe's Convention 108+ for the protection of individuals with regard to the processing of personal data.[2057]

The Data Protection Inspectorate[2058] is the national supervisory authority in Estonia. Despite being a member of the Global Privacy Assembly (GPA) since 2006, the Inspectorate has not endorsed the 2018 GPA Declaration on Ethics and Data Protection in Artificial Intelligence;[2059] the 2020 GPA Resolution on AI Accountability;[2060] the 2022 GPA Resolution on Facial Recognition Technology[2061] or the 2023 GPA Resolution on Generative AI.[2062]

Algorithmic Transparency

Estonia is subject to the GDPR and Convention 108+. Estonians have a general right to obtain access to information about automated decision-making and to the factors and logic of an algorithm.[2063] The 2020 Recommendation of the Council of Europe Committee of Ministers clarifies that the right to algorithmic transparency extends to contestation,[2064] including "an opportunity to be heard, a thorough review of the decision and the possibility to obtain a non-automated decision. This right may not be waived, and should be affordable and easily enforceable before, during and

---

[2057] Council of Europe, *Convention 108 and Protocols* (May 18, 2018) https://www.coe.int/en/web/data-protection/convention108-and-protocol

[2058] Data Protection Inspectorate, https://www.aki.ee/en

[2059] Global Privacy Assembly, *Declaration on Ethics and Data Protection in Artificial Intelligence* (Oct. 23, 2018), https://globalprivacyassembly.org/wp-content/uploads/2018/10/20180922_ICDPPC-40th_AI-Declaration_ADOPTED.pdf

[2060] Global Privacy Assembly, *Resolution on Accountability in the Development and Use of Artificial Intelligence* (Oct. 2020), https://globalprivacyassembly.org/wp-content/uploads/2020/10/FINAL-GPA-Resolution-on-Accountability-in-the-Development-and-Use-of-AI-EN-1.pdf

[2061] Global Privacy Assembly, *Resolution on Principles and Expectations for the Appropriate Use of Personal Information in Facial Recognition Technology* (Oct. 2022), https://globalprivacyassembly.org/wp-content/uploads/2022/11/15.1.c.Resolution-on-Principles-and-Expectations-for-the-Appropriate-Use-of-Personal-Information-in-Facial-Recognition-Technolog.pdf

[2062] Global Privacy Assembly, *Resolution on Generative Artificial Intelligence Systems* (Oct. 2023), https://globalprivacyassembly.org/wp-content/uploads/2023/10/5.-Resolution-on-Generative-AI-Systems-101023.pdf

[2063] See Recital 63 and Article 22 of the GDPR. Article 9 c) of the Convention 108+ as well as Recital 77, *Explanatory Report, Convention 108+*, p. 24, https://rm.coe.int/convention-108-convention-for-the-protection-of-individuals-with-regar/16808b36f1

[2064] *Recommendation CM/Rec(2020)1 of the Committee of Ministers to member States on the human rights impacts of algorithmic systems* (Apr. 8, 2020), https://search.coe.int/cm/pages/result_details.aspx?objectid=09000016809e1154

after deployment, including through the provision of easily accessible contact points and hotlines."[2065]

Estonia's Ministry of Justice announced plans in 2020 to draft legislation addressing high-risk algorithmic systems that will require the creators of AI (both public and private) to provide transparency regarding when AI communicates with an individual, processes an individual's data, or makes a decision on the basis of the individual's data.[2066] However, as of 2025, the legislation has not been drafted.

Estonia's National AI Strategy for 2022–2023, which is currently being updated for the 2024–2026 period, underscores the importance of transparency in AI systems. The strategy outlines measures to ensure that AI applications, including algorithms used in public services, are transparent, accountable, and aligned with ethical standards.[2067] Estonia demonstrated a commitment to transparency by creating public repositories for open-source AI components. The Kratid repository,[2068] for example, is open source to provide visibility into the AI components that have been developed or used by the public sector.

Digitization of Public Services

According to Estonia's Digital Decade Country Report of 2025,[2069] the country is committed to increasing efficiency of public services in line with the Estonian Digital Agenda 2030[2070] and the National AI strategy. Indeed, 93% of Estonian internet users were reported to use the internet to interact with public services. In light of its commitment to e-government, Estonia emphasizes the use of AI for government services. Indeed, *KrattAI* refers to "the vision of how digital public services should work in the age of artificial intelligence"; or more specifically, KrattAI is described as an "interoperable network of AI applications, which enable citizens to use public services with virtual assistants through voice-based interaction."[2071]

2065 Ibid

2066 Baltic Course, *Estonian Ministry: Use of AI Must Respect Fundamental Rights* (Aug. 19, 2020) www.baltic-course.com/eng/Technology/?doc=158411&output=d

2067 Data Guidance, *Estonia: DPI Publishes Data Security Recommendations* (Apr. 8, 2024), https://www.dataguidance.com/news/estonia-dpi-publishes-data-security-recommendations

2068 Ministry of Justice and Digital Affairs, *Reusable AI Components* (2024), https://www.kratid.ee/en/kratijupid

2069 Ministry of Justice and Digital Affairs, *Europe's Digital Decade Strategic Roadmap: Estonia* (2025), https://www.justdigi.ee/sites/default/files/documents/2025-03/Estonian%20National%20Digital%20Decade%20Strategic%20Roadmap%202025.pdf

2070 Ministry of Economic Affairs and Communications, *Estonia's Digital Agenda 2030* (2021), https://www.justdigi.ee/en/digital-communications-and-cyber/digital-agenda-2030

2071 KRATT Artificial Intelligence Programme of #Estonia, *#KrattAI: Roadmap for 2020* (2020), https://f98cc689-5814-47ec-86b3-db505a7c3978.filesusr.com/ugd/7df26f_19625e00a7b84900b99e952b1ce7d21a.pdf; Republic of Estonia, Ministry of Economic Affairs and Communications, *Report of Estonia's AI Taskforce* (May 2019), https://f98cc689-5814-47ec-86b3-

The Estonian government makes use of automated decision-making in many different contexts.[2072] For example, the Tax and Customs Board uses automated decision-making to facilitate tax refunds following the submission of an online income tax return. Other examples include the use of tachographs on lorries and automated speed checks on motorways to issue cautionary fines and the use of automated decision-making for the determination of a child's school on the basis of their registered residence.[2073] There has been international coverage of Estonia's ambitious plans for AI in the public sector – including on the issue of "Robot Judges."[2074] The Estonian court system embraces digitalization and started an e-File system in 2005. The use of AI to tackle an immense backlog of cases has been considered, including the adoption of projects that can make "autonomous decisions within more common court procedures/tasks that would otherwise occupy judges and lawyers alike for hours."[2075]

The government has also launched an AI-based digital assistant (*Bürokratt*) with the aim to efficiently deliver public services and simplify communication with the State.[2076] Bürokratt is an interoperable network of public and private sector AI solutions, which from the user's point of view, act as a single channel for public services and information.[2077] According to the International Research Centre on Artificial Intelligence (IRCAI), operating under UNESCO, "all the aspects regarding KrattAI are fully transparent—starting from business strategy and roadmap to the technology side: architecture, technical solutions—which is Open Source and available to all users."[2078] Open source AI components are also used for the Kratid website.[2079]

The Estonian government provides a data tracker tool accessible through the state portal (eesti.ee) that enables anyone with an eID to keep track of which institutions have accessed their data and for what purposes.[2080] As pointed out on the

db505a7c3978.filesusr.com/ugd/7df26f_486454c9f32340b28206e140350159cf.pdf. The report of Estonia's AI Taskforce defined 'kratt' as being "a practical application that uses artificial intelligence and that fulfils a specific function."

[2072] See also *#KrattAI Roadmap for 2020* (2020), https://f98cc689-5814-47ec-86b3-db505a7c3978.filesusr.com/ugd/7df26f_19625e00a7b84900b99e952b1ce7d21a.pdf

[2073] Council of State of the Netherlands and ACA-Europe, *An Exploration of Technology and the Law* (May 14, 2018), http://www.aca-europe.eu/colloquia/2018/Estonia.pdf

[2074] Eric Niller, *Can AI Be a Fair Judge in Court? Estonia Thinks So,* Wired (Mar. 23, 2019) https://www.wired.com/story/can-ai-be-fair-judge-court-estonia-thinks-so/

[2075] Anett Numa, *Artificial Intelligence as the New Reality of E-justice,* e-estonia (Apr. 2020) https://e-estonia.com/artificial-intelligence-as-the-new-reality-of-e-justice/

[2076] Ministry of Justice and Digital Affairs, *Virtual Assistant Bürokratt* (2024), https://www.kratid.ee/en/burokratt

[2077] Riigi Infosüsteemi Amet, *Bureaucrat [Bürokratt]*, https://buerokratt.ee/

[2078] International Research Centre on Artificial Intelligence, *IRCAI Global 2021 Top 100 List* (2021), https://ircai.org/top100/entry/krattai/

[2079] Ministry of Justice and Digital Affairs, *Kratijupid*, https://www.kratid.ee/kratijupid

[2080] Federico Plantera, *Data Tracker—Tool that Builds Trust in Institutions,* e-estonia (Sept. 2019), https://e-estonia.com/data-tracker-build-citizen-trust/

e-estonia website, transparency is "fundamental to foster trust in the effective functioning of the whole system." Information is also provided regarding automated processing although Algorithm Watch states that it "is not always clear if data is used as a part of an automatic process or viewed by an official."[2081] In spite of the ambition of this tool, the Estonian Human Rights Center argue that the data provided is variable depending on the service and at times not detailed enough.

To assist transparency and understanding, the Estonian Human Rights Center suggests that visual depictions of data use should be provided.[2082] Similarly, Algorithm Watch states that the current tool does not provide a "clear understanding of what profiling is done by the state, which data is collected, how it is used, and for what purpose."[2083] "It is a duty of a country based on the rule of law to have foresight and prevent serious interferences with fundamental rights by means of setting out a relevant legislative framework," said Kai Härmand at the time Deputy Secretary General on Legal Policy at the Ministry of Justice.

Environmental Impact

Estonia, a nation with a highly digitized public sector, recognizes the potential environmental consequences of its digital infrastructure. To address these concerns, the Estonian government adopted the Digital Agenda 2030 in 2021.[2084] This long-term strategy aims to establish a "green digital government" by prioritizing environmentally responsible solutions and minimizing the ecological impact of digital services. To achieve these objectives, the Estonian Ministry of Economic Affairs and Communications commissioned a report to analyze the environmental footprint of Estonia's digital society comprehensively. The report assessed several variables, including the life cycle of ICT equipment, the energy consumption of ICT equipment, data centres and cloud services, software solutions, and digital trash across 12 public institutions and four local governments. Results demonstrate that ICT-related activities contribute to significant emissions: The total impact of the workstation equipment (laptops, desktop PCs and monitors) life cycle of all Estonian state agencies is 26,000 t CO2e—roughly the the annual environmental impact of 5,000

[2081] Algorithm Watch, *Automating Society Report 2020*, p. 75 (Oct. 2020), https://automatingsociety.algorithmwatch.org/wp-content/uploads/2020/10/Automating-Society-Report-2020.pdf

[2082] Kari Käsper and Liina Rajavee, *Inimõigused, Infoühiskond Ja Eesti: Esialgne Kaardistus*, Estonian Human Rights Centre (2019), https://humanrights.ee/app/uploads/2019/12/Inimõigused-infoühiskond-ja-Eesti.-Esialgne-kaardistus.pdf

[2083] Algorithm Watch, *Automating Society Report 2020*, p. 75 (Oct. 2020), https://automatingsociety.algorithmwatch.org/wp-content/uploads/2020/10/Automating-Society-Report-2020.pdf

[2084] Ministry of Economic Affairs and Communications, *Digital Agenda 2030* (Sept. 10, 2024), https://mkm.ee/en/e-state-and-connectivity/digital-agenda-2030

households. The report stressed "the need for emerging technologies to be developed with principles of sustainability in mind."[2085]

Lethal Autonomous Weapons

In a 2019 meeting of the Group of Governmental Experts on Lethal Autonomous Weapons Systems, Estonia expressed the view that "humans must retain ultimate control and responsibility in relation to the use of force in armed conflict" and "humans must exercise such control over a weapon system as may be necessary to ensure that the weapon system operates consistently with international law."[2086]

Estonia was one of the 70 countries that endorsed a joint statement on autonomous weapons systems at the 2022 United Nations General Assembly. The joint statement urged "the international community to further their understanding and address these risks and challenges by adopting appropriate rules and measures, such as principles, good practices, limitations and constraints. We are committed to upholding and strengthening compliance with International Law, in particular International Humanitarian Law, including through maintaining human responsibility and accountability in the use of force."[2087]

In February 2023, Estonia participated in an international summit on the responsible application of artificial intelligence in the military domain hosted by the Netherlands. At the end of the Summit, Estonia endorsed a joint call for action on the responsible development, deployment and use of artificial intelligence in the military domain.[2088] In this joint call, States "stress the paramount importance of the responsible use of AI in the military domain, employed in full accordance with international legal obligations and in a way that does not undermine international security, stability and accountability." Estonia also endorsed the resulting Political

---

[2085] Eesti Elu, *Is Estonia's Digital Government Bad for the Environment?* (Sept. 12, 2024), https://eestielu.ca/is-estonias-digital-government-bad-for-the-environment/

[2086] Permanent Mission of Estonia to the UN and other International Organisations in Geneva, Group of Governmental Experts on Lethal Autonomous Weapons Systems – First meeting, *Statement of Estonia* (March 25-29, 2019), https://unoda-documents-library.s3.amazonaws.com/Convention_on_Certain_Conventional_Weapons_-_Group_of_Governmental_Experts_(2019)/LAWS%2BGGE%2B2019%2BI%2B-%2BEstonia%2B-%2BAgenda%2Bitem%2B5%28c%29.pdf

[2087] United Nations (UN) General Assembly, First Committee, *Joint Statement on Lethal Autonomous Weapons Systems First Committee, 77th United Nations General Assembly Thematic Debate – Conventional Weapons* (Oct. 21, 2022), https://estatements.unmeetings.org/estatements/11.0010/20221021/A1jJ8bNfWGlL/KLw9WYcSnnAm_en.pdf

[2088] Government of Netherlands, *Call to Action on Responsible Use of AI in the Military Domain* (Feb. 16, 2023), https://www.government.nl/latest/news/2023/02/16/reaim-2023-call-to-action

Declaration on Responsible Military Use of AI and Autonomy issued in November 2023.[2089]

At the 78th UN General Assembly First Committee in 2023, Estonia voted in favor[2090] of resolution L.56[2091] on autonomous weapons systems, along with 163 other states. The Resolution emphasized the "urgent need for the international community to address the challenges and concerns raised by autonomous weapons systems," and mandated the UN Secretary-General to prepare a report reflecting the views of member and observer states on autonomous weapons systems.

Estonia's Ministry of Defence developed its first Defence AI Strategy in 2025 to guide the use of AI in logistics, cybersecurity, and military decision-support to promote responsible, interoperable AI aligned with NATO and EU standards.[2092]

Human Rights

Estonia is a member of the European Union and the Council of Europe and is, accordingly, committed to the upholding of the Charter of Fundamental Rights and the European Convention on Human Rights. Estonia ratified the Universal Declaration on Human Rights and has acceded to international human rights treaties, such as the International Covenant on Civil and Political Rights. The Estonian Constitution also grants fundamental rights to citizens.

In Freedom House's 2025 Country Report, Estonia's ranking rose from 95/100 in 2024 to 96/100.[2093] The 2025 report explains that "Estonia's democratic institutions are generally strong, and both political rights and civil liberties are widely respected. However, more than 5 percent of the population remains stateless and cannot participate in national elections. Corruption is a persistent challenge, as is discrimination against ethnic Russians, Roma, LGBT+ people, and others. Far-right and Euroskeptic forces have become increasingly vocal in Estonian politics in recent years."

In a 2020 Recommendation to member States on the human rights impacts of algorithmic systems, the Council of Europe Committee of Ministers reminded member states that their obligations to the Convention for the Protection of Human Rights and Fundamental Freedoms "stands throughout the continuous processes of

---

[2089] US Department of State, *Political Declaration on Responsible Military Use of Artificial Intelligence and Autonomy*, Endorsing States (Feb. 12, 2024), https://www.state.gov/political-declaration-on-responsible-military-use-of-artificial-intelligence-and-autonomy/

[2090] Stop Killer Robots, *164 states vote against the machine at the UN General Assembly*, https://www.stopkillerrobots.org/news/164-states-vote-against-the-machine/

[2091] General Assembly, *Resolution L56: Lethal Autonomous Weapons* (Oct.12, 2023), https://reachingcriticalwill.org/images/documents/Disarmament-fora/1com/1com23/resolutions/L56.pdf

[2092] Ministry of Defence, *Defence Artificial Intelligence Strategy for Estonia* (Mar. 24, 2025), https://kaitseministeerium.ee/en/defence-artificial-intelligence-strategy-estonia

[2093] Freedom House, *Freedom in the World 2025: Estonia* (2025), https://freedomhouse.org/country/estonia/freedom-world/2025

technological advancement and digital transformation that European societies are experiencing," meaning "member States must ensure that any design, development and ongoing deployment of algorithmic systems occur in compliance with human rights and fundamental freedoms, which are universal, indivisible, inter-dependent and interrelated, with a view to amplifying positive effects and preventing or minimizing possible adverse effects."[2094]

In a 2018 report, the Commissioner for Human Rights of the Council of Europe urged the Estonian authorities to give careful consideration "to the ethical, legal and human rights implications of using robots and artificial intelligence in the care of older persons" given Estonia's strong focus on digitalization, new technologies, and AI.[2095]

OECD / G20 AI Principles

Estonia endorsed the OECD Principles on Artificial Intelligence, agreeing to "promote the use of AI that is innovative and trustworthy and that respects human rights and democratic values."[2096]

Council of Europe AI Treaty

Estonia contributed as a Council of Europe and EU Member State in the negotiations of the Council of Europe Framework Convention on AI, Human Rights, Democracy, and the Rule of Law.[2097] The European Union on behalf of all members on September 5, 2024; however, Estonia has not signed individually.[2098]

UNESCO Recommendation on AI Ethics

Estonia has endorsed the UNESCO Recommendations on AI, the first ever global agreement on the ethics of AI.[2099] No available records explicitly address

---

[2094] *Recommendation CM/Rec(2020)1 of the Committee of Ministers to member States on the human rights impacts of algorithmic systems* (Apr. 8, 2020), https://search.coe.int/cm/pages/result_details.aspx?objectid=09000016809e1154

[2095] Council of Europe, Commissioner for Human Rights, *Report of the Commissioner for Human Rights of the Council of Europe Dunja Mijatović Following Her Visit to Estonia from 11 to 15 June 2018*, p. 21 (Sept. 28, 2018),https://rm.coe.int/report-of-the-council-of-europe-commissioner-for-human-rights-dunja-mi/16808d77f4

[2096] OECD AI Policy Observatory, *OECD AI Principles Overview, Countries Adhering to the AI Principles* (2024), https://oecd.ai/en/ai-principles

[2097] Council of Europe, *Framework Convention on Artificial Intelligence* (2026), https://www.coe.int/en/web/artificial-intelligence/the-framework-convention-on-artificial-intelligence

[2098] Council of Europe Treaty Office, *Chart of Signatures and Ratifications of Treaty 225* (Feb. 4, 2026), https://www.coe.int/en/web/Conventions/full-list/?module=signatures-by-treaty&treatynum=225

[2099] UNESCO, *UNESCO Member States Adopt the First Ever Global Agreement on the Ethics of Artificial Intelligence* (Nov. 2021), https://www.unesco.org/en/articles/unesco-member-states-adopt-first-ever-global-agreement-ethics-artificial-intelligence.

implementation of the recommendation, though Estonia's efforts to align with the EU AI Act and participation in negotiations of the Council of Europe Framework Convention on AI demonstrate the country's attention to the Recommendations. Estonia has not initiated the UNESCO Readiness Assessment Methodology (RAM) process, a tool designed to assist countries in implementing the Recommendation.[2100]

Evaluation

As a member of the European Union and the Council of Europe, Estonia is committed to the protection of human rights, ethics in AI, and algorithmic transparency. Estonia has also endorsed the OECD AI Principles and the UNESCO Recommendation on the Ethics of AI and is party to the Council of Europe Framework Convention on AI through the EU signature. At regional level, Estonia has signed the Declaration of Collaboration on AI in the Nordic-Baltic Region which includes a commitment "to develop ethical and transparent guidelines, standards, norms and principles that can be employed as a steering mechanism to guide AI programmes."[2101]

By contrast, Estonia's first AI Strategy did not consider the issues of ethics and human rights in significant depth. Nevertheless, Estonia opted for a short-term strategy in order to be able to adapt it to a rapidly evolving AI landscape. Based on Estonia's experience as a leading country in the world regarding the use of data[2102] and e-governance, its second national AI strategy, although still wary of "proactive over-regulation in a rapidly evolving area" and of "creating unnecessary obstacles," includes the bases for an ethical framework and provides for specific actions in this regard.

Reflective of this approach is Estonia's Prime Minister Kaja Kallas' statement: "We are already using artificial intelligence in our operations and services, but we see there is a huge potential to make our services more convenient for the people." "But at the same time, we are a rule-of-law country and every individual's privacy is a very important matter for us."[2103] This is a similar position that Estonia defended with regard to the draft EU AI Act. "We welcome the proposed AI Act because we have long supported the idea that European-wide harmonized regulation ensures a common market, and minimizing risks increases acceptance of AI technologies," said Mr. Velsberg, the Estonian Government Chief Data Officer. Estonia shall establish a

[2100] UNESCO Global AI Ethics and Governance Observatory, *Global Hub* (Oct. 2025), https://www.unesco.org/ethics-ai/en/global-hub

[2101] Nordic Co-operation, *AI in the Nordic-Baltic region, Declaration* (May 14, 2018), https://www.norden.org/en/declaration/ai-nordic-baltic-region

[2102] E-estonia, *Estonia Is Leading the World in the Use of Data* (May 23, 2022), https://e-estonia.com/estonia-is-leading-the-world-in-the-use-of-data/

[2103] Invest in Estonia, *Prime Minister Kaja Kallas: Artificial Intelligence Is the Next Big Thing for Estonia* (2022), https://investinestonia.com/prime-minister-kaja-kallas-artificial-intelligence-is-the-next-big-thing-for-estonia/

national supervisory mechanism as part of implementation of the AI Act. Thus far, only fundamental rights protection authorities have been officially designated.

Ethiopia

In 2025, Ethiopia deepened international cooperation and partnerships for developing and governing artificial intelligence through support for the Africa Declaration on Artificial Intelligence and the BRICS Leaders' Statement on the Global Governance of Artificial Intelligence.

National AI Strategy

Ethiopia's Council of Ministers created the Artificial Intelligence Institute[2104] to develop a National AI Strategy in 2022. After three years of consultations with key stakeholders, the Council approved the National AI Policy in 2024.[2105] The policy creates a framework to leverage AI for economic growth, social development, and national security through data management, human capital development, infrastructure development, and ethical considerations.[2106] Key areas of alignment with international standards include a human-centric approach, transparency and explainability, ethical AI development, inclusivity and non-discrimination, human rights protection, and regulatory frameworks. The AI policy charges the AI Institute with monitoring and implementing the AI Policy.

African Union

As a member of the African Union (AU), Ethiopia is committed to advancing the formulation and implementation of human-centered AI policies, in alignment with the goals of the AU digital transformation strategy[2107] and the Continental Data Policy Framework.[2108]

The Continental AI Strategy was adopted in 2024 during the 45th Ordinary Session of the African Union Executive Council in Accra, Ghana,[2109] following

---

[2104] Ethiopian Ministry of Justice, *Council of Ministers Regulation No. 510/2022 establishing Artificial Intelligence Institute* (2022), https://justice.gov.et/am/law/artificial-intelligence-institute-establishment-council-of-ministers-regulation-customs-regulation-council-of-ministers-regulation-no-510-2022/

[2105] Ethiopian Artificial Intelligence Institute, *Ethiopia Has Approved and Operationalized a National AI Policy*, LinkedIn (2025), https://www.linkedin.com/posts/etartificialintelligenceinstitute_fdre-national-artificial-intelligence-policy-activity-7376224568568348673-tsIx

[2106] Federal Democratic Republic of Ethiopia, *National Artificial Intelligence Policy* (Jun. 2024), https://www.lawethiopia.com/images/Policy_documents/Ethiopian%20ai%20policy.pdf

[2107] African Union, *The Digital Transformation Strategy for Africa (2020–2030)*, https://au.int/sites/default/files/documents/38507-doc-dts-english.pdf

[2108] African Union, *40th Ordinary Session of the Executive Council* (Feb. 2–3, 2022), https://au.int/sites/default/files/decisions/41584-EX_CL_Dec_1143-1167_XL_E.pdf

[2109] African Union, *Continental Artificial Intelligence Strategy* (2024), https://au.int/en/documents/20240809/continental-artificial-intelligence-strategy

endorsement by African ministers.[2110] The Strategy calls for unified national approaches among AU Member States for inclusive and responsible AI development The AU envisions a prosperous Africa where responsible and African-centric AI is pivotal in the inclusive growth and socio-economic development of Africans positioning the continent as a key player in the global AI landscape.[2111] The Union seeks to achieve this by leveraging AI to accelerate socio-economic development and promote African cultural renaissance in line with the AU Agenda 2063[2112] and Sustainable Development Goals.[2113] The Strategy was supported both technically and financially by UNESCO, ensuring that the African Union's approach to AI regulation and innovation aligned with the UNESCO Recommendation on the Ethics of Artificial Intelligence.[2114]

Along with the development of the Continental AI Strategy, the African Union engaged with other international regulatory frameworks on AI in 2024. In April 2024, African Union senior officials met in Addis Ababa with the G7 Presidency to strengthen collaboration on AI.[2115] Among other discussions, the meeting highlighted the plans for the Italian G7 Presidency alongside the United Nations Development Programme (UNDP) to co-design an AI Hub for Sustainable Development with African leaders in 2024. The Italian delegation explained, "The AI Hub aims to be a catalyzer for systemic change and accelerate the achievement of the Sustainable Development Goals. The benefits of an AI empowered Africa will have a positive impact not just within the African continent but across the world."[2116]

Ethiopia further committed to aligning governance frameworks with the African Union Continental AI Strategy by endorsing the Africa Declaration on Artificial Intelligence.[2117] The Declaration centers objectives to leverage AI to advance African economies, industries, and societies "to position Africa as a global leader in ethical, trustworthy, and inclusive AI adoption" through intergovernmental and regional cooperation on AI governance, development, and capacity building.

[2110] African Union, *African Ministers Adopt Landmark Continental Artificial Intelligence Strategy* (2024), https://au.int/en/pressreleases/20240617/african-ministers-adopt-landmark-continental-artificial-intelligence-strategy

[2111] African Union, *Continental Artificial Intelligence Strategy* (Aug. 9, 2024), https://au.int/sites/default/files/documents/44004-doc-EN-_Continental_AI_Strategy_July_2024.pdf

[2112] African Union, *AU Agenda 2063: The Africa We Want* (Jun. 10, 2013), https://au.int/Agenda2063/popular_version

[2113] United Nations, *The 17 Goals*, https://sdgs.un.org/goals

[2114] UNESCO, *UNESCO's Support to AU's Continental AI Framework and Strategy* (Aug. 28, 2024), https://www.unesco.org/en/articles/unescos-support-aus-continental-ai-framework-and-strategy

[2115] UN Development Programme (UNDP), *The G7 Presidency Partners with UNDP Africa to Advance AI for Sustainable Development* (Apr. 15, 2024), https://www.undp.org/africa/press-releases/g7-presidency-partners-undp-africa-advance-ai-sustainable-development

[2116] Ibid

[2117] Global AI Summit on Africa, *The Africa Declaration on Artificial Intelligence* (Apr. 4, 2025), https://c4ir.rw/docs/Africa%20Declaration%20on%20Artificial%20Intelligences.pdf

BRICS

Ethiopia, together with Iran, Egypt, Saudi Arabia and the United Arab Emirates, officially joined China, Russia, India, Brazil and South Africa as a new BRICS member in a 2023 vote. Russian President Putin who took over the rotating chairmanship of BRICS in January 2024 said that the bloc seeks "strengthening multilateralism for equitable global development and security."[2118] Egypt, Ethiopia, Iran, and the UAE participated in their first BRICS summit in 2024.[2119]

At the most recent BRICS Summit in 2025, Ethiopia endorsed the Leaders' Statement on the Global Governance of Artificial Intelligence.[2120] The statement emphasized digital and governance sovereignty to protect workers and citizens while emphasizing the leading role for the United Nations in facilitating discussions on how to leverage AI technology for the Sustainable Development Goals (SDGs).

BRICS also elaborated on the Digital Economy Working Group proposed in a concept note at the 2023 BRICS Summit,[2121] aiming to develop digital education cooperative mechanisms. Deepening technological cooperation through this Working Group may also encourage future BRICS cooperation on AI.

In January 2024, China also called for the development of a BRICS AI Development Cooperation Center to be constructed in China. This announcement was followed by the release in February 2024, of the AI Governance International Evaluation Index, which evaluates BRICS' AI development in contrast to G7 countries.[2122]

Data Protection

The 1995 Constitution of the Federal Democratic Republic of Ethiopia establishes a fundamental right to privacy.

Ethiopia made a significant stride in data protection with the enactment of the Personal Data Protection Proclamation in July 2024.[2123] This comprehensive law replaces Ethiopia's fragmented regulatory framework and empowers individuals by

---

[2118] Taarifa Rwanda, *Ethiopia Officially Confirmed as BRICS Member* (Jan. 8, 2024), https://furtherafrica.com/2024/01/08/ethiopia-officially-confirmed-as-brics-member/

[2119] BRICS-Russia 2024, *Address by President of the Russian Federation Vladimir Putin on the start of Russia's BRICS Chairmanship*, 2024 https://brics-russia2024.ru/en/index_07.php?sphrase_id=39447

[2120] BRICS Brasil 2025, *BRICS Leaders' Statement on the Global Governance of Artificial Intelligence*, Presidency Documents (2025), https://brics.br/en/documents/presidency-documents

[2121] BRICS 2022 China, *Concept Note: BRICS Digital Economy Cooperation* (2022), https://www.gov.br/memp/pt-br/assuntos/brics-sme-working-group-1/documents-on-smes/2022-china/concept-note-on-brics-digital-economy-2-china.pdf

[2122] Global Times, *China Researchers Release AI Governance Index, with China Ranking in First Echelon,* (Feb. 6, 2024), https://www.globaltimes.cn/page/202402/1306760.shtml

[2123] Federal Negarit Gazette, *Personal Data Protection Proclamation No. 1321/2024* (Jul. 24, 2024), https://www.dataguidance.com/sites/default/files/personal_data_protection_proclamation_1321-2024.pdf; see also https://justice.gov.et/en/law/personal-data-protection-proclamation/

granting them rights over their personal data, such as access, rectification, erasure, and objection to processing. The proclamation also imposes stringent obligations on organizations handling personal data, including implementing robust security measures, obtaining informed consent, and ensuring transparency. The Proclamation emphasizes data localization and regulates cross-border data transfers, emphasizing "sovereignty of data" to keep Ethiopian citizen data within national borders.[2124] Rules allow personal data transfers outside of Ethiopia only to countries with similar data protection standards.

The Proclamation charges the Ministry of Technology and Innovation with formulating policies and strategies on personal data and the Ethiopian Communications Authority with enforcing the Proclamation, including investigating complaints and imposing fines.[2125] The Communications Authority responsibility builds on the Communication Services Proclamation,[2126] which establishes the Communication Authority's responsibility to promote data privacy. The Authority's Consumers Rights and Protection Directive No. 832/2021 explicitly defines personal data as any information relating to an identified or identifiable natural person leading to identify such person, directly or indirectly in particular by reference to an identifier such as a name, an identification number, location data, telephone number, traffic and billing data, and other personal information in the context of Telecommunications Services.[2127] The Ethiopian Communications Authority is not a member of the Global Privacy Assembly (GPA).[2128]

The right to privacy is further reinforced in several legal instruments, including the Criminal Code of the Federal Democratic Republic of Ethiopia (Proclamation No. 414/2004), the Civil Code of the Empire of Ethiopia (Proclamation No. 165/1960), the Freedom of the Mass Media and Access to Information Proclamation (No. 590/2008), and the recently enacted Media Proclamation No. 1238/2021. The Media Proclamation of 2021 repealed portions of the Mass Media & Access to Information Proclamation specifically related to mass media.[2129]

[2124] International Trade Administration, *Ethiopia- Country Commercial Guide* (Sept. 2024), https://www.trade.gov/country-commercial-guides/ethiopia-digital-economy

[2125] Federal Negarit Gazette, *Personal Data Protection Proclamation No. 1321/2024*, pp. 15629–15631 (Jul. 24, 2024), https://www.dataguidance.com/sites/default/files/personal_data_protection_proclamation_1321-2024.pdf

[2126] Ministry of Finance of Ethiopia, *Proclamation No. 1148/20219: Proclamation for the Communications Service*, Federal Negarit Gazette (Aug. 12, 2019), https://www.mofed.gov.et/media/filer_public/7c/78/7c781a5b-1c06-4f6d-9128-6fa116cf71e8/communicationsserviceproclamationno1148-2019.pdf

[2127] Ethiopian Communications Authority, *Telecommunications Consumer Rights and Protection Directive No. 832/2021* (Aug. 2021), https://cyrilla.org/api/files/1689857290044ckib706mn44.pdf

[2128] Global Privacy Assembly (GPA), *List of Accredited Members* (2026), https://globalprivacyassembly.com/participation-in-the-assembly/list-of-accredited-members/

[2129] Bowmans Law, *Africa Guide to Data Protection* (2022), https://bowmanslaw.com/wp-content/uploads/2022/06/Data-Protection_01.06.2022.pdf

The Digital ID Proclamation[2130] was enacted in 2023 with the view to adopt a comprehensive legal framework for the regulation of the national digital identification system. The Proclamation provides the rules for collecting, processing, transferring, disclosing, modifying, and overall management of personal data of registrants. "Personal Data" for the said Proclamation means the biometric and demographic data collected with the 'digital identification system [Article (2)(17)]. Interestingly, according to Article 17(1) of the Proclamation, the owner of the personal data is the registrant, and they shall give their data to the registrar upon their consent for processing, transferring, disclosing, and modifying.

Ethiopia has not signed the African Union's Convention on Cyber Security and Personal Data Protection,[2131] known as the Malabo Convention.[2132] which entered into force in June 2023.[2133]

Algorithmic Transparency

Ethiopia has introduced initial safeguards for automated decision-making through Art. 31 Personal Data Protection Proclamation No. 1321/2024, which recognizes individuals' right to object to decisions made solely by automated means.[2134] The National AI Policy builds on these rights with ethical principles emphasizing transparency, explainability, and accountability in the use of AI systems.[2135] However, Ethiopia has yet to adopt comprehensive AI legislation or regulatory guidance on algorithmic accountability that outlines enforcement mechanisms and technical standards.

Biometric Identification

Ethiopia is currently implementing a Digital ID Program,[2136] Fayda, as part of the Digital Ethiopia 2025 Strategy. Given the ongoing civil war, there are fears that

---

[2130] Federal Negarit Gazette, *Proclamation No. 1284/2023: Ethiopian Digital Identification Proclamation*, Citizenship Rights in Africa Initiative (Apr. 18, 2023), https://citizenshiprightsafrica.org/ethiopian-digital-identity-proclamation-number-1284-2023/

[2131] African Union, *List of Countries which Have Signed, Ratified/Acceded to the Convention on Cyber Security and Personal Data Protection* (Jul. 8, 2024), https://au.int/sites/default/files/treaties/29560-sl-AFRICAN_UNION_CONVENTION_ON_CYBER_SECURITY_AND_PERSONAL_DATA_PROTECTION.pdf

[2132] African Union, *Convention on Cyber Security and Personal Data Protection* (2014), https://au.int/sites/default/files/treaties/29560-treaty-0048_-_african_union_convention_on_cyber_security_and_personal_data_protection_e.pdf

[2133] Data Protection Africa, *Africa: AU's Malabo Convention Set to Enter Force after Nine Years* (May 19, 2023), https://dataprotection.africa/malabo-convention-set-to- enter-force/

[2134] Ministry of Justice, *Personal Data Protection Proclamation* (2024), https://justice.gov.et/en/law/personal-data-protection-proclamation/

[2135] Federal Democratic Republic of Ethiopia, *National Artificial Intelligence Policy* (Jun. 2024), https://www.lawethiopia.com/images/Policy_documents/Ethiopian%20ai%20policy.pdf

[2136] *National Id Ethiopia* (2026), https://id.gov.et/

the Digital ID Program could reinforce discrimination against ethnic minorities.[2137] The Digital ID Proclamation 1284/2023[2138] instigated the Fayda program. Over 200 key stakeholders representing Federal ministries and government agencies, regional governments, civil society organizations, local human rights organizations, and international institutions participated in a 2022 consultation on the National Identity Program (NIP).[2139]

The government has taken several initiatives to promote Fayda. Fayda is the main identification credential for all civil servants[2140] and biometric ID serves as the official student ID across educational institutions,[2141] working with UNICEF to support digital ID registration for children.[2142]

Although government officials told parliament in November 2023 that digital ID registration is not mandatory and that people can continue to use their analogue Kebele ID card, the digital ID became mandatory for all transactions with financial institutions[2143] and access to public services in 2024.[2144]

The program is being implemented with $350 million funding from the World Bank's International Development Association (IDA), from which $10 million was disbursed as of August 2024.[2145] The program's goal is for 90 million Ethiopians and legal residents to hold the Fayda digital ID by January 2029.[2146] As of August 2024, Fayda has seen 8.2 million registrations. The Personal Data Protection Proclamation

---

[2137] Zecharias Zelalem, *Ethiopia Digital ID Prompts Fears of Ethnic Profiling*, Context (Jan. 31, 2023), https://www.context.news/surveillance/ethiopia-digital-id-prompts- fears-of-ethnic-profiling

[2138] *National Id Ethiopia* (2023), https://id.gov.et/law

[2139] Citizenship Rights in Africa Initiative, *First Ethiopian Digital Identification Draft Proclamation Public Consultation* (Mar. 15, 2022), https://citizenshiprightsafrica.org/first- ethiopian-digital-identification-draft-proclamation-public-consultation/

[2140] Biometric Update, *Ethiopia Makes Fayda the Main Credential for Civil Servants* (Sept. 18, 2023), https://www.biometricupdate.com/202309/ethiopia-makes-fayda-the-main-credential-for-civil-servants

[2141] Biometric Update, *Ethiopia Rolls Out Student IDs, Integrates Biometric Data to Issue Fayda* (Sept. 14, 2023), https://www.biometricupdate.com/202309/ethiopia-rolls-out-student-ids-integrates-biometric-data-to-issue-fayda

[2142] Biometric Update, *UNICEF Pens Deal to Support Ethiopia's Digital ID Registration for Children* (Oct. 16, 2023), https://www.biometricupdate.com/202310/unicef-pens-deal-to-support-ethiopias-digital-id-registration-for-children

[2143] Biometric Update, *Ethiopia to Make Digital ID Obligatory for Banking Operations* (Jul. 12, 2023), https://www.biometricupdate.com/202307/ethiopia-to-make- digital-id-obligatory-for-banking-operations

[2144] Biometric Update, *Ethiopia to Make Digital ID Compulsory for Access to Government Services* (Jan. 15, 2024), https://www.biometricupdate.com/202401/ethiopia-to-make-digital-id-compulsory-for-access-to-government-services

[2145] World Bank Group, *The Transformative Power of Ethiopia's Digital ID: Unlocking a Better Future for All* (Feb. 27, 2025), https://www.worldbank.org/en/news/feature/2025/02/27/the-transformative-power-of-ethiopia-afe-digital-id-unlocking-a-better-future-for-all

[2146] Biometric Update, *Ethiopia Hopes to Take Digital ID Issuance to 63M by Year-End* (Oct. 14, 2025), https://www.biometricupdate.com/202510/ethiopia-hopes-to-take-digital-id-issuance-to-63m-by-year-end

is an important step to safeguard data and digital privacy as this digital ID registration increases.

Ethiopia is also rolling out the Fayda program to the country's refugee population, currently focus on 77,000 refugees living in Addis Ababa.[2147] The program's goal is to extend digital ID access to the over 1 million refugees in Ethiopia, primarily individuals from South Sudan, Somalia, Eritrea, and Sudan. This digital ID expansion is supported in partnership with the National ID Program (NIDP) and UNHCR, the UN Refugee Agency. Access to this ID card will help refugees engage in financial transactions, enroll in schools, and obtain SIM cards.

Environmental Impact of AI

Ethiopia supports the African Union Continental AI Strategy, which emphasizes sustainable development. However, Ethiopia has not indicated specific policies or projects to curb the emissions and other environmental impacts of AI systems and infrastructure.

Lethal Autonomous Weapons

Ethiopia has not acceded to the Convention on Certain Conventional Weapons (CCW) nor to any of its Protocols.[2148] The country has a mixed voting record in the UN on resolutions related to Lethal Autonomous Weapons Systems (LAWS). However, the country was party to a working paper submitted on behalf of the Non-Aligned Movement and other Parties to the CCW to the GGE on LAWS, which stated that "a strengthened and reinforced multilateral approach, with new legally-binding provisions for addressing the humanitarian and international security challenges posed by emerging technologies in the area of LAWS, is vital. There is an urgent need to pursue a legally-binding instrument on LAWS."[2149]

At the 79th UN General Assembly First Committee in November 2024, Ethiopia voted in favor of Resolution L.77[2150] on autonomous weapons systems, in

---

[2147] The UN Refugee Agency, *Inclusion in Ethiopia's ID System Opens New Doors for Refugees,* (May 24, 2024), https://www.unhcr.org/us/news/stories/inclusion-ethiopia-s-id-system-opens-new-doors-refugees

[2148] UN Office for Disarmament Affairs, *High Contracting Parties and Signatories CCW* (1980), https://disarmament.unoda.org/en/our-work/conventional-arms/convention-certain-conventional-weapons/high-contracting-parties-and-signatories-ccw

[2149] Non-Aligned Movement Geneva Chapter Coordinator of the Group of NAM and other states to the CCW, *Convention on Prohibitions or Restrictions on the Use of Certain Conventional Weapons which May Be Deemed to Be Excessively Injurious or to Have Indiscriminate Effects* (2021), https://documents.unoda.org/wp-content/uploads/2021/06/NAM.pdf

[2150] UN General Assembly, *Lethal autonomous weapons systems draft resolution L.77* (Oct. 18, 2024), https://docs.un.org/en/A/C.1/79/L.77

coherence with its vote in favor of Resolution L.56[2151] in 2023.[2152] Ethiopia did not contribute to the Secretary General report on LAWS[2153] mandated by Resolution L.56. Resolution L.77 emphasizes the urgency of addressing the challenges and concerns outlined in the report as well as the necessity to make the process as inclusive and participative as possible. Ethiopia was among only 6 member states to abstain[2154] from voting Resolution L.43[2155] on Artificial intelligence in the military domain and its implications for international peace and security in 2025.

AI Literacy

Ethiopia's National AI Policy commits to broad-based AI literacy and to building institutional capacity for oversight. It mandates activities to "increase the AI awareness of the society" and to build citizens' ability to use AI through practice-oriented education, alongside lifelong learning and short-/medium-term training programs to expand a skilled talent pool.[2156]

Human Rights

Ethiopia has supported or is party to a number of international and regional human rights instruments, including the Universal Declaration of Human Rights, the International Covenant on Civil and Political Rights,[2157] the Convention of the Rights of the Child,[2158] and the African Charter on Rights and Welfare of the Child.[2159] These human rights instruments ratified by Ethiopia form an 'integral part' of the laws of the country, according to Article 9 of its Constitution.

[2151] UN General Assembly, *Lethal Autonomous Weapons, Resolution L56* (Oct. 12, 2023), https://reachingcriticalwill.org/images/documents/Disarmament-fora/1com/1com23/resolutions/L56.pdf
[2152] Automated Decision Research, *State Position – Ethiopia*, https://automatedresearch.org/news/state_position/ethiopia/
[2153] UN General Assembly, *Lethal Autonomous Weapons Systems, Report of the Secretary General* (Jul. 1, 2024), https://docs.un.org/en/A/79/88
[2154] United Nations General Assembly, *General and Complete Disarmament Report of the First Committee*, pp. 45–54 (Nov. 15, 2024), https://docs.un.org/en/A/79/408
[2155] United Nations General Assembly, *Lethal Autonomous Weapons Systems Draft Resolution L.43* (Oct. 16, 2024), https://docs.un.org/en/A/C.1/79/L.43
[2156] Federal Democratic Republic of Ethiopia, *National Artificial Intelligence Policy*, Section 4.1, Human Resource Development, Articles 4.1.1–4.1.3 (2024), https://www.lawethiopia.com/images/Policy_documents/Ethiopian%20ai%20policy.pdf
[2157] United Nations (1966), https://www.ohchr.org/en/instruments-mechanisms/instruments/international-covenant-civil-and-political-rights
[2158] United Nations (1989), https://www.ohchr.org/en/instruments-mechanisms/instruments/convention-rights-child
[2159] African Union, *African Charter on the Rights and Welfare of the Child* (1990), https://au.int/sites/default/files/treaties/36804-treaty-african_charter_on_rights_welfare_of_the_child.pdf

The human rights landscape in Ethiopia is marked by significant challenges, as reflected in the Freedom House 2025 Freedom in the World report, which assigns Ethiopia a score of 18/100.[2160] The "Not Free" classification comes from a lack of political rights and civil liberties given ongoing political, social, and ethnic tensions. This score underscores the urgent need for reform and accountability in the country, particularly around political repression, ethnic tensions and violence, media freedom, and restrictions on humanitarian aid. The ongoing conflict in Amhara and Oromia along with the declaration of the state of emergency between august 2023 and June 2024 resulted, according to the Office of the High Commission for Human Rights, in "actions by security forces in violation of Ethiopia's international human rights obligations."[2161] Human Rights Watch reports[2162] that both parties have committed war crimes and abuses resulting in serious civilian losses as well as destruction of health facilities in Amhara.

OECD / G20 AI Principles

Ethiopia is not listed among the countries that have formally endorsed the OECD AI Principles and G20 AI Guidelines.[2163] However, as a member of the African Union, which became a permanent member of the G20 in 2023,[2164] Ethiopia participates in emerging global dialogues on responsible AI governance.

Council of Europe AI Treaty

Ethiopia has not signed the Council of Europe Framework Convention on Artificial Intelligence and Human Rights, Democracy, and the Rule of Law.[2165]

[2160] Freedom House, *Freedom in the World 2025: Ethiopia* (2025), https://freedomhouse.org/country/ethiopia/freedom-world/2025

[2161] United Nations Office of the High Commissioner for Human Rights, *UN Human Rights Chief calls for sustained efforts to halt violations and abuses* (Jun. 14, 2024), https://www.ohchr.org/en/press-releases/2024/06/ethiopia-un-human-rights-chief-calls-sustained-efforts-halt-violations-and

[2162] Human Rights Watch, *World Report 2025, Ethiopia Events of 2024*, https://www.hrw.org/world-report/2025/country-chapters/ethiopia

[2163] OECD AI Policy Observatory, *AI Principles: Countries adhering to the AI Principles* (2026), https://oecd.ai/en/ai-principles

[2164] G20, *New Delhi Leaders' Declaration, Welcoming the African Union as a Permanent Member* (Sept. 9–10, 2023), https://www.mea.gov.in/Images/CPV/G20-New-Delhi-Leaders-Declaration.pdf

[2165] Council of Europe Treaty Office, *Chart of Signatures and Ratifications of Treaty 225* (Mar. 12, 2025), https://www.coe.int/en/web/conventions/full-list?module=signatures-by-treaty&treatynum=225

UNESCO Recommendation on AI Ethics

As a member of UNESCO, Ethiopia along with 192 other member states, adopted the UNESCO Recommendation on the Ethics of AI, the first global standard on the ethics of AI.[2166]

The Deputy General Director, Research and Development Cluster, from the Ethiopian Artificial Intelligence Institute participated in the June 2024 UNESCO-Eastern Africa Sub-Regional Forum on Artificial Intelligence (EARFAI). EARFAI focused on the potential for AI to benefit sustainable development and societal change in Eastern Africa, including discussions on the landscape of AI governance.[2167] A major theme of EARFAI involved discussions of implementation of UNESCO's Recommendation in Eastern African states.

The Readiness Assessment Methodology (RAM) is a tool to guide Member States in identifying institutional and regulatory gaps that may inhibit development of an AI ecosystem aligned with the Recommendation. The RAM is currently in process in Ethiopia.[2168]

Evaluation

Ethiopia has initiated the development of a governance framework for AI, as outlined in its National AI Policy. The strategy aims to establish regulatory measures that align with international standards. However, the framework lacks comprehensive regulations to ensure accountability and transparency in AI applications and an effective organized roadmap for implementation. These gaps present risks related to misuse and ethical concerns. For example, while the National AI Strategy and AU Continental Strategy include commitments to fairness, accountability, and transparency, implementation of these ethical guidelines remains inconsistent, with limited mechanisms to monitor AI systems for bias or discrimination.

Ethiopia is beginning to address issues of privacy, data protection, and individual rights in the digital landscape. For example, the Ethiopian Communications Authority is charged with enforcing the data protection proclamation. However, legal frameworks are still evolving, and there is a pressing need to strengthen laws that safeguard citizens' rights in the face of increasing surveillance and data collection. Fears remain that the Digital ID Program holds the potential to become a tool of mass surveillance in a country with a historically challenging human rights landscape.

Ethiopia's progress in AI is noteworthy. Nevertheless, significant challenges remain in governance, ethical implementation, capacity building, public engagement, and the protection of digital rights.

[2166] UNESCO, *UNESCO Adopts First Global Standard on the Ethics of Artificial Intelligence* (2023), https://www.unesco.org/en/articles/unesco-adopts-first-global-standard-ethics-artificial-intelligence

[2167] UNESCO, *Programme: UNESCO-Eastern Africa Sub-Regional Forum on Artificial Intelligence (EARFAI)* (Jun. 24–26, 2024), https://www.unesco.org/en/earfai/programme?hub=149901

[2168] UNESCO Global AI Ethics and Governance Observatory, *Global Hub* (Oct. 2025), https://www.unesco.org/ethics-ai/en/global-hub

Finland

In 2025, Finland passed legislation designating supervisory authorities for implementation of the EU AI Act and released a proposal on regulatory sandboxes and an AI registry aligned to the second phase of AI Act implementation. The ethical recommendations for AI in education released in 2025 align with the UNESCO Recommendation and OECD AI Principles around digital competence.

National AI Strategy

In 2017, Finland was among the first countries to develop a national AI strategy, establishing proposed target dates and allocating public funds in furtherance of the country's AI-related business objectives. The vision put forth by the national strategy is that "In another five years time, artificial intelligence will be an active part of every Finn's daily life. Finland will make use of artificial intelligence boldly in all areas of society—from health care to the manufacturing industry—ethically and openly. Finland will be a safe and democratic society that produces the world's best services in the age of artificial intelligence. Finland will be a good place for citizens to live and a rewarding place for companies to develop and grow. Artificial intelligence will reform work as well as create wellbeing through growth and productivity."[2169] The AI strategy demonstrated Finland's objectives to: (1) enhance business competitiveness using AI, (2) ensure top-level expertise and attract top experts, (3) provide the world's best public services, and (4) make Finland a front runner in the age of AI.[2170]

To that end, in May 2017, Finland's Minister of Economic Affairs launched Finland's AI Programme,[2171] an operational program designed for "turning Finland into a leading country in the application of artificial intelligence." The Programme focused on three areas: an efficient public sector, a well-functioning society, and a competitive business and industry sector.[2172]

[2169] Ministry of Economic Affairs and Employment, *Finland's Age of Artificial Intelligence: Turning Finland into a Leading Country in the Application of Artificial Intelligence, Objective and Recommendations for Measures* (2017), https://julkaisut.valtioneuvosto.fi/bitstream/handle/10024/160391/TEMrap_47_2017_verkkojulkaisu.pdf

[2170] OECD, *State of Implementation of the OECD AI Principles* (Jun. 18, 2021), https://www.oecd.org/content/dam/oecd/en/publications/reports/2021/06/state-of-implementation-of-the-oecd-ai-principles_38a4a286/1cd40c44-en.pdf

[2171] Finnish Government, *Turning Finland into a Leading Country in the Age of Artificial Intelligence* (Mar. 14, 2018), https://valtioneuvosto.fi/en/-/1410877/raportti-suomi-ponnistaa-tekoalyajan-karkimaaksi

[2172] Ministry of Economic Affairs and Employment, *Leading the Way into the Age of Artificial Intelligence: Final Report of Finland's Artificial Intelligence Programme* (Jun. 12, 2019), https://julkaisut.valtioneuvosto.fi/bitstream/handle/10024/161688/41_19_Leading%20the%20way%20into%20the%20age%20of%20artificial%20intelligence.pdf

The Minister of Economic Affairs appointed a steering committee, which included representatives from the public, private, and research sectors, and charged the steering committee with publishing a report containing its recommendations regarding the operationalization of Finland's AI Programme. Within the steering committee, the Minister of Economic Affairs established four subgroups focused on four (4) key areas: (1) expertise and innovations; (2) data and platform economy; (3) transformation of work and society; and (4) ethics. The steering committee published three reports. The first, released in 2017, included recommendations to facilitate Finland's objective to adopt and benefit from AI.[2173] The second report examined (1) the impact of artificial intelligence on growth and employment; (2) labor market dynamics in a technological revolution; (3) learning and skills in a transition; and (4) good application of artificial intelligence technology and ethics.[2174] The subgroup identified three key "values of a good artificial intelligence society"—transparency, responsibility, and extensive societal benefits—and stresses that "humans [must] assume ultimate legal and moral responsibility for the decisions."[2175]

The third report detailed the steering committee's policy recommendations for Finland's AI Programme,[2176] setting forth the contours of Finland's vision of a country that, by 2025, "is competitive and able to attract talent and has the most relevantly educated population and where citizens are well-informed and independent."[2177]

Nordic-Baltic and Nordic Cooperation on AI

As for the regional landscape, the Finnish Minister in charge of digitalization signed the declaration on "AI in the Nordic-Baltic region" in 2018, establishing a collaborative framework on "developing ethical and transparent guidelines, standards, principles and values to guide when and how AI applications should be used" and "on the objective that infrastructure, hardware, software and data, all of which are central to the use of AI, are based on standards, enabling interoperability, privacy, security, trust, good usability, and portability. In 2024, a proposal for a shared vision for Nordic

---

[2173] Ministry of Economic Affairs and Employment, *Finland's Age of Artificial Intelligence: Turning Finland into a Leading Country in the Application of Artificial Intelligence, Objective and Recommendations for Measures* (2017), https://julkaisut.valtioneuvosto.fi/bitstream/handle/10024/160391/TEMrap_47_2017_verkkojulkaisu.pdf

[2174] The report cautioned that "The conclusions of the report do not necessarily represent the group's joint views," but "do represent a majority opinion."

[2175] Ministry of Economic Affairs and Employment, *Work in the Age of Artificial Intelligence: Four Perspectives on the Economy, Employment, Skills and Ethics*), p.50 (Sep. 10, 2018), https://julkaisut.valtioneuvosto.fi/bitstream/handle/10024/160980/TEMjul_21_2018_Work_in_the_age.pdf

[2176] Ministry of Economic Affairs and Employment of Finland, *Leading the Way into the Age of Artificial Intelligence: Final Report of Finland's Artificial Intelligence Programme* (Jun. 12, 2019), http://julkaisut.valtioneuvosto.fi/bitstream/handle/10024/161688/41_19_Leading%20the%20way%20into%20the%20age%20of%20artificial%20intelligence.pdf

[2177] Ibid, p. 12

AI was discussed by high-level decision-makers from the private and public sectors with Nordic and Baltic ministers in Copenhagen, where the vision outlined was for large-scale AI adoption, with tangible benefits for both citizens and businesses.[2178]

The ministerial declaration Digital North 2.0[2179] builds on the common priorities of the Nordic-Baltic countries, and follows the previous ministerial declaration, Digital North 2017-2020. "In order to promote work with digitalisation, co-ordinate efforts, and follow up on the goals of the declaration, a council of ministers for digitalisation (MR-DIGITAL) was established in 2017. The aim is to promote development in three areas: (1) Increase mobility and integration in the Nordic and Baltic region by building a common area for cross-border digital services; (2) Promote green economic growth and development in the Nordic-Baltic region through data-driven innovation and a fair data economy for efficient sharing and re-use of data; and (3) Promote Nordic-Baltic leadership in the EU/EEA and globally in a sustainable and inclusive digital transformation of our societies."[2180]

In November 2021, the Nordic and Baltic ministers for digitalization released another joint statement announcing a focus on digital inclusion, striving to implement measures to make digital services more accessible to all Finnish inhabitants and ensuring that those who do not possess the necessary level of skills get the opportunity to acquire them. [2181]

The Nordic and Baltic ministers of digitalization issued a common statement on the importance of cooperation on digital security in the Nordic-Baltic region in September 2022, following the COVID-19 pandemic and the war in Ukraine. In their common statement, the ministers stressed that this "rapid transformation has challenged everyone to adapt to new, digital ways of doing business, learning and accessing public authorities." The ministers declared that they "have committed to ensuring that our region maintains its position as a leader in digitalisation, and that everyone in the region benefit from digitalisation regardless of age, wealth, education or level of digital skills. One important factor that helps ensure a strong level of digitalisation in the region is the trust citizens put in digital services from the public sector—be it at regional, national or local level. In order to keep up this high level of trust, we need to continue our efforts to make our digital public services human centric and accessible. [...] Robust and secure digital services, safeguarding users' privacy

---

[2178] Nordic Cooperation, *A Shared Nordic Future with Artificial Intelligence* (Aug. 30, 2024), https://www.norden.org/en/news/shared-nordic-future-artificial-intelligence

[2179] Nordic and Baltic Ministers of Digitalization, *Ministerial Declaration Digital North 2.0* (Jul. 14, 2023), https://dk.usembassy.gov/u-s-nordic-leaders-summit-in-helsinki/

[2180] Nordic Co-operation, *Nordic-Baltic Co-operation on Digitalisation*, https://www.norden.org/en/information/nordic-baltic-co-operation-digitalisation

[2181] Nordic and Baltic Ministers of Digitalization, *Common Statement on the Importance of Promoting Digital Inclusion as a Central Part of the Digital Transformation in the Nordic-Baltic Region* (Nov. 26, 2021), https://www.norden.org/en/declaration/common-statement-importance-promoting-digital-inclusion-central-part-digital

and ensuring that personal data are stored and processed in a trustworthy way, are crucial to the citizens' sustained trust in digital services."[2182]

As part of its action plan for Vision 2030 (2021–2024), the Nordic Council of Ministers also identified innovation, digital integration, the safe use of artificial intelligence, data development and open data, education and digitalization as key objectives.[2183] A new initiative for research collaboration on ethical and responsible use of AI was documented in March 2024 as a discussion paper.[2184] There are suggestions for a regional funding scheme and to identify instruments and thematic scoping relevant to Nordic research funders. Further, the suggested Nordic initiative not just understands the consequences and potential of AI in Nordic societies (Nordic value added, EU AI Act, data biases, cross-cutting strategic priorities) but uses thematic calls to "identify, amplify and fully capitalize on Nordic strengths and provide grounds for transfer of knowledge and lessons learned between sectors."

A Finnish-Estonian Declaration introduced in February 2025 emphasized AI, digitalization, and media literacy as essential civic skills.[2185] The declaration aligns with OECD's digital competence goals, supporting AI-driven education, and upholding the UN's recognition of information access as a fundamental right.

Digital Services Act

As an EU member state, Finland shall apply the EU Digital Services Act (DSA).[2186] The DSA regulates online intermediaries and platforms. Its main objective is to prevent illegal and harmful activities online and the spread of disinformation.

[2182] Nordic and Baltic Ministers of Digitalization, *Common Statement on the Importance of Cooperation on Digital Security in the Nordic-Baltic Region* (Sept. 6, 2022), https://www.norden.org/en/declaration/common-statement-importance-cooperation-digital-security-nordic-baltic-region

[2183] Nordic Council of Ministers, *The Nordic Region: Toward Being the Most Sustainable and Integrated Region in the World, Action Plan for 2021–2024* (Dec. 14, 2020), https://www.norden.org/en/publication/nordic-region-towards-being-most-sustainable-and-integrated-region-world

[2184] NordForsk, *A Nordic Initiative for Research and Innovation on Responsible and Ethical Use of Artificial Intelligence* (May 8, 2024), https://www.norden.org/en/publication/nordic-initiative-research-and-innovation-responsible-and-ethical-use-artificial

[2185] Republic of Estonia Ministry of Education and Research, *Estonian—Finnish Declaration Embracing AI and Media Literacy in Education to Safeguard Democracy and the Rule of Law* (Feb. 18, 2025), https://hm.ee/sites/default/files/documents/2025-02/EE-FI%20hisavaldus%20final%2018.02.2025.pdf

[2186] EUR-Lex, *Regulation (EU) 2022/2065 of the European Parliament and of the Council of 19 October 2022 on a Single Market for Digital Services and amending Directive 2000/31/EC (Digital Services Act)* (Oct. 21, 2022), https://eur-lex.europa.eu/legal-content/EN/TXT/?uri=celex%3A32022R2065

The Finnish Transport and Communications Agency, Traficom, is the DSA main supervisor.[2187] The Consumer Ombudsman and the Data Protection Ombudsman supervise specific other obligations. The Finnish authorities received a total of 78 complaints under the DSA in 2024.[2188]

Signatories also commit to reducing the risk of online harms under the 2022 Strengthened Code of Practice on Disinformation,[2189] which counts as a mitigation measure under the DSA. Actions include demonetizing the dissemination of disinformation, ensuring the transparency of political advertising, empowering users, enhancing the cooperation with fact-checkers, and providing researchers with better access to data.

EU AI Act

As an EU member State, Finland is bound by the EU AI Act,[2190] which entered into force on August 1, 2024.[2191] The EU AI Act is a risk-based market regulation that supports the objective of promoting a human-centric approach to AI and making the EU a global leader in the development of secure, trustworthy, and ethical AI. Finland has appointed members to the European AI Board and identified 8 authorities protecting fundamental rights under the act.[2192] A Working Group on national implementation with a charge through June 2026 drafted national legislation to implement the act and determine necessary actions.[2193] A law designating ministries

---

[2187] Finnish Transport and Communications Agency (Traficom), *Questions and Answers about the Digital Services Act*, https://www.traficom.fi/en/communications/data-economy-and-digital-services/questions-and-answers-about-digital-services-act

[2188] Office of the Data Protection Ombudsman, *Digital Services Act in Force for a Year—Nearly 80 Complaints to Authorities in Finland in 2024* (Feb. 18, 2025), https://tietosuoja.fi/en/-/digital-services-act-in-force-for-a-year-nearly-80-complaints-to-authorities-in-finland-in-2024

[2189] European Commission, *The 2022 Code of Practice on Disinformation*, https://digital-strategy.ec.europa.eu/en/policies/code-practice-disinformation

[2190] European Parliament, *Artificial Intelligence Act, European Parliament legislative resolution of 13 March 2024 on the proposal for a regulation of the European Parliament and of the Council on laying down harmonised rules on Artificial Intelligence (Artificial Intelligence Act) and amending certain Union Legislative Acts (COM(2021)0206 – C9-0146/2021 – 2021/0106(COD))*, P9_TA(2024)0138, https://www.europarl.europa.eu/RegData/seance_pleniere/textes_adoptes/definitif/2024/03-13/0138/P9_TA(2024)0138_EN.pdf

[2191] European Commission, *European Artificial Intelligence Act Comes into Force*, Press Release (Jul. 31, 2024), https://ec.europa.eu/commission/presscorner/detail/en/ip_24_4123

[2192] Ministry of Economic Affairs and Employment of Finland, *National Implementation of EU Artificial Intelligence Regulation*, https://tem.fi/en/ai-regulation

[2193] Ministry of Employment and the Economy, *Working Group on the National Implementation of the EU Artificial Intelligence Regulation [EU:n tekoälyasetuksen kansallisen toimeenpanon työryhmä]* (2024), https://valtioneuvosto.fi/hanke?tunnus=TEM044:00/2024

and agencies as supervisory authorities according to applications[2194] entered into force at the start of 2026.[2195] Another proposal sandboxes and a national register underwent public consultation in 2025.[2196]

The Finnish Data Protection Authority is one of the market surveillance authorities for high-risk systems, though the Finnish Transport and Communications Agency (Traficom) will act as the single point of contact.[2197]

Public Participation

Finland's longstanding and broad commitment to an open democracy has traditionally been given expression by extensive consultation with established groups.[2198] The Finnish Constitution also states that "democracy entails the right of the individual to participate in and influence the development of society and his or her living conditions." Provisions on consultation and participation are given further weight in various laws and guidelines including the Act on the Openness of Government Activities.

In 2024, the Finnish government requested an opinion on the draft of the government's proposal for legislation on the implementation of the artificial intelligence regulation.[2199] The draft seeks to advance the implementation of the EU's artificial intelligence regulation, regarding the supervision of certain artificial intelligence systems. The proposed law sets the competent market surveillance authorities and the authorities with the competence to receive and process notifications. The bill also proposed the amendment of the law on the market control of certain products (Law 1137/2016) to include within its scope.

[2194] Ministry of Employment and the Economy, *Government Proposal to Parliament for Legislation Supplementing the EU Artificial Intelligence Regulation*, HE 46/2025vp (May 8, 2025), https://tem.fi/en/ai-regulation

[2195] Ministry of Economic Affairs and Employment, *National Supervision of EU Artificial Intelligence Act to Begin—Laws on Powers of Authorities to Take Effect at Start of the Year* (Dec. 22, 2025), https://tem.fi/en/-/national-supervision-of-eu-artificial-intelligence-act-to-begin-laws-on-powers-of-authorities-to-take-effect-at-start-of-the-year

[2196] Ministry of Employment and the Economy, *The Second Phase of the Implementation of the EU Artificial Intelligence Regulation Is Open for Comments* (Apr. 25, 2025), https://tem.fi/-/eu-n-tekoalyasetuksen-toimeenpanon-toinen-vaihe-lausuntokierrokselle

[2197] FinLex, *1377/2025 Act on the Supervision of Certain Artificial Intelligence Systems [Laki eräiden tekoälyjärjestelmien valvonnasta]* (Dec. 22, 2025), https://www.finlex.fi/fi/lainsaadanto/saadoskokoelma/2025/1377

[2198] OECD, *Better Regulation in Europe: Finland, Transparency through Consultation and Communication*, p. 71 (May 27, 2010), https://www.oecd-ilibrary.org/governance/better-regulation-in-europe-finland-2010_9789264085626-en

[2199] Out letande.fi, *Request for Opinion: Draft Government Bill with Proposed Legislation on the Implementation of the Artificial Intelligence Regulation [Begäran om utlåtande: utkast till regeringens proposition med förslag till lagstiftning om genomförande av förordningen om artificiell intelligens]* (Dec. 4, 2024), https://www.lausuntopalvelu.fi/SV/Proposal/Participation?proposalId=0e252297-c14b-4b6b-a0da-0a35756c9a90

Data Protection

Since Finland is an EU Member State, the General Data Protection Regulation (GDPR)[2200] is directly applicable in Finland and to Finnish people. The aim of the GDPR is to "strengthen individuals' fundamental rights in the digital age and facilitate business by clarifying rules for companies and public bodies in the digital single market."[2201] The Finnish Data Protection Act supplements the GDPR.[2202]

Regarding the activities of law enforcement authorities, Finland transposed the EU Data Protection Law Enforcement Directive (LED)[2203] through the Act on the Processing of Personal Data in Criminal Cases and in connection with Maintaining National Security, which entered into force on January 1, 2019, along with the Data Protection Act. The LED "will in particular ensure that the personal data of victims, witnesses, and suspects of crime are duly protected and will facilitate cross-border cooperation in the fight against crime and terrorism."[2204] The LED provides for the prohibition of any decision based solely on automated processing, unless it is provided by law, and of profiling that results in discrimination.[2205]

Finland is also a member of the Council of Europe and ratified the Council of Europe's Convention 108+ for the protection of individuals with regard to the processing of personal data.[2206]

The Data Protection Ombudsman is the national supervisory authority charged with overseeing compliance with data protection legislation and safeguards the rights and freedoms of individuals with regard to the processing of their personal data.[2207]

[2200] *Regulation (EU) 2016/679 of the European Parliament and of the Council of 27 April 2016 on the protection of natural persons with regard to the processing of personal data and on the free movement of such data*, https://eur-lex.europa.eu/legal-content/EN/TXT/PDF/?uri=CELEX:32016R0679

[2201] European Commission, *Legal Framework of EU Data Protection* (2025), https://commission.europa.eu/law/law-topic/data-protection/data-protection-eu_en

[2202] Ministry of Justice of Finland, *Data Protection Act (1050/2018)*, https://www.finlex.fi/en/legislation/translations/2018/eng/1050?language=eng; complete text https://www.finlex.fi/api/media/statute-foreign-language-translation/687709/mainPdf/main.pdf?timestamp=2018-12-04T22%3A00%3A00.000Z

[2203] *Directive (EU) 2016/680 of the European Parliament and of the Council of 27 April 2016 on the protection of natural persons with regard to the processing of personal data by competent authorities for the purposes of the prevention, investigation, detection or prosecution of criminal offences or the execution of criminal penalties, and on the free movement of such data, and repealing Council Framework Decision 2008/977/JHA*, https://eur-lex.europa.eu/legal-content/EN/TXT/?uri=celex%3A02016L0680-20160504

[2204] European Commission, *Legal Framework of EU Data Protection*, https://commission.europa.eu/law/law-topic/data-protection/legal-framework-eu-data-protection_en

[2205] Article 11 (1) and (2) of the LED, https://eur-lex.europa.eu/legal-content/EN/TXT/?uri=celex%3A02016L0680-20160504

[2206] Council of Europe, *Convention 108 and Protocols* (2026), https://www.coe.int/en/web/data-protection/convention108-and-protocol

[2207] Office of the Data Protection Ombudsman of Finland, https://tietosuoja.fi/en/home

The Ombudsman has made numerous important decisions,[2208] including with administrative fines amounting to several hundred thousand Euros.[2209] The institutional mission of the Data Protection Ombudsman is expected to be broadened in alignment with the draft proposal for the Finnish EU AI Act national regulation. The recent attributions encompass supervisory authority over prohibited AI systems, as well as numerous high-risk AI systems, addressing sensitive areas such as biometrics, law enforcement, migration, education, and employment.

Concerning this aspect, the Finnish Data Protection Ombudsman and current EDPB chair, Anu Talus, asserted the importance of the cooperation between authorities for the EU AI Act Enforcement, pointing out that "the debate on artificial intelligence in particular continued to be intense and is being even more so in 2024. It is important to ensure that cooperation between authorities in the enforcement of the EU's AI Act is smooth at both national and European level."[2210]

Despite being a member of the Global Privacy Assembly (GPA) since 2002, the Data Protection Ombudsman has not endorsed the 2018 GPA Declaration on Ethics and Data Protection in Artificial Intelligence,[2211] 2020 GPA Resolution on AI Accountability,[2212] 2022 GPA Resolution on Facial Recognition Technology,[2213] or 2023 GPA Resolution on Generative AI.[2214]

[2208] European Data Protection Board, *News on Finland*, https://edpb.europa.eu/news/news_en?news_type=All&field_edpb_member_states_target_id=76

[2209] European Data Protection Board, *Administrative Fine Imposed On Psychotherapy Centre Vastaamo For Data Protection Violations* (Dec 7, 2021), https://edpb.europa.eu/news/national-news/2022/administrative-fine-imposed-psychotherapy-centre-vastaamo-data-protection_en European Data Protection Board, *Finnish SA: Administrative Fine on Viking Line for Unlawful Processing of Employees' Health Data* (Dec. 9, 2022), https://edpb.europa.eu/news/national-news/2023/finnish-sa-administrative-fine-viking-line-unlawful-processing-employees_en

[2210] Data Protection Ombudsman, *Annual Report of the Office of the Data Protection Ombudsman 2023: Data Protection Work in a Changing Digital World*, Press Release (Apr. 14, 2024), https://tietosuoja.fi/en/-/annual-report-of-the-office-of-the-data-protection-ombudsman-2023-data-protection-work-in-a-changing-digital-world

[2211] Global Privacy Assembly, *Declaration on Ethics and Data Protection in Artificial Intelligence* (Oct. 23, 2018), https://globalprivacyassembly.org/wp-content/uploads/2018/10/20180922_ICDPPC-40th_AI-Declaration_ADOPTED.pdf

[2212] Global Privacy Assembly, *Resolution on Accountability in the Development and Use of Artificial Intelligence* (Oct. 2020), https://globalprivacyassembly.org/wp-content/uploads/2020/10/FINAL-GPA-Resolution-on-Accountability-in-the-Development-and-Use-of-AI-EN-1.pdf

[2213] Global Privacy Assembly, *Resolution on Principles and Expectations for the Appropriate Use of Personal Information in Facial Recognition Technology* (Oct. 2022), https://globalprivacyassembly.org/wp-content/uploads/2022/11/15.1.c.Resolution-on-Principles-and-Expectations-for-the-Appropriate-Use-of-Personal-Information-in-Facial-Recognition-Technolog.pdf

[2214] Global Privacy Assembly, *Resolution on Generative Artificial Intelligence Systems* (Oct. 2023), https://globalprivacyassembly.org/wp-content/uploads/2023/10/5.-Resolution-on-Generative-AI-Systems-101023.pdf.

Algorithmic Transparency

Finland is subject to the GDPR-like Finnish Law and Convention 108+. Finns have a general right to obtain access to information about automated decision-making and to the factors and logic of an algorithm.[2215] Per a 2020 Recommendation of the Council of Europe Committee of Ministers on human rights impacts of algorithm systems, those rights ought to extend to the ability to contest a decision with adverse effects.[2216] In 2023, OECD recognized Finland and Helsinki for the Open AI Register of AI Systems use in the public sector as an example of OECD AI Principle 1.3 on transparency and explainability.[2217]

Finland, through the European Centre for Algorithmic Transparency (ECAT),[2218] has been actively participating in supporting and enforcing these standards, with particular emphasis on monitoring algorithmic impact across various sectors, from social media to public administration. The EU's DSA Implementing Regulation standardizes the format, content, and reporting periods for transparency reports, and providers will begin to collect data according to the Implementing Regulation as of July 1, 2025, with the first harmonized reports due at the beginning of 2026.[2219]

Digitization of Public Services

In 2020, Finland launched a National Artificial Intelligence Programme—Aurora AI, a decentralized open network and data-based model for smart public services and applications.[2220] However, core components and production of the service ended on December 31, 2023, owing to limited usage.[2221]

[2215] See Recital 63 and Article 22 of the GDPR. Article 9 c) of the Convention 108+ as well as Recital 77, *Explanatory Report, Convention 108+*, p. 24, https://rm.coe.int/convention-108-convention-for-the-protection-of-individuals-with-regar/16808b36f1

[2216] *Recommendation CM/Rec(2020)1 of the Committee of Ministers to Member States on the Human Rights Impacts of Algorithmic Systems* (Apr. 8, 2020), https://search.coe.int/cm/pages/result_details.aspx?objectid=09000016809e1154

[2217] OECD, *The State of Implementation of the OECD AI Principles Four Years On,* p. 38 (Oct. 2023), https://www.oecd-ilibrary.org/docserver/835641c9-en.pdf

[2218] European Commission, *About*: *European Centre for Algorithmic Transparency (ECAT)* (2024), https://algorithmic-transparency.ec.europa.eu/about_en

[2219] European Commission, *Implementing Regulation Laying Down Templates Concerning the Transparency Reporting Obligations of Providers of Online Pplatforms* (Nov. 4, 2024), https://digital-strategy.ec.europa.eu/en/library/implementing-regulation-laying-down-templates-concerning-transparency-reporting-obligations

[2220] United Nations, Department of Social and Economic Affairs, *AuroraAI Finland*, https://publicadministration.desa.un.org/good-practices-for-digital-government/compendium/aurora-ai-finland#

[2221] Finnish Government, *Maintenance of the AuroraAI Network's Core Components Will End on 31 December 2023* (Oct. 5, 2023), https://valtioneuvosto.fi/en/-/16079645/maintenance-of-the-auroraai-network-will-end

In September 2020, the city of Helsinki launched an AI registry in beta version to detail how city government uses algorithms to deliver services. The registry includes training datasets, "a description of how an algorithm is used, how humans utilize the prediction, and how algorithms were assessed for potential bias or risks" as well as a contact person responsible for deployment so citizens can provide feedback.[2222] The city of Helsinki describes it AI register as "a window into the artificial intelligence systems used by the City of Helsinki. Through the register, you can get acquainted with the quick overviews of the city's artificial intelligence systems or examine their more detailed information based on your own interests. You can also give feedback and thus participate in building human-centred AI in Helsinki."[2223] The city of Helsinki's public register for algorithmic transparency has eight use cases on the website, a majority of which are chatbots.[2224.]

In 2022, a new research project, Civic Agency in AI also started with the aim to help the public sector ensure that their AI tools are transparent, accountable, and equitable. Its purpose is to develop best practices and recommendations regarding AI governance.[2225]

In May 2023, general legislation on automatic decision-making in public administration entered into force, with an 18-month transition period.[2226] The legislation requires authorities to provide appropriate information about the use of an automated decision-making procedure. The legislation allows automated decision-making on administrative matters in so far as making a decision on the matter in question does not require individual consideration. Concerns exist with regard to its compatibility with Article 22(1) GDPR which provides for the right of the data subject not to be subject to a decision solely based on automated processing.

On the basis of this law, in October 2023, the Finnish Immigration Service adopted a decision to introduce automated decision making for students' first residence permits. One condition for the use of automated decision is that "the residence permit can be granted according to the application, and the matter does not require more extensive holistic deliberation."[2227]

[2222] Khari Johnson, *Amsterdam and Helsinki Launch Algorithm Registries to Bring Transparency to Public Deployments of AI* (Sept. 28, 2020), https://venturebeat.com/2020/09/28/amsterdam-and-helsinki-launch-algorithm-registries-to-bring-transparency-to-public-deployments-of-ai/

[2223] *City of Helsinki AI Register*, https://ai.hel.fi/en/get-to-know-ai-register/

[2224]City of Helsinki, *AI Register* (Nov. 15, 2024), https://ai.hel.fi/en/ai-register/

[2225] Finish Center for Artificial Intelligence (FCAI), *Participatory Research To Improve Artificial Intelligence Based Public Sector Services and Empower Citizens* (Feb 8, 2022), https://fcai.fi/news/2022/2/8/participatory-research-to-improve-artificial-intelligence-based-public-sector-services-and-empower-citizens

[2226] Ministry of Finance, *Assessment Criteria for Information Security in Public Administration: Recommendation and Criteria* (Nov. 14, 2022), https://julkaisut.valtioneuvosto.fi/items/4fee11a6-0f1a-45fc-9330-54efa9cc278f

[2227] Finnish Immigration Service, *Decision to Introduce Automated Decision Making: First Residence Permits of Students* (Oct. 18, 2023),

AI and Children

Finland and UNICEF have been collaborating to create internationally applicable policy guidance for the use and development of AI for children. The Ministry for Foreign Affairs supports the project, where practices are developed for the planning of safe and inclusive AI solutions that take the rights of the child into account.[2228]

Facial Recognition

According to news reports, Finland's National Bureau of Investigation has acknowledged using facial recognition technology in connection with certain law enforcement activities.[2229] After initially denying the use of facial recognition technology in media questioning, officials from the National Bureau of Investigation acknowledged that four members of the Child Exploitation Investigation Unit had conducted 120 searches of the Clearview AI system in 2019–2020.

The Deputy Data Protection Ombudsman issued a note to the National Bureau of Investigation regarding the controversial use of Clearview AI facial recognition technology. In September 2021, the Deputy Data Protection Ombudsman warned the National Bureau of Investigation that police officers had used a facial recognition technology system without first verifying that it complied with data security or data protection laws.[2230]

A Digital Travel Credentials biometric pilot funded by the European Union for border checks was launched in August 2023 at the Helsinki-Vantaa airport.[2231] Finland also began work on a national digital identity wallet in April 2024, in line with the EU regulation that introduces the European Digital Identity (EUDI)

---

https://migri.fi/documents/5202425/182258005/Decision+to+introduce+automated+decision+making+-+First+residence+permits+of+students.pdf

[2228] Ministry of Foreign Affairs, *Policy Guidance on AI for Children Piloted in Different Parts of the World* (Oct. 19, 2021), https://um.fi/current-affairs/-/asset_publisher/gc654PySnjTX/content/lapsiin-liittyvan-tekoalyn-pelisaantoja-pilotoitu-eri-puolilla-maailmaa

[2229] Finnish Government, *Testing of Facial Recognition Software by NBI Reported to Data Protection Ombudsman* (Apr. 9, 2021), https://valtioneuvosto.fi/en/-/25235045/testing-of-facial-recognition-software-by-nbi-reported-to-data-protection-ombudsman

[2230] European Data Protection Board, *Finnish SA: Police Reprimanded for Illegal Processing of Personal Data with Facial Recognition Software* (Oct. 7, 2021), https://www.edpb.europa.eu/news/national-news/2021/finnish-sa-police-reprimanded-illegal-processing-personal-data-facial_en

[2231] Biometric Update.com, *Tech5 Face Biometrics Deployed for Digital Travel Credential Pilot in Finland* (Mar. 12, 2024), https://www.biometricupdate.com/202403/tech5-face-biometrics-deployed-for-digital-travel-credential-pilot-in-finland

Wallet.[2232] Biometric data via facial recognition technologies are processed under the GDPR and the EU AI Act.[2233]

The Ministry of the Interior proposed amending the regulation on biometric data, including facial images, stored in the registers of the police and the Finnish Immigration Service to aid in crime prevention.[2234] The draft proposes to amend the provisions of the Police Personal Data Act, the Immigration Administration Personal Data Act, and the Aliens Act on the processing of biometric data to allow for the use of personal data beyond their original processing purpose. These amendments extend to biometric data of citizens of foreign nations. Public comments raised concerns over the potential for mass surveillance, especially real-time facial recognition, and the use of personal data beyond the original, consented purpose.[2235] The proposal passed to Parliament in September 2025.[2236]

Environmental Impact of AI

In November 2020, Finland launched its Artificial Intelligence 4.0 strategy, where the objective is to be a frontrunner in both green and digital transitions.[2237] According to an interim report[2238] issued in December 2021, Finland plans to be carbon neutral by 2035. Finland also plans to reinforce the EU's Green Deal,[2239] which

[2232]Biometric Update.com, *Finland Starts Work on National Digital Identity Wallet* (Apr. 30, 2024), https://www.biometricupdate.com/202404/finland-starts-work-on-national-digital-identity-wallet

[2233] Chambers & Partners, *Artificial Intelligence 2024: Finland* (2024), https://chambers.com/downloads/gpg/996/008_finland.pdf

[2234] Ministry of the Interior, *Proposal to Amend Regulation on Biometric Data in Crime Prevention Circulated for Comments* (Feb. 3, 2025), https://intermin.fi/en/-/proposal-to-amend-regulation-on-biometric-data-in-crime-prevention-circulated-for-comments

[2235] lausuntopalvelu.fi [Statement Service], *Request for Opinion on the Proposal to Amend the Regulation on Biometric Data Stored in the Registers of the Police and the Finnish Immigration Service [Lausuntopyyntö ehdotuksesta poliisin ja Maahanmuuttoviraston rekistereihin tallennettuja biometrisia tietoja koskevan sääntelyn muuttamiseksi]* (Mar. 17, 2025), https://www.lausuntopalvelu.fi/FI/Proposal/Participation?proposalId=9cefc822-fe12-4b34-ae30-1e6d1a9525b8

[2236] Ministry of the Interior, *Government Proposal on the Use of Biometric Data Will Enhance Prevention of the Most Serious Crimes* (Sept. 9, 2025), https://intermin.fi/en/-/government-proposal-on-the-use-of-biometric-data-will-enhance-prevention-of-the-most-serious-crimes

[2237] Ministry of Economic Affairs and Employment, *Artificial Intelligence 4.0 Report: Finland Has the Opportunity to Lead the Wy in the Ethical Digital and Green Transitions* (Oct. 24, 2022), https://valtioneuvosto.fi/en/-//1410877/artificial-intelligence-4.0-report-finland-has-the-opportunity-to-lead-the-way-in-the-ethical-digital-and-green-transitions

[2238] Finnish Government, *Artificial Intelligence 4.0 Programme. First Interim Report: From Launch to Implementation Stage* (Dec. 7, 2021), https://julkaisut.valtioneuvosto.fi/bitstream/handle/10024/163663/TEM_2021_53.pdf

[2239] European Commission, *The European Green Deal* (Sept. 11, 2024), https://commission.europa.eu/strategy-and-policy/priorities-2019-2024/european-green-deal_en

considers AI data centers significant consumers of energy[2240] and the challenge to European energy grids with the war in Ukraine.[2241] One of the four proposals in the Interim Report of the Artificial Intelligence 4.0 Programme published in December 2021 focuses on "Nature smartness and digital technology" to lead to the "triple victory of sustainable development": economic, ecological and social benefits.[2242] In February 2024, the prime minister, Petteri Orpo reiterated that "Work, security and skills: these are also at the heart of the Government Programme."[2243]

Along with Austria and Sweden, Finland has been evaluated at the top end, in the EU and globally, regarding overall sustainability (environmental, social and economic) as defined in Agenda 2030 in the Sustainable Development Report, 2022.[2244] The Climate Change Act offers Finland's goal to become carbon neutral.[2245] The report on UN 2030 Agenda for Sustainable Development states, "The Foreign and Security Policy the UN Sustainable Development Goals (SDGs) is to be achieved by Agenda 2030 have been jeopardised by, among other things, the lack of multilateral cooperation in the field of sustainable development. problems of multilateralism, the repercussions of the war of aggression in Russia, and the interest rate pandemic the consequences of the currency pandemic."[2246]

[2240] European Commission, *European Green Deal: Energy Efficiency Directive Adopted, Helping Make the EU 'Fit for 55'* (Jul. 25, 2023), https://energy.ec.europa.eu/news/european-green-deal-energy-efficiency-directive-adopted-helping-make-eu-fit-55-2023-07-25_en

[2241] European Commission, *State of the Energy Union Report* (Sept. 11, 2024), https://energy.ec.europa.eu/document/download/bd3e3460-2406-47a1-aa2e-c0a0ba52a75a_en?filename=State%20of%20the%20Energy%20Union%20Report%202024.pdf

[2242] Ministry of Economic Affairs and Employment, *How can Finland achieve the 2030 artificial intelligence vision? – Interim report of the Artificial Intelligence 4.0 programme compiles proposals for concrete aims and measures* (Dec. 12, 2021), https://tem.fi/en/-/how-can-finland-achieve-the-2030-artificial-intelligence-vision-interim-report-of-the-artificial-intelligence-4.0-programme-compiles-proposals-for-concrete-aims-and-measures

[2243] Valtioneuvosto Statstradet, *Prime Minister's Announcement on Government Policy in 2024 and the Most Important Proposals to Be Submitted to Parliament [Pääministerin ilmoitus hallituksen politiikasta vuonna 2024 ja keskeisimmistä eduskunnalle annettavista esityksistä]* (Feb. 13, 2024), https://valtioneuvosto.fi/-/10616/paaministerin-ilmoitus-hallituksen-politiikasta-vuonna-2024-ja-keskeisimmista-eduskunnalle-annettavista-esityksista

[2244] SDG Transformation Center, *Sustainable Development Report 2022* (Jun. 2022), https://sdgtransformationcenter.org/reports/sustainable-development-report-2022

[2245] Ministry of the Environment, *New Climate Change Act into Force in July* (Jun. 9, 2022), https://valtioneuvosto.fi/en/-/1410903/new-climate-change-act-into-force-in-july

[2246] Valtioneuvosto Statsradet, *Valtioneuvoston selonteko YK:n kestävän kehityksen toimintaohjelma Agenda2030:ntoimeenpanosta*, p. 10 (Nov. 14, 2024), https://julkaisut.valtioneuvosto.fi/bitstream/handle/10024/165921/VN_2024_49.pdf?sequence=1&isAllowed=y

Lethal Autonomous Weapons

Finnish officials called lethal autonomous weapons systems (LAWS) "a complex issue."[2247] Finnish officials cautioned that the "development of weapons and means of warfare where humans are completely out of the loop would pose serious risks from the ethical and legal viewpoint," stressing that "humans should always bear the ultimate responsibility when dealing with questions of life and death."[2248]

Finland reiterated support for a two-tier approach to an international treaty on LAWS in the submission to the UN General Secretary's report.[2249] Like a working paper Finland submitted with others to the 2022 Chair of the Group of Governmental Experts (GGE) on emerging technologies in the area of LAWS,[2250] the proposal suggests prohibitions and regulations to (1) prohibit autonomous weapons systems that operate with no human involvement and outside a human-responsible chain of command and (2) regulate the development and use of all other weapons systems having autonomous features or functions to ensure their compliance with the rules and principles of international law. Finland also supported the 2024 resolution for the high-level agenda on the general and complete disarmament of lethal autonomous weapons systems.[2251]

In February 2023, Finland participated in the 2023 REAIM Summit on the responsible application of artificial intelligence in the military domain hosted by the Netherlands. Finland endorsed a joint call for action on the responsible development, deployment and use of artificial intelligence in the military domain[2252] and the resulting Political Declaration on Responsible Military Use of AI and Autonomy

---

[2247] Government of Finland, *Statement to the UN General Assembly First Committee on Disarmament and International Security* (Oct. 22, 2014), https://reachingcriticalwill.org/images/documents/Disarmament-fora/1com/1com14/statements/22Oct_Finland.pdf

[2248] Human Rights Reports, *Stopping Killer Robots: Country Positions on Banning Fully Autonomous Weapons and Maintaining Human Control* (Aug. 10, 2020), https://www.hrw.org/report/2020/08/10/stopping-killer-robots/country-positions-banning-fully-autonomous-weapons-and

[2249] UN Office of Disarmament Affairs, *Submission by Finland concerning United Nations General Assembly resolution 78/241 of 22 December 2023 on Lethal Autonomous Weapons Systems to the United Secretary-General* (May 16, 2024), https://docs-library.unoda.org/General_Assembly_First_Committee_-Seventy-Ninth_session_(2024)/78-241-Finland-EN_0.pdf

[2250] UN Office of Disarmament Affairs, *Working Paper Submitted by Finland, France, Germany, the Netherlands, Norway, Spain, and Sweden to the 2022 Chair of the Group of Governmental Experts (GGE) on Emerging Technologies in the Area of Lethal Autonomous Weapons Systems (LAWS)*, Documents from the 2022 CCW GGE on LAWS (Jul. 13, 2022), https://documents.unoda.org/wp-content/uploads/2022/07/WP-LAWS_DE-ES-FI-FR-NL-NO-SE.pdf

[2251] UN Digital Library, *Lethal Autonomous Weapons Systems: Resolution / Adopted by the General Assembly*, A/79/62 (draft A/C.1/79/L77) (Dec. 2, 2024), https://digitallibrary.un.org/record/4068497?ln=en

[2252] Government of Netherlands, *Call to Action on Responsible Use of AI in the Military Domain* (Feb. 16, 2023), https://www.government.nl/latest/news/2023/02/16/reaim-2023-call-to-action

issued in November 2023.[2253] Finland also endorsed the Paris Declaration at the AI Action Summit, pledging to "not authorise the decision of life and death to be made by an autonomous weapon system" and "not develop, deploy, or use systems incapable of being used in accordance with [International Humanitarian Law]."[2254]

A new act on the export of dual-use items entered into force on September 1, 2024 to harmonize Finnish regulations with the EU Export Control Regulation.[2255] The list applies to software and technology.[2256] This Act can be extended to AI-based systems as per the EU AI Act, which states (Item 24), "Nonetheless, if an AI system developed, placed on the market, put into service, or used for military, defense, or national security purposes is used outside those temporarily or permanently for other purposes, for example, civilian or humanitarian purposes, law enforcement, or public security purposes, such a system would fall within the scope of this Regulation."[2257] However, the regulation, also mentions (Item 24) "An AI system placed on the market for civilian or law enforcement purposes which is used with or without modification for military, defense or national security purposes should not fall within the scope of this Regulation, regardless of the type of entity carrying out those activities."

AI Literacy

The Ministry of Education and Culture developed a Recommendation on AI with plans for AI literacy, legal and ethical issues.[2258] The Recommendation identifies competencies students can development through AI literacy education and the safe use of AI tools in learning. The recommendations also aim to encourage equal opportunities for education and training educators. The University of Helsinki's free Elements of AI course is internationally recognized and has been translated into a number of languages.[2259]

[2253] US Department of State, *Political Declaration on Responsible Military Use of Artificial Intelligence and Autonomy* (Nov. 9, 2023), endorsing States as of Nov. 27, 2024, https://www.state.gov/bureau-of-arms-control-deterrence-and-stability/political-declaration-on-responsible-military-use-of-artificial-intelligence-and-autonomy

[2254] Élysée, *Paris Declaration on Maintain Human Control in AI Enabled Weapon Systems* (Feb. 11, 2025), https://www.elysee.fr/emmanuel-macron/2025/02/11/paris-declaration-on-maintaining-human-control-in-ai-enabled-weapon-systems

[2255] Ministry of Foreign Affairs, *Legislation on Export Control of Dual-Use Items to be Reformed* (Aug. 16, 2024), https://valtioneuvosto.fi/en/-/legislation-on-export-control-of-dual-use-items-to-be-reformed

[2256] EU LEX, *Regulation (EU) 2021/821 of The European Parliament and of The Council* (May 20, 2023), https://eur-lex.europa.eu/legal-content/EN/TXT/?uri=celex%3A02021R0821-20230526

[2257] EU LEX, *Regulation (EU) 2024/1689 of The European Parliament and of The Council* (Jul. 7, 2024), https://eur-lex.europa.eu/legal-content/EN/TXT/?uri=CELEX%3A32024R1689&qid=1730891247545

[2258] Ministry of Education and Culture, *Recommendations for AI* (Sept. 18, 2024), https://okm.fi/en/project?tunnus=OKM021:00/2024

[2259] Elements of AI, *Welcome to the Elements of AI Free Online Course* (2026), https://www.elementsofai.com/

Human Rights

As one of the signatories to the Universal Declaration of Human Rights and several international human rights treaties and conventions,[2260] Finland is committed to protecting human rights, civil liberties, and political rights. Under Finnish law, these rights are guaranteed and subject to the rule of law as interpreted by an independent judiciary.

Freedom House gave Finland a top score (100/100) in the 2025 Freedom in the World Report for political rights and civil liberties, observing that "Finland's parliamentary system features free and fair elections and robust multiparty competition. Corruption is not a significant problem, and freedoms of speech, religion, and association are respected. The judiciary is independent under the constitution and in practice."[2261]

Finland's Ministry for Foreign Affairs' use of the AI-assisted tool, OpenEval, to share information about evaluations related to development policy reflects the Ministry's desire to "be a frontrunner in the appropriate use of artificial intelligence."[2262] The application reflects the Council of Europe advice to member states that their human-rights commitments require them to "ensure that any design, development and ongoing deployment of algorithmic systems occur in compliance with human rights and fundamental freedoms, which are universal, indivisible, inter-dependent and interrelated."[2263]

OECD / G20 AI Principles

Finland is a long-time member of the OECD and has adopted OECD AI Principles,[2264] committing "to uphold international standards that aim to ensure AI systems are designed to be robust, safe, fair and trustworthy."[2265]

The OECD recognizes Finland's progress implementing principle 1.3 transparency and explainability (in Helsinki), and for promoting international research

---

[2260] These include the International Covenant on Economic, Social and Cultural Rights, International Covenant on Civil and Political Rights, European Convention on Human Rights and Protocol amending the Convention for the Protection of Individuals with regard to Automatic Processing of Personal Data (108+)

[2261] Freedom House, *Freedom in the World 2025: Finland* (2025), https://freedomhouse.org/country/finland/freedom-world/2025

[2262] Ministry for Foreign Affairs, *Harness AI for Efficient Use of Evaluation Data* (Jun. 18, 2025), https://valtioneuvosto.fi/en/-/harnessing-ai-for-efficient-use-of-evaluation-evidence

[2263] *Recommendation CM/Rec(2020)1 of the Committee of Ministers to member States on the human rights impacts of algorithmic systems* (Apr. 8, 2020), https://search.coe.int/cm?i=09000016809e1154

[2264] OECD AI Policy Observatory, *AI in Finland*, https://oecd.ai/en/dashboards/national/finland

[2265] OECD AI Policy Observatory, *OECD AI Principles Overview, Countries Adhering to the AI Principles* (2026), https://oecd.ai/en/ai-principles

collaboration.[2266] Finland was also among the first countries to have a National AI Strategy.

Council of Europe AI Treaty

Finland contributed as a Council of Europe Member State in the negotiations of the Council of Europe Framework Convention on AI, Human Rights, Democracy, and the Rule of Law.[2267] The European Commission signed on behalf of the members of the European Union (including Finland) on September 5, 2024.[2268] Finland has not signed independently.[2269]

UNESCO Recommendation on AI Ethics

Finland has endorsed the UNESCO Recommendations on AI, the first ever global agreement on the ethics of AI.[2270] Finland has not yet initiated the Readiness Assessment Methodology (RAM).

Finland has taken other steps to implement the UNESCO Recommendation. For example, The Finnish National Agency for Education (*Opetushallitus*) ethical principles for AI in education draws from the UNESCO Recommendation and the organization's guidance on AI in education.[2271]

The Finnish National Commission for UNESCO "act[s] as liaison between UNESCO and national authorities," serving as an advisory body to the Ministry of Education and Culture and the Government.[2272] The

Evaluation

Finland remains one of the most digital countries in the EU in 2025. In the national AI strategy, one of the first in the world, Finland adopted an approach based

---

[2266] OECD, *State of Implementation of the OECD AI Principles Four Years On* (Oct. 27, 2023), https://www.oecd.org/en/publications/the-state-of-implementation-of-the-oecd-ai-principles-four-years-on_835641c9-en.html

[2267] Council of Europe, *Framework Convention on Artificial Intelligence* (2026), https://www.coe.int/en/web/artificial-intelligence/the-framework-convention-on-artificial-intelligence

[2268] Europe Union External Action, *The European Commission Signs Historic Council of Europe Framework Convention on Artificial Intelligence and Human Rights* (Sept. 10, 2024), https://www.eeas.europa.eu/delegations/council-europe/european-commission-signs-historic-council-europe-framework-convention-artificial-intelligence-and_en

[2269] Council of Europe Treaty Office, *Chart of Signatures and Ratifications of Treaty 225* (Jan. 18, 2026), https://www.coe.int/en/web/Conventions/full-list/?module=signatures-by-treaty&treatynum=225

[2270] UNESCO, *UNESCO Member States Adopt the First Ever Global Agreement on the Ethics of Artificial Intelligence* (Nov. 2021), https://www.unesco.org/en/articles/unesco-member-states-adopt-first-ever-global-agreement-ethics-artificial-intelligence

[2271] Finnish National Agency for Education, *Background Material: AI and Ethics in Education* (2026), https://www.oph.fi/en/teemat-ja-kehittaminen/backround-material-ai-and-ethics-education

[2272] Ministry of Education and Culture, *Finnish National Commission for UNESCO*, https://okm.fi/en/national-commission-for-unesco

on its vision of "a good artificial intelligence society." Risks posed by AI do not occupy center stage in the strategy, but Finland has tried to develop best practices and hands-on solutions for ensuring trustworthy AI by design. The capital, Helsinki, is the first city to have developed and adopted an AI register to foster trust in AI and ensure the best service to citizens.

Finland benefits from strong European and national data protection frameworks and the EU AI Act. Finland has identified authorities responsible for enforcing the EU AI Act and legislation to implement the next phase of the Act is pending. The Data Inspectorate's reprimand of police for the use of Clearview technology reflects the authority's independence. However, the introduction of a new law authorizing automated decisions in administration raises questions regarding its compatibility with data subjects' rights not to be subjected to automated decision making enshrined in the GDPR.

Finland has also worked closely with UNICEF to develop internationally applicable policy guidance for the use AI by children and used international models from UNESCO and OECD to develop recommendations for ethical AI use in education, including AI literacy competencies. The country has endorsed both the OECD AI Principles and the UNESCO Recommendation on the Ethics of AI. These international standards align with many of Finland's actions related to AI governance, even when they are not explicitly referenced. Finland has yet to take independent action on the Council of Europe Framework Convention on Artificial Intelligence.

France

In 2025, France launched a new phase of the National AI Strategy and facilitated collaboration on AI governance as the host of the AI Action Summit. France also designated a single point of contact and the market surveillance and competent authorities who will implement and enforce the EU AI Act.

National AI Strategy

France's national Strategy on Artificial Intelligence entitled "AI for Humanity" aims to make France a world leader in AI. "AI will raise a lot of issues in ethics, in politics, it will question our democracy and our collective preferences," stated French President Emmanuel Macron in 2018.[2273] He continued: "If you want to manage your own choice of society, your choice of civilization, you have to be able to be an acting part of this AI revolution."[2274] The Ministry of Economy, Finance, and Industrial, Energy, and Digital Sovereignty (*Ministère de l'Économie, des Finances et de la Souveraineté industrielle, énergétique et numérique*) oversees the national AI strategy.

France's AI strategy sets out four objectives: (1) Reinforcing the AI ecosystem to attract the very best talent, (2) Developing an open data policy, especially in sectors where France already has the potential for excellence, such as healthcare, (3) Creating a regulatory and financial framework favoring the emergence of "AI champions," and (4) Promoting AI regulation and ethics to ensure to high standard and acceptability for citizens. This high standard includes supporting human sciences research on the ethics of AI use, making all algorithms used by the State public, including admission to higher education, and encouraging AI's openness to diversity.

The second phase of the national AI strategy launched in 2022 focused on priority areas such as trusted AI and generative AI.[2275] The government initiated the third phase dedicated to "Make France an AI Powerhouse" at the AI Action Summit in 2025.[2276] This phase emphasizes strategic infrastructure and widespread AI adoption across various sectors while building trust through AI Cafés to encourage democratic discourse about AI and establishment of the National Institute for AI

[2273] Élysée, *France's New National Strategy for Artificial Intelligence—Speech of Emmanuel Macron* (Mar. 29, 2018), https://www.elysee.fr/emmanuel-macron/2018/03/29/frances-new-national-strategy-for-artificial-intelligence-speech-of-emmanuel-macron.en

[2274] Nicholas Thompson, *Emmanuel Macron Talks to WIRED about France's AI Strategy* (Mar. 31, 2018), https://www.wired.com/story/emmanuel-macron-talks-to-wired-about-frances-ai-strategy

[2275] Ministère de l'Économie, des Finances, *The National Strategy for Artificial Intelligence [la stratégie nationale pour l'intelligence artificielle* (Feb. 7, 2025), https://www.economie.gouv.fr/actualites/strategie-nationale-intelligence-artificielle

[2276] Présidence de la République, *Make France an AI Powerhouse* (Feb. 10, 2025), https://www.elysee.fr/admin/upload/default/0001/17/d9c1462e7337d353f918aac7d654b896b77c5349.pdf

Evaluation and Security (*l'Institut National pour l'Évaluation et la Sécurité de l'Intelligence Artificielle*, INESIA).[2277]

The national AI strategy built on a report by Cedric Villani, a French mathematician and member of parliament, who dedicated a chapter to ethical considerations and included proposals to open the "black box," implement ethics by design, and set up an AI Ethics Committee.[2278] The strategy also referenced report by the Strategy department under the Prime Minister and national data protection authority (*Commission nationale de l'informatique et des libertés*, CNIL).[2279]

The National Coordinator for AI, tasked with the implementation of the national AI strategy, works with all administrations, centers and research laboratories dedicated to AI.[2280]

National Pilot Committee for Digital Ethics

With regard to AI regulation and ethics (objective 4 of the National AI Strategy), in July 2019, the Prime Minister asked the French National Consultative Committee on Bioethics (CCNE) to launch a pilot initiative dedicated to Digital Ethics. The National Pilot Committee for Digital Ethics (NPCDE), created in December 2019, "shall submit initial contributions on the ethics of digital sciences, technologies, uses and innovations and determine relevant equilibria for the organization of public debate on digital ethics and artificial intelligence.'' It is also tasked with maintaining ethical oversight and to raise awareness, inform and assist individuals, companies, administrations, institutions, etc., in their decision-making process.[2281] The committee has been seized by the Prime Minister to give opinions on the ethical issues concerning three specific topics of digital applications using in particular machine learning: 1) Conversational agents (chatbots); 2) Autonomous cars; and 3) Medical diagnosis and health AI.

However, civil society groups such as Access Now have objected to government studies that simply propose ethical guidelines rather than hard law.

[2277] Ministère de L'Économie des Finances et de la Souveraineté Industrielle et Numérique, *The Government Announces the Creation of the National Institute for AI Evaluation and Security [Le Gouvernement annonce la création de l'Institut national pour l'évaluation et la sécurité de l'intelligence artificielle (INESIA)]* (Jan. 31, 2025), https://presse.economie.gouv.fr/le-gouvernement-annonce-la-creation-de-linstitut-national-pour-levaluation-et-la-securite-de-lintelligence-artificielle-inesia/

[2278] Cedric Villani, *For a Meaningful Artificial Intelligence: Toward a French and European Strategy* (Mar. 2018), https://www.jaist.ac.jp/~bao/AI/OtherAIstrategies/MissionVillani_Report_ENG-VF.pdf

[2279] CNIL, *Algorithms and Artificial Intelligence: CNIL's Report on the Ethical Issues* (May 25, 2018), https://www.cnil.fr/en/algorithms-and-artificial-intelligence-cnils-report-ethical-issues

[2280] Gouvernement, *Appointment of Mr. Renaud VEDEL as National Coordinator for Artificial Intelligence [Nomination de M. Renaud Vedel comme coordinateur national pour l'intelligence artificielle]* (Mar. 9, 2020), https://www.info.gouv.fr/communique/11429-nomination-de-m-renaud-vedel-comme-coordinateur-national-pour-l-intelligence-artificielle

[2281] Claude Kirchner, *The French National Committee for Digital Ethics* (Feb. 24, 2020), https://ai-regulation.com/the-french-national-committee-for-digital-ethics/

Access Now explains, "[t]here is solid and creative thinking in the advisory paper that informed the strategy around the ethical and regulatory challenges posed by AI, but at the moment the proposed solutions largely involve the creation of groups to study them rather than the proposal of new or modified norms."[2282] "France's AI strategy generally cleaves to the 'ethics' framework and makes scant reference to hard legal constraints on AI development." AccessNow notes that the "Villani report is considerably more detailed about the ethical and legal challenges posed by AI." While some civil society organizations oppose the need for the NPCDE, they continue to work towards their objectives and convened a conference with the European Research Consortium for Informatics and Mathematics (ERCIM) Ethics Working Group to discuss digital technology ethics and connected technologies in October 2022.[2283]

In September 2023, the Prime Minister launched a new Committee on generative AI.[2284] Its aim is to make concrete recommendations to adapt French AI strategy. Among the members are Joëlle Barral from Google, Yann le Cun from Meta and Arthur Mensch and Cedric O, both involved with Mistral AI. Concerns about a possible conflict of interests and influence on the French position regarding the draft EU AI Act have been raised.[2285]

EU Digital Services Act

As an EU member state, France shall apply the EU Digital Services Act (DSA).[2286] The DSA regulates online intermediaries and platforms. Its main objective is to prevent illegal and harmful activities online and the spread of disinformation. The DSA also bans targeted advertising based on a person's sexual orientation, religion, ethnicity, or political beliefs and targeted advertising to minors based on profiling. Arcom, France's regulatory authority for audiovisual and digital communication, or *le régulateur de la communication audiovisuelle et numérique*,[2287]

[2282] AccessNow, *Mapping Regulatory Proposals for Artificial Intelligence in Europe* (Nov. 18, 2018), https://www.accessnow.org/cms/assets/uploads/2018/11/mapping_regulatory_proposals_for_AI_in_EU.pdf

[2283] ERCIM, *Forum on Digital Ethics in Research: Beyond Compliance*, ERCIM News (Jan. 11, 2023), https://ercim-news.ercim.eu/en132/jea/forum-on-digital-ethics-in-research-beyond-compliance

[2284] French Ministry of Economy and Finance, *France Sets Up a Committee for Generative Artificial Intelligence* (Oct. 3, 2023), https://www.economie.gouv.fr/comite-intelligence-artificielle-generative

[2285] Théophane Hartmann, *AI Act: French Government Accused of Being Influenced by Lobbyist with Conflict of Interest* (Dec. 21, 2023), https://www.euractiv.com/section/artificial-intelligence/news/ai-act-french-government-accused-of-being-influenced-by-lobbyist-with-conflict-of-interests/

[2286] *Regulation (EU) 2022/2065 of the European Parliament and of the Council of 19 October 2022 on a Single Market for Digital Services and amending Directive 2000/31/EC (Digital Services Act)*, (Oct. 21, 2022), https://eur-lex.europa.eu/legal-content/EN/TXT/?uri=celex%3A32022R2065

[2287] Arcom, *5 Things to Remember about the Digital Services Act (DSA)* (Feb. 16, 2024), https://www.arcom.fr/en/news/5-things-remember-about-digital-services-act-dsa

is the digital services coordinator charged with overseeing the EU Digital Services Act.[2288]

EU AI Act

As an EU member State, France is bound by the EU AI Act,[2289] a risk-based market regulation that supports the objective of promoting a human-centric approach to AI and making the EU a global leader in the development of secure, trustworthy, and ethical AI. France designated authorities responsible for enforcing the AI Act in 2025. According to a draft designation, the Directorate-General for Competition, Consumer Affairs, and Fraud Control (*Direction Générale de la concurrence, de la Consommation et de la Répression des Fraudes*, DGCCRF) will coordinate market-surveillance authorities and act as the single point of contact. The Directorate-General for Enterprise (*Direction générale des Entreprises*, DGE) will represent France in the European AI Committee.[2290] Existing regulators CNIL (data protection), Arcom (audiovisual), ANSSI (cybersecurity), and PEReN (digital regulation) will serve as competent authorities in applicable sectors.[2291]

Public Participation

The Villani report, which has been a key source of inspiration for France's national strategy, relied on the work of multiple stakeholders but not the public at large. The data protection authority (CNIL) relied on extensive public outreach to inform the report on "the ethical matters raised by algorithms and artificial intelligence" that also informed the strategy. More than 3,000 people took part in 45

[2288] European Commission, *Digital Services Coordinators* (Mar. 10, 2026), https://digital-strategy.ec.europa.eu/en/policies/dsa-dscs#1720699867912-1

[2289] European Parliament, *Artificial Intelligence Act, European Parliament legislative resolution of 13 March 2024 on the proposal for a regulation of the European Parliament and of the Council on laying down harmonised rules on Artificial Intelligence (Artificial Intelligence Act) and amending certain Union Legislative Acts (COM(2021)0206 – C9-0146/2021 – 2021/0106(COD))*, P9_TA(2024)0138 (Mar. 13, 2024), https://www.europarl.europa.eu/RegData/seance_pleniere/textes_adoptes/definitif/2024/03-13/0138/P9_TA(2024)0138_EN.pdf

[2290] Direction générale des Entreprises, *DGE and DGCCRF Publish the Draft Designation of National Authorities Responsible for Implementing the European AI Regulation [La DGE et la DGCCRF publient le projet de désignation des autorités nationales en charge de la mise en œuvre du règlement européen sur l'IA]* (Sept. 9, 2025), https://www.entreprises.gouv.fr/espace-presse/la-dge-et-la-dgccrf-publient-le-projet-de-designation-des-autorites-nationales-en

[2291] Direction générale des Entreprises, *The Authorities Responsible for Implementing the European Regulation on Artificial intelligence [Les autorités compétentes pour la mise en œuvre du règlemente européen sur l'intelligence artificielle* (Sept. 9, 2025), https://www.entreprises.gouv.fr/priorites-et-actions/transition-numerique/soutenir-le-developpement-de-lia-au-service-de-0

debates and events, organized by 60 partners, including research centers, public institutions, trade unions, think tanks, and companies.[2292]

A 2020 survey by the European consumer organization BEUC found that more than 80% of those polled in France were familiar with Artificial Intelligence. Over 50% of respondents agreed that companies use AI to manipulate consumer decisions.[2293] BEUC also reported that there is little trust over authorities to exert effective control over organizations and companies using AI. More than 60% of respondents in France said users should be able to say "no" to automated decision-making.

Public engagement has increased. Open consultations around AI to Serve the Public Good? ahead of the AI Action Summit engaged over 10,000 citizens and 200 experts who "clearly expressed a strong demand for robust governance of emerging AI technologies a the national and international level."[2294]

Data Protection

The right to the protection of personal data falls within the scope of application of the right to respect for private life,[2295] which is constitutionally protected.[2296] The 1978 French Data Protection Act, which was amended in 2018 to adapt it to both the GDPR and the LED,[2297] further delineates these protections. The National Commission of Data and Rights (*Commission Nationale de l'Informatique et des Libertés*, CNIL) is the independent administrative authority for data protection with jurisdiction over ministers, public authorities, and private and public company directors.[2298]

Since France is an EU Member State, the General Data Protection Regulation (GDPR)[2299] is directly applicable in France and to French people. The aim of the

---

[2292] CNIL, *Algorithms and Artificial Intelligence: CNIL's Report on the Ethical Issues* (May 25, 2018), https://www.cnil.fr/en/algorithms-and-artificial-intelligence-cnils-report-ethical-issues

[2293] BEUC, *Artificial Intelligence, What Consumers Say: Findings and Policy Recommendations of a Multi-Country Survey on AI* (Sept. 7, 2020) https://www.beuc.eu/publications/beuc-x-2020-078_artificial_intelligence_what_consumers_say_report.pdf

[2294] Make.org, *Consultations ahead of the AI Summit: Citizens, Academics, and Civil Society Agree on Stronger Regulation and Perceive AI as Challenge and Opportunity* (Dec. 10, 2024), https://about.make.org/articles-en/consultations-ahead-of-the-ai-summit-citizens-academics-civil-society-agree-on-stronger-regulation-and-perceive-ai-as-challenge-and-opportunity; for results of the consultation, see https://make.org/FR/consultation/ai-action-summit/results

[2295] See Cassation, Civ. 1, (Nov. 5) 1996.

[2296] See Constitutional Council, decision n° 99-416 DC (July 23, 1999)

[2297] "Informatique et Libertés" Act, as modified by Bill n° 2018-493 (June 20, 2018), implemented by Decree n° 2018-687 (Aug. 1, 2018), and Order n° 2018-1125 (Dec. 12, 2018)

[2298] CNIL, *Status & Composition* (Apr. 29, 2025), https://www.cnil.fr/en/cnil/status-composition

[2299] *Regulation (EU) 2016/679 of the European Parliament and of the Council of 27 April 2016 on the protection of natural persons with regard to the processing of personal data and on the free movement of such data* (Apr. 27, 2016), https://eur-lex.europa.eu/EN/legal-content/summary/general-data-protection-regulation-gdpr.html

GDPR is to "strengthen individuals' fundamental rights in the digital age and facilitate business by clarifying rules for companies and public bodies in the digital single market. A single law will also do away with the current fragmentation in different national systems and unnecessary administrative burdens."[2300]

The EU Data Protection Law Enforcement Directive (LED)[2301] "will in particular ensure that the personal data of victims, witnesses, and suspects of crime are duly protected and will facilitate cross-border cooperation in the fight against crime and terrorism."[2302] The LED provides for the prohibition of any decision based solely on automated processing, unless it is provided by law, and of profiling that results in discrimination.[2303] The EU Charter of Fundamental Rights more generally provides that EU citizens have the right to protection of their personal data. Article 8 of the Charter states that: "Everyone has the right to the protection of personal data concerning him or her. Such data must be processed fairly for specified purposes and on the basis of the consent of the person concerned or some other legitimate basis laid down by law."

France is a member of the Council of Europe and ratified COE Convention 108+ for the protection of individuals with regard to the processing of personal data.[2304]

The CNIL has been a member of the Global Privacy Assembly (GPA) since 2002.[2305] The commission's work contributed to the Declaration on Ethics and Data Protection in AI, which was later adopted by the Global Privacy Assembly with CNIL's sponsorship.[2306] The CNIL did not endorse the 2020 GPA Resolution on AI

[2300] European Commission, *Legal Framework of EU Data Protection*, https://commission.europa.eu/law/law-topic/data-protection/legal-framework-eu-data-protection_en

[2301] *Directive (EU) 2016/680 of the European Parliament and of the Council of 27 April 2016 on the protection of natural persons with regard to the processing of personal data by competent authorities for the purposes of the prevention, investigation, detection or prosecution of criminal offences or the execution of criminal penalties, and on the free movement of such data, and repealing Council Framework Decision 2008/977/JHA* (Apr. 27, 2016), https://eur-lex.europa.eu/legal-content/EN/TXT/?uri=celex%3A02016L0680-20160504

[2302] European Commission, *Legal Framework of EU Data Protection*, https://commission.europa.eu/law/law-topic/data-protection/legal-framework-eu-data-protection_en

[2303] Article 11 (1) and (2) of the LED, https://eur-lex.europa.eu/legal-content/EN/TXT/?uri=celex%3A02016L0680-20160504

[2304] Council of Europe Treaty Office, *Chart of Signatures and Ratifications of Treaty 223* (Jan. 8, 2026), https://www.coe.int/en-GB/web/conventions/full-list?module=signatures-by-treaty&treatynum=223

[2305] Global Privacy Assembly (GPA), *List of Accredited Members* (2026), https://globalprivacyassembly.com/participation-in-the-assembly/list-of-accredited-members/

[2306] Global Privacy Assembly, *Declaration on Ethics and Data Protection in AI* (Oct. 23, 2018), http://globalprivacyassembly.org/wp-content/uploads/2019/04/20180922_ICDPPC-40th_AI-Declaration_ADOPTED.pdf

Accountability[2307] but co-sponsored the 2022 GPA Resolution on Facial Recognition Technology[2308] and the 2023 GPA Resolution on Generative AI.[2309]

Algorithmic Transparency

Following the assassination in October 2020 of history professor Samuel Paty, the Secretary of State for the Digital Transition and Electronic Communications, Cédric O, reiterated the importance of algorithmic transparency. He wrote in a blog post "the opacity of the functioning of (social media) algorithms and their moderation is a societal and democratic aberration." He added "it is also essential that full transparency be observed vis a vis the public authorities as regards the principles governing in detail the choices made by their moderation algorithms, whether it is about online hatred or dissemination of false information."[2310]

France is subject to the GDPR and Convention 108+ which both provide for a general right to obtain access to information about automated decision-making and to the factors and logic of an algorithm.[2311] These rights, including the right and access to contest automated decisions with adverse consequences to fundamental rights, align with the 2020Recommendation of the Council of Europe Committee of Ministers on human rights impacts of algorithm systems.[2312]

France has continued to advance concrete measures for algorithmic transparency. The CNIL's Recommendations on AI and GDPR advised that organizations inform individuals when their personal data may be used to train AI models and ensure rights such as access and erasure remain effective.[2313] The data protection agency's recommendations on the development of AI systems further

---

[2307] Global Privacy Assembly, *Resolution on Accountability in the Development and Use of Artificial Intelligence* (Oct. 2020), https://globalprivacyassembly.org/wp-content/uploads/2020/10/FINAL-GPA-Resolution-on-Accountability-in-the-Development-and-Use-of-AI-EN-1.pdf

[2308] Global Privacy Assembly, *Resolution on Principles and Expectations for the Appropriate Use of Personal Information in Facial Recognition Technology* (Oct. 2022), https://globalprivacyassembly.org/wp-content/uploads/2022/11/15.1.c.Resolution-on-Principles-and-Expectations-for-the-Appropriate-Use-of-Personal-Information-in-Facial-Recognition-Technolog.pdf

[2309] Global Privacy Assembly, *Resolution on Generative Artificial Intelligence Systems* (Oct. 2023), https://globalprivacyassembly.org/wp-content/uploads/2023/10/5.-Resolution-on-Generative-AI-Systems-101023.pdf.

[2310] Cédric O, *Régulations*, Medium (Oct. 20, 2020), https://medium.com/@cedric.o/r%C3%A9gulations-657189f5d9d2

[2311] See Recital 63 and Article 22 of the GDPR. Article 9 c) of the Convention 108+ as well as Recital 77, *Explanatory Report, Convention 108+*, p. 24, https://rm.coe.int/convention-108-convention-for-the-protection-of-individuals-with-regar/16808b36f1

[2312] *Recommendation CM/Rec(2020)1 of the Committee of Ministers to Member States on the Human Rights Impacts of Algorithmic Systems* (Apr. 8, 2020), https://search.coe.int/cm/pages/result_details.aspx?objectid=09000016809e1154

[2313] CNIL, *AI and GDPR: The CNIL Publishes New Recommendations to Support Responsible Innovation* (Feb. 7, 2025), https://www.cnil.fr/en/ai-and-gdpr-cnil-publishes-new-recommendations-support-responsible-innovation

clarified that models trained on personal data fall under the GDPR and encouraged documentation and governance measures to strengthen transparency.[2314]

AI Oversight

The French data protection authority (CNIL) has published several papers on AI, including a 2018 report on the ethical issues raised by AI.[2315] This report set out two founding principles—fairness and vigilance—six recommendations, and six concerns. Following this report, a joint 2020 paper by the CNIL and the Defender of Rights detailed concerns around the transparency obligations of those responsible for AI systems.[2316]

The CNIL imposed a €20 million fine on Clearview AI for unlawful use of facial recognition technology.[2317] In April 2023, the French DPA considered that the company had not complied with the order and consequently imposed an overdue penalty payment of €5,200,000 on Clearview AI.[2318]

In January 2023, the CNIL created an AI Department (AID) to improve the authority's understanding of the risks posed by AI systems, prepare for EU AI Act enforcement, and develop further relationships within the AI ecosystem.[2319]

Generative AI was one of the key topics in the 2023 action plan. The CNIL's work on Generative AI was supported by the AID[2320] and the Digital Innovation Laboratory (LINC), which released a study to better understand the functioning and risks of generative AI.[2321] The CNIL issued initial recommendations on the development of AI systems in April 2024 following a public consultation. The recommendations seek to support AI ecosystem players in complying with the GDPR,

---

[2314] CNIL, *AI: The CNIL Finalises Its Recommendations on the Development of Artificial Intelligence Systems and Announces Its Upcoming Work* (Jul. 22, 2025), https://www.cnil.fr/en/ai-cnil-finalises-its-recommendations-development-artificial-intelligence-systems

[2315] CNIL, *Algorithms and Artificial Intelligence: CNIL's Report on the Ethical Issues* (May 25, 2018), https://www.cnil.fr/en/algorithms-and-artificial-intelligence-cnils-report-ethical-issues

[2316] CNIL, *Algorithmes et discriminations : le Défenseur des droits, avec la CNIL, appelle à une mobilisation collective* (May 2020), https://www.cnil.fr/fr/algorithmes-et-discriminations-le-defenseur-des-droits-avec-la-cnil-appelle-une-mobilisation

[2317] Secrétaire générale du Gouvernement, *Deliberation San-2023-005 [Délibération SAN-2023-005]*, LegiFrance (Apr. 2023), https://www.legifrance.gouv.fr/cnil/id/CNILTEXT000047527412/

[2318] EDPB, *Facial Recognition: The French SA Imposes a Penalty Payment on CLEARVIEW AI* (May 11, 2023), https://www.edpb.europa.eu/news/national-news/2023/facial-recognition-french-sa-imposes-penalty-payment-clearview-ai_en

[2319] CNIL, *The CNIL Creates an Artificial Intelligence Department and Begins to Work on Learning Databases* (Jan. 26. 2023), https://www.cnil.fr/en/cnil-creates-artificial-intelligence-department-and-begins-work-learning-databases; https://www.freevacy.com/news/cnil/cnil-adds-ai-department-commences-machine-learning-study/3278

[2320] CNIL, *AI: The Action Plan of the CNIL* (May 2023), https://www.cnil.fr/en/artificial-intelligence-action-plan-cnil

[2321] LINC, *Dossier IA generative – ChatGPT : un beau parleur bien entraîné* (Apr. 26, 2023), https://linc.cnil.fr/dossier-ia-generative-chatgpt-un-beau-parleur-bien-entraine

focusing on the purpose of data processing, data retention, and data minimization.[2322] A second public consultation focusing on the legal basis of legitimate interest to develop AI systems, data subject rights, and safe development of AI systems closed on October 1, 2024.[2323] The CNIL's finalized recommendations offer practical guidance to ensure compliance with the GDPR during the development phase of AI systems regarding AI model training, data annotation, and secure system development.[2324]

Use of AI in Public Administration

The CNIL selected four AI projects aimed at improving public services for incubation in its sandbox in 2023. One project led by the French employment agency (France Travail) is to equip job agents with AI conversational assistants to help them offer a personalized path adapted to the needs of job seekers.[2325] Another project, the "Albert" project of the DINUM (interministerial directorate for digital affairs), is aimed at civil servants. The Albert project would assist civil servants in the search for information and formulating specific responses to users. The project is based on an open language model and considered a potential "lever for the deployment of AI in administrations."

The CNIL published the results of its AI sandbox for public services in 2025.[2326] The organization recommended, among other measures, the principle of data minimization, meaningful human intervention, transparency of AI systems to users, and the assessment and mitigation of potential bias that could lead to discrimination.

Mass Surveillance

Facial recognition technology (FRT) processes sensitive personal data prohibited, at least in principle, by the GDPR and the French data protection law, subject to exceptions such as individual's consent or important public interest. In the latter case, FRT can be authorized by a Decree of the Conseil d'État informed by an opinion from the CNIL

[2322] CNIL, *AI: CNIL Publishes Its First Recommendations on the Development of Artificial Intelligence Systems* (Jun. 7, 2024), https://www.cnil.fr/en/ai-cnil-publishes-its-first-recommendations-development-artificial-intelligence-systems

[2323] CNIL, *[Closed] Artificial Intelligence: The CNIL Opens a New Public Consultation on the Development of AI Systems* (Jul. 2. 2024), https://www.cnil.fr/en/closed-artificial-intelligence-cnil-opens-new-public-consultation-development-ai-systems

[2324] CNIL, *AI: The CNIL Finalises Its Recommendations on the Development of Artificial Intelligence Systems and Announces Its Upcoming Work* (Jul. 22, 2025), https://www.cnil.fr/en/ai-cnil-finalises-its-recommendations-development-artificial-intelligence-systems

[2325] CNIL, *Artificial Intelligence and Public Services "Sandbox": The CNIL Supports 8 Innovative Projects* (Dec. 4, 2023), https://www.cnil.fr/en/artificial-intelligence-and-public-services-sandbox-cnil-supports-8-innovative-projects

[2326] CNIL, *Sandbox "AI and Public Services [Bac à sable «IA et services publics»*] (Mar. 2025), https://www.cnil.fr/sites/cnil/files/2025-04/bac_a_sable_recommandations.pdf

FRT has long been used in France on a voluntary basis with regard to passport control in airports, bank security, and has also been tested in several colleges. The deployment of an FRT-based ID program, Alicem,[2327] was scheduled for November 2019, despite a very critical opinion from the CNIL. Deployment was delayed after a group of NGOs appealed to the Conseil d'État requesting the annulment of the decree that authorized Alicem. In early November 2019, the Conseil d'État dismissed the appeal.[2328]

The CNIL published guidance on the use of facial recognition in 2019.[2329] The document, primarily directed at public authorities in France that want to experiment with facial recognition, presents the technical, legal and ethical elements that need to be considered. After recalling that facial recognition, experimental or not, must comply with the GDPR and the LED, the CNIL sets out three general requirements: (1) facial recognition can only be used if there is an established need to implement an authentication mechanism that ensures a high level of reliability, and there are no other less intrusive means that would be appropriate; (2) the experimental use of facial recognition must respect the rights of individuals (including consent and control, transparency and security); and (3) the use of facial recognition on an experimental basis must have a precise timeline and be based on a rigorous methodology setting out the objectives pursued and the criteria for success.

In December 2019, the Observatoire des Libertés Numériques[2330] and 80 organizations signed an open letter calling on the French Government and Parliament to ban any present and future use of facial recognition for security and surveillance purposes.[2331] In 2020, the administrative tribunal of Marseille rendered a decision on facial recognition that ruled illegal a decision by the South-East Region of France (Provence-Alpes-Côte d'Azur) to test facial recognition at the entrance of two High

---

[2327] Charlotte Jee, *France Plans to Use Facial Recognition to Let Citizens Access Government Services,* MIT Technology Review (Oct. 3, 2019), https://www.technologyreview.com/2019/10/03/132776/france-plans-to-use-facial-recognition-to-let-citizens-access-government-services/

[2328] Marion Garreau, *Le ministère de l'Intérieur va pouvoir lancer l'application Alicem, basée sur la reconnaissance faciale*, L'Usine Nouvelle (Nov. 5, 2020), https://www.usinenouvelle.com/editorial/le-ministere-de-l-interieur-va-pouvoir-lancer-l-application-alicem-basee-sur-la-reconnaissance-faciale.N1024754

[2329] CNIL, *Reconnaissance faciale: Pour un débat à la hauteur des enjeux* (Nov. 2020), https://www.cnil.fr/sites/default/files/atoms/files/reconnaissance_faciale.pdf

[2330] The Observatoire des Libertés Numériques federates several French NGOs monitoring legislation impacting digital freedoms: Le CECIL, Creis-Terminal, Globenet, La Ligue des Droits de l'Homme (LDH), La Quadrature du Net (LQDN), Le Syndicat des Avocats de France (SAF), Le Syndicat de la Magistrature (SM)

[2331] La Quadrature du Net, *Joint Letter from 80 Organisations: Ban Security and Surveillance Facial Recognition* (Dec. 19, 2019), https://www.laquadrature.net/en/2019/12/19/joint-letter-from-80-organisations-ban-security-and-surveillance-facial-recognition/

schools.[2332] Following an analysis from the CNIL,[2333] the court ruled that there was no opportunity for free and informed consent and also that there were other, less intrusive means to manage entrance to high schools.[2334] This was the first decision ever by a court applying the General Data Protection Regulation (GDPR) to Facial Recognition Technologies (FRTs).[2335]

In November 2023, investigative media Disclose revealed that in 2015, law enforcement secretly acquired surveillance video image analysis software from the Israeli company Briefcam. For eight years, the Ministry of the Interior concealed the use of this tool, which enables facial recognition. According to Disclose, the algorithmic video-surveillance option was unlawfully used by municipal police forces across the nation.[2336] The Grenoble administrative court ruled in 2025 that the deployment of Briefcam's algorithmic video surveillance technology in Moirans, Isère is unlawful, following a 3-year legal challenge by civil rights organization La Quadrature du Net. [2337]

Despite rulings about facial recognition and GDPR, many experiments took place leading up to the Paris Games in 2024 as companies positioned themselves for a share of seven-billion euro market.[2338] In May 2023, the French Government enacted the Olympic law legitimizing automated algorithmic image processing at the Olympic Games[2339] as an experiment until March 31, 2025.[2340] An automated surveillance system was put in place to detect "suspect" behavior or objects. According to Human Rights Watch, "the surveillance provision of the proposed bill would constitute a serious threat to civic freedoms and democratic principles." Human Rights Watch also

---

[2332] Tribunal Administratif de Marseille, *La Quadrature du Net, No. 1901249* (Nov. 27, 2020), https://forum.technopolice.fr/assets/uploads/files/1582802422930-1090394890_1901249.pdf

[2333] CNIL, *Expérimentation de la reconnaissance faciale dans deux lycées : La CNIL précise sa position* (Oct. 29, 2019), https://www.cnil.fr/fr/experimentation-de-la-reconnaissance-faciale-dans-deux-lycees-la-cnil-precise-sa-position

[2334] La Quadrature du Net, *First Success against Facial Recognition in France* (Feb. 27, 2020), https://www.laquadrature.net/en/2020/02/27/first-success-against-facial-recognition/

[2335] AI Regulation, *First Decision of a French Court Applying GDPR to Facial Recognition* (Feb. 27, 2020), https://ai-regulation.com/first-decision-ever-of-a-french-court-applying-gdpr-to-facial-recognition/

[2336] Disclose, *The French National Police Is Unlawfully Using an Israeli Facial Recognition Software* (Nov. 14, 2023) https://disclose.ngo/en/article/the-french-national-police-is-unlawfully-using-an-israeli-facial-recognition-software

[2337] Tribunal Administratif de Grenoble, *Decision N° 2105328* (Jan. 24, 2025), https://www.laquadrature.net/wp-content/uploads/sites/8/2025/01/2105328_24012025.pdf

[2338] France Culture, *Quand la reconnaissance faciale en France avance masquée* (Sept. 4, 2020), https://www.franceculture.fr/societe/quand-la-reconnaissance-faciale-en-france-avance-masquee

[2339] Jennifer Krueckeberg, *Let the Games Begin: France's Controversial Olympic Law Legitimizes Automated Surveillance Testing at Sporting Events*, AlgorithmWatch (May 30, 2023), https://algorithmwatch.org/en/let-the-games-begin-frances-controversial-olympic-law-legitimizes-automated-surveillance-testing-at-sporting-events/

[2340] La Galaxie Sénat, *Bill on the 2024 Olympic and Paralympic Games, Law No. 2023-380* (May 19, 2023) https://www.senat.fr/travaux-parlementaires/textes-legislatifs/la-loi-en-clair/projet-de-loi-jeux-olympiques-et-paralympiques-de-2024.html

cited the European Data Protection Board and European Data Protection Supervisor stance that "biometric surveillance stifles people's reasonable expectation of anonymity in public spaces and reduces their will and ability to exercise their civic freedoms, for fear of being identified, profiled or even wrongly persecuted."[2341] La Quadrature du Net also launched a campaign against biometric surveillance.[2342] For its part, the CNIL identified the use of augmented cameras in the framework of the Olympic games as a priority topic for investigation in 2024.[2343]

French MPs unsuccessfully challenged the law before the French Constitutional Council. The French Constitutional Council noted that facial recognition will not be used and the AI-camera system was set for a legitimate purpose and only temporarily. As long as the necessary safeguards are put into place, there is no reason to invalidate the law at this stage.[2344]

Enforcement against facial recognition technology tightened in 2025 with the recognition that untargeted extraction of facial images for recognition databases is prohibited by principle.[2345] Real-time biometric identification systems can only be deployed under strict temporal and geographic limits, requiring judicial or administrative authorization, mainly for locating missing persons or counterterrorism purposes. Still, the French government launched a working group in May 2025 to legalize controlled real-time facial recognition use, particularly for criminal investigations, emphasizing high regulatory scrutiny.[2346]

Environmental Impact of AI

France's AI Strategy includes raising international awareness about the environmental impact of AI, promoting research to make AI infrastructure "more environmentally friendly," and making ecological data open to the public among its

---

[2341] Human Rights Watch, *France: Reject Surveillance in Olympic Games Law - Algorithm-Driven System Would Violate Rights* (Mar. 7, 2023) https://www.hrw.org/news/2023/03/07/france-reject-surveillance-olympic-games-law

[2342] La Quadrature du Net, *France Becomes the First European Country to Legalise Biometric Surveillance* (Mar. 29, 2023), https://www.laquadrature.net/en/2023/03/29/france-becomes-the-first-european-country-to-legalize-biometric-surveillance/

[2343] CNIL, *CNIL Investigations in 2024: Minors Data, Olympic Games, Right of Access and Digital Receipts* (Feb. 8, 2024), https://www.cnil.fr/en/cnil-investigations-2024-minors-data-olympic-games-right-access-and-digital-receipts

[2344] Conseil Constitutionnel, *Law relating to the Olympic and Paralympic Games of 2024 and Various Other Provisions - Decision No. 2023-850 DC* (May 7, 2023) https://www.conseil-constitutionnel.fr/decision/2023/2023850DC.htm

[2345] Assemblée Nationale, *Written Parliamentary Question No. 6750: Security, Algorithmic Video Surveillance, and Facial Recognition Software [Question écrite n° 6750: Sécurités Vidéosurveillance algorithmique et logiciels de reconnaissance faciale]* (Sept. 9, 2025), https://questions.assemblee-nationale.fr/q17/17-6750QE.htm

[2346] La Quadrature du Net, *In France, the Eternal Return of Facial Recognition* (Sept. 4, 2025), https://www.laquadrature.net/en/2025/09/04/in-france-the-eternal-return-of-facial-recognition/

key objectives.[2347] At the AI Action Summit in Paris, France spearheaded the launch of the Coalition for Environmentally Sustainable AI, alongside the UN Environmental Programme (UNEP) and the International Telecommunication Union (ITU) aimed at aligning AI development with Sustainable Development Goals (SDGs) and mitigating AI's environmental footprint.[2348]

The launch of the first global observatory dedicated to AI energy consumption and sustainability marks progress toward implementing France's goals. The observatory emerged from a partnership between International Energy Agency (IEA) and a coalition of 36 experts including scientists, companies, and institutions catalyzed by France's Ministry of Ecological Transition and the National Institute for Research in Digital Science and Technology (INRIA).[2349] The observatory aims to optimize data center energy use, better forecast energy needs, and promote innovations for carbon reduction.[2350]

Lethal Autonomous Weapons

President Macron declared in an interview that he is "dead against" the deployment of lethal autonomous weapons: "You always need responsibility and assertion of responsibility."[2351] In the submission to the UN Secretary General 2024 report on member States' views on lethal autonomous weapons systems (LAWS), France emphasized support for a two-tier approach to address the challenges of LAWS. France advocates for the prohibition of fully autonomous lethal systems that are inherently indiscriminate and supports regulation of partially autonomous systems to ensure compliance with International Humanitarian Law (IHL) while maintaining human control and accountability.[2352] The Paris Declaration on Maintaining Human Control in AI-Enabled Weapon Systems released during the AI Action Summit in Paris reiterates this position, explicitly stating France's commitment to retaining

---

2347 Digital Trade & Data Governance Hub, *France AI Strategy, Overview of the Strategy* (2018), https://datagovhub.elliott.gwu.edu/france-ai-strategy/

2348 SDG Knowledge Hub, *Paris Summit Launches Coalition for Sustainable AI* (Feb. 26, 2025), https://sdg.iisd.org/news/paris-summit-launches-coalition-for-sustainable-ai/

2349 Ministère Transition écologique, Aménagement du territoire, Transports, Ville et Logement, *New Coalition Aims to Put Artificial Intelligence on a More Sustainable Path* (Feb. 11, 2025), https://www.ecologie.gouv.fr/en/press/new-coalition-aims-put-artificial-intelligence-more-sustainable-path

2350 International Energy Agency (IEA), *Energy and AI Observatory* (Jun. 16, 2025), https://www.iea.org/data-and-statistics/data-tools/energy-and-ai-observatory?tab=Energy+for+AI

2351 Nicholas Thompson, *Emmanuel Macron Talks to Wired About France's AI Strategy*, Wired (Mar. 31, 2018), https://www.wired.com/story/emmanuel-macron-talks-to-wired-about-frances-ai-strategy/

2352 France, *Submission to the United Nations Secretary-General on Lethal Autonomous Weapons Systems, Resolution 78/241* (May 2024), https://docs-library.unoda.org/General_Assembly_First_Committee_-Seventy-Ninth_session_%282024%29/78-241-France-EN.pdf

meaningful human involvement in decisions where lethal force may be used and calling for international norms upholding this principle.[2353]

France has continually advocated for human oversight and the need for an international framework to guide constraints on the use of LAWS in the UN and international summits such as REAIM. For example, France endorsed a joint statement to the UN General Assembly in 2022, committing to "upholding and strengthening compliance with International Law, in particular International Humanitarian Law (IHL), including through maintaining human responsibility and accountability in the use of force."[2354] France also voted in favor of resolution L.56[2355] on autonomous weapons systems, along with 163 other states.

Following the international summit on the Responsible application of Artificial Intelligence in the Military domain hosted by the Netherlands (REAIM 2023), France agreed on a joint call for action on the responsible development, deployment and use of artificial intelligence in the military domain.[2356] France also endorsed the resulting Political Declaration on Responsible Military Use of AI and Autonomy.[2357] France contributed to the 2024-2025 REAIM Summit outcomes, which reaffirm the imperative to keep humans "in the loop" and prevent proliferation of autonomous weapons contrary to international peace and security."[2358]

AI Literacy

France's national AI strategy emphasizes human development and digital skills key components across the three phases. However, the strategy emphasizes the development of AI experts and programs rather than inclusion and democratic participation. A report on National Strategy for Digital Inclusion notes plans for a

---

[2353] Présidence de la République, *Paris Declaration on Maintaining Human Control in AI-enabled Weapon Systems* (Feb. 11, 2025), https://www.elysee.fr/emmanuel-macron/2025/02/11/paris-declaration-on-maintaining-human-control-in-ai-enabled-weapon-systems

[2354] United Nations (UN) General Assembly, First Committee, *Joint Statement on Lethal Autonomous Weapons Systems First Committee, 77th United Nations General Assembly Thematic Debate – Conventional Weapons* (Oct. 21, 2022), https://estatements.unmeetings.org/estatements/11.0010/20221021/A1jJ8bNfWGlL/KLw9WYcSnnAm_en.pdf

[2355] General Assembly, *Resolution L56: Lethal Autonomous Weapons* (Oct. 12, 2023), https://reachingcriticalwill.org/images/documents/Disarmament-fora/1com/1com23/resolutions/L56.pdf

[2356] Responsible AI in the Military domain Summit, *REAIM Call to Action* (Feb. 16, 2023), https://www.government.nl/documents/publications/2023/02/16/reaim-2023-call-to-action

[2357] US Department of State, *Political Declaration on Responsible Military Use of Artificial Intelligence and Autonomy* (Nov. 9, 2023), endorsing States as of Nov. 27, 2024, https://www.state.gov/political-declaration-on-responsible-military-use-of-artificial-intelligence-and-autonomy/

[2358] Global Commission on Responsible AI in the Military Domain, *Responsible by Design Strategic Guidance Report on the Risks, Opportunities, and Governance of Artificial Intelligence in the Military Domain* (Sept. 2025), https://hcss.nl/wp-content/uploads/2025/09/GC-REAIM-Strategic-Guidance-Report-Final-WEB.pdf

Grand Beginners skills course with a certification platform,[2359] aligned with the objectives to develop Digital Skills for Citizens and Digital Skills for the Labor Force.[2360]

Human Rights

In France, human rights are understood as the fundamental freedoms and protections guaranteed to all individuals based on the principles of "liberty, equality and fraternity" (liberté, égalité, fraternité). These rights are enshrined in key legal documents such as the Declaration of the Rights of Man and of the Citizen[2361] and the French Constitution.[2362] France is also subject to the Charter of Fundamental Rights of the European Union[2363] and the Council of Europe Convention for the Protection of Human Rights (1950).[2364] As part of the EU, France upholds human rights standards related to dignity, freedoms, equality, solidarity, citizenship, and justice.

France is also a signatory to many international human rights treaties and conventions. France typically ranks among the top nations in the world for the protection of human rights and transparency. Freedom House ranked France "Free" in 2025 with a score of 89/100, given the country's democratic processes and generally strong protections for civil liberties and political rights.[2365]

The French Ombudsman and the CNIL have "both, in their own area of expertise, voiced their concerns regarding the impact of algorithmic systems on fundamental rights."[2366] Following a joint expert seminar in May 2020, they called for a collective mobilization to prevent and address discriminatory biases of algorithms.[2367] The report called on the government and relevant actors to take

---

[2359] République Française, *National Strategy for Digital Inclusion: Key Recommendations* (Sept. 26, 2022), https://labo.societenumerique.gouv.fr/en/articles/national-strategy-for-an-inclusive-digital-main-recommendations/

[2360] European Union Digital Skills & Jobs Platform, *France—National Plan for Digital Inclusion* (Dec. 23, 2024), https://digital-skills-jobs.europa.eu/en/actions/national-initiatives/national-strategies/france-national-plan-digital-inclusion

[2361] Élysée, *Declaration of the Rights of Man and of the Citizen 1789* (Dec. 14, 2022), https://www.elysee.fr/en/french-presidency/the-declaration-of-the-rights-of-man-and-of-the-citizen

[2362] Conseil Constitutionnel, *Constitution of 4 October 1958*, https://www.conseil-constitutionnel.fr/en/constitution-of-4-october-1958

[2363] Official Journal of the European Union, *Charter of Fundamental Rights of the European Union* (2012), https://eur-lex.europa.eu/legal-content/EN/TXT/HTML/?uri=CELEX:12012P/TXT

[2364] Council of Europe, *European Convention on Human Rights* (2021), https://www.echr.coe.int/documents/d/echr/convention_ENG

[2365] Freedom House, *Freedom in the World 2025: France* (2025), https://freedomhouse.org/country/france/freedom-world/2025

[2366] Défenseur des droits and CNIL, *Algorithms: Preventing Automated Discrimination* (2020), https://www.defenseurdesdroits.fr/sites/default/files/2023-07/ddd_rapport_algorithmes_2020_EN_20200531.pdf

[2367] CNIL, *Algorithms and Discrimination: Defender of Rights, with the CNIL, Calls for Collective Mobilization* (Jun. 2, 2020), https://www.cnil.fr/fr/algorithmes-et-discriminations-le-defenseur-des-droits-avec-la-cnil-appelle-une-mobilisation

appropriate measures to avoid algorithms that replicate and amplify discrimination.[2368] In particular, the Ombudsman recommended to: i) support research to develop studies to measure, and methods to prevent bias; ii) reinforce information, transparency and explainability requirements with regard to algorithms; and iii) perform impact assessments to anticipate discriminatory effects of algorithms.[2369]

In April 2022, the French National Consultative Commission on Human Rights also adopted an opinion on the impact of artificial intelligence on fundamental rights. The consultative Commission recommends prohibiting certain uses of AI that are considered too prejudicial to fundamental rights, such as social scoring or remote biometric identification of people in public spaces and places accessible to the public.[2370]

AI Safety Summit

In November 2023, France participated in the first AI Safety Summit and endorsed the Bletchley Declaration.[2371] France thus committed to participate in international cooperation efforts on AI "to promote inclusive economic growth, sustainable development and innovation, to protect human rights and fundamental freedoms, and to foster public trust and confidence in AI systems to fully realise their potential." Endorsing parties affirmed that for the good of all, AI should be designed, developed, deployed, and used, in a manner that is safe, in such a way as to be human-centric, trustworthy and responsible.

France hosted the third iteration of the AI Safety Summit, rebranded as the AI Action Summit.[2372] The AI Action Summit emphasized moving from principles to implementation, complementing the EU AI Act rollout with international coordination on AI safety and accountability and a Joint Statement on Data Governance for AI Systems by data protection authorities reaffirming human-rights-

---

[2368] Defender of Rights, *Algorithms: Preventing Automated Discrimination*, p. 19 (May 2020), https://www.defenseurdesdroits.fr/sites/default/files/2023-07/ddd_rapport_algorithmes_2020_EN_20200531.pdf

[2369] Inside Tech Media, *French CNIL Publishes Paper on Algorithmic Discrimination* (Jun. 9, 2020), https://www.insideprivacy.com/artificial-intelligence/french-cnil-publishes-paper-on-algorithmic-discrimination/

[2370] CNCDH, *Avis relatif à l'impact de l'intelligence artificielle sur les droits fondamentaux (A-2022-6)* (May 20, 2022), https://www.cncdh.fr/publications/avis-relatif-limpact-de-lintelligence-artificielle-sur-les-droits-fondamentaux-2022-6

[2371] UK Department for Science, Innovation & Technology, Foreign, Commonwealth & Development Office, *Prime Minister's Office, The Bletchley Declaration by Countries Attending the AI Safety Summit* (Nov. 2023), https://www.gov.uk/government/publications/ai-safety-summit-2023-the-bletchley-declaration/the-bletchley-declaration-by-countries-attending-the-ai-safety-summit-1-2-november-2023

[2372] Élysée, *Artificial Intelligence Action Summit*, https://www.elysee.fr/en/sommet-pour-l-action-sur-l-ia

based data governance, transparency, and accountability commitments.[2373] More than 60 nations endorsed the resulting Statement on Inclusive and Sustainable Artificial Intelligence for People and the Planet.[2374]

OECD / G20 AI Principles

France endorsed the OECD and the G20 AI Principles.[2375] In 2020, France and Canada, and a dozen other countries announced the Global Partnership on Artificial Intelligence (GPAI) to "support the responsible and human-centric development and use of AI in a manner consistent with human rights, fundamental freedoms, and our shared democratic values."[2376]

The OECD praised France for its commitment to following OECD recommendations on public administration by publishing a guide for public administrations on the responsible use of algorithms in the public sector.

During a 2023 interview, President Emmanual Macron stated, "The G7 and the Organisation for Economic Co-operation and Development (OECD), which includes 38 countries, would be a 'good platform' to develop global regulation" on artificial intelligence and other technology.[2377]

Council of Europe AI Treaty

France contributed as a Council of Europe and EU Member State in the negotiations of the Council of Europe Framework Convention on AI, Human Rights, Democracy and the Rule of Law.[2378] The European Commission signed this first

[2373] CNIL, *Joint Statement on Building Trustworthy Data Governance Frameworks* (Sept. 18, 2025), https://www.cnil.fr/en/joint-statement-on-trustworthy-governance-for-ai-signed-by-twenty-dpa
[2374] Élysée, *Statement on Inclusive and Sustainable Artificial Intelligence for People and the Planet* (Feb. 11, 2025), https://www.elysee.fr/en/emmanuel-macron/2025/02/11/statement-on-inclusive-and-sustainable-artificial-intelligence-for-people-and-the-planet
[2375] OECD Legal Instruments, *Recommendation of the Council on Artificial Intelligence, Adherents* (May 3, 2024), https://legalinstruments.oecd.org/en/instruments/OECD-LEGAL-0449#adherents
[2376] Government of Canada, *Joint Statement from Founding Members of the Global Partnership on Artificial Intelligence* (Jun. 15, 2020), https://www.canada.ca/en/innovation-science-economic-development/news/2020/06/joint-statement-from-founding-members-of-the-global-partnership-on-artificial-intelligence.html
[2377] French President Emmanuel Macron, *France Sees Global AI Regulation Ideas by the End of This Year, Wants to Work with US on Tech Laws - CNBC interview* (Jun 15, 2023), https://www.cnbc.com/2023/06/15/ai-regulation-france-sees-ideas-on-global-laws-by-end-of-year.html
[2378] Council of Europe, *Framework Convention on Artificial Intelligence* (2026), https://www.coe.int/en/web/artificial-intelligence/the-framework-convention-on-artificial-intelligence

legally binding international AI treaty on behalf of EU Member States.[2379] France has not acted independently to sign or ratify the treaty.[2380]

UNESCO Recommendation on AI Ethics

France endorsed the 2021 UNESCO Recommendation on AI Ethics, the first ever global agreement on the ethics of AI.[2381] France's attention to the environmental impact of AI and the need for capacity building in digital and AI literacy for citizens and civil servants mark some progress toward implementing the recommendations, as does the proposed governance plan related to the EU AI Act and the founding of INESIA to monitor AI security. France is not implementing or preparing the UNESCO Readiness Assessment Methodology (RAM), a tool to facilitate implementation of the AI ethics recommendations by helping countries identify gaps in their AI governance and infrastructure.[2382]

Evaluation

France continues as a leader in national AI policies. France has endorsed the OECD/G20 AI Principles and contributed to the creation of the Global Partnership on AI. French authorities in charge of human rights, data protection and ethics are actively involved in AI policy and have published practical guidance regarding facial recognition and algorithmic transparency. France anticipated the entry into force of the EU AI Act by creating a dedicated AI unit with its data protection authority and has since taken steps to clarify reporting and enforcement of the EU AI Act.

However, citizens, civil society, and authorities continue to debate France's approach to the use of AI by law enforcement authorities and for security purposes and the potential negative impact on fundamental rights. France successfully advocated for a national security exemption in both the EU AI Act and the Council of Europe Framework Convention on AI, human rights, democracy and the rule of law. France also played a key role in imposing a separate and lighter regime for GPAI models under the EU AI Act.

[2379] Council of Europe, *Framework Convention on Artificial Intelligence and Human Rights, Democracy and the Rule of Law* (Sept. 5, 2024), https://rm.coe.int/1680afae3c

[2380] Council of Europe Treaty Office, *Chart of Signatures and Ratifications of Treaty 225* (Jan. 8, 2026), https://www.coe.int/en/web/conventions/full-list?module=signatures-by-treaty&treatynum=225

[2381] UNESCO, *Member States*, https://www.unesco.org/en/countries

[2382] UNESCO Global AI Ethics and Governance Observatory, *Global Hub* (Oct. 2025), https://www.unesco.org/ethics-ai/en/global-hub

Germany

In 2025, Germany centralized AI policy under the new Federal Ministry for Digital Transformation and Government Modernisation. The first Digital Minister framed the Ministry's mission around digital sovereignty, regulatory simplification, and local business competitiveness.

National AI Strategy

The German government published an initial national AI strategy in November 2018.[2383] The three main goals are 1) "to make Germany and Europe a leading centre for AI and thus help safeguard Germany's competitiveness in the future," 2) to ensure "a responsible development and use of AI which serves the good of society," and 3) to "integrate AI in society in ethical, legal, cultural and institutional terms in the context of a broad societal dialogue and active political measures." Germany updated the AI Strategy[2384] to respond to a new AI principles and frameworks developed between 2018 and 2020. The updated Strategy also included a discussion on how AI can help with "pandemic control" or environmental and climate protection.

The Federal Ministry of Education and Research (*Bundesministerium für Bildung und Forschung*) launched an Action Plan[2385] to implement the AI Strategy in 2023. The Federal Ministry of Research, Technology and Space (*Bundesministerium für Forschung, Technologie und Raumfahrt*) now oversees the AI Action Plan. The AI Action Plan outlines targeted initiatives to "ensure that Germany remains a leading AI nation" through the Made in Germany motto.[2386] The High-Tech Agenda Germany[2387] furthers the Made in Germany motto through the AI initiative, which centers emphasizes AI as one of 5 key technologies that will help Germany "become

[2383] Federal Ministry of Research, Technology and Space, *Artificial Intelligence*, https://www.bmftr.bund.de/EN/Research/EmergingTechnologies/ArtificialIntelligence/artificialintelligence_node.html

[2384] German Federal Government, *Artificial Intelligence Strategy* (Dec. 2020), https://www.ki-strategie-deutschland.de/files/downloads/Fortschreibung_KI-Strategie_engl.pdf

[2385] Federal Ministry of Education and Research, *BMBF—Aktionsplan "Künstliche Intelligenz"* (Nov. 2023), https://www.bmftr.bund.de/SharedDocs/Downloads/DE/2023/230823-executive-summary-ki-aktionsplan.pdf?__blob=publicationFile&v=1; to download in English, Federal Government, *Action Plan Artificial Intelligence 2023 (BMBF)(English)*, Downloads (Nov. 2023), https://www.ki-strategie-deutschland.de/

[2386] Federal Ministry of Research, Technology and Space, *BMFTR—"Artificial Intelligence" Action Plan [BMFTR—Aktionsplan "Künstliche Intelligenz"]*, https://www.bmftr.bund.de/DE/Forschung/Schluesseltechnologien/KuenstlicheIntelligenz/KiAktionsplan/kiaktionsplan_node.html

[2387] Federal Ministry of Research, Technology and Space, *High-Tech Agenda Germany* (Jul. 2025), https://www.bmftr.bund.de/SharedDocs/Publikationen/DE/FS/1118830_Hightech_Agenda_Germany.pdf?__blob=publicationFile&v=5

a leading technology nation again."[2388] The new Federal Ministry for Digital Transformation and Government Modernisation established in 2025 has extensive power and consolidates responsibilities, including over artificial intelligence, that were formerly divided among six departments.[2389] The new Ministry is charged with implementing the binding political agenda of the coalition government agreement, which emphasizes digital sovereignty, GPU infrastructure, and innovation-friendly EU AI Act implementation.[2390]

Still, AI ethics is a core component of the ongoing AI Strategy.[2391] The German government further emphasizes transparency for the development of AI to ensure the protection of civil rights and maintain trust in businesses and institutions. The AI Strategy suggests that "government agencies or private-sector auditing institutions" should "verify algorithmic decision-making in order to prevent improper use, discrimination and negative impacts on society."

Germany launched several projects to implement its National AI Strategy. According to the OECD, there are approximately 29 initiatives on AI across several topics and institutions[2392] ranging from the ethical guidelines to initiatives that foster fruitful business environments. Three specifically focus on ethical questions related to AI. First, the Ethical Guidelines for Automated and Connected Driving set out 20 ethical principles for autonomous and semi-autonomous vehicles.[2393] This was among the first guidelines worldwide to establish ethical principles for connected vehicular traffic. The Ethical Guidelines led to an action plan and the "creation of ethical rules for self-driving cars" that was adopted by the Federal Government.[2394] Second, the German AI Observatory forecasts and assesses AI technologies' impact on society. The AI Observatory also develops regulatory frameworks that help deal with the rapidly changing labor market to ensure that social aspects of these changes are not

---

[2388] Ibid, p. 2

[2389] EuroNews, *Merz's Government Decree Gives German Digital Ministry Extensive Powers* (May 7, 2025), https://www.euronews.com/next/2025/05/07/merzs-government-decree-gives-german-digital-ministry-extensive-powers

[2390] Christian Democratic Union, Christian Social Union of Bavaria, and Social Democratic Party, *Responsibility for Germany: Coalition Agreement between CDU, CSU, and SPD, 21st Legislative Period of the German Bundestag* (May 5, 2025), https://www.spd.de/fileadmin/Dokumente/Koalitionsvertrag2025_bf.pdf

[2391] The Federal Government of Germany, *Artificial Intelligence Strategy* (Nov. 2018), https://www.ki-strategie-deutschland.de/home.html?file=files/downloads/Nationale_KI-Strategie_engl.pdf

[2392] OECD AI Policy Observatory, *AI Initiatives from Germany* (2026), https://oecd.ai/en/dashboards/national/germany

[2393] Federal Ministry of Transport and Digital Infrastructure, *Ethics Commission: Automated and Connected Driving* (Jun. 2017), https://www.bmv.de/SharedDocs/EN/publications/report-ethics-commission.pdf?__blob=publicationFile

[2394] Federal Ministry of Transport, *Automated and Connected Driving* (Jul. 30, 2020), https://www.bmv.de/SharedDocs/EN/Articles/StV/automated-and-connected-driving.html

neglected.[2395] Finally, the Federal Ministry for Economic Cooperation and Development's Development Cooperation initiative FAIR Forward[2396] aims to promote a more "open, inclusive and sustainable approach to AI on an international level" by "working together with five partner countries: Ghana, Rwanda, South Africa, Uganda and India." Several projects underway in partner countries work to Strengthen Technical Know-How on AI, Remove Entry Barriers to AI, and Develop Policy Frameworks ready for AI.

Further, the Federal Ministry for Economic Affairs and Energy launched a Regulatory Sandbox initiative in 2018. This initiative, although not specifically dedicated to AI, focuses on "testing innovation and regulation which enable digital innovations to be tested under real-life conditions and experience to be gathered."[2397]

In response to the European Commission's White Paper on AI,[2398] Germany called for tighter regulation of AI on the EU level in 2020. The German government stated it welcomes new regulations but wants more specific definitions and stricter requirements for data storage, more focus on information security and more elaborate definitions of when human supervision is needed.[2399]

Data Ethics Commission

In 2018, the German federal government established a Data Ethics Commission to "build on scientific and technical expertise in developing ethical guidelines for the protection of the individual, the preservation of social cohesion, and the safeguarding and promotion of prosperity in the information age."[2400] In 2020, the Commission recommended to the German parliament that sustainability, justice and solidarity, democracy, security, privacy, self-determination and human dignity should

---

[2395] Denkfabrik: Digitale Arbeitsgesellschaft, *About the AI Observatory [Über das KI-Observatorium]* (Accessed Mar. 16, 2025), https://www.ki-observatorium.de/das-ki-observatorium/ueber-das-ki-observatorium

[2396] Toolkit Digitalisierung, *FAIR Forward – Artificial Intelligence for All*, https://www.giz.de/expertise/html/61982.html

[2397] Federal Ministry for Economic Affairs and Energy, *Regulatory Sandboxes—Testing Environments for Innovation and Regulation* (2025), https://www.bmwk.de/Redaktion/EN/Dossier/regulatory-sandboxes.html

[2398] European Commission, *White Paper on Artificial Intelligence: a European Approach to Excellence and Trust* (Feb. 19, 2020), https://commission.europa.eu/publications/white-paper-artificial-intelligence-european-approach-excellence-and-trust_en

[2399] Federal Government, Statement by the Federal Government of the Federal Republic of Germany on the White Paper on Artificial Intelligence – A European Approach to Excellence and Trust [Stellungnahme der Bundesregierung der Bundesrepublik Deutschland zum Weißbuch zur Künstlichen Intelligenz—ein europäisches Konzept für Exzellenz und Vertrauen (Jun. 30, 2020), https://www.bmi.bund.de/SharedDocs/downloads/DE/veroeffentlichungen/2020/stellungnahme-breg-weissbuch-ki.html

[2400] Bundesministerium der Justiz und für Verbraucherschutz, *Report of the Data Ethics Commission Strengthens Data Protection* (Oct. 23, 2019), https://www.bfdi.bund.de/SharedDocs/Pressemitteilungen/EN/2019/24_Datenethikkommission.html

be the ethical and legal principles that guide the regulation of AI.[2401] The Data Ethics Commission suggested a risk-based approach to the regulation of AI, which distinguishes five levels of criticality in a "criticality pyramid" and respective measures in its risk-adapted regulatory system for the use of algorithmic systems. The Commission also recommended the establishment of "*ex ante* approval mechanisms and continuous supervision by oversight bodies."[2402]

The German consumer organization federation *Verbraucherzentrale Bundesverband* (vzbv) favored the creation of the Commission and strongly supported the recommendations, as did the main German industry body *Bundesverband der Deutschen Industrie* (BDI).[2403] The vzbv further emphasized that the aim of Automated Decision Making (ADM) regulation must be to ensure compliance with existing laws. Toward that goal, "it must be possible for supervisory authorities to scrutinise and verify the legality of ADM systems and their compliance with existing laws so that they can impose penalties if the law is infringed."[2404] Vzbv also noted it is "important to ensure consumers' self-determination when making decisions, to strengthen consumers' confidence in ADM systems by creating transparency and to foster competition and innovation."

EU Digital Services Act

As an EU member state, Germany shall apply the EU Digital Services Act (DSA).[2405] The DSA regulates online intermediaries and platforms. Its main objective is to prevent illegal and harmful activities online and the spread of disinformation. The DSA also bans targeted advertising based on a person's sexual orientation, religion, ethnicity, or political beliefs. The DSA also bans targeted advertising to minors based on profiling.

Through the 2022 Strengthened Code of Practice on Disinformation,[2406] which counts as a mitigation measure under the DSA, signatories commit to take action in several domains, such as demonetizing the dissemination of disinformation; ensuring the transparency of political advertising; empowering users; enhancing the cooperation with fact-checkers; and providing researchers with better access to data.

[2401] Datenethikkommission, *Opinion of the Data Ethics Commission—Executive Summary* (Oct. 23, 2019), https://www.bfdi.bund.de/SharedDocs/Downloads/EN/Datenschutz/Data-Ethics-Commission_Opinion.pdf

[2402] Louisa Well, *Algorithmwatch Automating Society Report - Germany* (2020), https://automatingsociety.algorithmwatch.org/report2020/germany/

[2403] Vzbz, Communication between the Editor and Isabelle Buscke (Nov. 27, 2020)

[2404] Vzbv, *Artificial Intelligence: Trust is Good, Control is Better* (2019), https://www.vzbv.de/sites/default/files/2019_vzbv_factsheet_artificial_intelligence.pdf

[2405] EUR-Lex, Regulation (EU) 2022/2065 of the European Parliament and of the Council of 19 October 2022 on a Single Market for Digital Services and amending Directive 2000/31/EC (Digital Services Act) (Oct. 21, 2022), https://eur-lex.europa.eu/legal-content/EN/TXT/?uri=celex%3A32022R2065

[2406] European Commission, *The 2022 Code of Practice on Disinformation*, https://digital-strategy.ec.europa.eu/en/policies/code-practice-disinformation

EU AI Act

As an EU member State, Germany is bound by the EU AI Act.[2407] The EU AI Act is a risk-based market regulation which supports the objective of promoting a human-centric approach to AI and making the EU a global leader in the development of secure, trustworthy and ethical AI. The first provisions of the AI Act entered into force on August 1, 2024.[2408] Germany missed the August 2025 deadline to designate competent authorities but released a draft bill to designate authorities in late 2025. The draft AI Market Surveillance and Innovation Promotion Act[2409] would designate the Federal Network Agency (*Bundesnetzagentur*) as the central market surveillance and notifying authority, providing a one-stop shop for businesses. The Agency launched an AI Service Desk to guide businesses through compliance requirements.[2410]

The Digital Minister established in 2025 expressed disagreements with the EU AI Act in its current form, seeing it as "overloaded and too complex," and wanting to make it more "innovation-friendly"[2411] by loosening the rules. He noted, "You should regulate when you have a market, when you have products," otherwise "companies develop elsewhere."[2412]

Public Participation

Germany's national AI strategy points out that the 2019 data strategy is based on "broad public consultation and a host of expert discussions," though participation

---

[2407] European Parliament, Regulation (EU) 2024/1689 of the European Parliament and of the Council of 13 June 2024 laying down harmonised rules on artificial intelligence and amending Regulations (EC) No 300/2008, (EU) No 167/2013, (EU) No 168/2013, (EU) 2018/858, (EU) 2018/1139 and (EU) 2019/2144 and Directives 2014/90/EU, (EU) 2016/797 and (EU) 2020/1828 (Artificial Intelligence Act) (Jul. 12, 2024), https://eur-lex.europa.eu/legal-content/EN/TXT/?uri=CELEX%3A32024R1689

[2408] European Commission, *Artificial Intelligence - Questions and Answers*, Press Corner (Aug. 1, 2024), https://ec.europa.eu/commission/presscorner/detail/en/qanda_21_1683

[2409] Federal Ministry for Digital Transformation and Government Modernisation, *Draft Law for Implementation of EU Regulation on Artificial Intelligence [Gesetz zur Durchführung der KI-Verordnung]* (Sept. 12, 2025), https://bmds.bund.de/service/gesetzgebungsverfahren/gesetz-zur-durchfuehrung-der-ki-verordnung

[2410] Federal Network Agency, *Bundesnetzagentur's AI Service Desk Launched* (Aug. 3, 2025), https://www.bundesnetzagentur.de/SharedDocs/Pressemitteilungen/EN/2025/20250703_KI_ServiceDesk.html?nn=691794

[2411] Euroactiv, *German Digital Minister Attacks AI Act, Defends Gigafactories* (Aug. 8, 2025), https://www.euractiv.com/news/german-digital-minister-attacks-ai-act-defends-gigafactories/

[2412] Business Insider, *The Head of Germany's Answer to DOGE Wants to Scale Back Regulation and "Open up the Gates"* (Sept. 25, 2025), https://www.businessinsider.com/germany-karsten-wildberger-ai-regulation-doge-2025-9

in developing and updating its AI strategy was primarily limited to "experts in expert forums."[2413]

One initiative, Platform for Self-Learning Systems (*Plattform Lernende Systeme*),[2414] sometimes known as Platform for AI, focuses on fostering dialogue between stakeholders including civil society, government, and business on the topic of self-learning systems. The Platform also aims to "shape self-learning systems to ensure positive, fair and responsible social coexistence" as well as strengthen skills for developing and using self-learning systems. The IT Security, Privacy, Legal and Ethical Framework working group has published several papers on AI security and ethical issues.[2415]

To inform the public about AI policy, the German government created a website (KI-Strategie-Deutschland.de) to provide information on AI strategy implementation and new policy developments.[2416] *Plattform Lernende Systeme* also offers a map that shows AI developments across Germany by region.[2417]

Germany's federal parliament also set up the Study Commission on Artificial Intelligence, Social Responsibility, and Economic, Social, and Ecological Potential, which comprised equal numbers of parliamentary representatives and experts.[2418] The Commission's aim was to develop recommendations on AI and examine its impact on "our value systems, fundamental and human rights, and [its] benefits for society and the economy." Some of their meetings were broadcast on parliamentary television or could be attended in person. After two years of work, the Study Commission presented its final report to the federal parliament (*Bundestag*) on October 28, 2020. The Commission's findings were then debated in the Bundestag.[2419]

Germany's Federal Ministry for Digital and State Modernization opened consultation on a draft federal law implementing the EU AI Act, the AI Market Surveillance and Innovation Promotion Act,[2420] in 2025. The Ministry published a

---

[2413] German Federal Government, *Artificial Intelligence Strategy of the German Federal Government* (Dec. 2020), https://www.ki-strategie-deutschland.de/files/downloads/Fortschreibung_KI-Strategie_engl.pdf

[2414] Lernende Systeme, *Mission Statement*, https://www.plattform-lernende-systeme.de/mission-statement.html

[2415] Lernende Systeme, *WG 3: IT Security, Privacy, Legal and Ethical Framework*, https://www.plattform-lernende-systeme.de/wg-3.html

[2416] German Federal Government, *KI-Strategie*, https://www.ki-strategie-deutschland.de/home.html

[2417] Lernende Systeme, *Artificial Intelligence in Germany*, https://www.plattform-lernende-systeme.de/map-on-ai.html

[2418] German Parliament, Enquete-Kommission, *Künstliche Intelligenz – Gesellschaftliche Verantwortung und wirtschaftliche, soziale und ökologische Potenziale*, https://www.bundestag.de/webarchiv/Ausschuesse/ausschuesse19/weitere_gremien/enquete_ki

[2419] German Parliament, *Executive Summary of the Final Report of the Study Commission on Artificial Intelligence* (Oct. 27, 2020), https://www.bundestag.de/resource/blob/804184/f31eb697deef36fc271c0587e85e5b19/Kurzfassung-des-Gesamtberichts-englische-Uebersetzung-data.pdf

[2420] Federal Ministry for Digital Transformation and Government Modernisation, *Draft Law for Implementation of EU Regulation on Artificial Intelligence [Gesetz zur Durchführung der KI-*

spreadsheet with more than 50 responses to the call for comments, along with individual files for additional comments received outside the consultation period. The Cabinet approved the government draft in February 2026.

Data Protection

Since Germany is an EU Member State, the General Data Protection Regulation (GDPR)[2421] is directly applicable in Germany and to Germans. The aim of the GDPR is to "strengthen individuals' fundamental rights in the digital age and facilitate business by clarifying rules for companies and public bodies in the digital single market. A single law will also do away with the current fragmentation in different national systems and unnecessary administrative burdens."[2422] The GDPR entered into force in May 2016 and has applied since May 2018.

The activities of law enforcement authorities have been addressed at the EU level by the EU Data Protection Law Enforcement Directive (LED).[2423] The directive ensures that the "personal data of victims, witnesses, and suspects of crime are duly protected and will facilitate cross-border cooperation in the fight against crime and terrorism."[2424] The LED prohibits any decision based solely on automated processing, unless it is provided by law, and of profiling that results in discrimination.[2425]

Germany's Federal Data Protection Act (*Bundesdatenschutzgesetz*, BDSG) was designed to bring the German privacy law on par with the GDPR and the LED. It is usually referred to as the "BDSG-new" since it replaced the former BDSG.[2426] The German Federal Cabinet approved the Draft Law[2427] to amend the BDSG in February 2024. After the legislative body's (*Bundesrat*) opinion, the Draft Law was discussed

---

Verordnung] (Sept. 12, 2025), https://bmds.bund.de/service/gesetzgebungsverfahren/gesetz-zur-durchfuehrung-der-ki-verordnung

[2421] EUR-Lex, *General Data Protection Regulation (GDPR) Summary* (Jan. 7, 2022), https://eur-lex.europa.eu/EN/legal-content/summary/general-data-protection-regulation-gdpr.html

[2422] European Commission, *Legal Framework of EU Data Protection*, https://commission.europa.eu/law/law-topic/data-protection/legal-framework-eu-data-protection_en

[2423] Directive (EU) 2016/680 of the European Parliament and of the Council of 27 April 2016 on the protection of natural persons with regard to the processing of personal data by competent authorities for the purposes of the prevention, investigation, detection or prosecution of criminal offences or the execution of criminal penalties, and on the free movement of such data, and repealing Council Framework Decision 2008/977/JHA, https://eur-lex.europa.eu/legal-content/EN/TXT/?uri=celex%3A02016L0680-20160504

[2424] European Commission, *Legal Framework of EU Data Protection*, https://commission.europa.eu/law/law-topic/data-protection/legal-framework-eu-data-protection_en

[2425] Article 11 (1) and (2) of the LED, https://eur-lex.europa.eu/legal-content/EN/TXT/?uri=celex%3A02016L0680-20160504

[2426] *Federal Data Protection Act* (Bundesdatenschutzgesetz, BDSG), https://www.gesetze-im-internet.de/bdsg_2018/

[2427] Bundesregierung, Draft of a First Act to Amend the Federal Data Protection Act [Entwurf eines Ersten Gesetzes zur Änderung des Bundesdatenschutz- gesetzes] (Feb. 6, 2024), https://www.bmi.bund.de/SharedDocs/gesetzgebungsverfahren/DE/Downloads/kabinettsfassung/VII4/aendg-bdsg.pdf

by the Federal Parliament (*Bundestag*).[2428] The Draft Law would institutionalize a committee of Independent German Federal and State Data Protection Supervisory Authorities (Data Protection Conference—DSK), whose decisions would be legally non-binding. No updates on further action on the BDSG-new are available.

The German Federal Commissioner for Data Protection and Freedom of Information (*Der Bundesbeauftragte für den Datenschutz und die Informationsfreiheit*) is the national data protection authority for Germany. The Federal Commissioner is, however, in charge of only federal government authorities and private telecoms and postal services. [2429] Any other private entity and all other authorities in Germany are regulated by the relevant state DPA. Bavaria has two DPAs, one responsible for the private sector and one for the public sector.[2430] In other states, such as Hessian,[2431] one authority is responsible for all data protection supervision and enforcement in the state. The Federal Data Protection Commissioner serves as the joint representative of the German data protection authorities within the European Data Protection Board.[2432]

The Digital Minister established in 2025 championed simplification and centralization of the data protection supervisory authority over the various data protection commissioners at the federal and state levels.[2433] The Minister explained that making Germany attractive to businesses "requires consistent data protection and data security. [...] however, data protection must not become an innovation brake, because we also have to be able to try digital business models.[2434]

In addition to the BDSG-new, there are various sector-specific data protection regulations, such as those governing financial and energy industries. For example, the Telecommunications Telemedia Data Protection Act (TTDSG) set to be enforced in 2023, contains regulations about cookie management and Personal Information

---

[2428] Bundesministerium des Innern und für Heimat, *Erstes Gesetz zur Änderung des Bundesdatenschutzgesetzes*, Gesetzgebungverfahren (Aug. 9, 2023), https://www.bmi.bund.de/SharedDocs/gesetzgebungsverfahren/DE/aendg_bdsg.html

[2429] Federal Commissioner for Data Protection and Freedom of Information, *Tasks and Powers, BfDI*, https://www.bfdi.bund.de/EN/BfDI/UeberUns/DieBehoerde/diebehoerde_node.html

[2430] Bayerisches Landesamt für Datenschutzaufsicht, *Zuständigkeit des BayLDA, Aufgaben*, https://www.lda.bayern.de/de/aufgaben.html

[2431] Hessische Beauftragte für Datenschutz und Informationsfreiheit , *Zuständigkeit: Aufgaben und Organisation*, Über Uns, https://datenschutz.hessen.de/ueber-uns/aufgaben-und-organisation

[2432] Federal Commissioner for Data Protection and Freedom of Information, *The European Data Protection Board (EDPB)*, BfDI, https://www.bfdi.bund.de/EN/Fachthemen/Gremienarbeit/EuropaeischerDatenschutzausschuss/europaeischerdatenschutzausschuss_node.html

[2433] IT-Journal, *Digital Minister Criticizes "Patchwork" on Data Protection* (Aug. 2, 2025), https://www.it-journal.de/220186-digitalminister-kritisiert-flickenteppich-beim-datenschutz.html

[2434] Federal Government, *Speech by the Federal Minister for Digital and State Modernisation, Dr. Karsten Wildberger, at the German Savings Bank Day 2025 on the 21st May 2025 in Nuremberg* (May 24, 2025), https://www.bundesregierung.de/breg-de/service/newsletter-und-abos/bulletin/rede-des-bundesministers-fuer-digitales-und-staatsmodernisierung-dr-karsten-wildberger--2350424

Management Systems.[2435] The name of the regulation was changed to Telecommunications Digital Services Data Protection Act (TDDDG) in 2024 as a part of the introduction of the Digital Services Act. The contents are essentially identical.[2436] Further, the German Civil Code Article 327q is intended to protect user privacy in cases where a consumer gives their personal data to access a service.[2437]

The EU Charter of Fundamental Rights more generally provides that EU citizens have the right to protection of their personal data. Article 8 of the Charter states that: "Everyone has the right to the protection of personal data concerning him or her. Such data must be processed fairly for specified purposes and on the basis of the consent of the person concerned or some other legitimate basis laid down by law."

Germany is also a member of the Council of Europe and ratified the Council of Europe's Convention 108+ for the protection of individuals with regard to the processing of personal data.[2438] The German Federal Council introduced a draft law in July 2024 to protect personal rights from deepfakes under criminal law. The draft law addresses privacy concerns related to deepfake technologies and provides coverage for individuals whose rights to privacy are violated.[2439]

The Federal Data Protection Commissioner is a member of the Global Privacy Assembly since 2002. It co-sponsored the 2018 GPA Declaration on Ethics and Data Protection in AI[2440] and the 2023 GPA Resolution on Generative AI.[2441] The Federal Data Protection Commissioner also sponsored the 2020 GPA Resolution on AI Accountability.[2442] However, the Federal Data Protection Commissioner did not endorse the 2022 GPA Resolution on Principles and Expectations for the Appropriate Use of Personal Information in Facial Recognition Technology.[2443] Several State DPA

2435 TDDDG, *Telekommunikation-Digitale-Dienste-Datenschutz-Gesetz*, https://gesetz-ttdsg.de

2436 Ibid

2437 Federal Ministry of Justice, *German Civil Code,* https://www.gesetze-im-internet.de/englisch_bgb/

2438 Council of Europe, *Modernised Convention for the Protection of Individuals with Regard to the Processing of Personal Data* (May 18, 2018), https://www.coe.int/en/web/data-protection/convention108-and-protocol

2439 Bundesrat, *Entwurf eines Gesetzes zum strafrechtlichen Schutz von Persönlichkeitsrechten vor Deepfakes* (Mar. 13, 2025), https://www.bundesrat.de/SharedDocs/drucksachen/2024/0201-0300/222-24(B).pdf?__blob=publicationFile&v=1

2440 Global Privacy Assembly, *Declaration on Ethics and Data Protection in Artificial Intelligence* (Oct. 23, 2018), http://globalprivacyassembly.org/wp-content/uploads/2018/10/20180922_ICDPPC-40th_AI-Declaration_ADOPTED.pdf

2441 Global Privacy Assembly, *Resolution on Generative Artificial Intelligence Systems* (Oct. 2023), https://globalprivacyassembly.org/wp-content/uploads/2023/10/5.-Resolution-on-Generative-AI-Systems-101023.pdf.

2442 Global Privacy Assembly, *Adopted Resolution on Accountability in the Development and Use of Artificial Intelligence* (Oct. 2020), https://globalprivacyassembly.org/wp-content/uploads/2020/10/FINAL-GPA-Resolution-on-Accountability-in-the-Development-and-Use-of-AI-EN-1.pdf

2443 Global Privacy Assembly, *Resolution on Principles and Expectations for the Appropriate Use of Personal Information in Facial Recognition Technology* (Oct. 2022),

are members of the Global Privacy Assembly but none of them has endorsed GPA's AI-related resolutions.[2444]

AI Oversight

The Federal Data Protection Commissioner is also one of the members of the committee of Independent German Federal and State Data Protection Supervisory Authorities (Data Protection Conference—DSK). The DSK published a position paper in May 2024, ensuring that the member data protection authorities are prepared to be the designated national supervisory authority under the EU AI Act.[2445] In the following days, the DSK published guidelines for implementing and using AI systems in compliance with the EU's data protection regulations.[2446] The guidelines aim to clarify uncertainties in the use of AI systems and provide pre-implementation and implementation guidance with compliance requirements. However, the proposed AI Market Surveillance and Innovation Promotion Act, which would designate the Federal Network Agency as lead AI oversight authority under the EU AI Act,[2447] reflects the new Digital Minister's stance against distributed authority. All 17 data protection authorities from every state (two from Bayern) immediately opposed the proposal, citing concerns over fundamental rights and the possible unconstitutionality of supervisory power from the federal agency on the uses of AI by state authorities.[2448]

The Federal Data Protection Commissioner is a member of the G7-Roundtable of the data protection and privacy authorities. The Roundtable issued the Statement on Generative AI[2449] in 2023, the Statement on the Role of Data Protection Authorities

---

https://globalprivacyassembly.org/wp-content/uploads/2022/11/15.1.c.Resolution-on-Principles-and-Expectations-for-the-Appropriate-Use-of-Personal-Information-in-Facial-Recognition-Technolog.pdf

2444 Global Privacy Assembly, *List of Accredited Members*, https://globalprivacyassembly.org/wp-content/uploads/2022/11/15.1.c.Resolution-on-Principles-and-Expectations-for-the-Appropriate-Use-of-Personal-Information-in-Facial-Recognition-Technolog.pdf

2445 Datenschutzkonferenz, *Nationale Zuständigkeiten für die Verordnung zur Künstlichen Intelligenz (KI-VO)*, (May. 3, 2024), https://www.datenschutzkonferenz-online.de/media/dskb/20240503_DSK_Positionspapier_Zustaendigkeiten_KI_VO.pdf

2446 Datenschutzkonferenz, *Künstliche Intelligenz und Datenschutz* (May 3, 2024), https://www.datenschutzkonferenz-online.de/media/oh/20240506_DSK_Orientierungshilfe_KI_und_Datenschutz.pdf

2447 Federal Ministry for Digital Transformation and Government Modernisation, *Draft Law for Implementation of EU Regulation on Artificial Intelligence (Gesetz zur Durchführung der KI-Verordnung)* (Sept. 12, 2025), https://bmds.bund.de/service/gesetzgebungsverfahren/gesetz-zur-durchfuehrung-der-ki-verordnung

2448 Independent Data Protection Supervisory Authorities of the Federal States, State data protection authorities sharply criticize: The Ministry of Digital Affairs wants to weaken fundamental rights protection in AI oversight (Sept. 4, 2025), https://www.datenschutz-berlin.de/fileadmin/user_upload/pdf/pressemitteilungen/2025/20250904-Landesdatenschutzbehoerden-PM_KI-VO.pdf

2449 Roundtable of G7 Data Protection and Privacy Authorities, *Statement on Generative AI* (Jun. 21, 2023), https://www.bfdi.bund.de/SharedDocs/Downloads/EN/G7/2023_Statement-AI.pdf?__blob=publicationFile&v=1

in Fostering Trustworthy AI,[2450] and the Statement on AI and Children[2451] in 2024. The Federal Data Protection Commissioner played a significant role in the statements.[2452]

The G7 Data Protection Authorities (DPAs) adopted a statement on AI governance in October 2024. They pointed out the critical role of DPAs in the realm of AI, emphasizing the need for protecting the right to privacy and data protection more than ever. The adopted statement "calls on policymakers and regulators for approaches that recognise the critical role of the DPAs in ensuring that AI technologies are developed and deployed responsibly and in identifying and addressing AI issues at their source." The statement emphasized that "DPAs apply a human-centred lens to their mandate and have vast experience with data-driven processing and in operationalising many data protection overarching principles that can be transposed into broader AI governance frameworks, in particular fairness, accountability, transparency, and security."[2453]

In 2021, numerous German DPAs launched a collective investigation on the use by German companies of third-party providers outside the EU and their compliance with the 2020 European Court of Justice's *Schrems II* decision regarding international data transfers. The Court has made clear that DPAs shall intervene to suspend or prohibit transfers which do not match the *Schrems II* criteria.[2454]

The Hamburg Commissioner for Data Protection and Freedom of Information published an advisory in June 2024 on Meta's new privacy policy, planned AI model training protocol, and the utilization conditions of users' personal data from social media applications. The Hamburg Commissioner provided helpful information on users' options and what they can do, Meta's new privacy policy's aims and goals, and what happens next.[2455] Furthermore, the BayLDA published guidelines regarding the

---

[2450] Roundtable of G7 Data Protection and Privacy Authorities, *Statement on the Role of Data Protection Authorities in Fostering Trustworthy AI* (Oct. 11, 2024), https://www.bfdi.bund.de/SharedDocs/Downloads/EN/G7/2024_Statement-Trustworthy-AI.pdf?__blob=publicationFile&v=2

[2451] Roundtable of G7 Data Protection and Privacy Authorities, *Statement on AI and Children* (Oct. 11, 2024), https://www.bfdi.bund.de/SharedDocs/Downloads/EN/G7/2024_Statement-AI-Children.pdf?__blob=publicationFile&v=3

[2452] Federal Commissioner for Data Protection and Freedom of Information, *BfDI at G7 Roundtable in Rome*, BfDI, (Oct. 11, 2024), https://www.bfdi.bund.de/SharedDocs/Pressemitteilungen/EN/2024/13_G7-Roundtable-Rom.html

[2453] Roundtable of G7 Data Protection and Privacy Authorities, *G7 DPAs' Communiqué:Privacy in the Age of Data* (Oct. 11, 2024), https://www.bfdi.bund.de/SharedDocs/Downloads/EN/G7/2024_Communique.pdf?__blob=publicationFile&v=3

[2454] European Data Protection Board, *Coordinated German Investigation of International Data Transfers* (Jul. 2021), https://edpb.europa.eu/news/national-news/2021/coordinated-german-investigation-international-data-transfers_en

[2455] Hamburg Commissioner for Data Protection and Freedom of Information, *AI Training with Personal Data on Instagram and Facebook,* Press Release (Jun. 27, 2024), https://datenschutz-

application of the GDPR and other legal requirements to the processing of personal data by AI systems. The BayLDA also expressed its opinion on the significant role and responsibility of data protection supervisory authorities in the field of AI supervision.[2456]

The German Institute for Human Rights, founded in 2001 by the German Bundestag (Parliament) as an independent national institution, works to ensure the observation and promotion of human rights by the German government in Germany and abroad.[2457] The Institute published an interview on protecting human rights when applying AI in the context of elderly care in 2019[2458] and has emphasized the importance of assessing and preventing the human rights risks of Artificial Intelligence in 2021,[2459] though the Institute has not indicated that it sees a sustained oversight of AI-related human rights infringements as a priority.

The Federal Ministry of the Interior and Community is creating the Advisory Center for AI (BeKI) as a central point for contacting and coordinating for AI projects at the federal level. BeKI is still under development[2460] and aims to ensure a coordinated approach, advise on the legal, ethical, and technical topics for responsible and transparent use of AI, and transform administrative processes sustainably.[2461]

The State Commissioner for Data Protection in Lower Saxony announced the establishment of the Office for AI under his authority in 2024. The new unit will support the use of AI technologies in relation to data protection laws, serve as a competence center for AI-related questions, and collaborate with other relevant authorities. The Commissioner Lehmkemper pointed out: "The establishment of the staff unit is an important step to proactively meet the challenges that AI brings with it, to ensure the protection of personal data in an increasingly digitalized world, and at the same time to accompany the opportunities of technical development." The unit

---

hamburg.de/fileadmin/user_upload/HmbBfDI/Datenschutz/Informationen/240627_PM_AI_Training_with_personal_data_Insta_Facebook_EN.pdf

[2456] BayLDA, *KI & Datenschutz*, Künstliche Intelligenz (Jun. 2024), https://www.lda.bayern.de/de/ki.html?mkt_tok=MTM4LUVaTS0wNDIAAAGVQHhz4nfvGN0UqzOzdh9guQEZYr0BeSK-YiffEsdz8i0DT-iweup1RCfhQNvco0i4ynZFqggLzoW0oAQM0DXYnSV_02iMc1DxKxuVAD-pGoad

[2457] German Institute for Human Rights, *Das Institut*, https://www.institut-fuer-menschenrechte.de/das-institut

[2458] German Institute for Human Rights, *Miteinander von Mensch und Maschine* (Nov. 19, 2019), https://www.institut-fuer-menschenrechte.de/aktuelles/detail/miteinander-von-mensch-und-maschine

[2459] German Institute for Human Rights, *Algorithmische Entscheidungssysteme* (Jul. 2021), https://www.institut-fuer-menschenrechte.de/fileadmin/Redaktion/Publikationen/Information/Information_Algorithmische_Entscheidungssysteme.pdf

[2460] Deutscher Bundestag, *Beratungszentrum für Künstliche Intelligenz,* Presse (Jul. 19, 2024), https://www.bundestag.de/presse/hib/kurzmeldungen-1013594

[2461] Federal Ministry of the Interior and Community, *Beratungszentrum für Künstliche Intelligenz (BeKI),* https://www.digitale-verwaltung.de/SharedDocs/downloads/Webs/DV/DE/Transformation/akteurssteckbrief-beki.pdf?__blob=publicationFile&v=5

also aims to devise safeguards for the implementation and evaluation of AI, evaluate the potential hazards of AI-based systems, provide assistance for research, promote awareness among various stakeholders, and conclusively ensure the implementation of AI in accordance with data protection regulations.[2462]

Algorithmic Transparency

Germany is subject to the GDPR and Convention 108+. Germans have a general right to obtain access to information about automated decision-making and to the factors and logic of an algorithm.[2463] In line with a 2020 Recommendation of the Council of Europe Committee of Ministers on human rights impacts of algorithm systems,[2464] these rights include requirements on transparency, accountability, and effective remedies.

The Ministry of Education and Research started a funding priority for AI R&D projects on explainability and transparency in 2019. The Ministry stated that improving explainability and transparency are two of the Federal government's central research goals.[2465] Funding is "aimed at collaborative projects between science and industry in an interdisciplinary composition."[2466] The German consumer organization vzbv emphasized in 2019 that the aim of automated decision-making regulation must be to ensure compliance with existing laws.[2467]

The German Data Ethics Commission sees auditability, explainability, and redress possibilities as crucial for algorithm safety.[2468] However, even though Germany has had a significant impact on the drafting of the EU AI Act and its transparency regulations, AlgorithmWatch has noted that Germany (and specifically its Ministry of the Interior) has in recent years tried to water down transparency obligations for AI systems in the AI Act, most notably with regard to exemptions for

---

[2462] Der Landesbeauftragte für den Datenschutz Niedersachsen, *Landesbeauftragter für den Datenschutz richtet Stabsstelle für Künstliche Intelligenz ein*, Pressemitteilung Nr. 16/2024 (Sept. 30, 2024), https://www.lfd.niedersachsen.de/startseite/infothek/presseinformationen/landesbeauftragter-fur-den-datenschutz-richtet-stabsstelle-fur-kunstliche-intelligenz-ein-236012.html

[2463] See Recital 63 and Article 22 of the GDPR. Article 9 c) of the Convention 108+ as well as Recital 77, *Explanatory Report, Convention 108+*, p. 24, https://rm.coe.int/convention-108-convention-for-the-protection-of-individuals-with-regar/16808b36f1

[2464] Council of Europe, *Recommendation CM/Rec(2020)1 of the Committee of Ministers to Member States on the Human Rights Impacts of Algorithmic Systems* (Apr. 8, 2020), https://search.coe.int/cm/pages/result_details.aspx?objectid=09000016809e1154

[2465] Bundesministerium für Bildung und Forschung, *KI-Erklärbarkeit und Transparenz*, https://www.softwaresysteme.pt-dlr.de/de/ki-erkl-rbarkeit-und-transparenz.php

[2466] OECD G20 Digital Economy Task Force, *Examples of AI National Policies* (2020), https://www.mcit.gov.sa/sites/default/files/examples-of-ai-national-policies.pdf

[2467] Vzbv, *Artificial Intelligence: Trust is Good, Control is Better* (2019), https://www.vzbv.de/sites/default/files/2019_vzbv_factsheet_artificial_intelligence.pdf

[2468] Datenethikkommission, *Opinion of the Data Ethics Commission* (Oct. 2019), https://www.bfdi.bund.de/SharedDocs/Downloads/EN/Datenschutz/Data-Ethics-Commission_Opinion.pdf?__blob=publicationFile&v=2

AI use in "law enforcement, migration, asylum and border control," including remote biometric identification systems.[2469]

According to AlgorithmWatch,[2470] the data protection agencies of the federal government and eight German federal states stated that greater transparency in the implementation of algorithms in the administration was indispensable for the protection of fundamental rights.[2471] The agencies demanded that if automated systems are used in the public sector, it is crucial that processes are intelligible and can be audited and controlled. In addition, public administration officials must be able to provide an explanation of the logic of the systems used and the consequences of their use. Self-learning systems must also be accompanied by technical tools to analyze and explain their methods. An audit trail should be created, and the software code should be made available to the administration and, if possible, to the public. According to the position paper, there need to be mechanisms for citizens to demand redress or reversal of decisions, and the processes must not be discriminating. In cases where there is a high risk for citizens, there needs to be a risk assessment done before deployment. Very sensitive systems should require authorization by a public agency that has yet to be created.

One of the State DPAs, the Berlin Commissioner for Data Protection and Freedom of Information (BlnBDI), rendered a decision reasserting the importance of algorithmic transparency. In May 2023, the Berlin Data Protection Authority fined a Berlin-based bank € 300,000. The Bank used an online form to request various data about the applicant's income, occupation and personal details for a credit card application. Based on the information requested and additional data from external sources, the bank's algorithm rejected the customer's application without any justification. The algorithm is based on criteria and rules previously defined by the bank. Since the client had a good credit rating and a regular high income, he doubted the automated rejection and complained to the Berlin data protection commissioner. The lack of transparency regarding the automated decision led to the imposition of a fine by the State DPA.[2472]

Mannheim, in collaboration with eight other cities across Europe and with the help of Eurocities' Digital Forum, adopted an algorithm register, the Algorithmic

---

[2469] Nikolett Aszódi and Matthias Spielkamp, *How the German Government Decided Not to Protect People against the Risks of AI* (Dec. 6, 2022), https://algorithmwatch.org/en/german-government-risks-of-ai/

[2470] Algorithm Watch, *Automating Society: Germany* (Jan. 29, 2019), https://algorithmwatch.org/en/automating-society-2019/germany

[2471] Freedom of Information Commissioners in Germany, *Transparenz der Verwaltung beim Einsatz von Algorithmen für gelebten Grundrechtsschutz unabdingbar* (Oct. 16, 2018), https://www.datenschutzzentrum.de/uploads/informationsfreiheit/2018_Positionspapier-Transparenz-von-Algorithmen.pdf

[2472] Berliner Beauftragte für Datenschutz und Informationsfreiheit (BlnBDI), *Computer Sagt Nein, Pressemitteilung* (May 31, 2023), https://www.datenschutz-berlin.de/pressemitteilung/computer-sagt-nein

Transparency Standard in January 2023.[2473] The aim is to provide more information for residents about the use of algorithm by municipalities and their impact. The register includes a range of information such as the type and purpose of an algorithm, the department using the algorithm, the geographical area and domain it relates to and a risk category. It also includes details on the data source and training data, any bias and mitigation, and human oversight. This initiative builds on similar algorithm registers launched in Amsterdam and Helsinki.[2474] The Federal Ministry of the Interior and Community opened the AI Opportunity Market[2475] in 2025. The AI Opportunity Market offers an overview of existing and planned AI systems in the German federal, state, and local governments. According to the Ministry website, the AI Opportunity Market "already meets many of the transparency requirements' set out in the EU AI Act."[2476]

The German Institution for Standardization (DIN) issued the DIN SPEC 92001-3 in 2023 as a guide for appropriate approaches and methods to enhance explainability throughout the life cycle of AI systems. The guide describes quality criteria, which promote explainable AI, for all types of AI systems and for all sectors. These criteria form the specific requirements for explainability and provide proof of technical operation correctness, allowing organizations to embed transparency principles in their AI-based applications. The guide is the outcome of the CERTIFIED AI project, which was financed by the Ministry of Economic Affairs, Industry, Climate Protection and Energy of the State of North Rhine-Westphalia.[2477]

The Federal Office for Information Security (BSI) published a whitepaper on transparency of AI systems in 2024. In the whitepaper, the BSI highlighted the importance of transparency for AI system's trustworthiness and defined transparency in the context of AI systems for different stakeholders. The definition has been made in a technology-neutral and future-proof manner and covers the entire ecosystem related to AI, considering transparency beyond the actual system. The whitepaper also enumerates the transparency requirements of EU AI Act and examines the opportunities and risks of transparent AI systems.[2478]

[2473] Algorithm Register, *Algorithmic Transparency Standard*, https://www.algorithmregister.org

[2474] Eurocities, *Nine Cities Set Standards for the Transparent Use of Artificial Intelligence* (Jan. 19, 2023), https://eurocities.eu/latest/nine-cities-set-standards-for-the-transparent-use-of-artificial-intelligence/

[2475] Federal Ministry for Digital Transformation and Government Modernisation, *AI Opportunity Market* (Jul. 24, 2025), https://www.kimarktplatz.bund.de/

[2476] Federal Ministry of the Interior and Community, *Artificial Intelligence in the Federal Administration: AI Opportunity Market and Transparency Database Now Open* (Jan. 27, 2025), https://www.bmi.bund.de/SharedDocs/pressemitteilungen/EN/2025/01/maki-pm.html

[2477] DIN, *Künstliche Intelligenz transparenter machen*, Presse (Jul. 19, 2023), https://www.din.de/de/din-und-seine-partner/presse/mitteilungen/kuenstliche-intelligenz-transparenter-machen-931058

[2478] Bundesamt für Sicherheit in der Informationstechnik, *Transparenz von KI-Systemen* (Jul. 1, 2024), https://www.bsi.bund.de/SharedDocs/Downloads/DE/BSI/KI/Whitepaper-Transparenz-KI-Systeme.pdf?__blob=publicationFile&v=3

The Federal Ministry for Digital and Transport (BMDV) and Federal Ministry of the Interior and Community (BMI) issued the AI guidelines in June 2024. The guidelines outline the framework for utilizing AI within the BMDV and BMI. It has been emphasized that the BMDV "use and design AI systems in such a way that results, modes of operation, decision-making principles and paths are presented as comprehensibly as possible for all those involved (explainable AI). This enables people to consciously acknowledge the results of AI, take responsibility for them and intervene if necessary. We make the use of AI recognizable." Furthermore, transparency regarding existing data, AI literacy, and broad participation opportunities in the AI environment are promoted.[2479] The BMI elucidated the manner and rationale for the utilization of AI, as well as their guiding principles regarding its application.[2480]

Facial Recognition

German governments have launched several projects on facial recognition technology, but these have been met with considerable public resistance. In 2017, Hamburg police deployed facial recognition technology in the wake of the G20 protests, which led to a three-year legal battle involving the Hamburg DPA and several courts that ended with the police deleting its biometric database in 2020.[2481] In 2018, the German Ministry of the Interior deployed facial recognition technology at a large train station in Berlin, sparking opposition from civil society.[2482] There was further outcry in 2020, when *Der Spiegel* reported that Germany planned to set up cameras capable of identifying people at 134 train stations and 14 airports.[2483] In 2021, Germany's incoming coalition government said it would exclude biometric recognition in public spaces as well as automated state scoring systems by AI.[2484] A

[2479] Bundesministerium für Digitales und Verkehr, *BMDV-KI-Leitlinien* (Jun. 17, 2024), https://bmdv.bund.de/SharedDocs/DE/Anlage/K/presse/pm-047-ki-richtlinie.pdf?__blob=publicationFile

[2480] Bundesministerium des Innern und für Heimat, *KI-Leitbild für das Ressort BMI* (Jun. 2024), https://www.bmi.bund.de/SharedDocs/downloads/DE/publikationen/themen/it-digitalpolitik/BMI24014.pdf?__blob=publicationFile&v=3

[2481] Identity Week, *Hamburg Deletes Facial Database for G20* (Jun. 2, 2020), https://identityweek.net/hamburg-deletes-facial-database-for-g20/

[2482] Janosch Delcker, *Big Brother in Berlin,* Politico (Sept. 13, 2018), https://www.politico.eu/article/berlin-big-brother-state-surveillance-facial-recognition-technology/

[2483] Phillipp Grüll, *Germany's Plans for Automatic Facial Recognition Meet Fierce Criticism,* Euractiv (Jan. 10, 2020), https://www.euractiv.com/section/data-protection/news/german-ministers-plan-to-expand-automatic-facial-recognition-meets-fierce-criticism/

[2484] POLITICO, *German Coalition Backs Ban on Facial Recognition in Public Places* (Nov. 24, 2021), https://www.politico.eu/article/german-coalition-backs-ban-on-facial-recognition-in-public-places/; Alliance for Freedom, Justice, and Sustainability, *Dare More Progress: Coalition agreement 2021–2025 between the Social Democratic Party of Germany (SPD), ALLIANCE 90 / THE GREENS and the Free Democrats (FDP),* https://www.welt.de/bin/Koalitionsvertrag%202021-2025.pdf_bn-235257672.pdf

January 2023 study showed that German citizens, when asked about facial recognition technology, generally "call for strong regulations to address associated risks," although citizens' trust in the government is correlated with their attitudes towards facial recognition technology.[2485]

In late 2022 and early 2023, Germany made clear the country's reservations on the AI Act, although the position seems to mesh well with the Act's general approach.[2486] The German government favored a total ban on remote biometric identification technology in 2021 as per its coalition agreement[2487] while, in 2023, it only supported a ban on *real-time* biometric recognition identification and would allow ex-post identification systems.[2488] The German Data Protection Conference (DSK) released a resolution on the use of facial recognition systems by health authorities in 2024 that reiterated Germany's resolve to uphold a ban on facial recognition in public spaces. The DSK asserts that the use of automated facial recognition—if permissible at all—must be absolutely necessary for the protection of high-ranking legal interests and must be subject to appropriate conditions depending on the deployment scenario and the level of intrusion. The legal basis for its use must include sufficient safeguards for fundamental rights and additional protective mechanisms.[2489]

The European Data Protection Board (EDPB) produced guidelines on the use of facial recognition technologies in law enforcement in 2022.[2490] The EDPB stresses that facial recognition tools should only be used in strict compliance with the Law Enforcement Directive (LED). Moreover, such tools should only be used if necessary and proportionate, as laid down in the Charter of Fundamental Rights."[2491] After

---

[2485] Genia Kostka, Léa Steinacker and Miriam Meckel, *Under Big Brother's Watchful Eye: Cross-Country Attitudes Toward Facial Recognition Technology*, Government Information Quarterly, 40(1) (Jan. 2023), https://doi.org/10.1016/j.giq.2022.101761

[2486] Luca Bertuzzi and Molly Killeen, *Tech Brief: Germany's AI Reservations, Fair Share Moves in the Metaverse* (Jan. 13, 2023), https://www.euractiv.com/section/digital/news/tech-brief-germanys-ai-reservations-fair-share-moves-in-the-metaverse/

[2487] SPD, Bündnis 90/Die Grünen and FDP, *Mehr Fortschritt Wagen* (2021), https://cms.gruene.de/uploads/assets/Koalitionsvertrag-SPD-GRUENE-FDP-2021-2025.pdf

[2488] Luca Bertuzzi, *Germany Could Become MEPs' Ally in AI Act Negotiations*, Euractiv (Jan. 9, 2023), https://www.euractiv.com/section/artificial-intelligence/news/germany-could-become-meps-ally-in-ai-act-negotiations/

[2489] German Data Protection Conference (DSK), *Caution Advised for Security Authorities Using Facial Recognition [Vorsicht bei dem Einsatz von Gesichtserkennungssystemen durch Sicherheitsbehörden]* (Sept. 20, 2024), https://www.datenschutzkonferenz-online.de/media/en/2024-09-20_Entschliessung_DSK_Gesichtserkennung.pdf

[2490] European Data Protection Board, *Guidelines 05/2022 on the use of facial recognition technology in the area of law enforcement* [obsolete] (May12, 2022), https://edpb.europa.eu/system/files/2022-05/edpb-guidelines_202205_frtlawenforcement_en_1.pdf

[2491] European Data Protection Board, *EDPB Adopts Guidelines on Calculation of Fines & Guidelines on the Use of Facial Recognition Technology in the Area of Law Enforcement, Press release* (May 16, 2022), https://edpb.europa.eu/news/news/2022/edpb-adopts-guidelines-calculation-fines-guidelines-use-facial-recognition_en

public consultation, the EDPB adopted guidelines for the use of facial recognition technologies in 2023. In version 2.0, practical examples and guidance were provided regarding facial recognition scenarios.[2492]

In 2024, the EDPB urged airline companies and airport operators to opt for less intrusive ways to streamline passenger flows, when possible. In the view of the EDPB Chair, "individuals should have maximum control over their own biometric data."[2493]

In 2024, after a terrorist attack in Solingen, the Federal Government presented two draft bills, the so-called "security package," to tighten security measures. The first draft bill, which aims to enhance internal security and the asylum system, empowers the Federal Office for Migration and Refugees to utilize biometric data to ascertain the identity of asylum seekers that do not carry IDs by comparing biometrics with publicly accessible data from the Internet.[2494] The second draft bill, which aims to improve the fight against terrorism, allows the use of biometric comparison for facial recognition to identify suspects easier.[2495] The security package has been subjected to serious criticism.[2496] The two draft bills were terminated with the end of the 20th legislative period. The Federal Ministry of the Interior of the new federal government, however, reportedly wants to renew these proposals in new draft legislation in 2025.[2497] Digital rights organizations have strongly criticized the renewed effort, stating that the measures will "enable total surveillance of public space."[2498]

Predictive Policing

The German police has launched several projects using AI to assist in predictive policing both on the federal and state level. The Federal Crime Agency has used risk scoring for "militant Salafists"[2499] while several state police forces have

---

[2492] European Data Protection Board, *Guidelines 05/2022 on the Use of Facial Recognition Technology in the Area of Law Enforcement*, Version 2.0 (Apr. 26, 2023), https://www.edpb.europa.eu/system/files/2023-05/edpb_guidelines_202304_frtlawenforcement_v2_en.pdf

[2493] European Data Protection Board, *Facial Recognition at Airports: Individuals Should Have Maximum Control Over Biometric Data* (Mar. 25, 2024), https://www.edpb.europa.eu/news/news/2024/facial-recognition-airports-individuals-should-have-maximum-control-over-biometric_en

[2494] Deutscher Bundestag, *Entwurf eines Gesetzes zur Verbesserung der inneren Sicherheit und des Asylsystems,* (Sept. 9, 2024), https://dserver.bundestag.de/btd/20/128/2012805.pdf

[2495] Ibid

[2496] Svea Windwehr, *Germany Rushes to Expand Biometric Surveillance,* EFF (Oct. 7, 2024), https://www.eff.org/deeplinks/2024/10/germany-rushes-expand-biometric-surveillance

[2497] netzpolitik.org, *Interior Minister Dobrindt Plans New Security Package* (Jul. 23, 2025), https://netzpolitik.org/2025/gesichtserkennung-und-ki-innenminister-dobrindt-plant-neues-sicherheitspaket/

[2498] Ibid

[2499] Louisa Well, *AlgorithmWatch Automating Society Report 2020: Germany* (2020), https://automatingsociety.algorithmwatch.org/report2020/germany/

deployed PRECOBS, an anti-burglary system, with varying success. The police of Baden-Württemberg, for instance, started a pilot project with PRECOBS in 2015[2500] that was ended in 2019 "due to data quality issues."[2501] North Rhine-Westphalia developed and tested the predictive policing tool SKALA from 2015 to 2018.[2502] In late 2022, Germany's Constitutional Court started a legal review of surveillance software deployed by police in the state of Hesse since 2017. The software, dubbed 'Hessendata', is based on the US company Palantir's Gotham program and its use has been met with public criticism.[2503] In the case, the court found that provisions in the laws of the states of Hesse and Hamburg, which enable the police to process data by matching data from various databases and to carry out automatic data analysis, were unconstitutional.[2504]

In the EU AI Act negotiations, Germany called for a ban on systems that would replace human judges in crime and recidivism risk assessment as well, but it also pushed for exemptions for law enforcement in other areas.[2505]

According to Germany's "security package," the Federal Criminal Police Office and the Federal Police will be able to compare biometric data with online sources and evaluate large volumes of data using AI. AI-based analysis can be used to establish connections between information held by law enforcement and the Internet, which is often used for predictive policing. The authority granted to law enforcement could be utilized in the event of a "particularly serious" crime suspicion.[2506] The widely criticized draft has not yet been enacted into law.

Environmental Impact of AI

The German Federal Government has acknowledged the importance of developing AI that is sustainable for the planet. In the 2020 update of Germany's Artificial Intelligence Strategy, the Federal Government says that it "will work to

---

[2500] Dominik Gerstner, *Predictive Policing in the Context of Residential Burglary: An Empirical Illustration on the Basis of a Pilot Project in Baden-Württemberg, Germany*, European Journal for Security Research, 3 (2018), https://doi.org/10.1007/s41125-018-0033-0

[2501] Louisa Well, *AlgorithmWatch Automating Society Report 2020: Germany* (2020), https://automatingsociety.algorithmwatch.org/report2020/germany/

[2502] Police North Rhine-Westphalia, *Projekt SKALA* (Jun. 8, 2018) https://polizei.nrw/artikel/projekt-skala-predictive-policing-in-nrw-ergebnisse

[2503] Ben Knight, *Germany: Police Surveillance Software a Legal Headache* (Dec. 22, 2022), https://www.dw.com/en/germany-police-surveillance-software-a-legal-headache/a-64186870

[2504] First Senate of 16 Feb. 2023 *Headnotes to the Judgment of 16 February 2023*, 1 BvR 1547/19, 1 BvR 2634/20, *Automated Data Analysis* (Feb. 16, 2023), https://www.bundesverfassungsgericht.de/SharedDocs/Entscheidungen/EN/2023/02/rs20230216_1bvr154719en.html

[2505] Luca Bertuzzi, *Germany Could Become MEPs' Ally in AI Act Negotiations,* Euractiv (Jan. 9, 2023), https://www.euractiv.com/section/artificial-intelligence/news/germany-could-become-meps-ally-in-ai-act-negotiations

[2506] Deutscher Bundestag, *Entwurf eines Gesetzes zur Verbesserung der Terrorismusbekämpfung,* (Sept. 9, 2024), https://dserver.bundestag.de/btd/20/128/2012806.pdf

encourage that the technology is designed to be both energy and resource-efficient and is used as an instrument for environmental conservation. Direct and indirect environmental impacts will also be taken into account in all of this, so as to include rebound effects and shifting of environmental problems."[2507] There are currently no policies that explicitly refer to both AI and sustainability. However, in April 2024, the G7 Ministers of Climate, Energy, and the Environment committed to "promote and share emerging transformative capacity using artificial intelligence and machine learning to improve the complementary use of remote and in situ environmental data for climate, energy and environment decision making" in the Climate, Energy and Environment Ministers' Meeting Communiqué.[2508]

The German Federal Ministry for Economic Cooperation (BMZ) and Development and the United Nations Development Programme (UNDP) announced a new platform designed to help ensure that responsible artificial intelligence is used for sustainable development in October 2024.[2509] The platform includes a list of five Sustainable Development Goals (SDG) as well as a way for participants to submit proposals for their initiatives. The initiatives should meet three criteria: (1) the initiative should be international, regional, or global, (2) the initiative should demonstrate its impact of its ability to achieve significant impact, and (3) at least one of the initiatives main goals should be to achieving one of the SDG in a developing country.[2510]

Lethal Autonomous Weapons

The German government's 2018 coalition agreement stated that it "rejects autonomous weapon systems devoid of human control" and called for a global ban.[2511] Also in 2018, in cooperation with the French government, the German government published a joint statement on Lethal Autonomous Weapons at the Meeting of the Group of Governmental Experts on Lethal Autonomous Weapons

---

[2507] German Federal Government, *Artificial Intelligence Strategy of the German Federal Government, 2020 Update* (Dec. 2020), https://www.ki-strategie-deutschland.de/files/downloads/Fortschreibung_KI-Strategie_engl.pdf

[2508] G7 Ministers of Climate, Energy, and the Environment, *Climate, Energy and Environment Ministers' Meeting Communiqué* (Apr. 29–30, 2024), https://www.g7italy.it/wp-content/uploads/G7-Climate-Energy-Environment-Ministerial-Communique_Final.pdf

[2509] UN Development Programme (UNDP), *Global Collective Action Platform for Responsible AI for Sustainable Development Initiated by UN Development Programme and Germany's Ministry for Economic Cooperation and Development* (Oct. 2024), https://www.undp.org/press-releases/global-collective-action-platform-responsible-ai-sustainable-development-initiated-un-development-programme-and-germanys-ministry

[2510] BMZ & UNDP, *AI for the Sustainable Development Goals* (2024), https://www.sdgaicompendium.org/

[2511] Konrad Adenauer Stiftung Europe, *A New Awakening for Europe. A New Dynamic for Germany. A New Solidarity for Our Country: Coalition Agreement between CDU, CSU, and SPD* (2018), https://www.kas.de/c/document_library/get_file?uuid=bd41f012-1a71-9129-8170-8189a1d06757&groupId=284153

Systems (LAWS). They wrote, "At the heart of our proposal is the recommendation for a political declaration, which should affirm that State parties share the conviction that humans should continue to be able to make ultimate decisions with regard to the use of lethal force and should continue to exert sufficient control over lethal weapons systems they use."[2512]

In 2019, the then-Foreign Minister of Germany, Heiko Maas, reasserted the German position as being in favor of a total ban on Lethal Autonomous Weapons Systems,[2513] and the Foreign Ministry organized a virtual forum on LAWS in 2020 to move closer to a "collective normative framework."[2514] Observers have argued, however, that the 2021 German coalition agreement is "ambivalent about legally binding action on autonomous weapons" and that it merely "rejects" these systems instead of pushing for regulation.[2515]

The German government's 2021 coalition agreement states, "We reject lethal autonomous weapon systems that are completely removed from human control. We actively promote their international outlawing. We want the peaceful use of space and cyberspace. For weapons technology developments in biotech, hypersonics, space, cyber and AI, we will take early arms control initiatives. We want to contribute to strengthening norms for responsible state behavior in cyberspace."[2516]

In October 2022, Germany was one of 70 states that endorsed a joint statement on autonomous weapons systems at the United Nations General Assembly. The statement called for the recognition of the dangers of autonomous weapons systems, acknowledged the need for human oversight and accountability, and emphasized the importance of an international framework of rules and constraints.[2517] In this joint statement, States declared that they "are committed to upholding and strengthening compliance with International Law, in particular International Humanitarian Law,

---

[2512] Permanent Representation of the Federal Republic of Germany to the Conference on Disarmament in Geneva & Représentation Permanente de la France auprès de la Conférence du Désarmement, Meeting of the Group of Governmental Experts on Lethal Autonomous Weapons Systems, *Statement by France and Germany* (Apr. 2018), http://perma.cc/2FQB-W8FX)

[2513] Federal Foreign Office, *Foreign Minister Maas on Agreement of Guiding Principles Relating to the Use of Fully Autonomous Weapons Systems* (Nov. 15, 2019), https://www.auswaertiges-amt.de/en/newsroom/news/maas-autonomous-weapons-systems/2277194

[2514] Federal Foreign Office, *Forum on Lethal Autonomous Weapons Systems* (Apr. 2, 2020), https://www.auswaertiges-amt.de/en/aussenpolitik/themen/forum-laws/2330682

[2515] Autonomous Weapons, *New German "Traffic Light Coalition" Ambivalent About Legally Binding Action on Autonomous Weapons* (Nov. 26, 2021), https://autonomousweapons.org/new-german-traffic-light-coalition-ambivalent-about-legally-binding-action-on-autonomous-weapons/

[2516] SPD, Bündnis 90/Die Grünen and FDP, *Mehr Fortschritt Wagen*, p. 45 (2021), https://cms.gruene.de/uploads/assets/Koalitionsvertrag-SPD-GRUENE-FDP-2021-2025.pdf

[2517] Ousman Noor, *70 states deliver joint statement on autonomous weapons systems at UN General Assembly*, Stop Killer Robots (Oct. 21, 2022) https://www.stopkillerrobots.org/news/70-states-deliver-joint-statement-on-autonomous-weapons-systems-at-un-general-assembly/

including through maintaining human responsibility and accountability in the use of force."[2518]

At the 78th UN General Assembly First Committee in 2023, Germany voted in favor[2519] of resolution L.56[2520] on autonomous weapons systems, along with 163 other states. The Resolution emphasized the "urgent need for the international community to address the challenges and concerns raised by autonomous weapons systems," and mandated the UN Secretary-General to prepare a report reflecting the views of member and observer states on autonomous weapons systems. Germany's submission reiterated support for "a legally binding instrument in the framework of the Group of Governmental Experts that prohibits the development, fielding or deployment of lethal autonomous weapons systems that cannot comply with international humanitarian law and are ipso facto prohibited." Germany also supported "a set of regulations of weapons systems with autonomous functions, to ensure that human control is retained at all times [and] the creation of an additional protocol to the Convention on Certain Conventional Weapons that entails the prohibition of lethal autonomous weapons systems operating outside of human control and a responsible chain of command."[2521]

Germany supported of Draft Resolution L.77 on lethal autonomous weapons systems at the 79th UN General Assembly First Committee. The resolution raises concerns about the "negative consequences and impact of autonomous weapon systems on global security and regional and international stability" and stresses "the importance of the role of humans in the use of force to ensure responsibility and accountability and for States to comply with international law."[2522]

Germany also co-facilitated the Pact for the Future adopted by UN Member States in September 2024 that provides concrete measures to ensure a world that is safer, more peaceful, sustainable, and inclusive for the generations to come. [2523] As per the provisions of the Pact, the Member States have taken the decision to "advance

[2518] UN General Assembly, First Committee, *Joint Statement on Lethal Autonomous Weapons Systems First Committee,* 77th United Nations General Assembly Thematic Debate – Conventional Weapons (Oct. 21, 2022), https://estatements.unmeetings.org/estatements/11.0010/20221021/A1jJ8bNfWGlL/KLw9WYcSnnAm_en.pdf

[2519] Isabelle Jones, *164 states vote against the machine at the UN General Assembly*, Stop Killer Robots (Nov. 1, 2023), https://www.stopkillerrobots.org/news/164-states-vote-against-the-machine/

[2520] General Assembly, *Resolution L56: Lethal Autonomous Weapons* (Oct.12, 2023), https://reachingcriticalwill.org/images/documents/Disarmament-fora/1com/1com23/resolutions/L56.pdf

[2521] UN Secretary-General, *Lethal Autonomous Weapons Systems* (Jul. 1, 2024), https://docs-library.unoda.org/General_Assembly_First_Committee_-Seventy-Ninth_session_(2024)/A-79-88-LAWS.pdf

[2522] UN General Assembly, *Seventy Ninth Session, First Committee, Agenda 98, Lethal Autonomous Weapons System* (Oct. 17, 2024), https://documents.un.org/doc/undoc/ltd/n24/305/45/pdf/n2430545.pdf

[2523] Vibhu Mishra, *Pact for the Future: World Leaders Pledge Action for Peace, Sustainable Development*, UN News (Sept. 22, 2024), https://news.un.org/en/story/2024/09/1154671

with urgency discussions on lethal autonomous weapons systems through the Group of Governmental Experts on Emerging Technologies in the Area of Lethal Autonomous Weapons Systems with the aim to develop an instrument, without prejudging its nature, and other possible measures to address emerging technologies in the area of lethal autonomous weapons systems, recognizing that international humanitarian law continues to apply fully to all weapons systems, including the potential development and use of lethal autonomous weapons systems."[2524]

Germany participated in an international summit on the responsible application of artificial intelligence in the military domain hosted by the Netherlands in February 2023. At the end of the Summit, Germany, together with other countries, agreed on a joint call for action[2525] on the responsible development, deployment and use of artificial intelligence in the military domain to "stress the paramount importance of the responsible use of AI in the military domain, employed in full accordance with international legal obligations and in a way that does not undermine international security, stability and accountability." Germany also endorsed the resulting Political Declaration on Responsible Military Use of AI and Autonomy issued in November 2023.[2526]

The second REAIM summit took place in Seoul on September 9–10, 2024. The REAIM Blueprint for Action was the outcome document for the summit. The blueprint encourages states, industry, civil society, and regional and international organizations to recognize and affirm the impact of AI applications on the military domain, especially on international peace and security. The blueprint exhorts stakeholders to implement ethical and human-centric AI applications while considering national and international laws. The importance of maintaining a global and regional dialogue and fostering active participation among stakeholders has been emphasized. Germany is one of the supporting countries for this blueprint.[2527]

AI Literacy

Germany's AI Action Plan recognizes education, skills, and public awareness as prerequisites for responsible innovation.[2528] Most of these initiatives have centered

---

[2524] United Nations, *Pact for the Future, Global Digital Compact and Declaration on Future Generations* (Sept. 2024), https://www.un.org/sites/un2.un.org/files/sotf-pact_for_the_future_adopted.pdf

[2525] Responsible AI in the Military Domain Summit, *REAIM Call to Action* (Feb. 16, 2023), https://www.government.nl/documents/publications/2023/02/16/reaim-2023-call-to-action

[2526] US Department of State, *Political Declaration on Responsible Military Use of Artificial Intelligence and Autonomy*, Endorsing States (Nov. 9, 2023), https://www.state.gov/political-declaration-on-responsible-military-use-of-artificial-intelligence-and-autonomy/

[2527] Responsible AI in the Military Domain Summit, *REAIM Blueprint for Action,* DigWatch (Sept. 11, 2024), https://dig.watch/resource/responsible-ai-in-the-military-domain-reaim-blueprint-for-action#REAIM_Blueprint_for_Action

[2528] Federal Ministry of Research, Technology and Space, *Artificial Intelligence, AI Action Plan* (Nov. 2023),

on building expertise in AI research and STEM, although some, including the Data Literacy Toolbox, focus on more general literacy skills, which would facilitate participation in policymaking related to digital and AI concerns.

Human Rights

According to Freedom House, Germany is one of the top countries in the world for the protection of political rights and civil liberties, receiving a score of 95/100[2529] and a ranking as "Free" in 2025. Freedom House reports that, "Germany is a representative democracy with a vibrant political culture and civil society. Political rights and civil liberties are largely upheld by law and in practice."

In a 2020 Recommendation to member States on the human rights impacts of algorithmic systems, the Council of Europe Committee of Ministers reminded member States of commitment to "ensuring the rights and freedoms enshrined in the Convention for the Protection of Human Rights and Fundamental Freedoms to everyone within their jurisdiction" and that "member States must ensure that any design, development and ongoing deployment of algorithmic systems occur in compliance with human rights and fundamental freedoms, which are universal, indivisible, inter-dependent and interrelated, with a view to amplifying positive effects and preventing or minimising possible adverse effects."[2530]

AI Safety Summit

In November 2023, Germany participated in the first AI Safety Summit and endorsed the Bletchley Declaration.[2531] Germany thus committed to participate in international cooperation efforts on AI "to promote inclusive economic growth, sustainable development and innovation, to protect human rights and fundamental freedoms, and to foster public trust and confidence in AI systems to fully realise their potential." Endorsing parties affirmed that for the good of all, AI should be designed, developed, deployed, and used in a manner that is safe, in such a way as to be human-centric, trustworthy and responsible."

---

https://www.bmftr.bund.de/EN/Research/EmergingTechnologies/ArtificialIntelligence/artificialintelligence_node.html

[2529] Freedom House, *Freedom in the World 2025: Germany* (2025), https://freedomhouse.org/country/germany/freedom-world/2025

[2530] *Recommendation CM/Rec(2020)1 of the Committee of Ministers to member States on the human rights impacts of algorithmic systems* (Apr. 8, 2020), https://search.coe.int/cm/pages/result_details.aspx?objectid=09000016809e1154

[2531] UK Department for Science, Innovation & Technology, Foreign, Commonwealth & Development Office, Prime Minister's Office, *The Bletchley Declaration by Countries Attending the AI Safety Summit* (Nov. 2023), https://www.gov.uk/government/publications/ai-safety-summit-2023-the-bletchley-declaration/the-bletchley-declaration-by-countries-attending-the-ai-safety-summit-1-2-november-2023

Germany participated in the AI Seoul Summit in May 2024, which built on the AI Safety Summit hosted by the UK at Bletchley Park.[2532]

OECD / G20 AI Principles

Germany is a member of the OECD and endorsed the OECD and the G20 AI Principles.[2533] In 2020, Germany joined 14 other countries to announce the Global Partnership on Artificial Intelligence to "support and accompany the responsible use of AI on the basis of human rights, fundamental freedoms, inclusion, diversity, innovation, economic growth and the common good, as well as to achieve the United Nations Sustainable Development Goals (SDGs)."[2534]

In 2021, the OECD noted several examples of Germany's implementation of the OECD AI Principles, including guidelines for trustworthy AI that are largely in line with the OECD AI Principles (Germany's Data Ethics Commission ethics recommendations), the establishment of a dedicated body to coordinate and evaluate AI strategies, and the development of partnerships between public and private research organizations.[2535] The OECD's 2024 review noted Germany's strong research base and public investment in AI infrastructure, as well as the ethical focus reflected in the Data Ethics Commission's recommendations. Persistent challenges include fragmented AI governance, limited AI adoption by small and medium-sized enterprises, and skills shortages.[2536]

Council of Europe AI Treaty

Germany contributed as a Council of Europe and EU Member State in the negotiations of the Council of Europe Framework Convention on AI, Human Rights, Democracy and the Rule of Law.[2537] On September 2024, the EU signed this first

---

[2532] UK Department for Science, Innovation & Technology, Foreign, Commonwealth & Development Office, *Seoul Declaration for Safe, Innovative, Inclusive AI by Participants Attending the Leaders' Session: AI Seoul Summit* (May 21, 2024), https://www.gov.uk/government/publications/seoul-declaration-for-safe-innovative-and-inclusive-ai-ai-seoul-summit-2024/seoul-declaration-for-safe-innovative-and-inclusive-ai-by-participants-attending-the-leaders-session-ai-seoul-summit-21-may-2024

[2533] OECD Legal Instruments, *Recommendation of the Council on Artificial Intelligence, Adherents* (May 3, 2024), https://legalinstruments.oecd.org/en/instruments/OECD-LEGAL-0449#adherents

[2534] Federal Ministry of Labour and Social Affairs, *Germany Is a Founding Member of the "Global Partnership on Artificial Intelligence"* (Jun. 15, 2020), https://www.bmas.de/DE/Service/Presse/Pressemitteilungen/2020/deutschland-ist-gruendungsmitglied-der-global-partnership-on-articifial-intelligence.html

[2535] OECD, *State of Implementation of the OECD AI Principles Insights from National AI Policies* pp. 10, 14, 15, 65 (Jun. 2021), https://www.oecd-ilibrary.org/science-and-technology/state-of-implementation-of-the-oecd-ai-principles_1cd40c44-en

[2536] OECD, *OECD Artificial Intelligence Review of Germany* (Jun. 2024), https://doi.org/10.1787/609808d6-en

[2537] Council of Europe, *Framework Convention on Artificial Intelligence* (2026), https://www.coe.int/en/web/artificial-intelligence/the-framework-convention-on-artificial-intelligence

legally binding international AI treaty on behalf of its members, including Germany. Germany has not taken independent action to sign or ratify the treaty.[2538]

UNESCO Recommendation on AI Ethics

Germany is a signatory to the UNESCO Recommendation on the Ethics of Artificial Intelligence. While its AI plan was published before the development of the UNESCO Recommendation, the German UNESCO Commission released a report in 2022 which showed that Germany "has taken important steps to regulate AI in a way that corresponds to human rights and the public good," although more work is needed in some areas.[2539]

The German Bundestag held discussions over the UNESCO Recommendation on the Ethics of AI in 2022. However, no action has followed despite the discussions proposing action within 4 years of the release of the UNESCO Recommendation.[2540]

UNESCO has provided Member States with the Readiness Assessment Methodology (RAM) to help countries assess how prepared they are to implement UNESCO's Recommendation on AI Ethics. The RAM uses both quantitative and qualitative questions to assess a country's readiness. Germany has not yet implemented the RAM.[2541] However, the country responded to a UNESCO survey reporting on progress implementing the Recommendations in 2025.[2542]

Evaluation

Germany's national AI strategy emphasizes AI ethics, and Germany has called for regulating AI at EU level. However, while generally in favor of regulation and transparency, the German position on the risks of AI systems (particularly on biometric identification and predictive policing) is difficult to pin down, as different ministries, coalition members, and state governments may have different priorities

---

[2538] Council of Europe Treaty Office, *Chart of Signatures and Ratifications of Treaty 225* (Mar. 18, 2026), https://www.coe.int/en/web/conventions/full-list?module=signatures-by-treaty&treatynum=225

[2539] UNESCO Deutsche UNESCO-Kommission, *Ethical Aspects of Artificial Intelligence* (Mar. 21, 2022), https://www.unesco.de/wissen/wissenschaft/ethik-und-philosophie/studie-umsetzung-ki-ethik-empfehlung

[2540] German Bundestag, *Meeting Protocol* (Sept. 23, 2022), https://www.bundestag.de/resource/blob/911624/0b4a2e35fc56a4f1197970a0684e8afd/to018-data.pdf

[2541] UNESCO Global AI Ethics and Governance Observatory, *Global Hub* (Oct. 2025), https://www.unesco.org/ethics-ai/en/global-hub

[2542] German Commission for UNESCO, *First German State Report (2025) on the UNESCO Recommendation of 2021 on the Ethics of Artificial Intelligence [Erster deutscher Staatenbericht (2025) zur UNESCO-Empfehlung von 2021 zur Ethik der Künstlichen Intelligenz]* (Feb. 2025), https://www.unesco.de/assets/dokumente/Digitalisierung_und_KI/01_Digitalisierung_KI_allgemein/2025_DE_Staatenbericht_No1_UNESCO_Empfehlung_KI_Ethik_FIN.pdf

and opinions.[2543] The same could be said about the regulation of GPAI models. Germany has made progress to establish a national supervisory mechanism to take the protection of human rights seriously. However, the draft law appointment of the Federal Network Agency and emphasis on innovation-friendly regulation raises concerns in this regard, as articulated by federal and state data protection agencies.

Germany has progressed on implementing the UNESCO Recommendation on the Ethics of AI and OECD AI Principles, through actions such as the establishment of the Data Ethics Commission. The data protection authorities have also demonstrated a willingness to challenge companies in violation of data protection. However, the lack of clarity on sanctions and even applications across the federated agencies remains a challenge.

[2543] Luca Bertuzzi, *Germany Could Become MEPs' Ally in AI Act Negotiations*, Euractiv (Jan. 9, 2023), https://www.euractiv.com/section/artificial-intelligence/news/germany-could-become-meps-ally-in-ai-act-negotiations/

Ghana

In 2025, Ghana official launched the National AI Strategy 2025–2030 and turned toward implementation through initiatives in education, training, and governance. Drafts of an updated Data Protection Bill and other technology-related bills would establish a legal framework to protect citizens from well-known risks of AI and other emerging technologies.

National AI Strategy

Ghana's Ministry of Communication, Digital Technology and Innovations (MoCDTI) officially established a national framework for ethical, inclusive, and transparent AI governance with the launch of the National Artificial Intelligence Strategy 2025–2033.[2544] The Strategy, which was developed in 2022 with the support of The Future Society, GIZ FAIR Forward, and Smart Africa,[2545] sets eight priority pillars: education and skills, data governance, digital infrastructure, applied research, sectoral AI adoption, ecosystem coordination, youth employment, and public-sector innovation, aligned with UNESCO's Recommendation on the Ethics of AI. The strategy emphasizes ethical AI practices, international collaboration, AI adoption in key sectors, skills development, and the establishment of a Natural Language Processing Centre of Excellence to promote local language integration.[2546] The Strategy is expected to ensure better conditions for the development and use of this innovative technology for the benefit of citizens.

Flagship programmes such as the One Million Coders Ghana initiative[2547] and a legislative package including the draft Data Protection Bill and Emerging Technologies Bill[2548] mark early activation of the Strategy. Minister Samuel Nartey George unveiled Ghana's plan to become Africa's leading AI hub by 2028, emphasizing AI innovation, digital transformation, and public-private

---

[2544] Ministry of Communication, Digital Technology and Innovations, *Ghana National Artificial Intelligence Strategy 2025–2033* (May 2025), https://www.africadataprotection.org/Ghana-AI-Strat.pdf

[2545] Ministry of Communications and Digitisation with Smart Africa, GIZ FAIR Forward, and The Future Society (TFS), *Republic of Ghana National Artificial Intelligence Strategy, 2023–2033*, Africa Data Protection (Oct. 2022), https://www.africadataprotection.org/Ghana-AI-Strat.pdf

[2546] Povo News, *Ghana's National Artificial Intelligence Strategy 2023–2033*, Slideshare (Aug. 14, 2024), https://www.slideshare.net/slideshow/ghana-s-national-artificial-intelligence-strategy-2023-2033-pdf/270928634

[2547] Ministry of Communication, Digital Technology and Innovations, *President Mahama Launches One Million Coders Program* (May 2, 2025), https://moc.gov.gh/2025/05/02/president-mahama-launches-one-million-coders-program/

[2548] Ministry of Communication, Digital Technology and Innovations, *Legislative Instruments: Legislative Review Exercise and Call for Comments*, https://moc.gov.gh/legislative_instruments/

collaboration.[2549] Ghana's Ministry of Communications, Digital Technology, and Innovation signed an MoU with MTN at the Mobile World Congress in Barcelona, focusing on training youth in AI, cybersecurity, and data governance.[2550]

The Ministry of Communications and Digitalization, through the Data Protection Commission (DPC), launched the UNESCO Readiness Methodology (RAM) in October 2024 to help guide AI governance and development.[2551]

Regional AI Governance Collaboration

As member of the African Union (AU), Ghana is working to align with the AU Digital Transformation Strategy[2552] as well as the Artificial Intelligence Continental Strategy for Africa. Ghana's Ministry of Communications and Digitalization participated in a consultative workshop organized by the AU High-Level Panel on Emerging Technologies (APET).[2553] The APET discussions aimed at gathering input about "myths, challenges, and benefits of Artificial Intelligence (AI) in Africa," urging African countries to invest in AI literacy, cooperate internationally for AI innovation, enhance data protection, invest in infrastructure, and review policy implementation frameworks governing AI.

The AU finalized the Artificial Intelligence Continental Strategy for Africa[2554] at its 45th Ordinary Session in Accra, Ghana, in July 2024. The AU introduced the African Digital Compact (ADC)[2555] in the same session as an initiative to leverage digital technologies and promote inclusivity in the drive for sustainable development and innovation across the continent. The ADC, aligned with Africa's Agenda 2063[2556] and the UN Global Digital Compact,[2557] calls for collaborative efforts from

---

[2549] Ministry of Communications, Digital Technology and Innovations Ghana, *Embracing the Future: Ghana Launches National AI Strategy Stakeholder Consultation Forum* (May 2, 2025), https://moc.gov.gh/2025/05/02/embracing-the-future-ghana-launches-national-ai-strategy-to-drive-innovation-2/

[2550] The Herald Ghana, *Ghana, MTN Sign MoU to Boost Digital Skills Training* (Mar. 4, 2025), https://theheraldghana.com/ghana-mtn-sign-mou-to-boost-digital-skills-training/

[2551] Ministry of Communications and Digitalisation, *Readiness Assessment Measurement (RAM)—Ethical Use of Artificial Intelligence Launched* (Oct. 1, 2024), https://moc.gov.gh/2024/10/01/readiness-assessment-measurement-ram-ethical-use-of-artificial-intelligence-launched/

[2552] African Union, *The Digital Transformation Strategy for Africa (2020–2030)*, https://au.int/sites/default/files/documents/38507-doc-dts-english.pdf

[2553] AUDA-NEPAD, *The African Union Artificial Intelligence Continental Strategy for Africa* (May 30, 2022), https://www.nepad.org/news/african-union-artificial-intelligence-continental-strategy-africa

[2554] African Union, *Continental Artificial Intelligence Strategy* (Jul. 2024), https://au.int/sites/default/files/documents/44004-doc-EN-_Continental_AI_Strategy_July_2024.pdf

[2555] African Union, *African Digital Compact* (Jul. 2024), https://au.int/en/documents/20240809/african-digital-compact-adc

[2556] African Union, *Agenda 2063: The Africa We Want* (Jun. 10, 2013), https://au.int/Agenda2063/popular_version

[2557] United Nations, *Global Digital Compact* (Sept. 22, 2024), https://www.un.org/techenvoy/global-digital-compact

governments, businesses, and civil society to bridge digital divides and secure digital rights, fostering a resilient, digitally empowered Africa. With major AU policy developments taking place in Accra itself, Ghana could see a renewed push towards AU's Digital Transformation Strategy.[2558]

Public Participation

Stakeholder engagement for public policies is an established practice in Ghana, led by the Ministry of Communication, Digital Technology and Innovation (MoCDTI). Annual budgets support the planning sessions to formulation of new policies.[2559] Stakeholder engagements took place for Ghana's Digital Economy Policy at the Accra Digital Centre in November 2022.[2560] Minister Ursula Owusu-Ekuful highlighted the significance of the ministerial consultation as a means to ensure inclusiveness and comprehensiveness of the policy, covering data governance, emerging tech and regulation, data classification, data sharing and open data.[2561]

In October 2025, the MoCDTI launched a nationwide public consultation on a package of draft digital governance bills, including the Emerging Technologies Bill, Data Protection Bill, and Cybersecurity (Amendment) Bill. Draft texts and public comment forms were published on the MoCDTI portal and submission deadlines extended to mid-November 2025 to invite feedback from civil society, academia, and the private sector.[2562] The Misinformation, Disinformation and Hate Speech Bill further introduces transparency and human rights safeguards in algorithmic content moderation, risk assessments, and fact-checking requirements, reinforcing participatory oversight of digital systems.

The National AI Strategy is the result of the collaborative efforts of Ghana's Ministry of Communications and Digitalisation, Ghana's Data Protection Commission, GIZ FAIR Forward,[2563] Smart Africa[2564] and The Future Society.[2565] In

---

[2558] African Union, *The Digital Transformation Strategy for Africa (2020–2030)*, https://au.int/sites/default/files/documents/38507-doc-dts-english.pdf

[2559] Ministry of Communication and Digitalization, *Programme Based Budget Estimates for 2022*, pp. 11 and 28 (2022), https://mofep.gov.gh/sites/default/files/pbb-estimates/2022/2022-PBB-MOCD.pdf

[2560] Graphic Online, *Ministry Develops Policy for Digital Economy* (Dec. 13, 2022), https://www.graphic.com.gh/business/business-news/ministry-develops-policy-for-digital-economy.html

[2561] The Future Society, *Stakeholder Consultation Workshops Drive Insights for National AI Strategies in Tunisia and Ghana* (Jun. 9, 2022), https://thefuturesociety.org/stakeholder-consultation-workshops-drive-insights-for-national-ai-strategies-in-tunisia-and-ghana/

[2562] Ministry of Communication, Digital Technology and Innovations (MoCDTI), *Legislative Review Exercise—Call for Comments* (Oct. 2025), https://moc.gov.gh/legislative_instruments/

[2563] Digital Global, *Artificial Intelligence for Development Cooperation—FAIR Forward*, https://www.bmz-digital.global/en/overview-of-initiatives/fair-forward/

[2564] Smart Africa, *SADA Launches Its National Digital Academy in Ghana* (May 20, 2022), https://smartafrica.org/sada-launches-its-national-digital-academy-in-ghana/

[2565] The Future Society, *Aligning Artificial Intelligence through Better Governance* (Jun. 9, 2022), https://thefuturesociety.org

total, the strategy is based on 40+ local stakeholder consultations, in-depth AI policy landscape mapping, and SWOT Analysis of Ghana's AI ecosystem, and 4 high-level public sector consultation workshops to iterate the mission and vision, recommendations and action plan, and a detailed booklet of AI use cases across key sectors.[2566]

For example, in May 2022, The Future Society co-led a stakeholder consultation workshop in Ghana to support the development of Ghana's AI strategy.[2567] The workshop sessions addressed AI governance and frameworks, policies, implementation plans, SWOT, local AI ecosystems, AI ethical guidelines and recommendations for the establishment of program offices to drive implementation. The Future Society together with Ghana's Data Protection Commission, GIZ FAIR Forward, and Smart Africa held the third high-level public sector consultation workshop in August 2022 with representatives from private and public sector, academia, and civil society, to discuss the establishment of precautionary guardrails for AI across sectors.

As a member of the African Union, Ghana also took part in the planning for the Continental Artificial Intelligence Strategy.[2568] The document, in section 2.4.1, AI Governance and Regulation, cites the need for a multi-tiered governance approach to ensure transparency and accountability of the systems used and developed in Africa. Such an approach would include open public consultations comprising various sectors.

The MoCD, in partnership with Smart Africa, held a workshop aimed at finalizing Ghana's National Data Strategy in June 2024.[2569] Supported by the Data Protection Commission and the GIZ, the workshop brought together experts and government officials to align on strategic objectives for data governance, infrastructure, and capacity building. A collaborative approach was recognized as vital for empowering regulatory bodies, fostering partnerships across sectors, and ensuring effective data management and compliance to enhance AI integration and cybersecurity.

While stakeholder inputs have been gathered in the past for the National AI Strategy in Ghana, not all policies are publicly accessible through government sources. Increasing transparency by making these contributions available could strengthen public trust and engagement in AI policy development.

[2566] The Future Society, *National AI Strategies for Inclusive and Sustainable Development* (Apr. 30, 2022), https://thefuturesociety.org/2022/04/03/policies-ai-sustainable-development/

[2567] The Future Society, *Stakeholder Consultation Workshops Drive Insights for National AI Strategies in Tunisia and Ghana* (Jun. 9, 2022), https://thefuturesociety.org/stakeholder-consultation-workshops-drive-insights-for-national-ai-strategies-in-tunisia-and-ghana/

[2568] African Union, *Continental Artificial Intelligence Strategy* (Jul. 2024), https://au.int/sites/default/files/documents/44004-doc-EN-_Continental_AI_Strategy_July_2024.pdf

[2569] Ministry of Communications and Digitalisation Ghana, *Ghana National Data Strategy Validation Workshop Kicks Off in Accra* (Jun. 27, 2024), https://moc.gov.gh/2024/06/27/ghana-national-data-strategy-validation-workshop-kicks-off-in-accra/

Data Protection

Ghana enacted the Data Protection Act in 2012.[2570] Section 41 of the Data Protection Act provides for the right of the data subject to object for decisions that significantly affect him to not be based solely on automated processing of personal data. The Data Protection Act establishes the Data Protection Commission,[2571] which is an independent statutory body in charge of enforcing the Data Protection Act and protecting data subjects' rights.

The Data Protection Commission and GIZ supported a workshop to finalize the National Data Strategy in June 2024, in partnership with the Ministry of Communication and Digitization (now MoCDTI) and Smart Africa.[2572] Key discussions among the experts and government officials centered on updating the legal framework to keep pace with technological advancements, enhancing technical infrastructure, improving data privacy and protection, and fostering inclusivity in data use, including the promotion of AI and cloud services.

The MoCDTI inaugurated a new Governing Board of the Data Protection Commission in 2025.[2573] Chaired by digital rights expert Teki Akuetteh, the Board is tasked with developing a National Data Governance Framework to advance Ghana's digital economy and AI goals. In October 2025, MoCDTI launched a public consultation on the Data Protection Bill 2025,[2574] proposing to repeal and replace the 2012 Act with a new Data Protection Authority, codify transparency in automated decision-making, require data protection impact assessments for AI systems, and strengthen cross-border data transfer and localization rules. The draft Data Protection Bill would strengthen enforcement mechanisms and citizens' rights in relation to AI systems, automated decision-making, and cross-border data transfers.

Ghana ratified (in 2019) the African Union Convention on Cyber Security and Personal Data Protection (Malabo Convention).[2575] The Convention, drafted in 2011 and adopted in 2014 is an initial step to the establishment of a regulatory framework

---

[2570] Data Protection Commission, *The Data Protection Act (Act 843)* (2012), https://nita.gov.gh/wp-content/uploads/2017/12/Data-Protection-Act-2012-Act-843.pdf

[2571] Data Protection Commission, *Home* (2026), https://dataprotection.org.gh/

[2572] Ministry of Communications and Digitalisation Ghana, *Ghana National Data Strategy Validation Workshop kicks off in Accra* (Jun. 27, 2024), https://moc.gov.gh/2024/06/27/ghana-national-data-strategy-validation-workshop-kicks-off-in-accra/

[2573] Akim Benamara, *Ghana Takes Major Step towards Digital Future with New Data Protection Board*, Tech Africa News (Aug. 1, 2025), https://techafricanews.com/2025/08/01/ghana-takes-major-step-towards-secure-digital-future-with-new-data-protection-board/

[2574] Data Protection Commission, *Draft Data Protection Bill, 2025* (2025), https://dataprotection.org.gh/wp-content/uploads/2025/11/DATA_PROTECTION_BILL1_DRAFT-1.pdf

[2575] African Union. *List of Countries which have signed, ratified/acceded to the African Union Convention on Cyber Security and Personal Data Protection* (Jul. 8, 2022), https://au.int/sites/default/files/treaties/29560-sl-AFRICAN_UNION_CONVENTION_ON_CYBER_SECURITY_AND_PERSONAL_DATA_PROTECTION.pdf

for data protection in the African region.[2576] The agreement emphasized that each country is to develop its own legislative framework, observing the African Charter on Human and People's Rights, respecting privacy and freedoms while enhancing the promotion and development of ICT.[2577]

Despite being a member of the Global Privacy Assembly since 2014, the Data Protection Commission has not endorsed the 2018 GPA resolution on AI and Ethics,[2578] the 2020 GPA Resolution on AI and Accountability,[2579] the 2022 Resolution on Facial Recognition Technology,[2580] or the 2023 GPA Resolution on Generative AI.[2581]

Algorithmic Transparency

The widespread use of social media and online tools in African countries makes regulating algorithmic transparency critical to protect against misuse.[2582] The Data Protection Act of 2012 establishes the principle of "fair and transparent processing of a data subject's personal data"[2583] and provides that subjects need to be informed by the data processor when a decision has been made by automated means. Section 17 of the DPA sets out principles that data controllers must follow when processing personal data. Section 17 (f) provides for openness as a principle for processing personal data. This section could be interpreted as including algorithmic transparency, but it would need confirmation either by the Data Protection Commission or by court. The Malabo Convention does not explicitly provide for

---

[2576] African Union, *African Union Convention on Cybersecurity (Malabo Convention)* (Jun. 27, 2014), https://au.int/en/treaties/african-union-convention-cyber-security-and-personal-data-protection

[2577] CCDCOE, *Mixed Feedback on the African Union Convention on Cyber Security and Personal Data Protection*, https://ccdcoe.org/incyder-articles/mixed-feedback-on-the-african-union-convention-on-cyber-security-and-personal-data-protection/

[2578] International Conference of Data Protection & Privacy Commissioners, *Declaration on Ethics and Data Protection in Artificial Intelligence* (Oct. 23, 2018), https://globalprivacyassembly.org/wp-content/uploads/2018/10/20180922_ICDPPC-40th_AI-Declaration_ADOPTED.pdf

[2579] Global Privacy Assembly (GPA), *Adopted Resolution on Accountability in the Development and Use of Artificial Intelligence* (Oct 2020), https://globalprivacyassembly.org/wp-content/uploads/2020/11/GPA-Resolution-on-Accountability-in-the-Development-and-Use-of-AI-EN.pdf

[2580] Global Privacy Assembly (GPA), *Resolution on Principles and Expectations for the Appropriate Use of Personal Information in Facial Recognition Technology* (Oct 2022), https://globalprivacyassembly.org/wp-content/uploads/2022/11/15.1.c.Resolution-on-Principles-and-Expectations-for-the-Appropriate-Use-of-Personal-Information-in-Facial-Recognition-Technolog.pdf

[2581] Global Privacy Assembly, *Resolution on Generative Artificial Intelligence Systems* (Oct. 2023), https://globalprivacyassembly.org/wp-content/uploads/2023/10/5.-Resolution-on-Generative-AI-Systems-101023.pdf.

[2582] African Policy Research Institute (APRI), *AI in Africa: Key Concerns and Policy Considerations for the Future of the Continent* (May 30, 2022), https://afripoli.org/ai-in-africa-key-concerns-and-policy-considerations-for-the-future-of-the-continent

[2583] Data Protection Commission, *The Data Protection Act, 2012 (ACT 843)* (2012), https://nita.gov.gh/wp-content/uploads/2017/12/Data-Protection-Act-2012-Act-843.pdf

algorithmic transparency. Both data protection regimes could more explicitly include algorithmic transparency to ensure that individuals are informed about how algorithms influence decisions that affect them.

The draft Data Protection Bill 2025[2584] lays out specific rights in relation to automated decisions, including the right to opt-out of fully automated decisions and to request a review of automated decisions. Under Application of Emerging Technologies,[2585] the bill also requires that "Technology-driven decisions impacting data subjects shall be explainable, contestable, and subject to human oversight to prevent bias and harm." The bill would also require mandatory data protection impact assessments prior to launch for "any service that utilises personal data to provide tailored experiences."

The Emerging Technologies Bill 2025 released by the Ministry of Communication, Digital Technology and Innovations (MoCDTI) for public comment proposes standards for accountability, human oversight, and explainability in automated decision-making.[2586] MoCDTI also introduced the Misinformation, Disinformation and Hate Speech (MDHI) Bill 2025, requiring human rights due diligence for algorithmic content moderation, annual risk assessments, and transparency obligations for media institutions.[2587] Together, these draft laws represent Ghana's first effort to codify algorithmic transparency and redress mechanisms, aligned with the governance and ethics pillar of the National AI Strategy 2025–2033.[2588]

A study by International Telecommunications Society (ITS) on the use of artificial intelligence in the Fintech industry of Ghana and three other African countries found gender bias in financial services decisions. A lack of algorithmic transparency and explainability were areas of concern, contributing to the inability to make institutions accountable for biased decisions.[2589]

2584 Data Protection Commission, *Draft Data Protection Bill, 2025*, sec. *65* (2025), https://dataprotection.org.gh/wp-content/uploads/2025/11/DATA_PROTECTION_BILL1_DRAFT-1.pdf

2585 Ibid, 53(2), 53(7)

2586 Ministry of Communication, Digital Technology and Innovations (MoCDTI), *Legislative Review Exercise—Call for Comments* (Oct. 2025), https://moc.gov.gh/legislative_instruments/

2587 Ministry of Communication, Digital Technology and Innovations (MoCDTI), *Misinformation, Disinformation, Hate Speech and Publication of Other Information (MDHI) Bill* (2025), https://moc.gov.gh/legislative_instruments/

2588 Ministry of Communication, Digital Technology and Innovations, *Ghana National Artificial Intelligence Strategy 2025–2033* (May 2025), https://www.africadataprotection.org/Ghana-AI-Strat.pdf

2589 Shamira Ahmed, *A Gender Perspective on the Use of Artificial Intelligence in Africa's FinTech Industry: Case Studies from South Africa, Kenya, Nigeria and Ghana*, Research ICT Africa (Jun. 21, 2021), https://researchictafrica.net/research/a-gender-perspective-on-the-use-of-artificial-intelligence-in-africas-fintech-industry-case-studies-from-south-africa-kenya-nigeria-and-ghana/

Facial Recognition

Practices of surveillance and use of facial recognition are of concern in the country. Ghana's National Service Scheme has implemented facial recognition in its registration process to prevent fraud and identity thefts in the receipt of payments to public employees. The deployment, which included testing, staff training sessions, and public awareness, has achieved its desired outcome according to government reports.[2590]

A 2023 report reveals that the government has increased its possession of surveillance technologies. Ghana is implementing the Integrated National Security Communications Enhancement Network (ALPHA) project, a safe city project, which will incorporate the use of facial recognition CCTV cameras.[2591] The project is valued at approximately $184 million, involving multiple international companies, primarily incorporating facial recognition AI from Huawei and biometric systems requiring citizen data.[2592] Concerns exist that the Ghanaian government will use this technology to identify and target citizens expressing dissenting viewpoints, with reported cases of harassment and detention of individuals expressing dissent in violation of international human rights law.[2593]

Minister Samuel Nartey George announced in March 2026 that Ghana's SIM re-registration exercise will introduce a liveness test requiring individuals to undergo facial recognition to verify their identity, citing widespread use of fraudulent Ghana Cards in the first phase of registration.[2594]

Credit Scoring

In 2019, Algorithm Watch found citizen scoring mechanisms in Ghana by the National Identification Authority. The investigation identified six biometric databases owned by public authorities that provide citizen scores, including credit reference scores issued by the Credit Reference Bureau.[2595]

[2590] BiometricUpdate, *Ghana National Service Officials Praise Face Biometrics Onboarding* (Nov. 23, 2022), https://www.biometricupdate.com/202211/ghana-national-service-officials-praise-face-biometrics-onboarding

[2591] Institute of Development Studies, *Mapping the Supply of Surveillance Technologies to Africa: Case Studies from Nigeria, Ghana, Morocco, Malawi, and Zambia* (2023), https://opendocs.ids.ac.uk/opendocs/handle/20.500.12413/18120

[2592] Institute of Development Studies, *Ghana's Democracy at Risk Due to Use of Surveillance Technology, Warns New Report* (Sept. 27, 2023), https://www.ids.ac.uk/press-releases/ghanas-democracy-at-risk-due-to-use-of-surveillance-technology-warns-new-report/

[2593] Institute of Development Studies, *Mapping the Supply of Surveillance Technologies to Africa: Case Studies from Nigeria, Ghana, Morocco, Malawi, and Zambia* (2023), https://opendocs.ids.ac.uk/opendocs/handle/20.500.12413/18120

[2594] GBC Ghana Online, *Ghanaians to Undergo Facial Recognition in SIM Re-registration—Sam George* (Mar. 17, 2026), https://www.gbcghanaonline.com/technology/ghanaians-to-undergo-facial-recognition-in-sim-re-registration-sam-george/2026/

[2595] Nicolas Kayser-Bril, *Identity-Management and Citizen Scoring in Ghana, Rwanda, Tunisia, Uganda, Zimbabwe and China*, Algorithm Watch (Oct. 21, 2019), https://algorithmwatch.org/en/wp-

According to a report by the Bank of Ghana, it initiated discussions with credit bureaus to introduce credit scoring as a complement to the credit reports already offered by these bureaus in 2023. This initiative followed the introduction of the Ghana Card, which serves as a distinct identifier for individuals with credit information. The goal was to facilitate the implementation of the credit scoring system by ensuring that financial institutions update existing credit information with new details from the Ghana Card.[2596]

Ghana's Vice President, Dr. Mahamudu Bawumia, launched the initiative known as myCreditScore in November 2024.[2597] Thus, the country is the second on the African continent to have such a system, the first being South Africa. Within the government, the initiative is viewed with favor and any discussions about possible harms are dismissed as criticism from the opposition.[2598] Additionally, several financial institutions have made significant progress in updating their existing credit information with the new Ghana Card data. This effort, combined with the integration of credit bureaus with the National Identification Authority (NIA) database, is anticipated to provide a strong foundation for the introduction of the credit scoring regime in the Ghanaian credit market.[2599]

The Bank of Ghana formally approved credit bureaus to implement credit scoring in 2025. MyCredit Score Limited and Dun & Bradstreet Credit Bureau Limited rolled out their scoring products. The Bank of Ghana stated it would continue to oversee and monitor scoring methodologies to ensure the system remains fair and consumer interests are protected.[2600] Concerns about algorithmic transparency in the credit scoring system have been raised by users, who note that the basis for individual credit score ratings is not disclosed, limiting individuals' ability to understand or contest decisions affecting their access to credit.[2601]

---

content/uploads/2019/10/Identity-management-and-citizen-scoring-in-Ghana-Rwanda-Tunesia-Uganda-Zimbabwe-and-China-report-by-AlgorithmWatch-2019.pdf

[2596] Bank of Ghana (Financial Stability Department), *Credit Reporting Activity, Annual Report* (2022), https://www.bog.gov.gh/wp-content/uploads/2023/10/Credit-Reporting-Activity-Report-2022.pdf

[2597] Joy Online, *Bawumia launches Credit Scoring System for Ghana* (Nov. 7, 2024) https://www.myjoyonline.com/bawumia-launches-credit-scoring-system-for-ghana/

[2598] GhanaWeb, *They Criticize without Knowledge—Bawumia Replies Critics of His Credit Scoring System* (Jul. 27, 2024), https://www.ghanaweb.com/GhanaHomePage/NewsArchive/They-criticize-without-knowledge-Bawumia-replies-critics-of-his-credit-scoring-system-1942562

[2599] Bank of Ghana (Financial Stability Department), *Credit Reporting Activity, Annual Report* (2023), https://www.bog.gov.gh/wp-content/uploads/2024/09/2023-ANNUAL-CREDIT-REPORTING-ACTIVITY-REPORT-170924.pdf

[2600] Ghana News Agency, *Bank of Ghana approves Credit Scoring to Strengthen Ghana's Financial Sector* (Aug. 13, 2025), https://gna.org.gh/2025/08/bank-of-ghana-approves-credit-scoring-to-strengthen-ghanas-financial-sector/

[2601] myCreditScore App Store Reviews, *myCreditScore—Credit Check App*, Apple App Store, https://apps.apple.com/gh/app/mycreditscore-credit-check/id6677019193

Biometrics

The Ghanaian Police Service has adopted biometric devices as part of a digitization strategy with the objective to "dramatically improve policing in Ghana."[2602] The police uses biometrics to check wanted individuals against the National Identification Authority's (NIA) database. The process started in 2017 with the issuing of the new Ghana card, aimed to centralize the identity management systems.[2603] The NIA biometric database amounts to 17 million records, including Ghana ID card, car registration, and insurance information, allowing for identification in real time with biometric devices. The NIA National Identity System employs three biometric technologies for identification: unique fingerprints represented as digitized templates, facial templates depicted as digitized color photos of the cardholder, and iris recognition.[2604] The Minister of Interior commissioned a new governing board for the NIA in May 2025, calling on members to deepen oversight, strengthen public trust, and accelerate the delivery of secure and inclusive identity services.[2605]

A report by Privacy International identified practices of surveillance in Ghana during the phase of emergency response to the COVID pandemic.[2606] The PanaBIOS app endorsed by the African Union and deployed by the Ghanaian border enforcement, used algorithms to track and trace individuals that might pose a health threat.

The Ghanaian government announced in January 2024 that government employees who had not registered with the Ghana Card system would have their salaries frozen as of March. According to presidential correspondent Charles Takyi-Boadu, the initiative assumed that any citizen who did not register was deceased and freezing salaries would eliminate "ghost registrations" from the payroll.[2607]

Groups such as the People's National Convention (PNC), Great Consolidated Popular Party (GCPP), and Election Watch Ghana held demonstrations calling on the

---

[2602] BiometricUpdate, *Ghana Police Get Over 100 Biometric Devices for Fields Checks against Database* (Jan. 2, 2023), https://www.biometricupdate.com/202301/ghana-police-get-over-100-biometric-devices-for-field-checks-against-national-database

[2603] Nicolas Kayser-Bril, *Identity-Management and Citizen Scoring in Ghana, Rwanda, Tunisia, Uganda, Zimbabwe and China*, Algorithm Watch (Oct. 21, 2019), https://algorithmwatch.org/en/wp-content/uploads/2019/10/Identity-management-and-citizen-scoring-in-Ghana-Rwanda-Tunesia-Uganda-Zimbabwe-and-China-report-by-AlgorithmWatch-2019.pdf

[2604] National Identification Authority, *Introduction to the GhanaCard*, https://nia.gov.gh/the-ghanacard-introduction/

[2605] Biometric Update, *Ghana's New Identification Authority Board to Focus on Public Trust, Inclusive Service Delivery* (May 12, 2025), https://www.biometricupdate.com/202505/ghanas-new-identification-authority-board-to-focus-on-public-trust-inclusive-service-delivery

[2606] Privacy International (PI), *Under Surveillance: (Mis)use of Technologies in Emergency Reponses. Global Lessons from the COVID-19 Pandemic* (Dec. 2022), https://privacyinternational.org/report/5003/under-surveillance-misuse-technologies-emergency-responses-global-lessons-covid-19

[2607] Daily Guide Network, *Govt Rolls Out New Payroll No Ghana Card, No Pay* (Jan. 11, 2024), https://dailyguidenetwork.com/govt-rolls-out-new-payroll-no-ghana-card-no-pay/

Electoral Commission of Ghana (EC) to be more transparent about biometric voter registration and the delivery of device kits to all political parties in the country after cases of the disappearance and destruction of several of these devices were reported.[2608] The following month, Deputy Minister for Foreign Affairs and Regional Integration Kwaku Ampratwum-Sarpong announced that a new model of digital passport would be introduced in the country, with an electronic chip that has a biometric identifier, which would be used to authenticate the holder's identity.[2609] President Akufo-Addo officially launched the chip-embedded biometric passport compliant with ICAO regulations on December 2, 2024.

Ghana launched the blockchain-based digital platform CitizenApp Data Interoperability System (CADIS) in October 2024 to facilitate access to digital government services. Identity verification is done through the Ghana Card and the citizen's phone number. Information such as biometric passport applications is available for consultation on the app.[2610] The mobile app was designed by the government to improve access to public services in Ghana, described as "Ghana's Roadmap to Intelligent Governance."[2611] Notably, this shift resonates with earlier initiatives, such as Ghana's adoption of biometrics for driver's licenses and vehicle registration.[2612]

The Minister of Communication, Digital Technology and Innovation announced a new legislative framework to link SIM cards to the Ghana Card national digital ID in 2025.[2613] The new framework involves biometric SIM registration and a Centralized Equipment Identity Register linked to the NIA database. The new framework aiming to combat mobile money fraud was operationalized through a mandatory facial recognition liveness test announced in March 2026. The Minister of Interior activated a biometric passenger monitoring system in February 2026 at Ghana's borders, designed to strengthen national security and facilitate trade under the

---

[2608] GhanaWeb, *Share Details of All Your Biometric Voter Registration Kits, Devices with All Political Parties—EC Told* (Apr. 16, 2024), https://www.ghanaweb.com/GhanaHomePage/NewsArchive/Share-details-of-all-your-biometric-voter-registration-kits-devices-with-all-political-parties-EC-told-1926363

[2609] GBC Ghana Online, *Ghana to Introduce Chip-Embedded Passports in Six Months—Kwaku Ampratwum-Sarpong* (Apr. 4, 2024), https://www.gbcghanaonline.com/general/ghana-chip-passport/2024/

[2610] Biometric Update, *Ghana Launches CitizenApp to Facilitate Access to Digital Govt Services* (Oct. 8, 2024), https://www.biometricupdate.com/202410/ghana-launches-citizenapp-to-facilitate-access-to-digital-govt-services

[2611] Lezeth Khoza, *Ghana Launches App to Improve Access to Public Services,* ITWeb Press Council FAIR (Oct. 9, 2024), https://itweb.africa/content/G98YdqLG8DaMX2PD; see also Ghana, *CitizenApp is Ghana's Roadmap to Intelligent Governance,* CitizenApp (2024), www.citizen.gov.gh/

[2612] Ayang Macdonald, *Benin, Ghana Adopt Biometrics for Driver's Licenses, Vehicle Registration,* Biometric Update (Dec. 21, 2022), www.biometricupdate.com/202212/benin-ghana-adopt-biometrics-for-drivers-licenses-vehicle-registration

[2613] Biometric Update, *Ghana Plans New Framework for SIM Card-Digital ID Linkage* (Sept. 8, 2025), https://www.biometricupdate.com/202509/ghana-plans-new-framework-for-sim-card-digital-id-linkage

African Continental Free Trade Area, replacing manual border checks with intelligence-driven biometric technology.[2614]

Generative AI was linked to over a third of new biometric fraud cases in Africa in 2024, with deepfake videos increasing sevenfold in the second half of the year.[2615] The rise of AI-driven digital identity fraud presents both challenges and investment opportunities in Ghana, with companies securing significant venture funding to counter these threats.

EdTech

Ghana was a subject of a study by Human Rights Watch about the use of government-endorsed Ed Tech tools for online learning during the COVID-19 pandemic across 49 countries.[2616] The findings show that the government of Ghana endorsed platforms that have the capability to identify, tag, and track users, including children. Further, learning apps endorsed by Ghana tracked and collected data from children and teachers for advertising and revenue purpose, transmitting data to AdTech companies. For Ghana, the apps in this category were Edmodo and Ghana Electronic Library. Ghana was one of only nine countries that disclosed in their privacy policies that they collected and used children's data for "behavioral advertising purposes."

The Cyber Security Authority issued the National Child Online Protection Framework in October 2024,[2617] outlining plans to address threats children face in the digital space including cyberbullying, harassment, and online exploitation, and announcing the development of guidelines for the removal of child abuse materials and reporting procedures.

In the same month, the Ministry of Education introduced a new Senior High School curriculum focused on competencies, character development, and Ghanaian identity, requiring the retraining of 68,000 teachers to deliver the programme to 1.4 million students, with subsidized AI support from AWS and Anthropic.[2618] Ghana is preparing for a national rollout in October 2025, with several African countries expressing interest in replicating the model.

[2614] GhanaWeb, *What You Need to Know about Ghana's Newly Activated Biometric Tracking System* (Feb. 7, 2026), https://www.ghanaweb.com/GhanaHomePage/NewsArchive/What-you-need-to-know-about-Ghana-s-newly-activated-biometric-tracking-system-2020667

[2615] Alexander Onukwue, *Surge in Deepfakes Heightens Fraud Risk for African Businesses*, Semafor (Jan. 29, 2025), https://www.semafor.com/article/01/28/2025/surge-in-deepfakes-heightens-fraud-risk-for-african-businesses

[2616] Human Rights Watch, *How Dare They Peep into My Private Life?* (May 25, 2022), https://www.hrw.org/report/2022/05/25/how-dare-they-peep-my-private-life/childrens-rights-violations-governments

[2617] Cyber Security Authority Ghana, *National Child Online Protection Framework* (Oct. 2024), https://digitalpolicyalert.org/digest/dpa-digital-digest-ghana

[2618] Ghana News Agency, *Ghana Charts a Bold Path for Ethical AI in Education* (Nov. 26, 2025), https://gna.org.gh/2025/11/ghana-charts-a-bold-path-for-ethical-ai-in-education-a-journey-begins-in-a-classroom/

UNICEF launched the Learning Pioneers Programme in 2024, a two-year initiative in which six countries, including Ghana, are partnering with the Office of Innovation Learning Innovation Hub "to pilot and accelerate digital learning tools to address the learning crisis and shape the future of learning." Under the slogan "Map, Match, and Make it Happen," the countries will be mapping, testing, and developing scalable EdTech tools.[2619]

Environmental Impact of AI

Ghana responded to the global call to action to end poverty, protect the earth's environment and climate, and ensure that people everywhere can enjoy peace and prosperity, captured in 17 UN Sustainable Development Goals.[2620] According to the 2025 Voluntary National Review, Ghana has made progress toward SDG 12 on responsible consumption and production and SDG 13 on climate action, while acknowledging significant challenges remain before the 2030 deadline.[2621] Notably, the environmental impact of Artificial Intelligence, including energy consumption and its carbon footprint poses further challenges to achieve these goals. While academic research purports that Artificial Intelligence can help the Ghanaian Government plan for a renewable energy future,[2622] the research also acknowledges that AI poses risks to the environment.

Lethal Autonomous Weapons

Ghana has not acceded to the Convention on Certain Conventional Weapons (CCW).[2623] However, the country has participated in several CCW meetings since 2014.[2624] In a statement to the 2016 CCW informal meeting of experts on LAWS, Ghana called for the "promotion and preservation of human dignity for humanity as a whole [...] In our view, fully automated lethal systems must be proscribed before they are fully developed."[2625]

[2619] UNICEF, *UNICEF's Learning Pioneers Programme: Shaping the Future of Learning* (Aug. 12, 2024), https://www.unicef.org/digitaleducation/stories/unicefs-learning-pioneers-programme-shaping-future-learning

[2620] United Nations Ghana, *Our Work on the Sustainable Development Goals in Ghana* (2024), https://ghana.un.org/en/sdgs

[2621] National Development Planning Commission, *Ghana's Voluntary National Review Report on the Implementation of the 2030 Agenda for Sustainable Development* (Jul. 2025), https://ndpc.gov.gh/media/2025_VNR_Report.pdf

[2622] Dom Byrne, *How Artificial Intelligence Is Helping Ghana Plan for a Renewable Energy Future,* Nature (May 7, 2024), https://doi.org/10.1038/d41586-024-01316-w

[2623] UN Office of Disarmament Affairs, *High Contracting Parties and Signatories CCW* (Mar. 18, 2025), https://disarmament.unoda.org/en/our-work/conventional-arms/convention-certain-conventional-weapons/high-contracting-parties-and-signatories-ccw

[2624] Automated Research, *State Positions: Ghana*, https://automatedresearch.org/news/state_position/ghana/

[2625] Ghana, *Comments and Questions by Ghana at the CCW Informal Meeting of Experts on Lethal Autonomous Weapons Systems*, Reaching Critical Will (Apr. 16, 2015),

Ghana is a member of the African Group within the United Nations, which supports the negotiation of a legally binding instrument on autonomous weapons systems. The African Group issued a statement in 2021 CCW calling attention to the "ethical, legal, moral and technical questions" in the use of autonomous weapons systems and urging concrete policy recommendations, including prohibitions and regulations.[2626]
At the 78th UN General Assembly First Committee in 2023, Ghana voted in favor[2627] of resolution L.56[2628] on autonomous weapons systems, along with 163 other states. The Resolution emphasized the "urgent need for the international community to address the challenges and concerns raised by autonomous weapons systems," and mandated the UN Secretary-General to prepare a report reflecting the views of member and observer states on autonomous weapons systems.

At the 79th Session of the First Committee in 2024, Ghana voted in favor of resolution L.77 on lethal autonomous weapon systems.[2629] At the thematic debate, Ghana's delegation stated that LAWS "raise serious humanitarian, legal, and ethical questions, which must be addressed through clear regulations,"[2630] aligning with the joint call of the UN Secretary-General and the ICRC for new prohibitions and restrictions on autonomous weapons. The resolution was formally adopted by the UN General Assembly on December 2, 2024 as Resolution 79/62, with 166 votes in favor.[2631] While the earlier resolution recognized potential risks of these systems, L.77 emphasizes the urgency of addressing the challenges posed by lethal autonomous weapons systems (LAWS), highlighting the need for a comprehensive approach to address diverse perspectives related to these systems. The resolution includes specific calls for action, such as convening open informal consultations in 2025 and inviting participation from various stakeholders, including civil society and the scientific community.

---

https://www.reachingcriticalwill.org/images/documents/Disarmament-fora/ccw/2015/meeting-experts-laws/statements/16April_Ghana.pdf

[2626] *Statement by the African Group, CCW Group of Governmental Experts Meeting on LAWS* (Dec 3, 2021), http://149.202.215.129:8080/s2t/UNOG/LAWS3-03-12-2021-AM_mp3_en.html

[2627] Isabelle Jones, *164 States Vote against the Machine at the UN General Assembly*, Stop Killer Robots (Nov. 1, 2023), https://www.stopkillerrobots.org/news/164-states-vote-against-the-machine/

[2628] General Assembly, *Lethal Autonomous Weapons*, Resolution L56 (Oct. 12, 2023), https://reachingcriticalwill.org/images/documents/Disarmament-fora/1com/1com23/resolutions/L56.pdf

[2629] Isabelle Jones, *161 States Vote against the Machine at the UN General Assembly*, Stop Killer Robots (Nov. 5, 2024), https://www.stopkillerrobots.org/news/161-states-vote-against-the-machine-at-the-un-general-assembly/

[2630] Ghana Permanent Mission to the United Nations, *First Committee, 18th Plenary Meeting on Thematic Debate on Conventional Weapons* (Oct. 24, 2024), https://www.ghanamissionun.org/10242024/

[2631] UN General Assembly, *Resolution 79/62: Lethal Autonomous Weapons Systems* (Dec. 2, 2024), https://documents.un.org/doc/undoc/ltd/n24/305/45/pdf/n2430545.pdf

AI Literacy

Ghana's Digital Economy Policy 2024 identifies digital skills as a core strategic pillar, integrating digital literacy into the national education curriculum and setting a target to train at least one million citizens in basic digital literacy through community-based programmes and digital centers.[2632] President Mahama launched the One Million Coders Ghana Programme in May 2025[2633] as a flagship pillar of the National AI Strategy 2025–2033, with a focus on youth and women. The program includes modules on ethical AI, data protection, and responsible innovation delivered through public universities and regional innovation hubs. The program complements MoCDTI's public consultation efforts by fostering an informed citizenry capable of engaging in AI governance discussions.[2634] A three-day Ministerial Bootcamp on AI run in partnership with UNDP aimed to equip Ministers and senior officials with foundational AI knowledge, including ethical implications and public service delivery applications. The bootcamp aligned with preparations to submit the National AI Strategy to Parliament.[2635]

Ghana's inclusive AI literacy efforts extend to gender equity. The Ministry of Communication's Ms. Geek programme aims to increase gender diversity in STEM, while Ghana Tech Lab supports women's enrollment in AI training programmes.[2636] In 2025, Heritors Labs and GIZ FAIR Forward published the Ghana Artificial Intelligence Practitioners' Guide, providing a multi-stakeholder framework for responsible AI adoption and a dedicated section on ethical AI governance for government, civil society, and the private sector.[2637]

The 2025 UNDP Human Development Report, launched in Accra in July 2025, found that more than 70% of Ghanaian respondents expressed optimism that AI will improve their productivity, while underscoring the urgency of closing digital

---

[2632] Ministry of Communication, Digital Technology and Innovations, *Programme-Based Budget Estimates for 2024* (2024–2027 Medium Term Expenditure Framework), https://mofep.gov.gh/sites/default/files/pbb-estimates/2024/2024-PBB-MOCD_.pdf

[2633] Ministry of Communication, Digital Technology and Innovations, *President Mahama Launches One Million Coders Program* (May 2, 2025), https://moc.gov.gh/2025/05/02/ president-mahama-launches-one-million-coders-program/

[2634] Ibid

[2635] UNDP Ghana, *Government of Ghana and UNDP Kicks Off Ministerial Bootcamp to Strengthen AI Capacity and Advance AI-enabled Public Services* (Jul. 25, 2025), https://www.undp.org/ghana/press-releases/government-ghana-and-undp-kicks-ministerial-bootcamp-strengthen-ai-capacity-and-advance-ai-enabled-public-services

[2636] Wilson Center, *Regulating Artificial Intelligence in Africa: Strategies and Insights from Kenya, Ghana, and the African Union* (2024), https://www.wilsoncenter.org/blog-post/regulating-artificial-intelligence-africa-strategies-and-insights-kenya-ghana-and-african

[2637] Heritors Labs and GIZ FAIR Forward, *Ghana Artificial Intelligence Practitioners' Guide* (2025), https://www.bmz-digital.global/wp-content/uploads/2025/10/GhanaArtificialIntelligence PractitionersGuide.pdf

access gaps to ensure AI benefits reach all segments of Ghanaian society, particularly youth and underserved populations.[2638]

Human Rights

Ghana is a signatory of the Universal Declaration of Human Rights (UDHR) and has enshrined the key provisions of the UDHR in its Constitution. The 1992 Constitution of Ghana[2639] provides for the protection of fundamental human rights and freedoms. The UDHR provides for the right to privacy under Article 12. The right to privacy is guaranteed under Article 18 of the Constitution: "No person shall be subjected to interference with the privacy of his home, property, correspondence or communication except in accordance with law and as may be necessary in a free and democratic society for public safety or economic well-being of the country, for the protection of health or morals, for the prevention of disorder or crime or for the protection of rights and freedoms of others."

Ghana is a party to the African Charter on Human and Peoples' Rights. In its 2019 Declaration of Principles on Freedom of Expression and Access to Information in Africa, the African Commission on Human and Peoples' Rights (ACHPR), in charge of interpreting the Charter,[2640] called on states to ensure that the "development, use and application of AI, algorithms and other similar technologies by internet intermediaries are compatible with international human rights law and standards, and do not infringe on the rights to freedom of expression, access to information and other human rights."[2641]

The ACHPR adopted Resolution 473 in February 2021, having recognized that emerging technologies such as AI have a bearing on the enjoyment of human rights under the African Charter.[2642] The ACHPR called on state parties to ensure that the

---

[2638] UNDP Ghana, *The Human Development and AI Paradox: A 35-Year Low with New Horizons for Ghana* (Jul. 9, 2025), https://www.undp.org/ghana/press-releases/human-development-and-ai-paradox-35-year-low-new-horizons-ghana

[2639] Ghana, *Constitution of Ghana*, Constitute Project (1996), https://www.constituteproject.org/constitution/Ghana_1996.pdf

[2640] African Union, *African Charter on Human and Peoples' Rights*, Articles 30, 45(3) (Oct. 21, 1986), https://au.int/en/treaties/african-charter-human-and-peoples-rights

[2641] Diplo, *Artificial Intelligence in Africa: Continental Policies and Initiatives* (2021), https://www.diplomacy.edu/resource/report-stronger-digital-voices-from-africa/ai-africa-continental-policies/

[2642] African Commission on Human and Peoples' Rights, *Resolution on the need to undertake a Study on human and peoples' rights and artificial intelligence (AI), robotics and other new and emerging technologies in Africa - ACHPR/Res. 473 (EXT.OS/ XXXI) 2021* (Mar. 10, 2021), https://achpr.au.int/en/adopted-resolutions/473-resolution-need-undertake-study-human-and-peoples-rights-and-art; University of Pretoria, *Centre for Human Rights Welcomes African Commission Resolution on Emerging Technologies*, Centre for Human Rights News (Mar. 18, 2021), https://www.chr.up.ac.za/tech4rights-news/2451-press-statement-centre-for-humanrights-welcomes-african-commission-resolution-on-emerging-technologies; CAIDP, *African Commission on Human and People's Rights Resolution 473: Statement of The Center for AI and Digital Policy (CAIDP)*

development and use of AI and other emerging technologies is compatible with the rights and duties in the African Charter and other regional and international human rights instruments to uphold human dignity, privacy, equality, non-discrimination, inclusion, diversity, safety, fairness, transparency, accountability, and economic development.

The Special Rapporteur on Freedom of Expression and Access to Information in Africa recalled in 2022 that, with Resolution 473, the Commission, "observed that whilst making government services and information digital enhances transparency and accessibility and artificial intelligence allows for a number of benefits in the society, it has to be accompanied by human rights considerations and a bridging of the digital divide." The Special Rapporteur also declared that "State Parties are encouraged to develop domestic legal frameworks regulating AI and e-governance; ensure these technologies are developed and used transparently; and ensure that imported AI and e-governance systems align with the African Charter."[2643] The ACHPR opened a call for inputs in April to its study mandated by Resolution 473 on AI, robotics, and emerging technologies' human rights impacts.[2644]

Freedom House's 2025 Freedom in the World report[2645] rates Ghana as "Free" with a score of 80/100. Ghana is rated "Partly Free" with a score of 65/100 on the Freedom on the Net report.[2646] According to Freedom House, Ghanaian political rights and civil liberties are backed by a stable democracy, with competitive multi-party elections since 1992 and peaceful transfer of power between the two major political parties. Discrimination against women persists, and LGBT+ people face widespread discrimination, intimidation, and violence. Areas of weakness concern judicial independence and the rule of law. Corruption also poses challenges to the effective performance of the government. The Freedom House report also highlights political violence as a growing concern.

(Apr. 5, 2021), https://www.caidp.org/app/download/8308244063/CAIDP-ACHPR-Res473-04052021.pdf

[2643] African Commission on Human and Peoples' Rights, *Press Statement by the Special Rapporteur on Freedom of Expression and Access to Information in Africa, on the Occasion of International Day for Universal Access to Information* (Sept. 28, 2022), https://achpr.au.int/en/news/press-releases/2022-09-28/special-rapporteur-freedom-expression-access-international-day

[2644] African Commission on Human and Peoples' Rights, *Call for Input on the Draft Study on Human and Peoples' Rights and Artificial Intelligence (AI), Robotics, and Other New and Emerging Technologies in Africa* (Apr. 8, 2025), https://achpr.au.int/en/news/announcements/2025-04-08/call-input-study-human-and-peoples-rights-artificial-intelligence-ai

[2645] Freedom House, *Freedom in the World 2025: Ghana* (2025), https://freedomhouse.org/country/ghana/freedom-world/2025

[2646] Freedom House, *Ghana: Freedom on the Net 2024* (2024), https://freedomhouse.org/country/ghana/freedom-net/2024

OECD / G20 AI Principles

Ghana is not a member of the OECD and has not officially endorsed the OECD principles.[2647] However, Ghana is a member of the African Union, which joined the G20 in 2023 and endorsed the G20 AI Principles.[2648] Ghana is not a member of the Global Partnership on Artificial Intelligence (GPAI).[2649]

Ghana is a member of the OECD Development Centre since 2015, collaborating with members and non-member countries in policy to improve living conditions in developing and emerging economies.[2650] In the development of Ghana's national AI strategy, the government of Ghana and The Future Society collaborated with the OECD AI Policy Observatory among other organizations that champion human-centric AI.[2651]

Council of Europe AI Treaty

Ghana is not a member state of the Council of Europe and has not signed this first legally binding international AI treaty.[2652]

UNESCO Recommendation on AI Ethics

Ghana is a member of UNESCO and adopted the Recommendation on the Ethics of AI in 2021.[2653]

Ghana's former Minister for Communications and Digitalisation, Ursula Owusu-Ekuful, reaffirmed the government's commitment to ensuring the ethical and responsible use of Artificial Intelligence (AI) technologies in Ghana at a public workshop launching the UNESCO Readiness Assessment Methodology (RAM) study in October 2024.[2654] The minister emphasized the role of public workshops as a

---

[2647] OECD Legal Instruments, *Recommendation of the Council on Artificial Intelligence, Adherents* (May 3, 2024), https://legalinstruments.oecd.org/en/instruments/OECD-LEGAL-0449

[2648] G20, *G20 Members* (2024), https://g20.org/about-g20/g20-members/

[2649] OECD AI Policy Observatory, *About the Global Partnership on Artificial Intelligence (GPAI)* (2026), https://oecd.ai/en/about/about-gpai

[2650] Modern Ghana, *Ghana Becomes the 50th Member of the OECD Development Centre* (Oct. 6, 2015), https://www.modernghana.com/news/647449/ghana-becomes-the-50th-member-of-the-oecd-development-centre.html

[2651] Yolanda Lannquist and Nicolas Miailhe, *National AI Strategies for Inclusive and Sustainable Development*, The Future Society (Apr. 30, 2022), https://thefuturesociety.org/2022/04/03/policies-ai-sustainable-development/

[2652] Council of Europe Treaty Office, *Chart of Signatures and Ratifications of Treaty 225* (Mar. 29, 2026), https://www.coe.int/en/web/conventions/full-list?module=signatures-by-treaty&treatynum=225

[2653] UNESCO, *UNESCO Member States Adopt the First Ever Global Agreement on the Ethics of Artificial Intelligence* (Apr. 21, 2022), https://www.unesco.org/en/articles/unesco-member-states-adopt-first-ever-global-agreement-ethics-artificial-intelligence

[2654] Ministry for Communications and Digitalisation, *Readiness Assessment Measurement (RAM) Ethical Use of Artificial Intelligence Launched* (Oct. 1, 2024), https://moc.gov.gh/2024/10/01/readiness-assessment-measurement-ram-ethical-use-of-artificial-

significant step to leverage AI responsibly, to follow guidelines and principles outlined by UNESCO's Recommendation on AI Ethics, and to play a pivotal role in achieving AU's Agenda 2063[2655] and the UN Sustainable Development Goals.[2656] Ghana has initiated the RAM process[2657] but has not yet implemented the UNESCO Ethical Impact Assessment tool.

Evaluation

Ghana has made significant progress toward ethical AI governance by endorsing the Continental AI strategy, launching a national steering committee to oversee AI readiness, and advancing data governance reforms including the 2025 Data Protection Bill. Ghana formalized the National AI Strategy in May 2025 and contributed to the development of the Continental AI Strategy for Africa, which was finalized in 2024. The Continental Strategy informs national implementation. The country has taken the first step to meaningfully implement the UNESCO Recommendation on the Ethics of AI with the initiation of RAM.

Efforts are underway to establish partnerships to further the digital strategy. While public addresses emphasize the importance of public participation and transparent policymaking, these principles are primarily showcased at events involving international partners, with limited practical engagement of Ghanaian citizens.

Ghana benefits from an independent Data Protection Commission and developed its own data protection legal framework and ratified the Malabo Convention. Amid the rapid deployment of AI in the country, the proposed amendments in the Data Protection Bill 2025 would add transparency and accountability. The regulation of practices of biometrics identification, facial recognition, and individual scoring is essential to realize the benefits of AI for the Ghanaian society.

---

intelligence-launched/; Maragaret Adjeley Sosah, *Govt Launches AI Readiness Assessment Measurement Tool*, Modern Ghana (Oct. 3, 2024), https://www.modernghana.com/news/1345918/govt-launches-ai-readiness-assessment-measurement.html

2655 African Union, *Agenda 2063: The Africa We Want*, https://au.int/en/agenda2063/overview

2656 Ministry for Communications and Digitalisation, *Readiness Assessment Measurement (RAM) Ethical Use of Artificial Intelligence Launched* (Oct. 1, 2024), https://moc.gov.gh/2024/10/01/readiness-assessment-measurement-ram-ethical-use-of-artificial-intelligence-launched/

2657 Ministry for Communications and Digitalisation, *Readiness Assessment Measurement (RAM) Ethical Use of Artificial Intelligence Launched* (Oct. 1, 2024), https://moc.gov.gh/2024/10/01/readiness-assessment-measurement-ram-ethical-use-of-artificial-intelligence-launched/

Greece

In 2025, Greece continued efforts to implement the Blueprint for AI Transformation and designated rights-protecting authorities under the EU AI Act. The country also ratified the Council of Europe Convention 108+.

National AI Strategy

Greece created A High Level Advisory Committee for Artificial Intelligence in October 2023 to develop a national AI strategy.[2658] The Committee published the Blueprint for Greece's AI Transformation that envisions a comprehensive plan to leverage AI to foster economic growth, enhance public services, and reinforce democratic values, while establishing safeguards for national security and resilience.[2659] Priority areas also include preparing citizens for the AI transition, promoting quality healthcare for all, enhancing access to and the quality of education, enriching cultural heritage, and mitigating and adapting to climate change.[2660]

The AI Blueprint promotes principles and values including respect for human dignity, human flourishing, pluralism, participation, transparency, oversight, adaptability and international multilateral cooperation.[2661] To advance its leadership in AI governance and ethics, the AI Blueprint proposes launching a Global AI Forum to increase AI literacy, convene high-level meetings on AI ethics and policy, and track the impact of AI on democratic processes globally.[2662] The AI Blueprint proposes establishing new regulatory agencies: the National AI Supervisory Authority as an independent oversight body aligned with the EU AI Act mandates, a Chief AI Strategy Officer for planning the AI Strategy, and a Data and AI Office for data curation, data governance, and AI implementation.[2663] The Blueprint also proposes transforming the Data Protection Authority into the Data Protection and Access to Information Authority to reconcile GDPR mandates for the right to personal data protection and the right to freedom of expression, noting "Data Controllers or AI providers/deployers would not be able to use data protection as a pretext to refuse access to crucial data."[2664]

[2658] Office of the Prime Minister of the Hellenic Republic, *Announcement—ßPrime Minister's Press Office* (Oct. 19, 2023), https://www.primeminister.gr/2023/10/19/32817

[2659] High-Level Advisory Committee on Artificial Intelligence, *A Blueprint for Greece's AI Transformation* (Nov. 2024), https://foresight.gov.gr/wp-content/uploads/2024/11/Blueprint_GREECES_AI_TRANSFORMATION.pdf

[2660] Ibid, pp. 10–11

[2661] Ibid, p. 13

[2662] Ibid, p. 22

[2663] Ibid, pp. 81–83

[2664] Ibid, p. 82

EU Digital Services Act

As an EU member state, Greece is fully and directly bound by the EU Digital Services Act.[2665] The Greek Parliament adopted Law 5099/2024 for the implementation of the EU Digital Services Act in Greece in April 2024.[2666] The law designates the Hellenic Telecommunications and Post Commission (EETT) as the Digital Services Coordinator. The National Council for Radio and Television (NCRTV) and the Hellenic Data Protection Authority (HDPA) are the competent authorities responsible for supervising intermediary service providers and enforcing the DSA in Greece.[2667]

EU AI Act

As part of implementing the EU AI Act, Greece has designated four existing national authorities to collectively supervise the protection of fundamental rights in relation to high-risk AI systems. The entities include the:[2668]

- Hellenic Data Protection Authority
- Greek Ombudsman
- Hellenic Authority for Communication Security and Privacy
- National Commission for Human Rights.

While the AI Blueprint proposes a National AI Supervisory Authority as the independent supervisory authority and for national and EU level coordination on AI, the office has not yet been officially announced as the designated market surveillance authority despite the EU AI Act deadline of August 2, 2025.[2669]

Public Participation

Members in the High-Level Advisory Committee on Artificial Intelligence that drafted Greece's AI Blueprint were predominantly academics from universities

---

[2665] European Union, *Digital Services Act* (Nov. 11, 2022), https://www.eu-digital-services-act.com/Digital_Services_Act_Articles.html

[2666] Government Gazette of the Hellenic Republic, *Law 5099/2024 [ΝΟΜΟΣ 5099, ΦΕΚ Α 48]*, Isokratis Legal Information Bank (Apr. 5, 2024), https://www.dsanet.gr/Epikairothta/Nomothesia/48.PDF

[2667] European Union, *Greece: Policy Monitor Country Profile* (Feb. 2025), https://better-internet-for-kids.europa.eu/en/knowledge-hub/greece-policy-monitor-country-profile; see also, Hellenic Telecommunications & Post Commission, *Who We Are* (2023), https://www.eett.gr/en/eett/about-eett/who-we-are/

[2668] Hellenic Republic Ministry of Digital Governance, *The Principles of Protection of Fundamental Rights in Relation to the Use of Artificial Intelligence in Greece [Οι Αρχές προστασίας των θεμελιωδών δικαιωμάτων σε σχέση με τη χρήση Τεχνητής Νοημοσύνης στην Ελλάδα]* (Nov. 12, 2024), https://www.mindigital.gr/archives/6901

[2669] European Commission, *Market Surveillance Authorities under the AI Act* (Sept. 26, 2025), https://digital-strategy.ec.europa.eu/en/policies/market-surveillance-authorities-under-ai-act#1720699867912-2

and research centers, with a few from the private and public sectors.[2670] The Committee engaged in interviews with stakeholders from various sectors, including government and state officials, researchers and stakeholders in education and research, representatives from the private sector, and international experts.[2671]

Public policies are generally accessible online through the relevant ministries and government agencies or ISOKRATIS Legal Information Bank.[2672]

Data Protection

Greece has progressively aligned its data protection framework with European standards, embedding personal data rights into both constitutional and legislative instruments. Article 9A of the Greek Constitution establishes the protection of personal data as a fundamental right.[2673] Prior to the EU GDPR, Greece enforced Law 2472/1997, which implemented Directive 95/46/EC and introduced foundational rules around consent, data minimization, and supervisory oversight.[2674] With the enactment of Law 4624/2019, Greece formally supplemented the GDPR, incorporating sector-specific provisions and further clarifying the mandate of the Hellenic Data Protection Authority (Hellenic DPA).[2675]

The Hellenic DPA is a member of the Global Privacy Assembly (GPA).[2676] The Hellenic DPA did not co-sponsor the resolutions and declarations related to AI and facial technology, however.[2677]

[2670] High-Level Advisory Committee on Artificial Intelligence, *A Blueprint for Greece's AI Transformation*, p. 1 (Nov. 2024), https://foresight.gov.gr/wp-content/uploads/2024/11/Blueprint_GREECES_AI_TRANSFORMATION.pdf

[2671] Ibid, p. 28

[2672] Athens Bar Association, *Isokratis Legal Information Bank* (2025), https://www.dsanet.gr/#

[2673] Hellenic Parliament, *The Constitution of Greece* (Nov. 25, 2019), https://www.hellenicparliament.gr/UserFiles/f3c70a23-7696-49db-9148-f24dce6a27c8/THE%20CONSTITUTION%20OF%20GREECE.pdf

[2674] Hellenic Data Protection Authority, *Law 2472/1997 on the Protection of Individuals with regard to the Processing of Personal Data* (Nov. 10, 1997), www.dpa.gr/sites/default/files/2019-10/law_2472-97-nov2013-en.pdf

[2675] Gazette of the Hellenic Republic, *Law 4624/2019* (Aug. 29, 2019), https://www.dpa.gr/sites/default/files/2020-08/LAW%204624_2019_EN_TRANSLATED%20BY%20THE%20HDPA.PDF

[2676] Global Privacy Assembly (GPA), *List of Accredited Members* (2025), https://globalprivacyassembly.com/participation-in-the-assembly/list-of-accredited-members/

[2677] International Conference of Data Protection & Privacy Commissioners, *Declaration on Ethics and Data Protection in Artificial Intelligence* (Oct. 23, 2018), https://globalprivacyassembly.com/wp-content/uploads/2018/10/20180922_ICDPPC-40th_AI-Declaration_ADOPTED.pdf; Global Privacy Assembly (GPA), *Adopted Resolution on Accountability in the Development and Use of Artificial Intelligence* (Oct. 2020), https://globalprivacyassembly.com/wp-content/uploads/2020/11/GPA-Resolution-on-Accountability-in-the-Development-and-Use-of-AI-EN.pdf; Global Privacy Assembly (GPA), *Resolution on Principles and Expectations for the Appropriate Use of Personal Information in Facial Recognition Technology* (Oct. 2022), https://globalprivacyassembly.com/wp-content/uploads/2022/11/15.1.c.Resolution-on-Principles-and-Expectations-for-the-Appropriate-Use-

Algorithmic Transparency

The GDPR and Greek Law No. 4624 establish algorithmic transparency as a legal right in Greece.[2678] Greece's ratification of Council of Europe Convention 108+ in 2025 clarified the extent of those rights.[2679] Greek citizens have the right to receive meaningful information about the logic involved in automated processing, and the right to contest a decision made solely by an algorithm that has a significant impact on them. This includes the right to obtain human intervention and challenge an automated conclusion. The Convention 108 + also ensures that citizens' views are taken into consideration, and they can obtain information about the logic behind the automated processing.

Law 4961/2022 explicitly addressed challenges posed by AI with the first dedicated legal framework addressing AI, algorithmic transparency, and emerging technologies.[2680] This law requires public entities to register the AI systems they deploy and perform algorithmic impact assessments (AIAs) for high-risk systems, especially those involving automated decision-making or profiling. Private companies using AI for personalization or prediction must also disclose key information about system operation, legal basis, and potential biases.

The AI Blueprint recommends transforming the Data Protection Authority into a Data Protection and Access to Information Authority to enhance algorithmic transparency rights.[2681] The proposed consolidated authority would ensure that access to data is not blocked by privacy claims by Data controllers and AI providers and deployers.[2682]

Clearview AI

The Hellenic DPA imposed a €20 million fine on Clearview AI, a facial recognition technology company that builds biometric profiles by scraping publicly available images from websites, including social media platforms without consent in

---

of-Personal-Information-in-Facial-Recognition-Technolog.pdf; Global Privacy Assembly (GPA), *Resolution on Generative Artificial Intelligence Systems* (Oct. 2023), https://globalprivacyassembly.com/wp-content/uploads/2023/10/5.-Resolution-on-Generative-AI-Systems-101023.pdf

[2678] Government Gazette of the Hellenic Republic, *Law 4624/2019* (Aug. 29, 2019), https://www.dpa.gr/sites/default/files/2020-08/LAW%204624_2019_EN_TRANSLATED%20BY%20THE%20HDPA.PDF

[2679] Council of Europe Treaty Office, *Chart of Signatures and Ratifications of Treaty 223* (Convention 108+) (Mar. 5, 2025), https://www.coe.int/en/web/conventions/full-list;?module=signatures-by-treaty&treatynum=223

[2680] National Printing House, *Gazette of the Hellenic Republic Law 4961/2022 [Νομοσ Υπ' Αριθμ. 4961]* (Jul. 27, 2022), https://search.et.gr/en/fek/?fekId=618783

[2681] High-Level Advisory Committee on Artificial Intelligence, *A Blueprint for Greece's AI Transformation*, p. 82 (Nov. 2024), https://foresight.gov.gr/wp-content/uploads/2024/11/Blueprint_GREECES_AI_TRANSFORMATION.pdf

[2682] Ibid, p. 22

2022. The ruling cited unlawful processing of biometric data, lack of transparency, and failure to comply with data subject rights.[2683] The Hellenic DPA found that the company violated "the principles of lawfulness and transparency (art. 5 paragraphs 1(a) and (2), 6, 9 GDPR) and its obligations under Articles 12, 14, 15 and 27 of the GDPR." The Hellenic DPA ordered the company to delete all biometric data of individuals in Greece and cease further processing operations within national territory.

AI in Border Control

The Hellenic DPA imposed an administrative fine of €175,000 on the Ministry of Migration and Asylum in 2024 for failing to cooperate fully with the Authority and for submitting impact assessments deemed "substantially incomplete and limited in scope."[2684] The Hellenic DPA received multiple inquiries in 2022—including from the European Parliament Committee on Civil Liberties, Justice and Home Affairs; UNHCR Greece; and civil society organizations—regarding the use of the Hyperion and Centaur systems in reception and accommodation facilities for asylum seekers. The Centaur system is an "integrated digital system for managing electronic and physical security," while Hyperion is an integrated entry-exit control system that uses an RFID reader and fingerprint for two-factor authentication. Both involve processing personal and biometric data and violated protections related to transparency and consent.

AI Literacy

Greece's AI Blueprint identifies improving AI literacy as a priority area to prepare citizens for the AI transition. As part of this mission, the Blueprint describes "introducing AI and its associated disciplines, including topics on the ethical use of AI, to the educational curriculum, starting at the primary level; introducing reskilling and upskilling programs for the general population; improving AI literacy and spreading its empowerment opportunities across the Greek population; strengthening Greece's AI innovation potential."[2685]

To carry out this strategy, the AI Blueprint proposes an AI Center of Excellence: a new graduate school and research center focusing on AI research, ethics,

---

[2683] European Data Protection Board, *Hellenic DPA Fines Clearview AI 20 Million Euros* (Jul. 22, 2022), https://www.edpb.europa.eu/news/national-news/2022/hellenic-dpa-fines-clearview-ai-20-million-euros_en

[2684] Hellenic Data Protection Authority, *Ministry of Migration and Asylum Receives Administrative Fine and GDPR Compliance Order Following an Own-Initiative Investigation by the Hellenic Data Protection Authority*
(Apr. 3, 2024), https://www.dpa.gr/en/enimerwtiko/press-releases/ministry-migration-and-asylum-receives-administrative-fine-and-gdpr

[2685] High-Level Advisory Committee on Artificial Intelligence, *A Blueprint for Greece's AI Transformation*, p. 10 (Nov. 2024), https://foresight.gov.gr/wp-content/uploads/2024/11/Blueprint_GREECES_AI_TRANSFORMATION.pdf

philosophy, social sciences, and executive education.[2686] It also proposes actions for lifelong learning, professional training, and vocational education to develop and support AI skills and talent and workforce.[2687]

Environmental Impact of AI

The EU Energy Efficiency Directive (2023/1791) introduces an obligation for the monitoring and reporting of the energy performance of large data centers (>500 kW).[2688] While the EU AI Act promotes standards and voluntary codes of conduct to improve energy efficiency,[2689] the limited focus overlooks the broader environmental impacts of AI's lifecycle, including intensive water use and the destructive extraction of rare earth minerals.[2690]

At the national level, Greece's AI Blueprint addresses some of these gaps by including appended guidelines on the responsible development and use of AI, highlighting environmental sustainability as a key consideration.[2691] The guidance extends beyond reducing carbon footprint to encompass responsible sourcing of materials, reducing electronic waste, and promoting recycling and reuse practices. However, government entities have yet to adopt formal regulation or legislation to hold developers or deployers accountable to these guidelines.

Lethal Autonomous Weapons

Greece has actively engaged in international discussions regarding lethal autonomous weapons systems (LAWS). Greece supports the UN General Assembly Resolution 78/241, which emphasized the urgent need to address the challenges posed by autonomous weapons systems.[2692] In comments to the UN Secretary-General for the report on Member States' views, Greece supported a two-tier system of

---

[2686] Ibid, pp. 17, 67

[2687] Ibid, pp. 72–79

[2688] European Union, *Directive (EU) 2023/1791 of the European Parliament and of the Council of 13 September 2023 on energy efficiency and amending Regulation (EU) 2023/955 (recast)*, Article 12 (Sept. 13, 2023), https://eur-lex.europa.eu/eli/dir/2023/1791/oj#art_12

[2689] European Commission, *Regulation (EU) 2024/1689 of the European Parliament and of the Council of 13 June 2024 laying down harmonised rules on artificial intelligence and amending Regulations (EC) No 300/2008, (EU) No 167/2013, (EU) No 168/2013, (EU) 2018/858, (EU) 2018/1139 and (EU) 2019/2144 and Directives 2014/90/EU, (EU) 2016/797 and (EU) 2020/1828 (Artificial Intelligence Act) OJ L2024/1689*, Article 40 and 95 (Aug. 1, 2024), https://eur-lex.europa.eu/eli/reg/2024/1689/oj/eng

[2690] UNEP, *AI Has an Environmental Problem. Here's What the World Can Do about That* (Sept. 21, 2024), https://www.unep.org/news-and-stories/story/ai-has-environmental-problem-heres-what-world-can-do-about

[2691] High-Level Advisory Committee on Artificial Intelligence, *A Blueprint for Greece's AI Transformation*, p. 147 (Nov. 2024), https://foresight.gov.gr/wp-content/uploads/2024/11/Blueprint_GREECES_AI_TRANSFORMATION.pdf

[2692] United Nations, *A/RES/78/241: Lethal Autonomous Weapons Systems* (Dec. 22, 2023), https://docs.un.org/en/A/RES/78/241

prohibitions and restrictions of LAWS and noted "compliance with fundamental principles and requirements of international humanitarian law, such as, the principles and requirements of target distinction, proportionality and precautions at the evolving of environment of a battlefield raises serious concerns."[2693]

Greece's position at the UN and Convention on Certain Conventional Weapons (CCW) Group of Governmental Experts (GGE) on LAWS underscores the importance of meaningful human control over the use of force and the necessity for accountability in military operations involving AI technologies.[2694]

Human Rights

Greece is a member of the United Nations and has ratified the Universal Declaration of Human Rights.[2695] Freedom House rates Greece as "Free" with a score of 85 of 100, commending the country's parliamentary democracy supported by active political competition and largely upheld civil liberties.[2696] However, persistent challenges remain, including corruption, government surveillance, discrimination against immigrants and religious and ethnic minority groups, and poor conditions for irregular migrants and asylum seekers.

Greece's approach to artificial intelligence governance emphasizes protecting human rights and democratic values. The country's AI strategy includes principles such as inclusive growth, respect for the rule of law, human rights, and democratic values, as well as transparency, explainability, robustness, security, safety, and accountability.[2697]

[2693] United Nations Office for Disarmament Affairs, *General Assembly First Committee, Resolution 78/241: Lethal Autonomous Weapons Systems* (Dec. 2023), https://docs-library.unoda.org/General_Assembly_First_Committee_-Seventy-Ninth_session_%282024%29/78-241-Greece-EN.pdf

[2694] Permanent Mission of Greece to the United Nations Office & the Other International Organizations, *Potential Challenges Posed by Emerging Technologies in the Area of Lethal Autonomous Weapons Systems to International Humanitarian Law,* Statement by Greece to the Group of Governmental Experts on Lethal Autonomous Weapon Systems (LAWS), UNODA (Mar. 2019), https://docs-library.unoda.org/Convention_on_Certain_Conventional_Weapons_-_Group_of_Governmental_Experts_%282019%29/GGE%2BLAWS%2BSTATEMENT%2Bby%2BGREECE-%2BChallenges%2Bto%2BIHL.pdf

[2695] Greece United Nations Security Council 2025–2026, *The Universal Declaration of Human Rights Turns 75* (Oct. 9, 2025), https://greeceforunsc.mfa.gr/on-human-rights-day

[2696] Freedom House, *Freedom in the World 2025: Greece* (2025), https://freedomhouse.org/country/greece

[2697] High-Level Advisory Committee on Artificial Intelligence, *A Blueprint for Greece's AI Transformation*, p. 22 (Nov. 2024), https://foresight.gov.gr/wp-content/uploads/2024/11/Blueprint_GREECES_AI_TRANSFORMATION.pdf

OECD / G20 AI Principles

As a member state of OECD, Greece is an adherent of the OECD AI Principles.[2698] Greece's AI Blueprint reflects many of these principles.

Council of Europe AI Treaty

Greece is party to the Council of Europe Framework Convention on Artificial Intelligence and Human Rights, Democracy, and the Rule of Law through its membership in the European Union,[2699] which endorsed the treaty on September 5, 2024.[2700] Endorsement signals a commitment to align national AI policies with the standards outlined in the treaty, however Greece has not acted independently to endorse or ratify the treaty.[2701]

UNESCO Recommendation on AI Ethics

As a member state of UNESCO, Greece endorses the UNESCO Recommendation on the Ethics of Artificial Intelligence.[2702] The guiding principles in Greece's AI Blueprint—human dignity, oversight, adaptability, and pluralism—and four domains of action—innovation and entrepreneurship, education and research, regulatory framework, public sector transformation—align with the Recommendation.[2703] However, these have not been implemented. Greece has not completed the UNESCO Readiness Assessment Methodology (RAM).[2704]

Evaluation

Greece's AI strategy and data protection regulations generally uphold fundamental rights, democratic values, and the rule of law. In addition to meeting EU-level obligations on AI and data protection, Greece's national AI strategy and international commitments—such as adherence to the OECD AI Principles, the Council of Europe AI Treaty, and the UNESCO Recommendation on the Ethics of

---

[2698] OECD Legal Instruments, *Recommendation of the Council on Artificial Intelligence*, Adherents (May 3, 2024), https://legalinstruments.oecd.org/en/instruments/oecd-legal-0449#adherents

[2699] CAIDP, *Council of Europe AI Treaty* (2025), https://www.caidp.org/resources/coe-ai-treaty/

[2700] Council of Europe, *The Framework Convention on Artificial Intelligence* (Sept. 5, 2024), https://www.coe.int/en/web/artificial-intelligence/the-framework-convention-on-artificial-intelligence

[2701] Council of Europe Treaty Office, *Chart of Signatures and Ratifications of Treaty 225* (Nov. 17, 2025), https://www.coe.int/en/web/Conventions/full-list/?module=signatures-by-treaty&treatynum=225

[2702] UNESCO, *Recommendation on the Ethics of Artificial Intelligence* (Sept. 26, 2024), https://www.unesco.org/en/articles/recommendation-ethics-artificial-intelligence

[2703] High-Level Advisory Committee on Artificial Intelligence, *A Blueprint for Greece's AI Transformation*, p. 22 (Nov. 2024), https://foresight.gov.gr/wp-content/uploads/2024/11/Blueprint_GREECES_AI_TRANSFORMATION.pdf

[2704] UNESCO Global AI Ethics and Governance Observatory, *Global Hub* (Oct. 2025), https://www.unesco.org/ethics-ai/en/global-hub

AI—reinforce a rights-based, democratic, and human-centric approach to AI governance. The decisions issued by the Hellenic Data Protection Authority demonstrate its effectiveness as an oversight body, ensuring compliance with data protection regulations and safeguarding individuals' fundamental rights.

AI governance is still nascent, given the recency of the AI Blueprint. Implementing the UNESCO RAM could provide the country "detailed and comprehensive insights into different dimensions of AI readiness"[2705] that could enhance Greece's ability to identify systematic institutional and regulatory gaps in its AI ecosystem. The RAM and further legislation could also provide the opportunity for greater public participation in AI governance.

[2705] UNESCO Global AI Ethics and Governance Observatory, *Readiness Assessment Methodology* (2025), https://www.unesco.org/ethics-ai/en/ram

Hong Kong

In 2025, Hong Kong's Digital Policy Office updated the Ethical Artificial Intelligence Framework and issued guidance on Generative AI. The Privacy Commission released guidelines on Generative AI for employees and joined 19 other data privacy authorities in a joint statement on data governance frameworks for privacy-protecting AI.

National AI Strategy

Hong Kong is making significant advances in AI development and policy implementation through the issuance of guidelines, policies, and AI technology applications across sectors. Although this special administrative region[2706] largely controlled by China does not have a national strategy for the regulation of AI, several sectoral and overall guidelines exist, along with more reforms underway to guide organizations in adopting accountable and ethical AI models and processes.

A 2021 Guideline on Ethical Development and Use of AI issued by the Office of the Privacy Commissioner for Personal Data (PCPD) aims to guide organizations in their adherence to personal data and privacy requirements.[2707] In 2022, the Government Chief Information Officer (now Digital Policy Office) expanded the seven original principles to twelve and categorized them into Performance and Generalized Principles.[2708] Performance Principles include (1) Transparency and Interpretability and (2) Reliability, Robustness, and Security while Generalized Principles are: (1) Fairness, (2) Diversity and Inclusion, (3) Human Oversight, (4) Lawfulness and Compliance, (5) Data Privacy, (6) Safety, (7) Accountability, (8) Beneficial AI, (9) Cooperation and Openness and (10) Sustainability and Just Transition.

Since the Hong Kong government officially merged the Government Chief Information Office and the Efficiency Office to create the Digital Policy Office (DPO) in 2024,[2709] the ethical principles were updated again, to Version 1.4.[2710] The updated

---

[2706] Referred to as the Hong Kong Special Administrative Region (HKSAR).

[2707] Office of the Privacy Commissioner for Personal Data PCPD), *Guidance on Ethical Development and Use of AI* (Aug. 18, 2021), https://www.pcpd.org.hk/english/resources_centre/publications/files/guidance_ethical_e.pdf; PCPD, *Inspection Report on Customers' Personal Data Systems* (Aug. 18, 2021), https://www.pcpd.org.hk/english/enforcement/commissioners_findings/files/r21_3099_e.pdf

[2708] Digital Policy Office, *Ethical Artificial Intelligence Framework* (Sept. 2022), https://www.digitalpolicy.gov.hk/en/our_work/data_governance/policies_standards/ethical_ai_framework/

[2709] Government of the Hong Kong Special Administrative Region, *Digital Policy Office Established Today* (Jul. 25, 2024), https://www.info.gov.hk/gia/general/202407/25/P2024072400394.htm

[2710] Government of the Hong Kong Special Administrative Region, *Ethical Artificial Intelligence Framework* (Version 1.4) (Jul. 2024), https://www.digitalpolicy.gov.hk/en/our_work/data_governance/policies_standards/ethical_ai_framework/doc/Ethical_AI_Framework.pdf

ethical principles include more detailed for the AI governance mechanism. The updated mechanism takes a three-prong defense approach. 1) Project Team, responsible for AI application development, risk evaluation, execution of actions to mitigate identified risks and documentation of AI Impact Assessments; 2) Project Steering Committee (PSC) and Project Assurance Team (PAT), responsible for ensuring project quality, defining acceptance criteria for AI applications, providing independent review, and approving AI applications; and 3) IT Board, or Chief Information Officer (CIO) if no IT Board is in place, with optional support by an Ethical AI Committee, which may consist of external advisors.

The DPO commissioned the Hong Kong Generative AI Research and Development Center (HKGAI) to research and formulate guidelines on generative AI for developers, service providers, and service users.[2711] The guidelines released in 2025 provides an introduction to Generative AI and lays out the technical limitations, dimensions of governance, and key principles for the governance of the technology. The final section details practical guidelines for each of Technology Developers, Service Providers, and Service Users, who are charged to be "proactive stewards of benevolent AI."

Sectoral initiatives are also relevant. The Hong Kong Monetary Authority (HKMA) issued guidelines on consumer protection in the use of AI in customer-facing applications in August 2024 to protect against user harms from hallucination and incomplete results from GenAI.[2712] However, in its circular, the HKMA acknowledges that adoption of GenAI in the banking sector is still at "an early stage, with most of the current applications focusing on improving banks' operational efficiency. Nonetheless, the ability of GenAI in content-creation means that GenAI could be more extensively adopted by the banking sector in customer- facing activities." The guidelines build on the 2019 HKMA guidelines on consumer protection and adds requirements.[2713] These include defining the scope, policies, using a human-in-the-loop approach, ensuring fairness, providing opt-out mechanisms, disclosing AI models' limitations, and abiding by Hong Kong's privacy regulations.

The HKMA is also taking steps to promote the use of AI for detecting suspicious activities by financial institutions. In September 2024, the HKMA released a circular on the Use of Artificial Intelligence for Monitoring of Suspicious Activities

---

[2711] Digital Policy Office, *Hong Kong Generative Artificial Intelligence Technical and Application Guideline* (Dec. 2025), https://www.digitalpolicy.gov.hk/en/our_work/data_governance/policies_standards/ethical_ai_framework/doc/HK_Generative_AI_Technical_and_Application_Guideline_en.pdf

[2712] Hong Kong Monetary Authority, *Consumer Protection in Respect of the Use of Generative Artificial Intelligence* (Aug. 2024), https://www.hkma.gov.hk/media/eng/doc/key-information/guidelines-and-circular/2024/20240819e1.pdf

[2713] Hong Kong Monetary Authority, *Consumer Protection in Respect of the Use of Big Data Analytics and Artificial Intelligence by Authorized Institutions* (Nov. 2019), https://www.hkma.gov.hk/media/eng/doc/key-information/guidelines-and-circular/2019/20191105e1.pdf

outlining several initiatives that the authority was undertaking to promote AI use.[2714] This includes an experience sharing forum, fintech sandbox,[2715] and guidance from the HKMA on innovating.

Hong Kong's technology development sector, specifically AI, is seeing increased integration with mainland China. In her 2021 policy address, the Chief Executive of Hong Kong, Carrie Lam stated that "the developments of Hong Kong and our country are closely related. Only by leveraging the Central Government's policies in support of Hong Kong can we give full play to our unique strengths, which will, in turn, bring continuous impetus to our economy."[2716] This, coupled with the passage of the Hong Kong National Security Law in 2020 by China's top legislature,[2717] cast some doubts on the future of the One Country, Two Systems model for the governance of Hong Kong.[2718]

Public Participation

Hong Kong does not have a structured process for public participation in the development of AI policy; however, government organizations have included citizen engagement in some significant AI-related developments. For instance, in July 2024, the Hong Kong Commerce and Economic Development Bureau began a two-month public consultation on required legal amendments to protect copyrights in the training and use of AI models.[2719] Contributions were accepted through various channels, including written submissions, briefing sessions, and a public forum. After the consultation process ended in February 2025, the Bureau published a document summarizing the public views and the Bureau's response. The document also presented agenda for the next steps on AI and copyright, including the future

---

[2714] Hong Kong Monetary Authority, *Use of Artificial Intelligence for Monitoring of Suspicious Activities* (Sept. 9, 2024), https://www.hkma.gov.hk/media/eng/doc/key-information/guidelines-and-circular/2024/20240909e1.pdf

[2715] Hong Kong Monetary Authority, *Generative Artificial Intelligence Sandbox* (Sept. 20, 2024), https://www.hkma.gov.hk/media/eng/doc/key-information/guidelines-and-circular/2024/20240920e1.pdf

[2716] Special Administrative Region, *The Chief Executive's 2021 Address* (Oct. 6, 2021), https://www.policyaddress.gov.hk/2021/eng/p38.html

[2717] Hong Kong Free Press, *Official English Translation of the Hong Kong National Security Law* (Jul. 1, 2020), https://hongkongfp.com/2020/07/01/in-full-english-translation-of-the-hong-kong-national-security-law/

[2718] William Overholt, *Hong Kong: The Rise and Fall of "One Country, Two Systems"* (Dec. 2019), https://ash.harvard.edu/files/ash/files/overholt_hong_kong_paper_final.pdf

[2719] Commerce and Economic Development Bureau Intellectual Property Department, *Copyright and Artificial Intelligence Public Consultation* (Jul. 2024), https://www.cedb.gov.hk/assets/resources/cedb/consultations-and-publications/Eng_Copyright%20and%20AI%20Consultation%20Paper%20(2024.07.08).pdf

formulation of Guidelines on Copyright Issues Relating to Generative AI and the need to amend the Copyright Ordinance to include a text and data mining exception.[2720]

Hong Kong's Department of Justice (DoJ) established the Consultation Group on LawTech Development to promote AI-driven legal solutions and document automation in 2025.[2721] The Consultation Group will provide advice and assistance to the DoJ regarding the promotion and development of law tech in Hong Kong.

Data Protection

Hong Kong passed the Personal Data (Privacy) Ordinance (PDPO) in 1995, which instilled a principles-based approach to data privacy and established the Office of the Privacy Commissioner for Personal Data (PCPD) as an independent data privacy regulator.

The PDPO saw amendments in 2012[2722] and 2021[2723] to address direct marketing and criminalize doxxing, respectively. The anti-doxxing amendments allow the PCPD to conduct investigations without a warrant, press charges independently, force content to be taken down, and charge non-compliant internet platforms, raising concerns that the amendment will be used to restrict dissenting opinions.[2724] The PCPD reported more than 1400 notices to remove messages on digital platforms in 2021–2022 and a man sentenced to eight months in prison.[2725] The efforts continued in 2025, with a number of reported investigations and sentences.[2726]

[2720] Commerce and Economic Development Bureau Intellectual Property Department, *Legislative Council Paper no. CB(2)240/2025(04): Enhancement of the Copyright Ordinance regarding Protection for Artificial Intelligence Technology Development—Outcomes of Public Consultation and Proposed Way Forward* (Feb. 2, 2025), https://www.legco.gov.hk/yr2025/english/panels/ci/papers/ci20250218cb2-240-4-e.pdf

[2721] Department of Justice, *Consultation Group on LawTech Development of DoJ Committed to Promoting Use of LawTech in Industry* (Feb . 2025), https://www.doj.gov.hk/en/community_engagement/press/20250210_pr2.html

[2722] Office of the Privacy Commissioner for Personal Data (PCPD), *Amendments 2012* (2012), https://www.pcpd.org.hk/english/data_privacy_law/amendments_2012/amendment_2012.html

[2723] Office of the Privacy Commissioner for Personal Data (PCPD), *The Personal Data (Privacy) (Amendment) Ordinance 2021 Takes Effect Today to Criminalise Doxxing Acts* (Oct. 8, 2021), https://www.pcpd.org.hk/english/news_events/media_statements/press_20211008.html

[2724] Law Society of Hong Kong, *Proposed Doxxing Offence - Personal Data (Privacy) (Amendment) Bill 2021* (Aug. 18, 2021), https://www.hklawsoc.org.hk/-/media/HKLS/pub_e/news/submissions/20210818.pdf

[2725] Office of the Privacy Commissioner for Personal Data (PCPD), *New Anti-Doxxing Regime Is Beginning to Bear Fruits* (Dec. 2022), https://www.pcpd.org.hk/english/news_events/newspaper/newspaper_20221228.html

[2726] Office of the Privacy Commissioner for Personal Data (PCPD), *A 54-Year-Old Male Arrested for Suspected Doxxing of a Former Employer* (Nov. 7, 2025), https://www.pcpd.org.hk/english/news_events/media_statements/press_20251107.html; PCPD, *A Female Arrested for Suspected Doxxing Arising from Online Shopping Dispute* (Mar. 10, 2025), https://www.pcpd.org.hk/english/news_events/media_statements/press_20250310.html

The PDPO applies to both private and public data usage. However, it allows for specific exemptions for criminal investigations, the performance of judicial functions, security and defense, and emergency situations.[2727] In the context of the recent Hong Kong National Security Law and associated protests, the broader implications of these exemptions on human rights are less clear. For example, police can request content be taken down or have online platforms provide information about users.[2728]

The PCPD released checklists and guidelines in 2025 to guide organizations on data protection related to artificial intelligence. The Checklist on Guidelines for the Use of Generative AI by Employees advises organizations to define permissible AI uses under the PDPO.[2729] The Guide to Getting Started with Anonymization enhanced data de-identification standards.[2730] The Chief Executive also proposed a review of laws to facilitate cross-border data flows with mainland China while safeguarding PDPO standards.[2731] This signaled legal modernization and regional integration for Hong Kong.

The PCPD has been an active participant in international discussions on data protection, algorithmic transparency, and many other key issues in the use of AI, especially in the General Privacy Assembly (GPA). The PCPD was a signatory to the GPA 2018 Declaration on Ethics and Data Protection in Artificial Intelligence,[2732] sponsored the 2020 Resolution on Accountability in the Development and Use of

---

[2727] Hong Kong e-legislation, *Cap. 486 Personal Data (Privacy) Ordinance, Part 8, Exemptions*, https://www.elegislation.gov.hk/hk/cap486

[2728] Government of the Hong Kong Special Administrative Region, *Implementation Rules for Article 43 of the Law of the People's Republic of China on Safeguarding National Security in the Hong Kong Special Administrative Region*, Press Release (Jul. 6, 2020), https://www.info.gov.hk/gia/general/202007/06/P2020070600784.htm

[2729] Office of the Privacy Commissioner for Personal Data (PCPD), *Checklist on Guidelines for the Use of Generative AI by Employees* (Mar. 31, 2025), https://www.pcpd.org.hk/english/resources_centre/publications/files/guidelines_ai_employees.pdf

[2730] Office of the Privacy Commissioner for Personal Data (PCPD), *Guide to Getting Started with Anonymization* (Jul. 31, 2025), https://www.pcpd.org.hk/english/resources_centre/publications/files/appa_anonymisation_guide072025.pdf

[2731] J. Lee, *2025 Policy Address of the Chief Executive—Building Hong Kong into an International AI Hub* (Sept. 17, 2025), https://www.policyaddress.gov.hk/2025/en/policy.html

[2732] International Conference of Data Protection & Privacy Commissioners, *Declaration on Ethics and Data Protection in Artificial Intelligence* (Oct. 23, 2018), https://globalprivacyassembly.com/wp-content/uploads/2018/10/20180922_ICDPPC-40th_AI-Declaration_ADOPTED.pdf

Artificial Intelligence,[2733] and co-sponsored the 2023 Resolution on Generative AI.[2734] The PCPD did not endorse the 2022 Resolution on Principles and Expectations for the Appropriate Use of Personal Information in Facial Recognition Technology.[2735] Hong Kong joined nineteen other national data protection authorities in signing the Joint Statement on Building Trustworthy Data Governance Frameworks, reaffirming privacy-by-design, following the GPA conference in 2025.[2736]

Algorithmic Transparency

The AI Guidance presented by the PCPD makes several recommendations to increase transparency around the use of AI, including "putting in controls that allow human oversight and intervention of the operations of the relevant AI system."[2737] Similarly, the guidelines for banks using AI from the Hong Kong Monetary Authority push banks to hold leadership accountable for AI decision-making, to ensure results from AI systems are explainable and auditable, and to provide transparency to consumers on the use of AI.[2738] These recommendations align with established principles such as the OECD AI Principles, and similar recommendations in the proposed EU AI Act; however, the recommendations are advisory and non-binding.

The PCPD published the Artificial Intelligence: Model Personal Data Protection Framework in 2024.[2739] The Framework provides a set of practical and detailed recommendations for local enterprises through a risk-based approach,

---

[2733] Global Privacy Assembly (GPA), *Adopted Resolution on Accountability in the Development and Use of Artificial Intelligence* (Oct. 2020), https://globalprivacyassembly.com/wp-content/uploads/2020/11/GPA-Resolution-on-Accountability-in-the-Development-and-Use-of-AI-EN.pdf

[2734] Global Privacy Assembly, *Resolution on Generative Artificial Intelligence Systems* (Oct. 2023), https://globalprivacyassembly.org/wp-content/uploads/2023/10/5.-Resolution-on-Generative-AI-Systems-101023.pdf

[2735] Global Privacy Assembly, *Resolution on Principles and Expectations for the Appropriate Use of Personal Information in Facial Recognition Technology* (Oct. 2022), https://globalprivacyassembly.org/wp-content/uploads/2022/11/15.1.c.Resolution-on-Principles-and-Expectations-for-the-Appropriate-Use-of-Personal-Information-in-Facial-Recognition-Technolog.pdf

[2736] Office of the Privacy Commissioner for Personal Data, *Joint Statement on Building Trustworthy Data Governance Frameworks to Encourage Development of Innovative and Privacy-Protecting AI* (Sept. 17, 2025), https://www.pcpd.org.hk/english/news_events/media_statements/files/New_JointStatementonDataGovernance_Signed.pdf

[2737] Office of the Privacy Commissioner for Personal Data (PCPD), *Guidance on the Ethical Development and Use of Artificial Intelligence* (Aug. 2021), https://www.pcpd.org.hk/english/resources_centre/publications/files/guidance_ethical_e.pdf

[2738] Hong Kong Monetary Authority, *High Level Principles on Artificial Intelligence* (Nov. 1, 2019), https://www.hkma.gov.hk/media/eng/doc/key-information/guidelines-and-circular/2019/20191101e1.pdf

[2739] Office of the Privacy Commissioner for Personal Data, *Artificial Intelligence: Model Personal Data Protection Framework* (Jun. 2024), https://www.pcpd.org.hk/english/resources_centre/publications/files/ai_protection_framework.pdf

offering key principles such as human oversight, data privacy, fairness, transparency, interpretability, reliability, robustness, and security to guide the procurement, implementation, and use of AI systems. The Generative Artificial Intelligence Technical and Application Guideline builds on this foundation by broadening the principle of transparency in generative AI.[2740] The Guideline requires that stakeholders maintain transparency across all stages of the AI system life cycle, from data collection and model development to deployment, usage, and maintenance, including through documentation of training data, disclosure of system capabilities and limitations, and traceability of outputs. The Guideline also discourages opaque or black box operations by promoting the use of explainable AI techniques.

Data Scraping

The PCPD and international data protection and privacy counterparts released a follow-up joint statement laying out expectations for industry related to data scraping and privacy in 2024.[2741] The initial statement in August 2023 established that "[social media companies] and other websites are responsible for protecting personal information from unlawful scraping."[2742] The follow-up, which incorporates feedback from engagement with companies such as Alphabet (YouTube), X Corp (formerly Twitter), and Meta Platforms (Facebook, Instagram, WhatsApp), adds that organizations must "comply with privacy and data protection laws when using personal information, including from their own platforms, to develop AI large langua The April 2025 Hong Kong Generative Artificial Intelligence Technical and Application Guideline[2743] builds on this foundation by broadening the principle of transparency in generative AI. It requires that stakeholders maintain transparency across all stages of the AI system life cycle, from data collection and model development to deployment, usage, and maintenance, including through documentation of training data, disclosure of system capabilities and limitations, and traceability of outputs. The Guideline also discourages opaque or "black box"

[2740] Digital Policy Office, *Generative Artificial Intelligence Technical and Application Guideline* (Dec. 2025), https://www.digitalpolicy.gov.hk/en/our_work/data_governance/policies_standards/ethical_ai_framework/doc/HK_Generative_AI_Technical_and_Application_Guideline_en.pdf

[2741] UK Information Commissioner's Office, *Global Privacy Authorities Issue Follow-Up Joint Statement on Data Scraping after Industry Engagement* (Oct. 28, 2024), https://ico.org.uk/about-the-ico/media-centre/news-and-blogs/2024/10/global-privacy-authorities-issue-follow-up-joint-statement-on-data-scraping-after-industry-engagement/

[2742] Regulatory Supervision Information Commissioner's Office of the United Kingdom, *Joint Statement on Data Scraping and the Protection of Privacy* (Aug. 24, 2023), https://ico.org.uk/media2/migrated/4026232/joint-statement-data-scraping-202308.pdf

[2743] Digital Policy Office, *Generative Artificial Intelligence Technical and Application Guideline* (Dec. 2025), https://www.digitalpolicy.gov.hk/en/our_work/data_governance/policies_standards/ethical_ai_framework/doc/HK_Generative_AI_Technical_and_Application_Guideline_en.pdf

operations by promoting the use of explainable AI techniques."ge models"; must regularly review and update privacy protection measures to keep pace with technological advances; and ensure that permissible data scraping, such as that for research, "is done lawfully and following strict contractual terms."[2744]

Data scraping generally involves the automated extraction of data from the web. Data protection authorities are seeing increasing incidents involving data scraping, particularly from social media companies and the operators of other websites that host publicly accessible data. Scraped personal information can be exploited for targeted cyberattacks, identity fraud, monitoring, profiling and surveillance purposes, unauthorized political or intelligence gathering purposes, unwanted direct marketing or spam.

Facial Recognition

Hong Kong has long-standing concerns about AI surveillance, especially in relation to mainland China and the preservation of democracy in Hong Kong. During the 2019 anti-government protests, Hong Kong Chief Executive Carrie Lam invoked the Emergency Regulations Ordinance, which gives the chief executive the power to to "make any regulations whatsoever which [they] may consider desirable in the public interest,"[2745] to ban the use of face masks. The ban aimed at preventing protestors from hiding their identities.[2746] Hong Kong's Court of Final Appeal largely upheld the application of the Emergency Regulations Ordinance for the facemask ban.[2747]

Hong Kong has also been a strong proponent of the use of technology in public spaces, laying out its future plans through the Hong Kong Smart Cities Blueprint 2.0, published.[2748] These plans include the use of the StayHomeSafe mobile app and companion tracking bracelet, whose use was required for all new entrants into Hong Kong.[2749] The blueprint and associated initiatives outline many plans such as the

---

[2744] Office of the Privacy Commissioner of Canada, *Concluding Joint Statement on Data Scraping and the Protection of Privacy* (Oct. 2024), https://ico.org.uk/about-the-ico/media-centre/news-and-blogs/2024/10/global-privacy-authorities-issue-follow-up-joint-statement-on-data-scraping-after-industry-engagement/

[2745] Hong Kong E-Legislation, *Cap. 241 Emergency Relations Ordinance* (2018), https://www.elegislation.gov.hk/hk/cap241

[2746] BBC News, *Hong Kong: Face Mask Ban Prompts Thousands to Protest* (Oct. 4, 2019), https://www.bbc.com/news/world-asia-china-49939173

[2747] Washington Post, *Hong Kong's Highest Court Upholds Ban on Masks at Protests* (Dec. 21, 2020), https://www.washingtonpost.com/world/hongkong-mask-ban-ruling/2020/12/20/f2722af0-4340-11eb-a277-49a6d1f9dff1_story.html

[2748] Innovation and Technology Bureau, *Hong Kong Smart Cities Blueprint 2.0* (Dec. 2020), https://www.smartcity.gov.hk/modules/custom/custom_global_js_css/assets/files/HKSmartCityBlueprint(ENG)v2.pdf

[2749] Government of Hong Kong Special Administrative Region, *StayHomeSafe User Guide*, https://www.coronavirus.gov.hk/eng/stay-home-safe.html

increased rollout of public wifi and contact tracing for COVID, as well as funding for robotic patrols of airport terminals and a new LawTech Fund.

Despite the clear potential upside in the Smart City Blueprint, there is no mention of potential concerns, such as data privacy, algorithmic bias, or the potential violation of human rights in the documents rolling out these new initiatives. Furthermore, there was no significant public engagement, despite concerns about negative impacts.[2750] The rollout of new technology in public places, the Hong Kong National Security Law, and the development and use of facial recognition technology by local companies all threaten the increased use of AI for surveillance purposes in Hong Kong.[2751]

In July 2024, Hong Kong's security chief, Chris Tang, informed local media that the police aim to install 2,000 new surveillance cameras and plan to incorporate facial recognition technology, with the potential for AI-assisted suspect tracking.[2752] However, it remains unclear whether Hong Kong will implement continuous live facial recognition scanning of public spaces or if it will limit facial recognition to retrospective analysis of footage following a crime or upon legal authorization. Some experts have expressed concern that Hong Kong may be moving closer to mainland China's pervasive surveillance practices.

Hong Kong launched a facial recognition pilot at the Chung Ying Street checkpoint in December 2024 to support future tourism development and simplify the process of crossing borders.[2753] The Hong Kong Police Force commissioned an independent consultancy to conduct a Privacy Impact Assessment for the use of facial recognition technology in the 'e-Corridor' system at the checkpoint. Legal advice was also sought to ensure compliance with privacy regulations and public notices inform individuals of ongoing CCTV surveillance. Additionally, Security Risk Assessments and independent audits were conducted to evaluate the system's security and data protection measures.[2754]

[2750] Neville Lai and Justin Chan, *People have to be at the heart of Hong Kong's smart city plans"* (Nov. 27, 2021), https://www.scmp.com/comment/opinion/hong-kong/article/3157263/people-have-be-heart-hong-kongs-smart-city-plans

[2751] Lachlan Markay, *Scoop: Chinese tech firm sidesteps sanctions* (Sept. 29, 2021), https://www.axios.com/chinese-tech-firm-sidesteps-sanctions-de43feaf-7df5-46ad-85bd-8a37ab468e2e.html

[2752] Jessie Yeung, CNN, *Hong Kong plans to install thousands of surveillance cameras. Critics say it's more proof the city is moving closer to China* (Oct. 2024), https://edition.cnn.com/2024/10/05/asia/hong-kong-police-cameras-facial-recognition-intl-hnk-dst/index.html

[2753] Government of Hong Kong Special Administrative Region, *Opening Ceremony of Reprovisioned Chung Ying Street Checkpoint in Sha Tau Kok Held Today* (Dec. 23, 2024), https://www.info.gov.hk/gia/general/202412/23/P2024122300484.htm?utm

[2754] Edith Lin, *Hong Kong Police Chief Raymond Siu Backs Government Plan to Install 2,000 Surveillance Cameras by End of Year* (Feb. 12, 2024), https://www.scmp.com/news/hong-kong/law-and-crime/article/3251776/hong-kong-police-chief-raymond-siu-backs-government-plan-install-2000-surveillance-cameras

Environmental Impact of AI

The Hong Kong Special Administrative Region has considered the environmental impact of AI. The Twelve Ethical AI Principles include "Sustainability and Just Transition." Any AI development should ensure that there are mitigation strategies in place to address potential impacts on society and the environment.[2755] The framework specifically acknowledges that AI can have an impact on planet preservation and cites examples of using AI to optimize power usage and reduce carbon footprint. Although no dedicated AI environmental impact assessment framework exists, the Digital Policy Office's 2025 Guideline encourages energy-efficient data processing and carbon footprint monitoring in government AI projects.[2756] Policies such as the Green Data Centre Guidelines issued by the Office of the Government Chief Information Officer seek to address the challenges expanded AI infrastructure poses on ambitions for carbon neutrality.[2757] Strategic efforts, including the Hydrogen Development Strategy 2024 and the Climate Action Plan 2050's target to increase renewable energy to 7.5-10% of the electricity mix by 2035, also aim to expand renewable energy use, ensuring AI growth aligns with broader climate objectives.[2758]

Lethal Autonomous Weapons

Given Hong Kong's status as a Special Administrative Region of China, policies and regulations formulated by the Chinese government, including those related to lethal autonomous weapons systems (LAWS), have the potential to influence Hong Kong's stance and policies on these matters.[2759] Hong Kong is not a signatory to international treaties such as the Convention on Certain Conventional Weapons (CCW) but adheres to the CCW provisions through China's ratification.[30]

[2755] Digital Policy Office, *Ethical Artificial Intelligence Framework* (Version 1.4) (Jul. 2024), https://www.digitalpolicy.gov.hk/en/our_work/data_governance/policies_standards/ethical_ai_framework/doc/Quick_Reference_Guide-Ethical_AI_Framework.pdf

[2756] Digital Policy Office, *Generative Artificial Intelligence Technical and Application Guideline* (Dec. 2025), https://www.digitalpolicy.gov.hk/en/our_work/data_governance/policies_standards/ethical_ai_framework/doc/HK_Generative_AI_Technical_and_Application_Guideline_en.pdf

[2757] Arendse Huld, *Hong Kong's Data Center Sector: Industry and Regulatory Landscape* (Sept. 7, 2023), https://www.china-briefing.com/news/hong-kong-data-center-sector-industry-regulatory-landscape/

[2758] Environment and Ecology Bureau, *The Strategy of Hydrogen Development in Hong Kong* (Jun. 2024), https://cnsd.gov.hk/strategy-of-hydrogen-development-in-hong-kong_booklet_en.pdf

[2759] United Nations Office for Disarmament Affairs (UNODA), *High Contracting Parties and Signatories to the CCW* (Mar. 18, 2025), https://disarmament.unoda.org/en/our-work/conventional-arms/convention-certain-conventional-weapons/high-contracting-parties-and-signatories-ccw

AI Literacy

Hong Kong mounted its most ambitious AI literacy campaign in 2025, combining government funding exceeding HK$4 billion with public-private partnerships targeting students, teachers, workers, and the public. The Education Bureau mandated 10–14 hours of AI education for all junior secondary level students (Junior Secondary Level) as part of the ICT curriculum implemented since 2023/24.[2760] The bureau aimed for 82% implementation across schools by June 2025 with a target of 95% by year-end. Teacher professional development programs targeting over 1,000 secondary ICT teachers and Microsoft Hong Kong's Partners in Learning Program 5.0 reached 450+ schools (70% participation), providing capacity-building for educators who could facilitate public understanding of AI governance issues, though no formal metrics on educator capacity to influence policy discussions have been established. The private-public partnership under the Just Start campaign set an ambitious goal to teach AI skills to 1 million people in Hong Kong.[2761] More than 30 organizations including Google, Microsoft, and Meta, and the government introduced the AI-Q Test, Hong Kong's first free, open-access AI self-assessment tool enabling personalized learning recommendations.

Human Rights

As a special administrative region of China, Hong Kong has not ratified the United Nations Universal Declaration of Human Rights (UDHR) but is protected by the convention through China's endorsement.[2762] The UN International Covenant on Economic, Social, and Cultural Rights (ICESCR) and UN International Covenant on Civil and Political Rights (ICCPR) apply to Hong Kong.[2763]

These human rights conventions are also complemented by other articles that protect the rights of citizens, including the Hong Kong Bill of Rights[2764] and the Basic Law.[2765] One of the key provisions in Article 30 of the Personal Data (Privacy) Ordinance protects citizens' right to privacy and clearly identifies principles to be

---

2760 Education Bureau, *Subjects under the Eight Key Learning Areas (2025/26 School Year)*, Technology Education (Sept. 16, 2025), https://www.edb.gov.hk/en/curriculum-development/kla/overview.html

2761 Hong Kong AI Literacy Initiative, *Just Start Campaign*, https://www.preface.ai/just-start/en/

2762 United Nations, *Member States*, https://www.un.org/en/about-us/member-states#gotoH

2763 United Nations Treaty Collection, *International Covenant on Civil and Political Rights, China*, fn6, https://treaties.un.org/Pages/ViewDetails.aspx?src=IND&mtdsg_no=IV-4&chapter=4&clang=_en#4; *International Covenant on Economic, Social, and Cultural Rights, China*, fn8 (with reservations), https://treaties.un.org/pages/ViewDetails.aspx?src=IND&mtdsg_no=IV-3&chapter=4&clang=_en#8

2764 Hong Kong e-Legislation, *Hong Kong Bill of Rights Ordinance of 1991*, CAP. 383 (Jun. 8, 1991), https://www.elegislation.gov.hk/hk/cap383

2765 Hong Kong e-Legislation, *Basic Law of Hong Kong Special Administrative Region of the People's Republic of China* (Apr. 4, 1990), https://www.elegislation.gov.hk/hk/A101!en@1997-07-01T00:00:00

adhered to in the use of personal data.[2766] The key regulatory bodies responsible for monitoring human rights implementation in Hong Kong include the Constitutional and Mainland Affairs Bureau (CMAB), the Equal Opportunities Commission (EOC), and the Office of the Privacy Commissioner for Personal Data, Hong Kong. These agencies are tasked with overseeing compliance with human rights standards and investigating violations, including issues concerning freedom of speech, assembly, and privacy.[2767]

The Legislative Council of the Hong Kong Special Administrative Region adopted the Safeguarding National Security Bill in March of 2024.[2768] The Article 23 National Security Law criminalizes secession, subversion, terrorism, and collusion with foreign forces, marking a significant step in the tightening of political and civil rights.[2769] Amnesty International has raised concerns that the implementation of Article 23 and the tightening restrictions freedoms of expression, association, and assembly will lead to the prosecution of individuals expressing dissenting views against the government.[2770]

The Freedom House 2025 Report ranks Hong Kong as "Partly Free" with a total score of 40/100.[2771] Political rights and civil liberties are scored very low on the report because of the prevalence of pro-Beijing interests in the country's political system and the limitations on freedom and autonomy because of political interventions from mainland China. Civil-society organizations continued to face restricted operating space under the Safeguarding National Security Ordinance (2024). Reports by Amnesty International[2772] and Human Rights Watch[2773] documented expanded surveillance and constraints on expression, de facto limiting public participation in AI policy dialogue and rights-impact assessments.

[2766] Hong Kong e-legislation, *Cap. 486 Personal Data (Privacy) Ordinance* (Aug. 1, 1996), https://www.elegislation.gov.hk/hk/cap486

[2767] Constitutional and Mainland Affairs Bureau (CMAB), *The Rights of the Individual* (2025), https://www.cmab.gov.hk/en/issues/individuals.htm

[2768] Security Bureau, *Basic Law Article 23 Legislation: Safeguarding National Security Bill* (Mar. 7, 2024), https://www.legco.gov.hk/yr2024/english/brief/sb20240308_20240308-e.pdf

[2769] Greg Torode and Jessie Pang, *Hong Kong's New National Security Law and Its impact*, Reuters (Mar. 19, 2024), https://www.reuters.com/world/asia-pacific/article-23-what-you-need-know-about-hong-kongs-new-national-security-laws-2024-03-19/

[2770] Amnesty International, *What Is Hong Kong's Article 23 Law? 10 Things You Need to Know* (Mar. 22, 2024), https://www.amnesty.org/en/latest/news/2024/03/what-is-hong-kongs-article-23-law-10-things-you-need-to-know/

[2771] Freedom House, *Freedom in the World 2025: Hong Kong* (2025), https://freedomhouse.org/country/hong-kong/freedom-world/2025

[2772] Amnesty International, *Hong Kong: Article 23 Law Used to "Normalize Repression"* (Mar. 27, 2025), https://www.amnesty.org/en/latest/news/2025/03/hong-kong-article-23-law-used-to-normalize-repression-one-year-since-enactment/

[2773] Human Rights Watch, *Hong Kong: Targeting of Exiled Activists' Families Escalates* (May 4, 2025), https://www.hrw.org/news/2025/05/04/hong-kong-targeting-exiled-activists-families-escalates

OECD / G20 AI Principles

Hong Kong is not a member of the OECD or the G20 and has not endorsed the OECD / G20 AI Principles.[2774]

Despite no official endorsement, Hong Kong's approach to AI governance such as the ethical framework includes many of the OECD's principles and implementation recommendations.[2775] Limitations, such as insufficient human capacity and incomplete alignment with human rights and democratic values, notably due to the use of facial recognition for surveillance purposes, undermine the region's implementation of the OECD Principles.

Council of Europe AI Treaty

Neither Hong Kong nor China have signed the Council of Europe AI treaty.[2776] Hong Kong and China also did not participate in the negotiations, which began in 2022, with input from 68 international representatives from civil society, academia, industry, and other international organizations was gathered.[2777]

UNESCO Recommendation on AI Ethics

As a member of UNESCO, Hong Kong adopted the Recommendation on the Ethics of Artificial Intelligence in November 2021.[2778] The AI Literacy initiatives to improve understanding of AI align with UNESCO's Education and Research policy action for implementing the Recommendation.[2779] Hong Kong's Ethical AI Framework includes an Application Impact Assessment.[2780] UNESCO offers a related Ethical Impact Assessment tool to implement the Recommendation.[2781]

[2774] OECD Legal Instruments, *Recommendation of the Council of Artificial Intelligence* (May 3, 2024), https://legalinstruments.oecd.org/en/instruments/OECD-LEGAL-0449

[2775] Office of the Privacy Commissioner of Hong Kong, *Guidance on the Ethical Development and Use of Artificial Intelligence* (Aug. 2021), https://www.pcpd.org.hk/english/resources_centre/publications/files/guidance_ethical_e.pdf

[2776] Council of Europe Treaty Office, *Chart of Signatures and Ratifications of Treaty 225* (Jan. 3, 2026), https://www.coe.int/en/web/Conventions/full-list/?module=signatures-by-treaty&treatynum=225

[2777] Council of Europe, *Framework Convention on Artificial Intelligence* (2026), https://www.coe.int/en/web/artificial-intelligence/the-framework-convention-on-artificial-intelligence

[2778] UNESCO, *Recommendation on the Ethics of Artificial Intelligence* (Sept. 26, 2024), https://www.unesco.org/en/articles/recommendation-ethics-artificial-intelligence

[2779] UNESCO, *UNESCO's Recommendation on the Ethics of Artificial Intelligence: Key Facts* (2023), https://www.unesco.org/en/artificial-intelligence/recommendation-ethics

[2780] Digital Policy Office, *Ethical Artificial Intelligence Framework* (Version 2.0) (Dec. 2025), https://www.digitalpolicy.gov.hk/en/our_work/data_governance/policies_standards/ethical_ai_framework/doc/Ethical_AI_Framework_en.pdf

[2781] UNESCO, *Ethics of Artificial Intelligence: The Recommendation* (2023), https://www.unesco.org/en/artificial-intelligence/recommendation-ethics

Hong Kong has not stated a plan or participated independently in UNESCO's Readiness Assessment Methodology (RAM) for artificial intelligence.[2782] Hong Kong's AI policies align with those of mainland China. China has not yet engaged in the RAM process and has not publicly indicated plans to do so.

Evaluation

Hong Kong has made significant progress in Artificial Intelligence policy, publishing frameworks and principles to enhance personal data privacy protection, to embed ethical considerations into the AI life cycle, and ensure consumer protection. Establishing the Digital Policy Office marked a development that brought with it the creation of an ethical framework in AI, as well as the establishment as an institution to monitor its development., although it could be safe to link this special administrative region with some of China's AI efforts and achievements due to the administrative relationship.

Hong Kong has not been a signatory to the OECD AI Principles; however, it has proposed and implemented similar principles and guidelines and endorsed the UNESCO Recommendation on the Ethics of AI. The leading role played by Hong Kong through the PCPD-sponsored resolution on greater accountability in the development and use of AI to the Global Privacy Assembly (GPA) shows that the region has the potential to play a prominent role in key global AI policy development and implementation. Hong Kong's policies around AI literacy, including children and students, support ethical AI initiatives in line with public engagement. However, as in other areas, public participation in policymaking is still limited

Irrespective of its complicated relationship with mainland China, especially as it relates to surveillance and data protection issues, there is some effort by the government to regulate and promote ethical AI development and use in the country. More effort is needed by the government to adopt a comprehensive national AI strategy that promotes democratic values and human rights, as well as alignment with international commitments to AI principles.

[2782] UNESCO, Global AI Ethics and Governance Observatory, *Global Hub* (Oct. 2025), https://www.unesco.org/ethics-ai/en/global-hub

www.ingramcontent.com/pod-product-compliance
Lightning Source LLC
LaVergne TN
LVHW081321110826
845149LV00007B/1562

* 9 7 9 8 9 9 5 6 9 5 8 1 3 *